ENGINEERING ORGANIZATION AND MANAGEMENT

BENJAMIN S. BLANCHARD

Virginia Polytechnic Institute and State University

PRENTICE-HALL, INC., *Englewood Cliffs, New Jersey*

Library of Congress Cataloging in Publication Data-

BLANCHARD, BENJAMIN S.
Engineering organization and management.

Bibliography: p.
Includes index.
1.-Engineering—Management. I.-Title.
TA190.B55 658'.91'62 75-38737
ISBN 0-13-279430-6

10 9 8 7 6 5 4 3 2

Printed in the United States of America

PRENTICE-HALL INTERNATIONAL, INC., *London*
PRENTICE-HALL OF AUSTRALIA PTY. LIMITED, *Sydney*
PRENTICE-HALL OF CANADA, LTD., *Toronto*
PRENTICE-HALL OF INDIA PRIVATE LIMITED, *New Delhi*
PRENTICE-HALL OF JAPAN, INC., *Tokyo*
PRENTICE-HALL OF SOUTHEAST ASIA PRIVATE LIMITED, *Singapore*

Dedicated to my wife, DOROTHY,
my children, BECKY, BENJIE, *and* LISA,
for their tolerance
and encouragement in this endeavor.

CONTENTS

PREFACE

The engineer in business, industry, and government today is faced with several major challenges. First, most products and systems are becoming more complex in nature, and there is an increasing demand relative to new technology applications at a time when our natural resources are dwindling. A pressing need exists to develop new products and processes that will effectively and efficiently respond to consumer requirements.

Second, successful engineering involves a "team" approach requiring the integration of many specialized disciplines. The engineer (as a member of the team) must not only be competent in his (or her) area of expertise, but must be thoroughly familiar with those related technical functions necessary for job accomplishment and know how they interface. Frequently, the engineer who initially is assigned highly specialized tasks may ultimately be required to accomplish a variety of functions covering a broad spectrum, which in turn necessitates an understanding of all facets of product/system development, production, operational use, and logistic support. This text addresses this issue by covering the product/system life cycle and engineering activities as they pertain to each phase of the life cycle.

Third, the engineer in accomplishing program functions requires the support of such nontechnical or organizational components as marketing, finance, budgeting and accounting, purchasing, personnel, etc. The interfaces here are numerous and varied, and the engineer should have a complete understanding of them if he (or she) is to operate effectively and efficiently in the fulfillment of objectives. This book provides the engineer with needed material to create this understanding.

Finally, the practicing engineer (regardless of technical background) often is pressed into the role of manager or supervisor responsible for the direction and control of some organizational entity. In such instances, the engineer may find himself (or herself) deficient in organizational principles, human relations, personnel management, budgeting, project accounting and control, and so on. This text is designed to cover some of the basic principles of organization and management with the objective of assisting the new engineering manager in the performance of his job.

The scope of the material included in this text is presented in five basic parts. Part One (Chapters 1 and 2) discusses engineering in modern times by introducing the engineering process and the fundamental concepts of engineering organization and management. Part Two (Chapters 3 through 8) describes engineering activities over the life cycle, commencing with the definition of product "need" and progressing through product planning, research, design and development, production or construction, consumer operational use, logistic support, material phase-out, and disposal. Part Three (Chapters 9 and 10) covers the principles of organization; organizational components and interfaces as they exist in corporations, businesses, and agencies; and engineering activities in the total organizational spectrum. Part Four (Chapters 11 through 16) addresses the major organizational and managerial interfaces that exist in most business enterprises and directly relate to the successful accomplishment of engineering functions and tasks (e.g., finance, budgeting and accounting, purchasing, personnel, legal, and so on). Part Five (Chapters 17 and 18) presents an overview of current environmental, social, political, economic, and related considerations that impact on and are in turn impacted by engineering activities and decisions.

The five parts of this text are mutually supportive, and the material is presented in a logical evolutionary sequence as the reader progresses from part to part. Emphasis is placed on the *life cycle approach,* with engineering activities occuring in all program phases. In addition, engineering as related to both large and small projects is represented as a team effort which involves the integration of many and varied technical disciplines and the support of numerous administrative components. In essence, this book is intended to provide engineering students, practicing engineers and/or engineering managers with a broad perspective of their functions, responsibilities, interfaces, and some of the experiences that they will encounter in the organizational arena. It certainly is not intended to be all-encompassing relative to the areas noted, but includes enough material of a varied nature to familiarize the engineer with the engineering process embedded in an organizational setting with associated management functions.

In the development of this text a number of real-world illustrations are used to enhance the material presentation. I wish to express my appreciation to the following individuals for their contributions in this area: J. Bussolini, Grumman Aerospace Corporation; D. Calhoun, Hughes Aircraft Company; H. Chambers, Rockwell International; J. Heater and S. LaFavor, McDonnell

Aircraft Company; E. Lowery, Xerox Corporation; R. Monroe, Martin Marietta Aerospace Corporation; and J. Snodgrass, The Cooper Tire and Rubber Company.

I also wish to thank Wolter J. Fabrycky, Associate Dean of Engineering, Virginia Polytechnic Institute and State University, for his encouragement in the writing of this text, and a special appreciation to my secretary, Ruth Hess, for her dedication and support in the preparation of the text materials.

BENJAMIN S. BLANCHARD

part one

Engineering in Modern Times

Chapter 1—Engineering and the Engineering Process
Chapter 2—Fundamental Concepts of Engineering Management

Part One is the focal point of engineering and responds to such questions as: What is it?; What does engineering do?; Why is engineering important?; What are the problems and challenges involved in engineering management?; and so on. It sets the stage for subsequent sections of the text and is represented by the inner circle in the illustrated conceptual outline.

Conceptual Outline

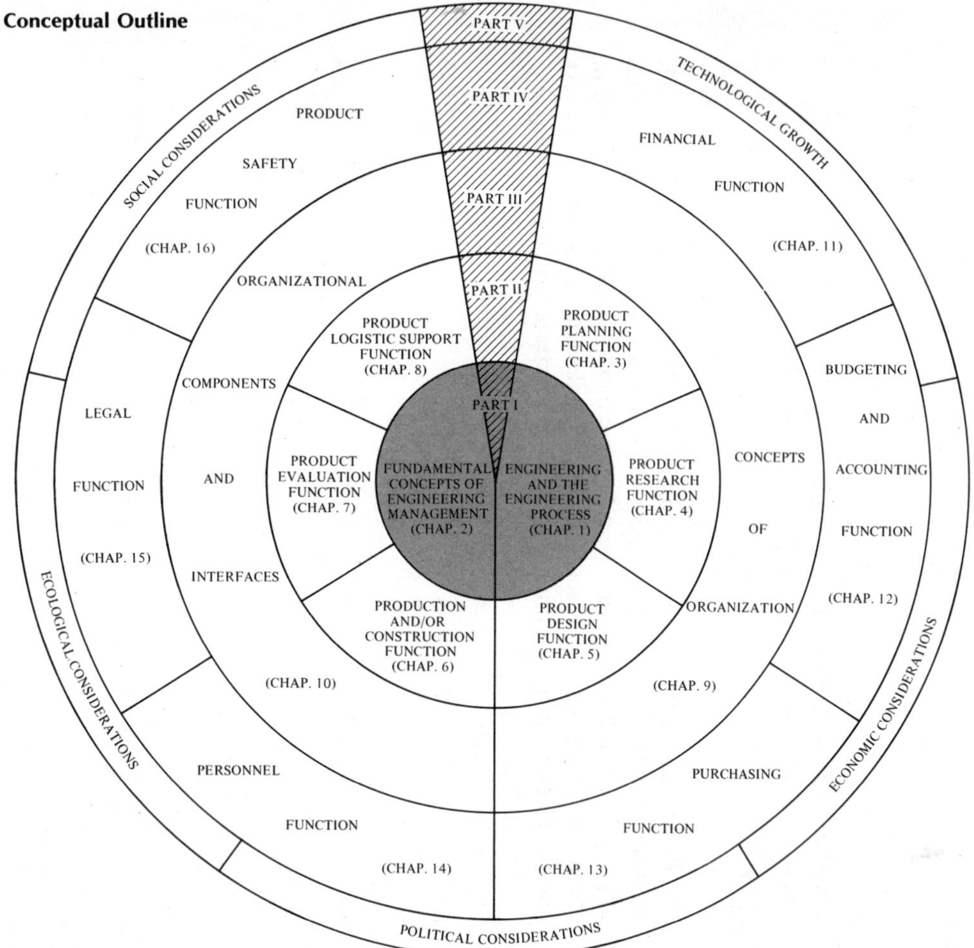

PART I	ENGINEERING IN MODERN TIMES
PART II	THE PRODUCT LIFE CYCLE
PART III	ORGANIZATION AND THE PRODUCT LIFE CYCLE
PART IV	RELATED FUNCTIONS AND INTERFACES
PART V	ENGINEERING AND THE ENVIRONMENT

1

ENGINEERING AND THE ENGINEERING PROCESS

"Engineering is the profession in which a knowledge of the mathematical and natural sciences gained by study, experience, and practice is applied with judgment to develop ways to utilize, economically, the materials and forces of nature for the benefit of mankind."[1] To define it another way, engineering constitutes the systematic application of physical and natural resources combined in such a manner as to create, develop, manufacture, and support a product or a process which economically provides some form of utility to man. Engineering basically involves the creation of something useful from a body of scientific knowledge by employing *technical expertise, individual skills* in the application of transformational methods, and personnel with the proper *attitudinal characteristics.*

1.1 INTRODUCTION TO ENGINEERING

Figure 1-1 depicts engineering as being concerned with creating things from a body of scientific knowledge. Science involves man's understanding of the physical and natural resources of the universe relative to both structure and behavior. Scientists direct their attentions to improving this understand-

[1]This definition is accepted by the Engineers' Council for Professional Development (ECPD).

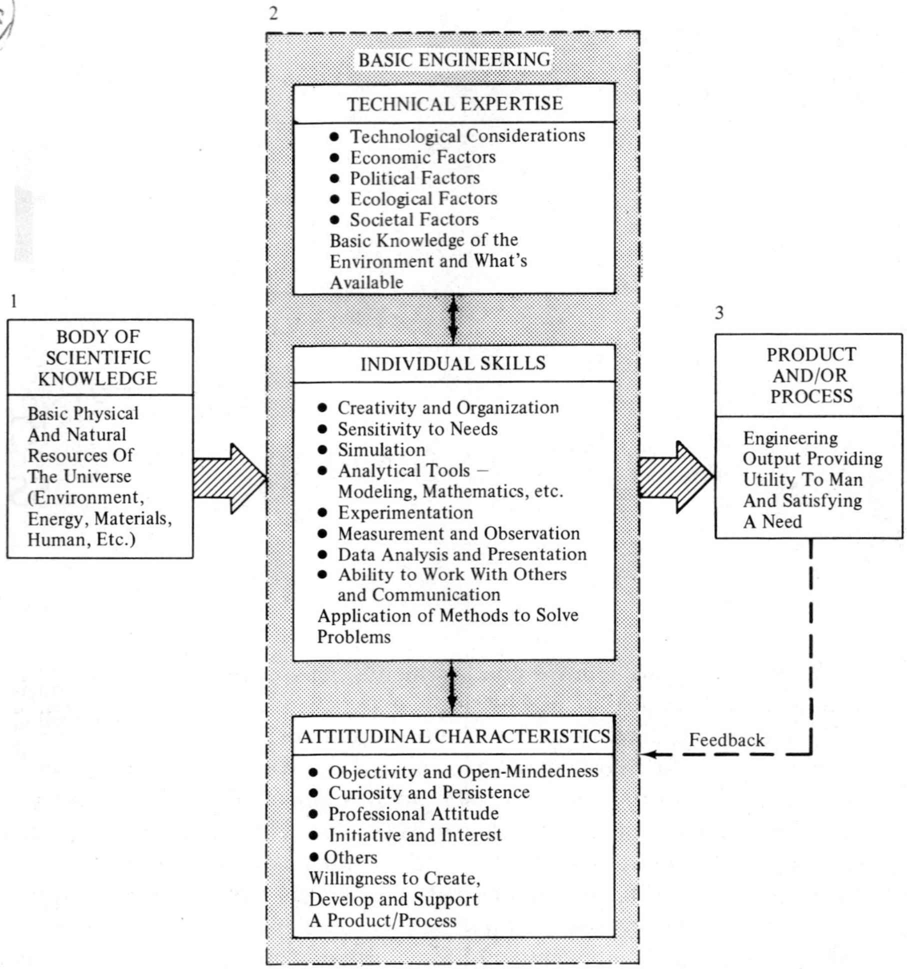

Figure 1-1. Basic engineering requisites.

ing through observing and assimilating what others have learned, developing theories and hypotheses, conducting research and experimentation, and translating the results into useful relationships (e.g., mathematical expressions, physical laws). The scientist is primarily oriented toward establishing theories and principles and generally has little concern for the application of those theories and principles in terms of a tangible product. The ultimate objective is to generate new knowledge.[2]

[2]The distinction between science and engineering is discussed in: Edward V. Krick, *An Introduction to Engineering and Engineering Design*, 2nd Edition, John Wiley & Sons, Inc., New York, 1969.

On the other hand, engineering utilizes this body of knowledge (represented by block 1, Figure 1-1) in creating tangible products or processes such as the airplane, bridge, washing machine, computer, industrial waste treatment facility, production capability, and the like. Typical products or processes may be categorized as aeronautical, biological, chemical, electrical, electronic, mechanical, nuclear, optical, and so on. Product/process characteristics will vary in different instances, depending on the intended function of the product. For example, if there is a need to provide electrical power to a specific area of the country, it will be necessary to develop and produce an electrical power system of some type that will provide an output of x kilowatts at a constant rate for a designated period of time and for a cost not to exceed y dollars.[3] In addition, the system must be supported throughout its planned life cycle such that it can perform its intended function in an effective and efficient manner on a sustaining basis. Typical product/process (or system) characteristics will include performance factors such as range, accuracy, and speed of performance; weight, size, and shape; operational effectiveness or availability; reliability, maintainability, and supportability; safety; cost; and comparable factors as applicable. In other words, a need is identified and engineering must create a tangible entity to satisfy that need.

In the accomplishment of engineering objectives, three basic ingredients must be present. They include a level of technical expertise, individual skills, and certain attitudinal characteristics (refer to block 2, Figure 1-1).

A. Technical Expertise

In actual practice, engineering is highly specialized as a result of the large, complex, and varied products/processes required to respond to the many different needs of our society. It is impossible for a single engineer to be entirely competent in designing an airplane, a dam, a radar system, and a chemical processing plant. Specialization is necessary, and a variety of individual engineering fields are required as discussed below and presented in Figure 1-2.[4]

---

[3]A system may be considered to be a nucleus of elements structured in such a manner as to accomplish a function designed to satisfy an identified need. A system may vary in form, fit, and function and is discussed at various levels. A system can be considered to be a product, but not all products are systems, since the above characteristics may not prevail (e.g., a radio without an operator). The material discussed throughout this book applies to both systems and products.

[4]The listing of engineering fields presented should be considered as being typical and not necessarily all-inclusive. In addition, the listing is not to be considered as representing any particular organizational structure within an industrial firm, a government agency, or an academic institution. A more in-depth discussion of the various engineering specialty areas is included in: G. C. Beakley and H. W. Leach, *Engineering—An Introduction to a Creative Profession*, 2nd edition, Macmillan Publishing Co., Inc., New York, 1972.

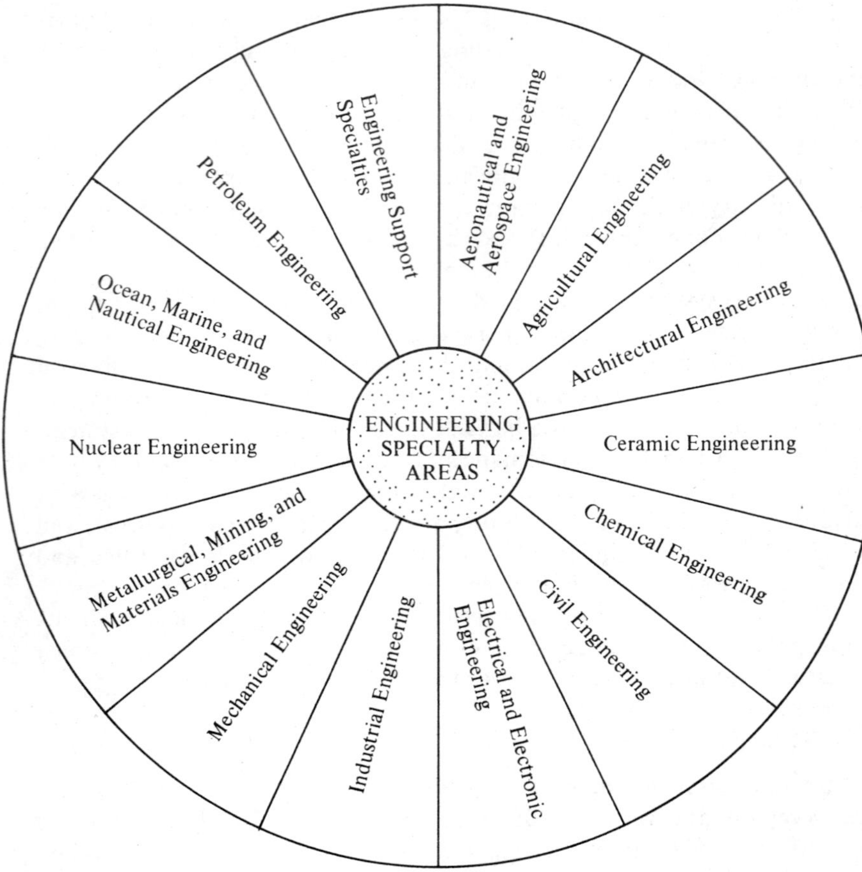

Figure 1-2. Engineering specialties.

*1. **Aeronautical and aerospace engineering.*** Aeronautical engineers are involved in the design and development of air transportation systems, supersonic and subsonic aircraft (jet, vertical takeoff, STOL), spacecraft, missiles, rockets, remotely piloted vehicles, and related systems. The need for faster methods of transportation and the advent of space exploration have created new and challenging opportunities in the field. An in-depth knowledge of aerodynamics and the principles of flight, vehicle stability and control (performance), structures, flight mechanics, boundary-layer and heat-transfer principles, fluid dynamics and flow characteristics, and so on, is required.

*2. **Agricultural engineering.*** Agricultural engineers deal with the production, processing, handling, and storage of food and related products.

This includes the design of farm machinery and power units, electric power and processing equipment, and farm structures (barns, silos, shelters, granaries, processing buildings, and the like). Agricultural engineers are also involved in soil and water conservation and control (drainage and irrigation systems, land surfacing, and reclamation).

3. Architectural engineering. Architectural engineers are primarily responsible for analysis and design activities associated with structural materials and their applications in the construction of large office buildings, manufacturing plants, sporting arenas, and so on. The architectural engineer, working closely with the architect, is trained in the sound principles of engineering and is concerned with safety and sound construction methods. The architect, in contrast, is more concerned with space arrangements, functional needs, and appearance.

4. Ceramic engineering. Ceramic engineers are specialists in the development of ceramic materials and their application to electrical and electronic systems, aerospace systems, chemical systems, and numerous components.

5. Chemical engineering. Chemical engineers design and develop new processes and processing facilities for the production of chemicals in large quantities. Such chemicals include acids, dyes, synthetics, alkalies, plastics, polymers, insecticides, fungicides, and many comparable products for domestic and industrial uses. The production process may involve chemical reactions of raw materials, such as oxidation and hydrogenation, chlorination, nitration, and reduction. Output products also include new fuels, medicines, plasmas, and textiles.

6. Civil engineering. Civil engineers design highways, bridges, tunnels, large buildings, dams, canals and waterways, water-supply systems (wells, pipelines, storage facilities), sanitary and sewage treatment and disposal systems, airports and transportation systems, and related projects. Civil engineers are also involved in city planning and management activities (street planning, zoning, industrial site development, and associated maintenance functions). Civil engineering is somewhat broader than most engineering fields and is often subdivided into more specialized areas, such as city management, construction engineering, highway engineering, hydraulic engineering, sanitary engineering, structural engineering, and transportation engineering. Depending on the field of specialty within civil engineering, knowledge of surveying, geology, structures, hydraulics, sanitation, air and water pollution control, soil mechanics, and/or environmental requirements is necessary.

7. Electrical and electronic engineering. Electrical and electronic engineers are responsible for the design and development of systems that create, transfer, and utilize energy. This includes electrical power systems, control systems, computer systems, communication systems, electronic systems (radar, navigation, fire control and missile guidance, signal processing equipment, etc.), electro-optical devices, instrumentation, appliances, and small electrical/electronic components. As is the case in civil engineering, this field is often subdivided into more specialized areas, such as electrical engineering, electronic engineering, communications engineering, power engineering, and computer engineering. Knowledge of electricity, electronic theory (e.g., analog and digital techniques, transistors, semiconductors, microelectronics, MSI, LSI, MOS), circuit design, power, electromagnetic wave theory, and related subjects is required.

8. Industrial engineering. The field of industrial engineering is a broad one and deals primarily with the design, installation, and operation of manufacturing systems (i.e., production and/or construction). The realm of industrial engineering may include parts of all engineering fields, since a person in this area may be involved in the production of an airplane, or an electronic device, or a motor, or almost any other type of system.

Industrial engineering is not only concerned with the production and manufacturing processes from a technical standpoint, but also encompasses the man-machine relationships in terms of safety and operational effectiveness. The field often involves related specialties such as plant engineering (facility design and plant layout), manufacturing engineering, tool design, methods engineering, production control, and quality control. Knowledge of statistical and operations research techniques, manufacturing techniques, process development and control methods, human factors engineering, industrial relations and organizational theory, and engineering economy is necessary.

9. Mechanical engineering. Mechanical engineers design and develop systems which primarily convert energy to useful mechanical forms. This may include both power-generating machines and machines that transform or consume this power in the completion of their particular functions. Mechanical engineers deal with such items as engines, turbines, motors, control mechanisms, transportation systems (automobiles, trains, space vehicles, etc.), refrigeration and air-conditioning systems, propulsion systems (steam, gas, nuclear), and cryogenic systems. The field requires a basic knowledge of mechanics, statics and dynamics, kinematics, turbulence and vibrations, materials, thermodynamics, heat transfer, fluidics and fluid flows, friction and wear, lubrication, power generation, and combustion.

10. Metallurgical, mining, and materials engineering. Metallurgical, mining, and materials engineers deal with the various forms and applications of metals, alloys, and materials in general. This includes the initial location and evaluation of materials, the methods associated with the mining and extraction of various materials from the earth, the accomplishment of land reclamation after the mining and extraction functions have been completed, and the techniques required in refining the extracted commodities (i.e., ores) into the basic metals or comparable alloys needed. This broad field also involves the design of mining and ore refineries and systems for changing the chemical and physical characteristics of metals (i.e., the extrusion, reforming, and/or hardening of metals). Engineers are often called upon to recommend the best materials to use for specific applications, which requires an in-depth knowledge of material characteristics (e.g., stress/strength features, fatigue characteristics).

11. Nuclear engineering. Nuclear engineering deals with the principles of fission and fusion reactions and the controlling of such reactions for power generation and medical applications. The energy dilemma has focused a great deal of attention on nuclear energy sources. Nuclear engineers also design and operate plants that control nuclear reactive materials, dispose of radioactive wastes, and decontaminate radioactive areas.

12. Ocean, marine, and nautical engineering. The field of ocean, marine, and nautical engineering encompasses the design and development of surface ships, submarines, hydrofoils, underwater sea laboratories, sea towers, and related marine structures. Knowledge of sea life, currents and fluid flows, structures, propulsion systems, material applications, marine biology, and the like is necessary.

13. Petroleum engineering. Petroleum engineers deal basically with the application of engineering principles to the exploration, location, development, and recovery of petroleum resources. This includes the drilling, separation and processing, transportation, and storage of crude oil, gases, and related products.

14. Engineering support specialties. The technical specialty areas that we have described are often supported, throughout the system design and development process, by design specialists in fields that generally are interdisciplinary in nature and cover many different types of systems—whether chemical, electrical, industrial, mechanical, or whatever. The objective of these specialists is to support basic engineering design and help to ensure that the end product/process contains certain inherent characteristics. Some of these specialties and their objectives are:

a. *Human factors engineering.* The human factors effort in the system design process is directed toward providing an optimum man-machine interface. Some system functions may be ideally accomplished through automation and others through manual means. Where manual activities are required (e.g., information handling, coordination, and decision making), the aspects of maximum manpower utilization in system operation and support, minimum personnel attrition and training, minimum personnel error rates, and maximum safety are prime considerations. Specialists in this area are directly oriented toward design and the analysis of system operational and maintenance functions from the standpoint of the "man" in the system.[5]

b. *Logistics engineering.* Although logistics in the past has primarily been treated as an afterthought, the lack of proper consideration of supportability in the early phases of system design and development has been costly in terms of total life-cycle cost (specifically, the high costs of sustaining system operation and maintenance throughout the planned life cycle). Thus in the system design and development process, it is necessary to consider (1) the design of prime mission equipment for supportability (i.e., reliability and maintainability), and (2) the design of the resulting system support capability in terms of optimum test and support equipment, supply support, personnel and training, transportation and handling, facilities, and technical data requirements. Design specialists in this area, usually constituting reliability and maintainability engineers, are oriented to optimum system design from the standpoint of the ultimate consumer utilization and support of that system in a true operational environment.[6]

c. *Maintainability engineering.* Maintainability is an inherent characteristic of system design and installation, which is concerned with ease, economy, safety, and accuracy in the performance of maintenance functions. It deals with maintenance times, supportability characteristics in design, and maintenance costs. Specialists in this area address such design features as accessibility, equipment packaging, performance monitoring and fault diagnostic provisions, mobility and transportability, standardization, and interchangeability.[7]

---

[5]Human factors engineering is discussed further in: E. J. McCormick, *Human Factors In Engineering And Design,* 4th ed., McGraw-Hill Book Company, New York, 1976; and D. Meister, *Human Factors: Theory and Practice,* John Wiley & Sons, Inc., New York, 1971.

[6]Logistics is presented in: B. Blanchard, *Logistics Engineering and Management,* Prentice-Hall, Inc., Englewood Cliffs, N.J., 1974.

[7]Maintainability engineering is covered in the following three sources: B. Blanchard and E. Lowery, *Maintainability—Principles and Practices,* McGraw-Hill Book Company, New York, 1969; A. Goldman and T. Slattery, *Maintainability—A Major Element of System Effectiveness,* John Wiley & Sons, Inc., New York, 1967; and C. Cunningham and W. Cox, *Applied Maintainability,* John Wiley & Sons, Inc., New York, 1972. Appendix A covers some of the quantitative measures of maintainability.

d. *Reliability engineering.* Reliability is an inherent characteristic of system design and installation, which is concerned with improving system operational availability and minimizing product/process failures. Specialists in this area are concerned with component selection and design application in terms of extending operational life (i.e., probability of survival) with respect to environmental conditions, stresses, tolerances, and other characteristics of the component in its intended application.[8]

e. *Systems engineering.* Systems engineering constitutes the application of efforts necessary to (1) transform an operational need into a description of system performance parameters and a preferred system configuration through the use of an iterative process of functional analysis, synthesis, optimization, definition, design, test, and evaluation; (2) integrate related technical parameters and assure compatibility of all physical, functional, and program interfaces in a manner that optimizes the total system definition and design; and (3) integrate performance, reliability, maintainability, logistic support, safety, producibility, survivability, human factors, and other related specialties into the total engineering effort.[9]

Systems engineering per se is not considered as an engineering discipline in the same context as the technical specialty areas described above. Actually, systems engineering is a *process* employed in the evolution of system development (particularly for large systems) from the point of when a need is identified through production and/or construction and ultimate deployment of that system in the field for consumer use. This process involves a series of steps accomplished in a logical manner and directed toward the development of an effective and efficient product/process output.

The requirement for systems engineering is brought about because many of the engineering specialists in one or more of the preceding areas (e.g., aeronautical engineering, civil engineering, electrical/electronic engineering) are not experienced to the depth necessary to ensure that *all* elements of the system are considered in a proper and timely manner. Thus, in some organizations, engineers possessing a broad technical background in a number of the specific technical specialties required in the development of large systems are assigned to perform the systems engineering role. This often involves the preparation of system specifications, the integration of the various engineer-

---

[8]Reliability is discussed further in the following four sources: S. Calalabro, ***Reliability Principles and Practices,*** McGraw-Hill Book Company, New York, 1962; W. Von Alven (ed), ***Reliability Engineering,*** Prentice-Hall, Inc., Englewood Cliffs, N.J., 1964; I. Bazovsky, ***Reliability Theory and Practice,*** Prentice-Hall, Inc., Englewood Cliffs, N.J., 1961; and W. Ireson (ed.), ***Reliability Handbook,*** McGraw-Hill Book Company, New York, 1966. Appendix A covers some of the quantitative measures of reliability.

[9]Systems engineering is discussed further in: B. Blanchard, ***Logistics Engineering and Management,*** Prentice-Hall, Inc., Englewood Cliffs, N.J., 1974.

ing design efforts, the test and evaluation of the system, and other related top-level engineering functions.

f. *Value engineering.* Value engineering objectives are directed toward the initial establishment of cost targets and the subsequent evaluation of proposed system and equipment configurations in terms of estimated production and/or construction costs. Specialists in this area should be knowledgeable in material applications and manufacturing processes and the economic considerations associated with various design and production alternatives.

Although there are other branches of engineering, the 14 basic areas described above represent the major ones. Successful accomplishment of engineering objectives (i.e., the development of a product or a process that provides utility to man) requires the technical expertise in any one or a combination of these specialized fields, depending on the output desired. In most instances, because of the complexity of modern systems, this expertise is provided by a team of engineers in varying fields integrated into a project organization and assigned the responsibility to develop the type of system that will be responsive to a need. If the objective is to design a dam, expertise is required from civil engineering, electrical engineering, mechanical engineering, industrial engineering, and perhaps other fields. In the event that an airplane is the desired output product, then aeronautical engineers, electrical and electronic engineers, mechanical engineers, mettallurgical engineers, industrial engineers, reliability and maintainability engineers, and others are required to accomplish the necessary development, production, and support functions.

In any event, engineering (particularly for large systems) is a *team effort*, and each member of the team must be an expert possessing the necessary technical knowledge in his or her field of specialty. In addition, the various individuals involved must be cognizant of the interfaces between their respective specialty areas and economical factors, political factors, ecological factors, societal factors, and the like. Engineering decisions require consideration of these factors in the early system design and development process, and the results of such decisions have a definite impact on these factors. Conversely, these factors usually impose constraints (i.e., input criteria) on the design process. Thus, technical expertise in this instance, as conveyed in Figure 1-1, includes not only the basic knowledge of individual specialty fields of engineering, but a knowledge of those fields in the context of the environment that we live in.

B. Individual Skills

In conjunction with a prescribed level of technical expertise available in the appropriate fields of engineering, it is necessary to cause the transformation of that knowledge into a useful output. This transformation step is dependent on the individual engineering skills assigned to a given project. As conveyed earlier, engineering is concerned with creating tangible products or processes. Thus, the success of any project is based on the creativeness and innovation of the engineers assigned to that project and their individual and collective abilities to employ the methodologies and techniques available in facilitating task accomplishment. Specifically, engineers:

1. Must be sensitive to consumer needs and have a basic understanding of what the product or process is to accomplish and how it is to be utilized. The development of mission scenarios, flow diagrams, or profiles of anticipated consumer operations is appropriate to ensure an understanding of the problem and need.
2. Must be creative enough to visualize the different means or alternatives by which they may be able to accomplish their objectives in developing and supporting the product or process desired. This should be a "blue-sky" type of effort to determine various possibilitites.
3. Must be able to assess what is currently available in terms of the state of the art in technology (i.e., the technical expertise discussed above). They should address such questions as: Will available "off-the-shelf" items perform the necessary function(s)? Can an existing item be modified to accomplish the purpose? Is new equipment development required? Is research required?
4. Must possess the necessary planning and organizational skills to identify the steps and resources required to ensure that product/process development and support are accomplished in a logical and efficient manner. In addition, they must schedule the tasks to be accomplished in a timely manner. What must be accomplished and when should it be accomplished?
5. Must be knowledgeable of the tools available to aid in the completion of their tasks, and must be able to apply these tools judiciously, effectively, and at the proper time. In other words, how should the task be accomplished and where should it be accomplished?

   Requirements in this area may vary considerably depending on the type of system and state of system development. When new systems are developed, engineers are generally involved in analysis, design, experimentation, measurement, and evaluation. In the performance of analytical functions, knowledge of mathematics, statistics, and operations research techniques are necessary. For design, engineers must be aware of available data (e.g., material and component sources, research and the results of experimentation in comparable areas, field data from similar systems already in the inventory, the design process, specifications and standards, computer methods, drafting practices, manufacturing methods, and the like. When involved in measurement and evaluation, engineers should be able to simulate the product or process in an operational environment and/or construct a prototype. They should be conversant with testing techniques, instrumentation, and test facilities and resources. These and other requirements and their applications will be discussed throughout subsequent chapters.

6. Must be able to communicate and work well with others. An engineer can be the best designer in the world; however, if he or she cannot adequately convey ideas and sell constructs, objectives cannot be met. In addition, engineers must be able to work effectively with others since, in most instances, they are not able to do a job alone. As mentioned earlier, engineering is generally a team effort, and individual specialists are highly dependent on each other. Thus, communication and the presentation of results both verbally and in writing are highly significant factors in the engineering process.

From the above, it should be evident that practicing engineers must not only be technically competent in a specialized field but must have certain skills if they are to put their technical know-how to use. These skills will become more evident through study of Part Two.

C. Attitudinal Characteristics

A third major factor necessary in engineering deals with the motivation and mental outlook of engineers in performing their tasks. Successful engineers must:

1. Be objective and open-minded. They should consider all possibilities and not be limited to traditional approaches or unduly compromised for other reasons. Quite often one hears the reasoning that "it should be done this way since we did it this way before." This approach precludes innovation and often promotes obsolescence in product/process design.
2. Be curious and persistent. They should remain current with state-of-the-art advances in technology and their fields of expertise and should pursue the possibility of new applications. They must be inventive and seek better ways of accomplishing things. As worthwhile outputs usually are difficult to attain without a great deal of effort, the attribute of persistence is certainly necessary and the key to success.

3. Assume the attitude of professionals in seeking to initially attain and subsequently maintain the standards of top-notch engineers. This is

acquired through day-to-day activities on the job and their approach to working with others, and is recognized in terms of the respect conveyed to them by their peers.

Attitudinal characteristics are definitely a major factor in the success (or failure) of engineers. Individuals interested in their profession, motivated, and with the proper initiative, will in all probability attain great success in life. Quite often their initiative on the job will far overshadow certain deficiencies in the technical lines and in the application of specific techniques. On the other hand, an individual may possess all the desired attributes required in the technical expertise and individual skill areas but lack the necessary motivation and initiative. Such a person is destined for stagnation and probable failure relative to career growth.

In summary, basic engineering as presented herein involves a number of characteristics (discussed above and identified in Figure 1-1) that must be properly integrated and managed to attain a product or process that will be responsive to a given need. These characteristics are inherent in our discussion from here on.

1.2 PRODUCT LIFE CYCLE

Fundamental to the application of engineering principles is a basic understanding of the *product/system life cycle.* The life cycle commences with the initial identification of a need by the consumer and extends through product planning, research, design, production or construction, evaluation, consumer use and product support in the field, and ultimate product phase-out. The product life cycle (sometimes known as the *consumer-to-consumer cycle*) and the major activities applicable to each phase are illustrated in Figure 1-3. Each step in the process is covered further in Part Two.[10] The steps illustrated in the figure follow a logical evolutionary flow, and the associated functions described for each step represent a full-scale effort. In reality, each product is different and the requirements in terms of "what has to be done" are different. A large system, such as an airplane or a transportation facility that requires new development, may evolve through all the steps illustrated by path A in Figure 1-4. That is, there is a product planning function (including a marketing analysis, a make-or-buy plan, a product design and production plan, etc.); a product research function to investigate

[10]The product life cycle depicted in Figure 1-3 represents a process involving the evolution of system development, production, use, and ultimate phase-out. The steps indicated are not necessarily to be considered in terms of an organizational structure. The activities described may be accomplished by different organizations, depending on the type and complexity of the system.

different methods by which specific objectives can be accomplished; a product design function involving detailed equipment, process, and software design; and so on. On the other hand, a relatively small item such as an electric toaster, a washing machine, or a motor, where design is relatively fixed need not be processed through all the steps but may phase directly from product planning into production (i.e., path C, Figure 1-4). Additional product development involving research and detailed design may not be necessary.[11]

THE PRODUCT LIFE CYCLE			
	CONSUMER	IDENTIFICATION OF NEED	"Wants or desires" for products (because of obvious deficiencies/problems or made evident through basic research results).
	PRODUCER	PRODUCT PLANNING FUNCTION	Marketing analysis; feasibility study; advanced product planning (product selection, specifications and plans, acquisition plan-research/design/production, evaluation plan, product use and logistic support plan); planning review; proposal.
	PRODUCER	PRODUCT RESEARCH FUNCTION	Basic research; applied research ("need" oriented); research methods; results of research; evolution from basic research to product design and development.
	PRODUCER	PRODUCT DESIGN FUNCTION	Design requirements; conceptual design; preliminary system design; detailed design; design support; engineering model/prototype development; transition from design to production.
	PRODUCER	PRODUCTION AND/OR CONSTRUCTION FUNCTION	Production and/or construction requirements; industrial engineering and operations analysis (plant engineering, manufacturing engineering, methods engineering, production control); quality control; production operations.
	PRODUCER	PRODUCT EVALUATION FUNCTION	Evaluation requirements; categories of test and evaluation; test preparation phase (planning, resource requirements, etc.); formal test and evaluation; data collection, analysis, reporting, and corrective action; retesting.
	CONSUMER*	PRODUCT USE AND LOGISTIC SUPPORT FUNCTION	Product distribution and operational use; elements of logistics and life cycle maintenance support; product evaluation; modifications; product phase-out; material disposal, reclamation, and/or recycling.

* Some of the specific supporting functions indicated may be accomplished by the producer throughout and/or at various stages in the product life cycle.

Figure 1-3. The product life cycle.

[11]This does not mean to infer that basic research directed toward new product lines and/or that new innovations concerning an existing product line are not being accomplished on a continuing basis. Research in this instance refers to that additional problem-oriented activity necessary to enable the development of an item that will satisfy a current need.

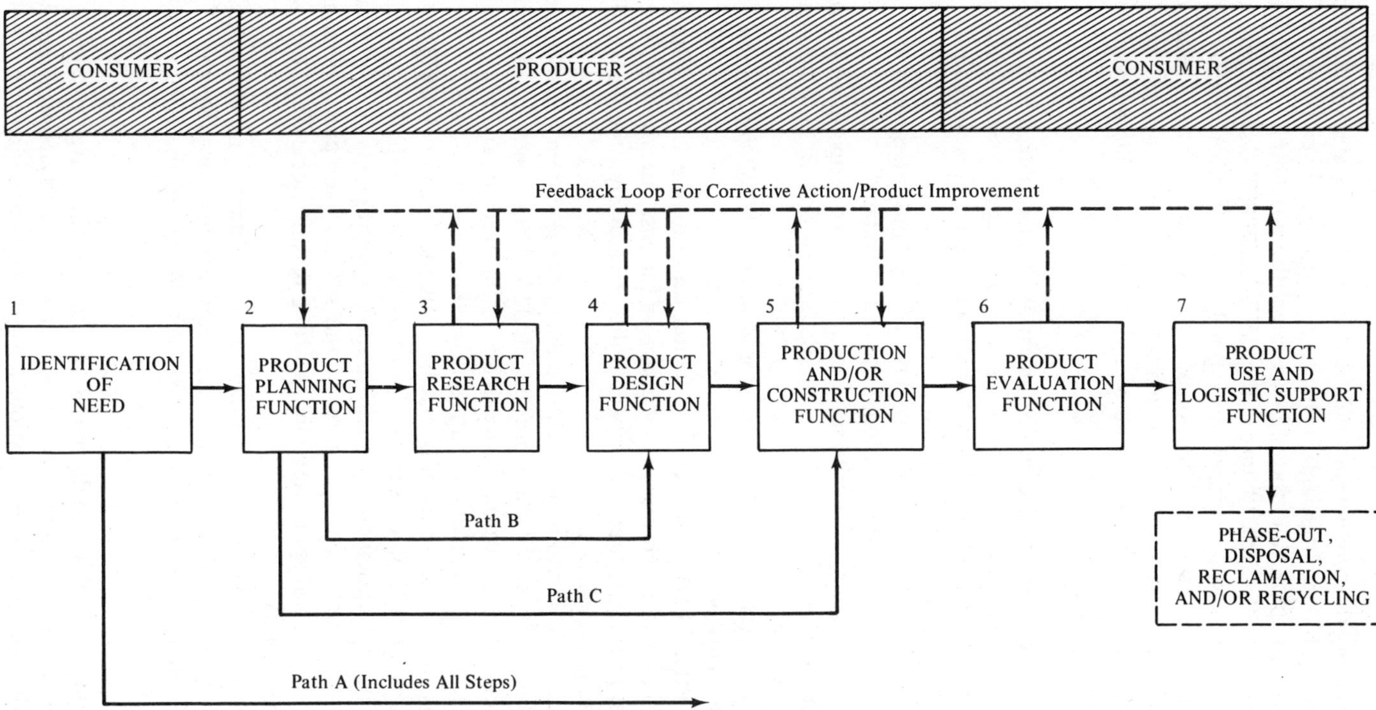

Note: The specific functions/tasks in each block are "tailored" to the individual system need. Only those activities that are actually necessary are accomplished.

Figure 1-4. Product life cycle application.

In discussing the product life cycle further, each of the steps identified in Figure 1-4 (i.e., blocks 1 through 7) should be addressed. Even though all the steps may not be applicable, the engineer must discipline himself to think in terms of the *total* life cycle to ensure that all aspects of the system are adequately and properly considered in a timely and effective manner. This approach stems from the identification of a need that develops as a result of a problem or deficiency and the subsequent "want" or "desire" for a product of some type. From the identification of a given need, one should respond to the following questions:

A. What is the system to accomplish in terms of operations and functional performance characteristics (i.e., range, accuracy, speed of performance, power output, "widgets" produced, liquid flow in gallons per hour, items processed per month, etc.)?
B. When is the system needed? What are the consumer requirements? What is the expected operational life of the system?
C. How is the system to be utilized in terms of hours of operation per day, quantity of "on-off" cycles per month, and so on? Define the system operational cycle.
D. How is the system to be distributed and deployed? Where are the various elements of the system to be located and for how long?
E. What effectiveness requirements should the system exhibit? Effectiveness figures of merit may include factors for cost effectiveness, system effectiveness, availability, dependability, reliability, maintainability, and supportability.
F. What are the environmental requirements for the system (i.e., temperature, humidity, shock and vibration, etc.)? Will the system be located in the arctic, tropical areas, mountainous or flat terrain, and what are the anticipated transportation, handling, and/or storage modes?
G. How is the system to be supported throughout its life cycle? This includes a definition of echelons of maintenance, functions at each echelon, and anticipated logistic support requirements (i.e., test and support equipment, supply support and spare/repair parts, personnel and training, transportation and handling requirements, facilities, and technical data).
H. When the system becomes obsolete and/or when items are removed from the inventory, what are the requirements for disposal? Can specific items be reclaimed and recycled? What are the effects on the environment?

Regardless of the size and type of system (whether large or small, or whether a mechanical system, an electrical or electronic system, a chemical system, or the like), the above questions apply to varying degrees and should be addressed. Answers to these questions generally evolve as part of the product planning function (represented by block 2, Figure 1-4) in the form of the development of system operational and maintenance concepts during a marketing analysis or a feasibility study. This, in turn, serves as the basis for both system design requirements and system planning requirements.

A. System Design Requirements

System performance parameters, operational and maintenance concepts, effectiveness factors, utilization, and environmental considerations basically define the requirements for the system in technical terms. Functional analyses are accomplished and design criteria (i.e., design guidelines and constraints) are established. An evaluation then determines the most appropriate engineering approach relative to acquiring an existing off-the-shelf item, modifying an item that is currently in the inventory, or developing a new item through research and detailed design. The results determine the extent of application of the steps in the product life cycle and the depth of activity required in each step (refer to Figure 1-3).

B. System Planning Requirements

The engineering approach selected (i.e., off-the-shelf versus newly developed items), combined with the consumer need(s) in terms of item quantities, schedules, distribution, and product support philosophies, will determine the requirements for planning. In addition, planning will be significantly affected by the market size and the competition. Planning factors should cover: the applicable product-life-cycle phases; program functions and organization by phase (i.e., what needs to be accomplished and how should it be accomplished?); scheduling; make-or-buy plan; design and production plan; evaluation plan; product distribution, use, and logistic support plan; and the plan for product phase-out, reclamation, and recycling (as applicable).

In essence, the product life cycle as illustrated in Figures 1-3 and 1-4 must be thoroughly understood and should be a major consideration by the engineer in the performance of his objectives. The engineer must intuitively think "life cycle" if the results of his efforts are to be effective. The particular need and resulting requirements developed through the product planning function will determine the specific application and depth of the life-cycle activities for a given system. Response to the questions stated above will aid in this objective. As each system is generally peculiar unto itself, the activity requirements will vary from instance to instance. Thus, some tailoring is required; that is, the appropriate tasks must be accomplished as needed and in a timely manner—not too much, or too little, or too early, or too late.[12] In any event, *all* facets of the life cycle must be addressed and appropriately

[12]Specification of the wrong requirements will result in a product that will not only be ineffective in accomplishing the desired consumer objectives, but will be costly in terms of acquisition and support.

tailored to the need. This approach is assumed throughout this text and discussed further in Part Two.

1.3 ROLE OF THE ENGINEER IN THE PRODUCT LIFE CYCLE

Thus far we have covered the basic requisites of engineering and some of the more conventional engineering disciplines. In addition, the product life cycle has been introduced to a limited degree. Now, we shall attempt to relate engineering functions directly to this life cycle.

There are many different engineering specialty areas which may be applied to varying degrees, depending on the type and complexity of the product or process in question. Also, there are variations in the application of engineering functions to the life cycle, depending on the size of the system and extent of new design and development required (refer to Figure 1-4). Therefore, the role of the engineer will usually be different from one situation to the next. However, in spite of these differences, it can safely be stated that engineering functions of one type or another *are* performed in *each* applicable phase in the product life cycle. Typical engineering activities by phase are discussed below.[13]

A. Product Planning Function (reference: Figure 1-4, block 2)

The product planning function often includes marketing and marketing analysis, the performance of feasibility studies, and advanced planning. The aspect of *marketing* can assume different proportions, depending on the nature of the product. If the product is relatively simple, the marketing function may not require direct engineering involvement except in the preparation of technical material or specifications which support the product in its promotion. On the other hand, for more complex products the marketing effort involves a team approach, constituting the nontechnical sales and promotional orientation, with the engineer in a role of technical support. In the past, many rather complex items have been sold on the basis of the pressures of nontechnical promotional activities alone. However, in today's society, where numerous questions are asked concerning product performance features, reliability and effectiveness characteristics, maintenance requirements, and so on, the need for engineering support is evident. In many instances, the engineer is therefore an active participant in product marketing activities that involve initial customer contacts, the preparation of technical sales information and proposals, and follow-up technical presentations for both internal producer management and customer consumption.

[13]The activities listed herein are presented as examples and must be appropriately tailored to meet the individual need.

The engineer must be responsive to those "what if?" type of questions that are technical in nature.

The accomplishment of *feasibility studies*, often considered as an inherent part of marketing analysis, involves such areas as needs analysis (i.e., size and location of the market, time of need, product performance requirements and characteristics, etc.), the identification of possible solutions to meet the need, the screening of alternatives and the selection of a preferred approach, and the preparation of a proposal. In this area, engineers are initially involved in assessing product performance characteristics and determining the various technical approaches that are possible in responding to the need. Second, given two or more alternatives, the engineer must evaluate each on the basis of technical performance, reliability and maintainability, size and weight, cost, and the like. As a final point, the engineer assists in the preparation of follow-up proposals and/or reports by providing the necessary technical inputs.

The third aspect of this program phase is that of *advanced planning*. Once a preferred product approach is selected, it is necessary to prepare a product specification; develop schedules; and prepare preliminary plans associated with product acquisition (design and production), product test and evaluation, product use and logistic support, and product phase-out and material disposition. In each area, engineering plays a significant role in response to such questions as those presented in Section 1.2. The engineer assists in the preparation of product specifications and ensures that the various plans and schedules are realistic and feasible from the technical standpoint. This activity is discussed further in Chapter 3.

B. Product Research Function (reference: Figure 1-4, block 3)

Research constitutes the "investigation or experimentation aimed at the discovery and interpretation of facts, revision of accepted theories or laws in the light of new facts, or practical application of such new or revised theories or laws."[14] For the purpose of discussion, research can be categorized as *basic research* and *applied or directed research*. *Basic research*, as defined herein, covers that scientific area of activity dedicated to advancing the state of the art in those technical fields of specialty considered significant by the internal management of an industrial firm or a government agency. In other words, the management of an industrial firm considers it necessary to expend resources in certain areas in anticipation of a future need. In this instance, the research activity is usually initiated from within.

Applied research, on the other hand, is defined as that scientific activity dedicated to advancing the state of the art in one or more technical fields of

---

[14] *Webster's Seventh New Collegiate Dictionary*, G. & C. Merriam Company, Springfield, Mass., 1963.

specialty, with the ultimate objective of solving a problem or correcting a deficiency. In this case, a problem has been identified and current technology is not available to respond to the need. Thus, specified goals are established and research is conducted accordingly. The requirement for research here is generally initiated from outside the industrial firm or government agency.

In either case, research involves engineering activity that is both extremely theoretical in nature and yet practical relative to direct product applications. Highly specialized engineers working with scientists in scientific fields are required when the objectives constitute advancing the state of the art in a given specialty area. As research results are generated, it is necessary to convert the new knowledge attained into a form that can lead to product development, production, and ultimate consumer use. In many firms/agencies, engineering activity has been consistent and productive in the first area; however, the transition of research results into product development has been a problem. Actually, both areas are equally important and engineering effort should be applied accordingly. Chapter 4 elaborates further on the functions of research.

C. Product Design Function (reference: Figure 1-4, block 4)

The design process stems from a set of stated requirements for a given system and evolves through (1) *conceptual design* (i.e., the establishment of performance parameters, operational requirements, and support policies), (2) *preliminary systems design* (sometimes called *advanced development*), and (3) *detailed design.* This process generally commences with a visualization of what is required and extends through the development test, and evaluation of an engineering or prototype model of the system. The output constitutes a configuration that can be directly produced or constructed from a specification, a set of drawings, and supporting documentation.

The engineer's role in this program phase involves a variety of functions that are dependent on the type of system and the extent of new development necessary. These functions may include all or any combination of the following:

1. Accomplishing functional analyses to identify the major operational and maintenance support functions that the system is to perform.
2. Establishing criteria (i.e., qualitative and quantitative technical parameters, bounds, and constraints) for system design.
3. Evaluating different alternative design approaches through the accomplishment of system/cost effectiveness analyses and trade-off studies.
4. Preparing system and component specifications.
5. Selecting components for the system and recommending supplier sources.

6. Assisting the purchasing and contracting functions in the preparation of supplier specifications and contractual documentation.
7. Preparing functional design layouts, engineering drawings, parts and material lists, standards, and so on, with the objective of thoroughly defining the product or process through documentation. This includes preliminary drawings, schematic diagrams, wiring and cable harness drawings, interface drawings, installation drawings, and manufacturing drawings. The engineer determines and specifies dimensions and design tolerances, input-output requirements, material applications, forms and finishes, and basic manufacturing assembly sequences.
8. Assessing the design through predictions, analyses, and the performance of periodic design reviews. This assessment is basically accomplished through a review of the engineering documentation described in item 7.
9. Developing breadboards, engineering models, and prototypes for system evaluation purposes.
10. Developing system software. Software refers to computer programs, associated data bases, and related documentation required to define, design, test, produce, operate, and maintain the system.
11. Developing system and component test specifications and procedures, and accomplishing specific tests to ensure that all design requirements are met.
12. Performing design modifications as necessary to correct deficiencies and/or to improve system design.

The performance of these functions is applicable to the design of prime mission equipment, design of the various elements of logistic support, and the design of associated software. The design function generally involves a team effort (particularly for large systems) which utilizes the services of engineering specialists from the appropriate technical fields, discussed in Section 1.1A, and supported primarily by draftsmen, technicians, and representatives of comparable technical trades. The engineer quite often visualizes himself only in the role of a "builder of hardware." However, the role of the engineer in product design is much more extensive, and his responsibilities are far greater. Full coverage of engineering design functions is presented in Chapter 5.

D. Production and/or Construction Function (reference: Figure 1-4, block 5)

This phase may constitute (1) the production of a multiple quantity of like items (i.e., mass production, or a *flop shop* type of operation), (2) the production of small quantities of a wide variety of different items (i.e., a *job-shop* type of operation), and/or (3) the construction of a single item, such as a large structure of some type. In production and/or construction operations, material and personnel resources must be combined in such a manner as to provide the necessary product/system output in an effective and efficient manner. Production actually stems from the point where system design is considered fixed and includes the total flow of materials, from the acquisition of raw materials to delivery of the finished product for evaluation and ultimate consumer use. The production flow process, regardless of the product type, involves inventories, material acquisition and control provisions, tooling and test equipment, transportation and handling methods, facilities, personnel, and data. Engineering is directly required in the design and development of a production capability and for defining the resources necessary for a large construction project. Engineering functions may entail:

1. Design of facilities for product fabrication, assembly, and test functions. This includes determining the capacity and location of both manufacturing and storage facilities, utility requirements, capital equipment needs, and material handling provisions.
2. Design of manufacturing processes (e.g., sequencing of tasks, man-machine operations, process specifications).
3. Selection of materials and the determination of inventory requirements.
4. Design of special tools, test equipment, transportation and handling equipment, and fixtures used in production/construction operations.
5. Establishment of work methods, time, and cost standards, and the subsequent evaluation of production/construction operations in terms of the established parameters.
6. Evaluation of production/construction operations to ensure that product performance, quality, reliability, maintainability, safety, and other desired features are maintained throughout the production/construction process. This also includes the initiation of corrective action as required.

These functions are generally categorized under industrial engineering, plant engineering, manufacturing engineering, methods engineering, production control, and/or quality control. Further discussion of this phase is included in Chapter 6.

E. Product Evaluation Function (reference: Figure 1-4, block 6)

Throughout product/system development, it is necessary to perform an evaluation or assessment to ensure that the end result will conform to the initially established requirements and meet the need(s) of the consumer. This evaluation is accomplished at periodic intervals and may take the form of an analysis of some type (i.e., an analytical evaluation) or a test of a system component or the entire system as an entity (i.e., a physical demonstration). The evaluation effort is iterative in nature and intended to provide the assurance that all is going well. If not, changes must be incorporated to correct those deficiencies which have been identified. Changes accomplished after the system is delivered to the consumer and in operational use are usually quite costly, whereas changes incorporated earlier in the life cycle to preclude a problem that may later occur can often be accomplished more economically. Thus, a key objective of the evaluation function is to detect possible problem areas as early as possible.

The evaluation function pertains to the design and development of prime mission equipment, the various elements of logistic support, associated software, and the production capability itself or the construction process, as applicable. Although this function is shown serially in Figure 1-4, evaluation is actually accomplished during design and throughout production and/or construction, involving some overlap of the functions depicted in the figure. Hence, one might consider blocks 4, 5 and 6 as an iterative loop within itself. This is supported by the feedback and corrective-action process shown in the figure.

Engineering plays an active role in the evaluation function by participating in the following activities:

1. Identification of test and evaluation requirements and the development of test plans for the system and its components. What needs to be tested? What are the system technical parameters to be measured? How is the test to be accomplished?
2. Determination of the logistic support resources required for test completion.
3. Accomplishment of specific test functions (with the support of qualified technicians).
4. Analysis of test data and the preparation of test reports.
5. Initiation of system modifications as required to correct deficiencies and/or for product-improvement purposes.

The specific objectives of product evaluation, the categories of test, and the procedures for test accomplishment are included in Chapter 7.

F. Product Use and Logistic Support Function (reference: Figure 1-4, block 7)

Functions during this phase constitute (1) consumer use of the product or system throughout its intended life cycle, (2) incorporation of product or system modifications for improvement, (3) the logistic support requirements necessary to ensure that the product or system is operationally available when needed, and (4) the ultimate phase-out and disposal of the product or system due to obsolescence or wear-out. These functions may be accomplished by the consumer alone, by the consumer with the support of the producer in certain areas, or by the consumer with the support of an outside organization (other than the producer) performing specific tasks. In any event, the engineering role (regardless of the organizations represented) may encompass the following:

1. Providing field service or customer service engineers at strategic geographical locations to assist the consumer in the day-to-day operation and maintenance support of the system.
2. Providing engineering support in the installation of system modifications and in the subsequent checkout of the system to ensure satisfactory operation.
3. Providing engineering support in the phase-out and disposal of the system at the end of the life cycle and in the subsequent reclamation and recycling of system components as appropriate.

This final phase of the product life cycle is discussed in Chapter 8.

As stated earlier, the engineering role in the product life cycle may vary considerably from system to system and should be tailored to meet the need. For small systems, one or two engineers may be assigned to accomplish all functions in each of the applicable program phases. An individual engineer may perform design tasks, production tasks, test and evaluation tasks, and others. In the development of large systems, any number of engineering specialists may be required in order to perform different tasks of a highly complex nature. In this instance, the engineering role is accomplished through a team effort.

In fulfilling the basic engineering requirements for various projects, the industrial firms, government agencies, or other organizations responsible for accomplishing these requirements will establish various categories of engineering specialists to suit their specific project applications and organizational structures. These categories may or may not directly coincide with the engineering specialist categories described in Section 1.1A. In fact, in most instances the industrial firm or government agency will develop new engineering classifications that will satisfy particular job needs. An example of such is presented in Figure 1-5. The engineering job classifications in the figure were taken in part from the want ads in several current news publications. This represents the market or need for engineers. It is readily evident that the more conventional classifications discussed in Section 1.1A and covered in most universities and colleges can be categorized further in terms of either the functions performed by the product life-cycle phase (e.g., sales engineer, tool engineer, manufacturing engineer) or by the type of system being developed (e.g., radar engineer, computer engineer). In any event, there are many classifications of engineers in our society today, and some of these classifications are continually changing as the state of the art in technology advances.

Figure 1-5. Typical engineering job-oriented classifications.

1.4 PROBLEMS AND CHALLENGES

The discussion up to this point has alluded to some problems and challenges for the future. The effective accomplishment of the many and varied engineering functions described earlier depends on the extent to which these problems and challenges are recognized, and the organization and management approaches employed in the fulfillment of engineering objectives. Engineering organization and management, in order to provide successful results, must address a number of areas, a few of which are highlighted below.

A. Large-Project Organizations

While relatively small products such as a portable radio, an electrical household appliance, or a small automobile may employ a limited amount of direct engineering resources, there are many large-scale systems that require the combined input of specialists representing a wide variety of engineering disciplines. Two typical examples include a ground mass-transit system and a commercial airplane.

1. Ground mass-transit system. Civil engineers are required for the layout and/or design of railroad tracks, tunnels, bridges, cables, and facilities. Electrical engineers are involved in the design of automatic train control provisions, traction power, substations and power distribution, automatic fare collection, digital data systems, and so on. Mechanical engineers are necessary in the design of passenger vehicles and related mechanical equipment. Architectural engineers may provide support in the construction of passenger terminals. Reliability and maintainability engineers would likely be involved in the design for system availability and the incorporation of supportability characteristics. Industrial engineers will deal with the production aspects of passenger vehicles and vehicle components. Test engineers will evaluate the system to ensure that all performance, effectiveness, and system support requirements are met. Engineers in the planning and marketing areas will be required to keep the public informed and to promote the technical aspects of the system (i.e., keep the politicians and local citizens happy). Additional engineering specialists of various categories will be required to perform specific project-related tasks on an as-required basis.

2. Commercial airplane. Aeronautical engineers determine aircraft performance (i.e., flight) requirements and design the overall airframe structure. Electrical engineers design the aircraft power distribution system and ground power requirements. Electronic engineers of various types are responsible for the development of subsystems, such as radar, navigation, communications, and data recording and handling. Mechanical engineers are called upon in the areas of mechanical structures, linkages, and hydraulics. Metallurgists are needed in the selection and application of materials for the aircraft structure. Reliability and maintainability engineers are concerned with availability, dependability, survivability, and system logistic support. Human factors engineers are interested in the man-machine functions, cockpit and cabin layout, and the design of the varous pilot controls panels. Systems engineering is concerned with the overall development of the airplane as a system and ensuring the proper integration of the numerous aircraft subsystems. Industrial engineers of various types are directly involved in the production of the aircraft itself and its many components. Test engineers are required to evaluate the system to ensure conformance to customer requirements. Other engineering specialties are employed on a task-by-task basis.

Although the above is not necessarily all-inclusive, it is apparent that many different engineering disciplines are directly involved. In fact, there are some large projects such as the development of a new airplane where the number of engineers assigned to perform engineering functions varies from 5,000 to 10,000. In addition, the quantity of different engineering types (refer to the job-oriented classifications in Figure 1-5) often ranges in the hundreds. These engineers, forming a part of a large organization, must not only be able to communicate with each other but must be conversant with such interface areas as purchasing, accounting, personnel, and legal.

Large projects of this type will usually fluctuate in terms of manpower loading. Depending on the functions to be performed on a project phase-by-phase basis (i.e., the product life cycle), some engineers will be assigned to a project until system development and production is completed, and others will be brought in for a short term to perform specific tasks. The requirements in terms of project man loading will also vary from phase to phase since the emphasis changes as system development progresses. During product planning, individuals with a broad systems engineering background are needed in greater quantities than detailed design specialists, whereas during the product design function, the reverse may be true. In any event, the needs for engineering will change as a system evolves through the product life cycle.

Another major factor associated with large projects is that a good deal of the activity that involves system development, production, evaluation, and support is often accomplished at supplier (sometimes known as subcontractor) facilities located throughout the world. In other words, there is a prime producer or contractor who is ultimately responsible for the development and production of the total system as an entity, and there are numerous suppliers providing different system components. For large systems, the percentage of supplier activity in terms of total system cost often ranges from 75 percent downward. That is, a great deal of the project work and associated engineering functions may be accomplished at dispersed locations.

These aspects of the many and varied quantities of engineering specialists required for large projects, combined with the shifting of needs by program phase and the integration effort involved in the prime producer-supplier activities, create certain inherent problems and present a definite challenge relative to engineering organization and management.

B. Technological Growth and Change

When discussing technological growth, there are trends that are somewhat of a long-term and continuous nature, and there are others that are more related to the immediate future. An example of some long-term possibilities is presented in Figure 1-6. Technological growth is occurring continuously and is stimulated through response to some unmet current needs and/or by attempting to accomplish ongoing activities in a more effective and efficient manner. In addition, stimulation in this area is being significantly influenced by social changes, political objectives, ecological factors, and economic considerations. For instance, concern for ecology (i.e., cleaning up the environment) has resulted in certain legislation which creates additional needs for developing new methods of reducing air and water pollution and the handling plus disposal of solid waste. The more recent concerns involving a possible energy shortage have caused a shift in emphasis toward the further development of other resources (i.e., solar energy, nuclear power, etc.) to supplement current oil and gas supplies. These and comparable situations, created either through properly planned programs or as a result of political or societal panic reactions, have definitely stimulated technological growth.

The response in fulfilling technology requirements is highly dependent on the scientists and engineers available in the needed fields of expertise, and whether they are up to date and creative in their respective specialty areas. In some fields, such as electronics and the medical profession (from the standpoint of engineering innovations), technological growth is rapid. Engineers in this instance have an extremely difficult time maintaining currency and keeping abreast of what's going on. In other areas, the pace is not as rapid. However, engineers in general tend to lose their technical competency within 3 to 10 years after college graduation unless they actively pursue some form of continuing education or happen to be involved in ongoing challenging project work which constitutes advancing the technical state of the art in some manner. Engineers who are assigned to projects and allow themselves to become obsolete not only inhibit technological growth but are a burden relative to both slowing down on-the-job progress and increasing project costs. This, in turn, adds to the overall system cost, which may be a particular detriment in a highly competitive market.

The challenge in this instance is to organize and manage an engineering effort which will not only stimulate technological growth but accomplish such in an effective and economical manner.

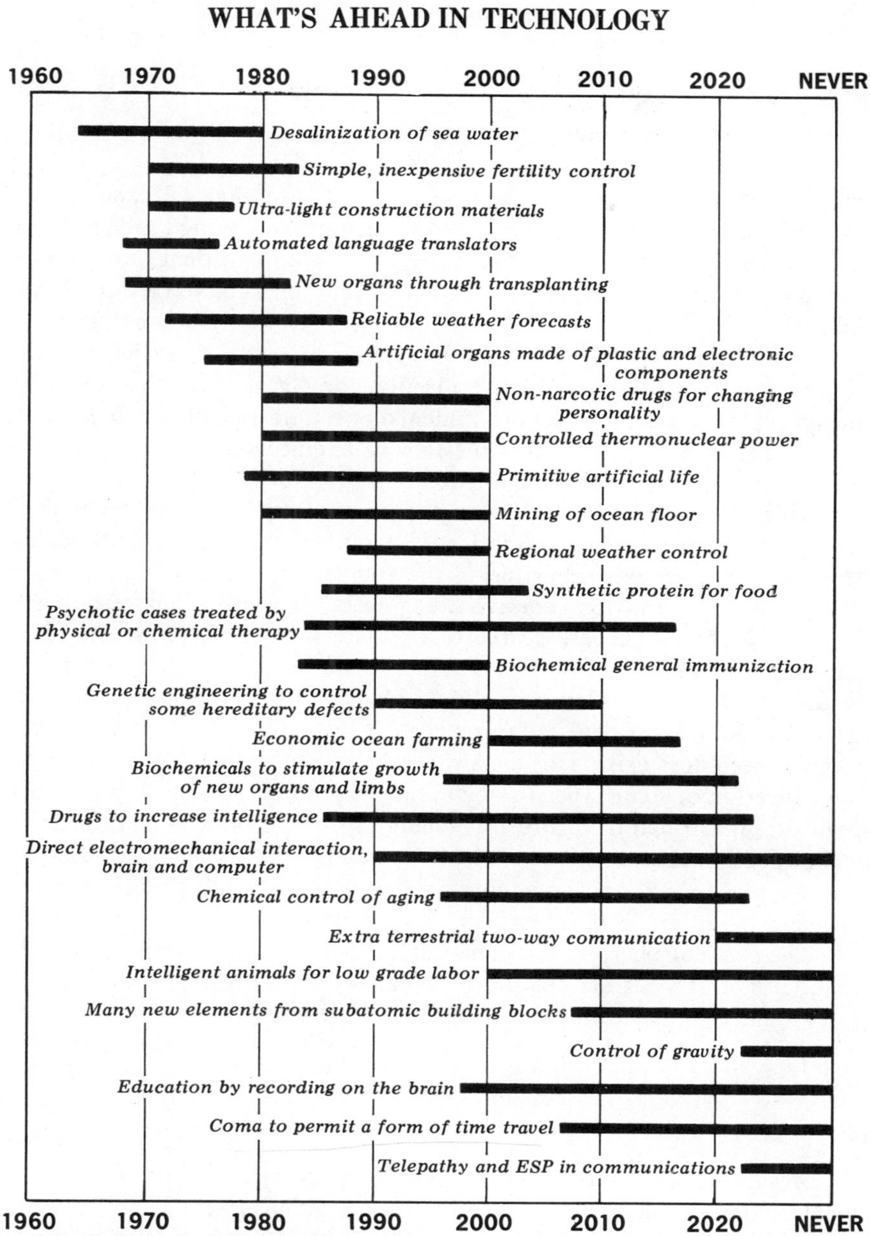

Figure 1-6. This chart on anticipated future breakthroughs is based on predictions by a panel of experts in a Rand Corporation study by Olaf Helmer, a mathematician. Bar lengths represent range of time estimates by "middle half" of the panel, with one-quarter of group predicting earlier and one-quarter later dates. Helmer's findings appear in his "Social Technology" (Basic Books, Inc., 1966). (*Source:* Daniel V. DeSimone, "Invention, the First Step, Is Often the Most Difficult," *The New York Times* (January 8, 1968. © by the New York Times Co. Reprinted by permission.)

C. Increased Demands[15]

A fairly large number of engineers in today's market shift their career objectives to less technical areas (such as business and related services) soon after graduation. Others are terminated involuntarily in their current jobs, owing to project cancellation or to the fact that their achieved levels of competence are no longer required. Some engineers change fields because they feel some degree of stagnation or frustration in terms of anticipated job growth. These situations, related through the "grapevine," cause future prospects (e.g., secondary school students) once interested in engineering to enter other career fields. Thus, the flow of engineers away from technical assignments often becomes significant.

This situation combined with the continuing trends in technological growth has created an increasing demand for engineers in many fields. Even with fluctuations occurring in the engineering job market, there will always be a demand for *competent* engineers. It is true that certain job-oriented positions become scarce or obsolete with time. However, if the engineer is at all astute, he should in most instances be able to detect trends in this direction and plan accordingly for a satisfactory job transition by acquiring knowledge and competence in some other closely related field where a need does exist. This usually can be accomplished through continuing education of some type. In any event, the engineer possessing the basic requisites illustrated in Figure 1-1, together with the interest and initiative, will be successful and will always be in demand. In fact, such attributes will soon be recognized, and an engineer possessing them will probably receive frequent job offers from other firms or agencies.

The challenge here is to organize and manage an engineering operation that will not only attract competent engineers to fulfill the necessary technical requirements but will also provide the motivation to promote individual job challenge, the feeling of personal accomplishment, the working environment, salary, and the like.

In summary, the basic aspects of engineering organization and management are not simple to understand nor are conditions fixed. There are small projects and large projects. Large-project organizations vary from system to system, and the type and variety of engineering specialties required are numerous. The demand for engineers shifts somewhat with technological

---

[15]A study of future engineering career opportunities is presented in *Career Outlook in Engineering* (IEEE Manpower Report 1973), The Institute of Electrical and Electronic Engineers, Inc., New York. This study projects increased technology growth and engineering manpower needs.

changes, and engineers in general are highly mobile by nature. These and related factors cause many problems and in turn create a definite challenge in the development, production, evaluation, operation, and support of a system in an effective and efficient manner.

1.5 TEXT OBJECTIVES

This chapter provided an introduction to engineering organization and management through a rather basic discussion of:

A. The definition of engineering in terms of the conventional engineering specialty areas plus the desired attributes or requisites of engineering (Section 1.1).
B. The definition of the product life cycle and the basic program functions accomplished in each phase of the life cycle (Section 1.2).
C. The role of the engineer in the product life cycle and applicable engineering job requirements for each life-cycle phase (Section 1.3).

The discussion thus far commenced with a response to such questions as "what is engineering?" and "what does engineering do?" and has been extended to include some coverage of the actual functions that engineers perform in industrial firms, government agencies, and other comparable organizations. In addition, a few major challenges associated with large-project organizations, technological growth, and future engineering demands have been identified. The ordering of the material up to this point is intended to cover the transition of engineering from the academic world to typical project applications that the engineer is likely to encounter after graduation. This, in turn, sets the stage for the material presented from here on.

As the reader progresses through subsequent chapters of this text, it is hoped that he or she will acquire a feeling for the role of the engineer in our society. The various aspects of the product life cycle for both small- and large-scale systems will be thoroughly covered in Part Two. This material will lead into the organization of engineering resources to accomplish product-life-cycle objectives in Part Three. In accomplishing these objectives, the typical engineer will have numerous interfaces with supporting nontechnical functions, such as finance, budgeting and accounting, purchasing, personnel, legal, and safety (Part Four). Finally, the engineer in fulfilling his role will be faced with external constraints of a technological, economic, political, ecological, and social nature (Part Five).

The objective of this book, through the material covered in Parts Two through Five, is to assist engineers to better understand the functions that they will likely perform in the real world and the many interfaces that they will encounter in the accomplishment of their objectives.

QUESTIONS AND PROBLEMS

1. Define engineering (in your own words).
2. What does engineering accomplish? What are some of the benefits to society? Name some of the detriments to society.
3. What are the basic personal attributes or requisites for engineering?
4. What are the basic differences between scientists and engineers?
5. Can engineers possessing the necessary technical expertise in their respective fields successfully accomplish their objectives? If so, how?
6. Once engineers graduate from college, they are considered experts in their respective technical fields of specialty. True or false? Why?
7. What is the definition of a system?
8. The various phases of the product life cycle shown in Figure 1-3 are applicable to all systems. True or false? Why?
9. Select a system of your choice and define the product life cycle for that system (construct a detailed flow diagram).
10. What is the scope of engineering activity in the product life cycle?
11. In determining system requirements, the engineer needs certain information. Describe that information.
12. All engineers who graduate from college in the same technical specialty area are likely to perform comparable functions after graduation. True or false? Why?
13. Does the consumer become involved in the system design process? If so, how?
14. How does engineering become involved in the marketing of a system? In the production of a system? In the evaluation of a system? In the support of a system throughout its life cycle? In the phase-out and disposal of a system?
15. What is the engineer's role in research?
16. Engineers entering the job market are likely to remain in the same jobs throughout their careers. True or false? Why?
17. Engineers are faced with certain challenges throughout their careers. Describe a few of the more significant ones.
18. In determining the specific engineering activities that are to be applied in each phase of the product life cycle, tailoring is required. What is meant by this?
19. The engineering job market fluctuates from time to time, and in periods of depression the need for engineers is nil. Is this true? Why?
20. What are some of the major challenges of engineering organization and management? List some objectives that you would establish in meeting these challenges.

2

FUNDAMENTAL CONCEPTS OF ENGINEERING MANAGEMENT

Within the context of the basic definition of engineering as presented in Chapter 1, it is necessary to set the stage for the material presented in subsequent sections by introducing a few fundamental concepts associated with engineering management. Goals and objectives must be established, an organization must be created, decisions are made, and management activities must prevail to ensure that the established goals and objectives are fulfilled.

2.1 GOALS AND OBJECTIVES OF ENGINEERING

A basic objective of engineering is to provide a system that will satisfactorily fulfill some identified need. Not only must the system *perform* in a specified manner, but it must do so both *effectively* and *efficiently*. That is, the development and production of a system must consider the appropriate combination of the following:

A. *System performance parameters*—capacity, energy output, delivery rate, range, accuracy, speed of performance, volume, weight, shape, and so on.

B. *System operational and support factors*—system effectiveness, availability, dependability, operational readiness, reliability, maintainability, supportability, transportability, and so on. Many of these terms are defined in Appendix A.

C. *Economic factors*—anticipated revenues, life-cycle cost (i.e., all future system costs to include research and development, investment, operation and support, and disposal costs), market sales price, producer cost, consumer cost, and so on. These factors are discussed further in subsequent sections.

Good system design reflects an optimum balance of these factors, which is attained through a sustaining trade-off and analysis effort accomplished at the early stages of system development. The results of this effort constitute engineering decisions that have a significant impact on what will occur later during system production or construction, operational use and support, and ultimate phase-out. The proper attention given to certain factors early in the life cycle may avoid problems later while ignoring other factors may prove to be quite costly. As one progresses through the life cycle, engineering decisions are made that could have a significant feedback effect on what has already been accomplished. For instance, if the consumer utilizes the system in a manner different from that which was intended initially in the design process, causing unanticipated stresses on the system, the results may be damaging. In essence, engineering decisions are made at various stages in the product life cycle, and the consequences, in most instances, will ultimately determine the success or failure of the system in meeting its objectives.

For most small- and large-scale systems, the accomplishment of engineering activities is the result of a team effort, involving a combination of:

A. Personnel with the technical expertise, individual skills, and the proper attitudinal characteristics necessary to perform required engineering functions (refer to Section 1.1).
B. Personnel trained as technicians, draftsmen, computer programmers, model builders, and other comparable tradesmen who directly support the professional engineering specialist (item A) in the performance of engineering functions.
C. Personnel representing nontechnical fields whose skills are necessary in providing direct support to ensure that engineering functions are accomplished in an effective and efficient manner. This includes financial management, budgeting and accounting, purchasing, personnel, legal, safety, and other related functions. These fields are discussed in Part Four.
D. The physical and natural resources (e.g., energy, materials) required for support in the accomplishment of manual tasks and in the production/construction of both hardware and software components of the system.

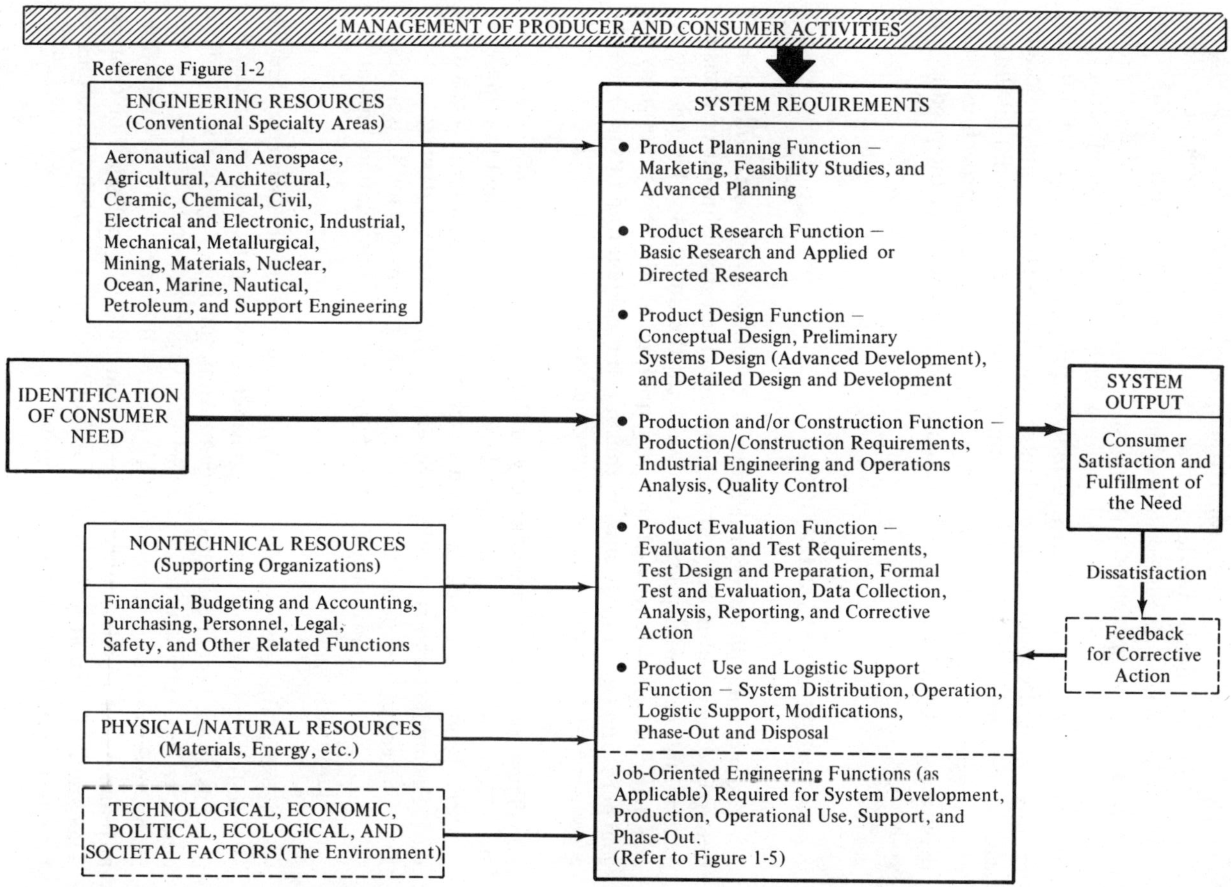

Figure 2-1. Integration of engineering and support resources.

These activities must be properly integrated (through effective organization and management) in order to meet the job-oriented requirements of the product life cycle as discussed in Section 1.3. Figure 2-1 illustrates this integration process from a broad standpoint. Thus, the objective of engineering is to produce a system that will perform satisfactorily in an effective and efficient manner. This, in turn, usually requires a team effort of different engineering and support personnel integrated into a structure of some form with the responsibility of accomplishing this objective. Successful fulfillment of this responsibility is dependent on the organizational approach used and the management of engineering decisions made throughout the product life cycle.

2.2 ENGINEERING ORGANIZATION

The purpose of an organization is to obtain the best utilization of resources (discussed in Section 2.1) through the establishment of decision-making and communications processes with the intent of achieving the full creativity and productivity of the personnel assigned. This is certainly our goal here, and from our discussion thus far, it is evident that this goal may be accomplished by varied means, depending on:

A. The type, size, and complexity of the system in question.

B. The extent of new design and development required, and the phases in the product life cycle through which the system is to be processed.

C. The budget and schedule constraints affecting the system.

D. The consumer, producer, and supplier requirements for a given system (e.g., the responsibility for functions and the interrelationships among the three).

Based on the needs, organization planning commences with a definition of specific requirements and the consideration of different organizational patterns (e.g., line/functional organization, project organization, and a mixed project-staff structure). The selected organizational pattern may vary from phase to phase in the product life cycle as the needs shift. Organizational planning also includes the consideration of levels and span of control, delegation of responsibilities and authority, and the basic principles of formal organization, which are as follows:[1]

A. Someone should be appointed and held responsible for supervising and managing all essential activities.

[1]These principles are taken in part from and discussed in: B. M. Bass, *Organizational Psychology*, Allyn and Bacon, Inc., Boston, 1968.

B. Project functions should not be duplicated or overlapping.
C. No single job position should have too many or too complex tasks. Tasks should be grouped in a meaningful manner such that position holders are not held responsible for a wide variety of nonrelated activities.
D. Position responsibilities should be written in a clear and concise manner and thoroughly understood by job occupants.
E. The authority for decision making should be commensurate with the responsibility associated with those decisions.
F. The delegation of authority for decision making should take place as close as possible to the point of action.
G. Managers should preferably have from 4 to 8 subordinates reporting to them. It is recommended that first-line supervisors have no more than 20 personnel.
H. Every position occupant should know who reports to him (or her) and to whom he should report. The chain of command should be clear, recognized, and followed. This chain of command may vary in different instances depending on the nature of information to be passed on, the problem type, and so on.

With preliminary planning accomplished, the next step constitutes structuring the organization, which includes determining the functions and tasks to be performed, defining the organization charter, and preparing position descriptions. Organization staffing follows through personnel recruiting (i.e., advertising, interviewing), hiring, and training. Once the organization is staffed and performing its assigned functions, it is necessary to accomplish an ongoing evaluation of organizational effectiveness. Consideration of such factors as personnel motivation, personnel development, salary growth, and working environment is critical if an organization is to be ultimately effective and productive.

These basic concepts, which are applicable to any organization, are presented at this point merely as an introduction to the subject area; this area is presented in depth in Part Three. The interest here (in the context of this book) pertains primarily to the organization of engineering and related functions.

2.3 ENGINEERING DECISION MAKING

Throughout the product life cycle, engineers are called upon to make many decisions or to provide the necessary technical information as an input to aid higher-level managers in making decisions. Decisions may range from establishing consumer operational deployment needs to the selection of a preferred system design configuration, the selection of appropriate supplier procurement sources for components, the determination of production facility requirements, the determination of the best logistic support policy for the system throughout its life cycle, and/or whether to incorporate a system modification. In each instance, the engineer or his manager is faced with the problem of deciding on a preferred approach.[2] There may be a number of feasible aternative approaches possible, each of which represents a configuration that can be evaluated in terms of some measure of performance, effectiveness, economics (revenues and total life-cycle cost), or an equivalent figure of merit. The intent is to select the best approach by conducting a series of individual evaluations where each evaluation stems from a specific problem definition.

Regardless of the type of problem at hand, the decision maker should initiate certain actions and follow an evaluation process that will ensure that all necessary factors are considered in a logical and methodical manner prior to the establishment of a firm approach. An example of such a process is illustrated in Figure 2-2 and described below.[3]

A. Definition of the Problem

The initial step in any analysis effort commences with the clarification of objectives, defining the issues of concern, and limiting the problem so that it can be studied in an efficient and timely manner. In many cases, the nature of the problem appears to be obvious, whereas the precise definition of the problem may be the most difficult part of the entire process. However, unless the problem is clearly and precisely defined, it is doubtful whether an analysis of any type will be meaningful.

B. Identification of Feasible Alternatives

The next step is to identify alternative solutions to the problem. All possible candidates must be initially considered, and yet the more alternatives that are considered, the more complex the analysis process becomes. Thus, it is desirable to list *all* possible candidates to ensure against inadvertent omissions, and then to eliminate those candidates which are clearly unattractive, leaving only a few for evaluation. Those few candidates may be presented in the form of a "decision tree" type of matrix, as illustrated in Figure 2-3, and then analyzed with the intent of selecting a preferred approach.

---

[2]The word "preferred" in this instance refers to the best approach possible within the constraints imposed by political, environmental, economical, time, or other factors. It does not imply "optimal," as optimization may not be possible because of these constraints.

[3]The analysis approach as described was taken in part from: B. Blanchard, *Logistics Engineering and Management,* Chapter 6, Prentice-Hall, Inc., Englewood Cliffs, N.J., 1974.

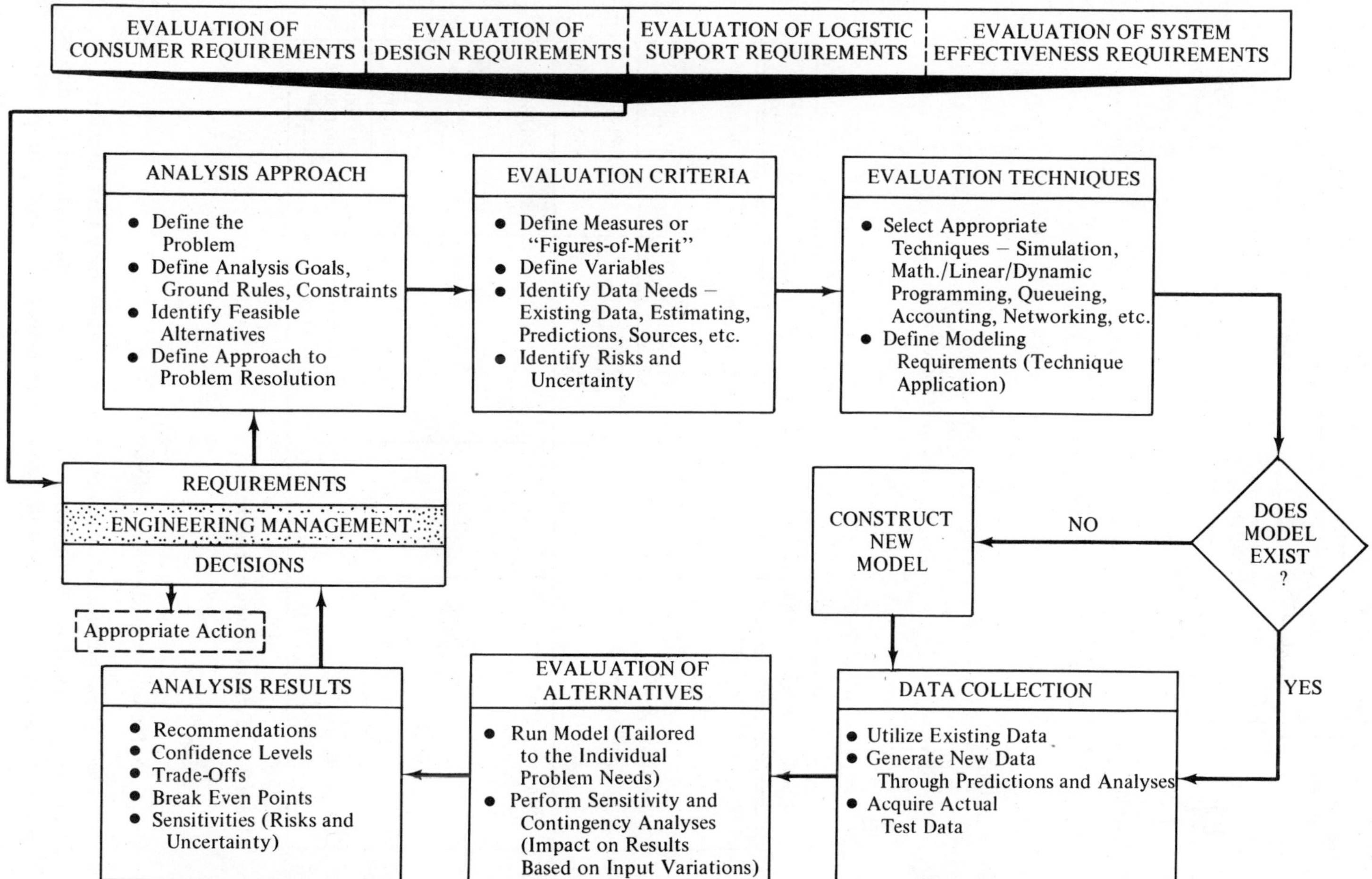

Figure 2-2. Engineering analysis process.

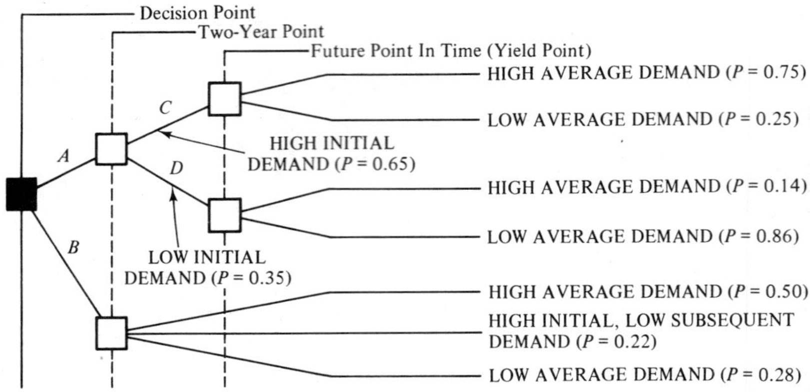

LEGEND

A PURCHASE SMALL FACILITY (INVESTMENT – $800,000)
B PURCHASE LARGE FACILITY (INVESTMENT – $2,000,000)
C EXPAND FACILITY (INVESTMENT TO BE DETERMINED)
D NO CHANGE IN FACILITY
■ DECISION POINT
□ CHANCE EVENT
P PROBABILITY OF OCCURRENCE

DECISION TREE WITH CHAIN OF EVENTS

CHOICE	CHANCE EVENT	PROBABILITY	VALUE OF YIELD ($)	DISCOUNTED VALUE OF YIELD ($)
Build Small Facility	High Initial Demand	0.65	2,670,000	1,735,500
A	Low Initial Demand	0.35	2,900,000	1,015,000
			TOTAL	2,750,500
			INVESTMENT	800,000
			NET VALUE	1,950,500
Build Large Facility	High Average Demand	0.50	8,700,000	4,350,000
B	High Initial, Low Subsequent Demand	0.22	2,200,000	484,000
	Low Average Demand	0.28	600,000	168,000
			TOTAL	5,002,000
			INVESTMENT	2,000,000
			NET VALUE	3,002,000

Figure 2-3. Typical decision tree matrix.

C. Selection of Evaluation Criteria

The criteria employed in the evaluation process may vary considerably, depending on the stated problem and the level and complexity of the analysis. For instance, at the system level, parameters of primary importance include cost effectiveness, system effectiveness, logistics effectiveness, life-cycle cost, operational availability, and performance. At the detailed level, the order of parameters will be different, as illustrated in Figure 2-4.

The parameters selected as evaluation criteria should relate directly to the problem statement. For instance, the problem may be to design a system that will perform a certain mission with a specific degree of effectiveness at minimum life-cycle cost. There may be several possible alternative design approaches, each of which is evaluated in terms of system effectiveness and life-cycle cost. On the other hand, the problem may entail the selection of the best among several alternative off-the-shelf equipment items, using supportability characteristics in design as criteria (e.g., accessibility, standardization of components, diagnostic aids). In this instance, there are a number of evaluation factors.

In the event that a number of parameters are involved, each parameter should be reviewed from the standpoint of relevancy or degree of importance. The degree of importance may be realized by applying parameter weighting factors (the most important items receiving the heaviest weighting). The application of weighting factors will depend on the evaluation technique employed. Sample applications in this area are illustrated in Part Two.

D. Application of Analytical Techniques

The next step in the analysis process involves the analytical phase. This entails the selection and combining of various analytical techniques in the form of a model or a series of models.[4] The model is a simplified representation of the real world which abstracts the features of the situation relative to the problem being analyzed. It is a tool employed by the analyst to assess the likely consequences of various alternative courses of action being examined. The model must be adapted to the problem at hand, and the output must be oriented to the selected evaluation criteria. The model, in itself, is not the decision maker but a tool that provides the necessary data, in a timely manner, in support of the decision-making process.

The model may be quite simple or very complex, highly mathematical or not at all mathematical, computerized or manually implemented, and so on. The extensiveness of the model will depend on the nature of the problem relative to the quantity of variables, input-parameter relationships, number of alternatives being evaluated, and the complexity of operation. The ultimate objective in the selection or development of a model is simplicity and usefulness. Evaluation techniques and models are discussed further in Appendix B.

---

[4]Models may constitute physical models, abstract models, symbolic models, mathematical models, and so on. Model, as defined here, refers to an analytical or mathematical model.

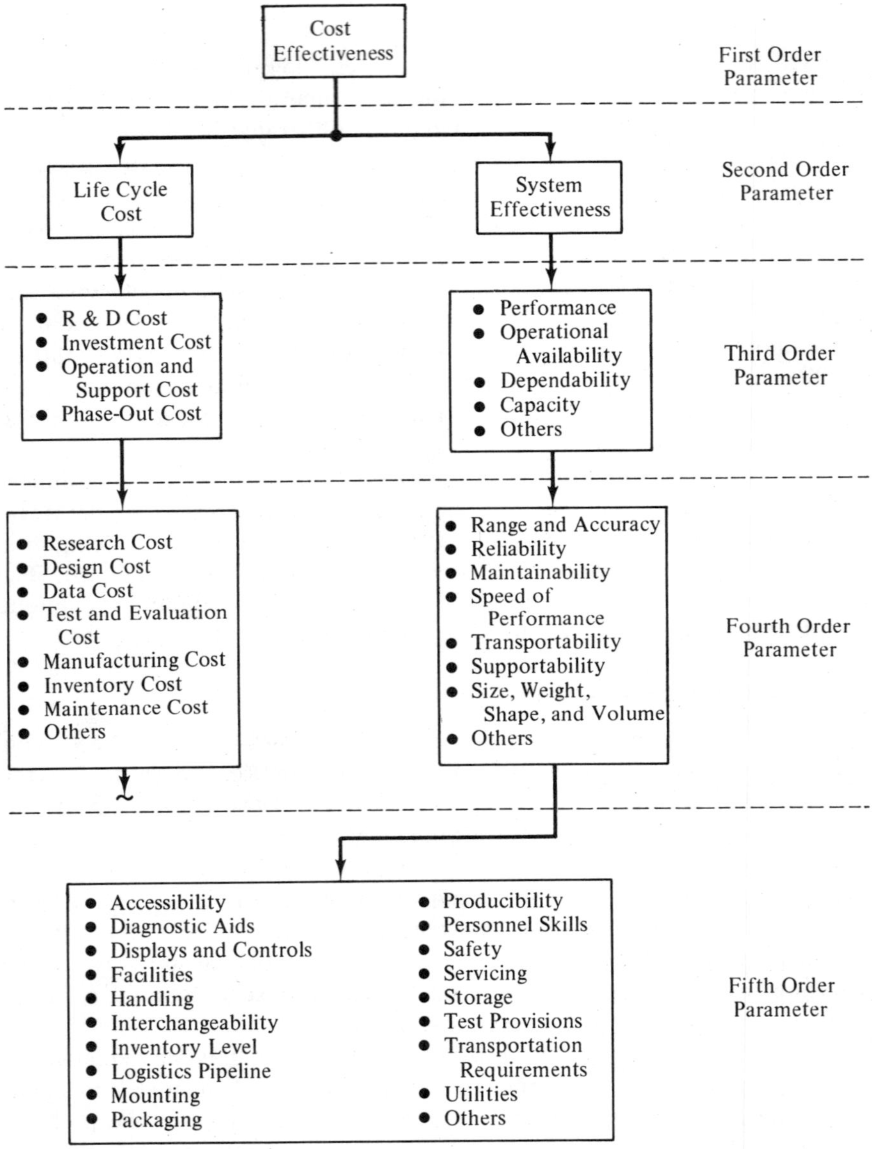

Figure 2-4. Order of evaluation parameters.

E. Data Generation and Application

One of the most important steps in the analysis process is to assemble the appropriate input data. The right type of data must be collected in a timely manner and must be presented in the proper format. Specific data requirements are identified from the evaluation criteria and from the input requirements of the model used for evaluation purposes.

F. Evaluation Results and Sensitivity Analysis

Data are gathered and inputed, and the model is exercised. The analysis results will lead to a recommendation for some type of action. However, the performance of a sensitivity analysis should be accomplished prior to making a final decision. In a given analysis, there may be a few key input parameters about which the analyst is very uncertain (owing to inadequate data, pushing the state of the art, etc.). The question is: How sensitive are the results of analysis to variations of these uncertain parameters? The analyst might wish to run the model using a baseline system configuration and then rerun the model, varying key input parameters to determine the effect on results. Variation is accomplished by applying various multiple factors to the input parameter being tested. If such variations cause a possible shift toward the consideration of another alternative (i.e., the analyst becomes unsure that his initial decision is adequately supported), then the analyst will readily be able to determine whether or not to probe further for better input data or to select a less risky alternative.

G. Risk and Uncertainty

The results of analyses generally influence decisions, having a significant impact on the future. Inherent in the decision-making process are the aspects of risk and uncertainty, since the future is of course unknown.[5] The basic question is: What are the chances of making the wrong decision? As the intent in any evaluation process is to minimize risk and uncertainty, the approach employed must address the following basic areas (not to be considered as being all-inclusive):

1. *Proper definition of the problem*—identification of the right problem and relationship of the evaluation technique employed to the problem selected. Is the right technique being used?
2. *Selection of evaluation criteria*—use of a single criterion measure versus a multiple criteria mix in the quantification of system performance characteristics, operational and logistic support parameters, and cost.
3. *Weighing of evaluation criteria*—the relative weighing (application of weighting factors) in terms of significance or level of importance of each criterion factor employed.

---

[5] In the interest of simplicity, the terms "risk" and "uncertainty" are often used jointly. Actually, risk implies the availability of explicit data in the form of a probability distribution around a certain parameter. Uncertainty implies a position that may be probabilistic in nature, but one that is not supported by explicit data. Evaluation criteria may be measurable in terms of risk or may be stated under conditions of uncertainty.

4. *Quantification of evaluation factors*—the inclusion versus exclusion of all applicable factors and the methods for quantification. What is quantifiable and how? What is left out?

5. *Use of ratios as a means of evaluation scoring*—the process of system selection based on the best ratio with and without consideration of capital budgeting constraints, effectiveness limits, the magnitude of the risk involved, and so on.

6. *Selection of an operational horizon*—consideration of variations in the anticipated operational life (time frame) for the alternatives being evaluated. A common time period should be used.

7. *System interaction effects*—the effect of various systems on each other.

8. *Evaluation data input*—the prediction of future events (probabilities, effectiveness figures of merit, etc.) and the estimation of costs. Valid and comprehensive data are required.

The aspects of decision making involving the use of analytical techniques, associated quantitative factors, and the areas of risk and uncertainty are discussed in subsequent chapters. The engineer must make decisions of varying magnitudes in the accomplishment of his functions, and the results of these decisions usually have a significant impact on the ultimate system performance and effectiveness. Thus, it is important that the engineer be thoroughly familiar with both the process illustrated in Figure 2-2 and some of the tools (i.e., methodologies and techniques) that are available for his use. These tools are an inherent part of the individual skills identified in block 2 of Figure 1-1.

2.4 ENGINEERING MANAGEMENT

As noted earlier, we live in a dynamic society. Technological advances are emanating on a perpetual basis, and new products are rapidly replacing old ones. New industries are being created while old ones phase out or fail. Economic factors and social needs are continually evolving. We exist in an environment in which many changes are occurring, and managing almost any type of enterprise is quite difficult in this environment.

The fulfillment of the objectives discussed in Section 2.1 involves both human and physical resources. The efforts of people must be coordinated, and each individual concerned must know the part that he is to play and his relationships with others. The manager of these resources is faced with many important decisions and often spends a great deal of time and effort analyzing and evaluating the various courses of action open to him. His decisions are of both a technical and an administrative nature, and he must decide *what to do, where to do it,* and *how it should be done.*

Management is basically a function of getting things done through others and deals primarily with the following aspects:

A. Planning

Planning involves the *process* of developing and formulating a course of action, and the output results constitute a plan. It is the systematic identification of program tasks, task schedules, and the resources required for task accomplishment. Planning is necessary to achieve some degree of order in completing activities scheduled in the immediate future and for long-range activities that are forecasted far out in time. In many firms and agencies, the planning activity is classified as follows:

1. Ongoing program planning. Current program activities that are scheduled in a relatively short-term period from the present out to 5 years in the future are covered. Plans for this period are fairly detailed in terms of specific scheduled events, personnel and material needs, cost, and so on.

2. Long-range planning. Included are activities that are anticipated (or forecasted) in the 5- to 10-year time frame and beyond. Plans in this area usually are more general in nature and are representative of the overall goals and objectives of the firm or agency involved.

Since the future is unknown, planning is based on forecasts and estimates of the market (i.e., consumer needs), competitive conditions with other firms developing similar or like products, changing technological and environmental factors, and so on. The planner predicts (based primarily on intuition, available data, and past experience) what needs to be accomplished in light of current goals and objectives. As conditions change from day to day, the planning activity must respond accordingly to determine the effects of the change(s) on plans already established. If these effects significantly impact the proposed methods for meeting the end objectives, then additional planning will be needed to develop alternative courses of action. Thus, planning is a dynamic and continuous process.

Planning addresses the "what if?" questions from the standpoint of managing the resources allocated and involves the aspects of decision making discussed in Section 2.3. Referring to Figure 2-3, there are various courses of action envisioned at the time that the plan is consumated. The various alternatives, represented by branches of the decision tree, are assigned probabilities that indicate the likelihood of occurrence. As chance events occur in time, these probabilities will change along with the possibility that, as a result, even some of the major alternatives will change.

Finally, planning is accomplished at all levels of organization. Top-level managers plan in terms of the overall objectives of the firm or agency, and their planning often extends further in the future than the planning accomplished at lower levels. Middle managers plan on how they will get the job done with the resources allocated. Even the individual engineer has to plan his approach in meeting a specific system design objective. In essence, planning is the starting point from which all other program activity evolves. Without it, the activities of a firm or agency may well become confused and ineffective. The importance of planning is discussed further in Chapter 3.

B. Organization

The second aspect of management is that of organizing and staffing. Organization is accomplished in response to the planning activity discussed above and consists of: (1) determining what activities need to be done, (2) grouping the identified activities in terms of a functionally oriented structure of some type (e.g., unit, group, department, division, section), and (3) staffing the structure with the appropriate personnel skills to perform the designated activities in a coordinated manner. The concepts of organization were introduced earlier in Section 2.2, and are discussed further in Part Three.

C. Directing (Program Implementation)

Directing in this instance consists of those day-to-day managerial functions associated with influencing, guiding, and supervising subordinates in the accomplishment of the organizational activities identified through planning. This requires a continuing awareness, on the part of the manager, of the specifics of the activities performed and whether the results are fulfilling the program objectives in an effective manner. Interaction between managers and their subordinates is complex and will be discussed in detail later.

D. Controlling

Controlling involves (1) the monitoring of program activities to ensure that the end objectives are being met (e.g., system design is in accordance with the specified requirements, program costs are within the established bounds, tasks are completed within the prescribed scheduled times), and (2) the initiation of corrective action as required to overcome deficiencies if the objectives are not being met. It is essential that program activities and their results be accomplished in compliance with the established plans and specifications. Thus, checks and balances are required on a periodic basis so that problem areas can be detected as early as possible in the system development process. Basically, this is accomplished through a combination of program reviews and system design reviews, discussed later.

In summary, engineering management is faced with a rapidly changing set of requirements. Response to these requirements is initiated through the aspects of planning, organizing, directing, and controlling a varied and mixed group of activities. Management is responsible for the accomplishment of certain designated goals and objectives in an effective and efficient manner. Management must not only accomplish this with the resources provided, but must do so in spite of certain imposed external constraints, many of which cannot be directly controlled. For instance, management is responsible to:

A. The firm's stockholders, who insist on making a profit and distributing high dividends.
B. The product consumer, who is "always right."
C. The employees of the firm or agency who must do the job and are entitled to continued growth (e.g., promotion, salary growth, more challenging jobs and improved fringe benefits).
D. The government, public, or citizens affected by the product.

In addition, there are legal constraints (i.e., laws, codes, etc.), social constraints, and political constraints which affect what is accomplished in terms of product development and use. These conditions, in conjunction with those described in Chapter 1, offer considerable challenge to engineering organization and management.

QUESTIONS AND PROBLEMS

1. Define (in your own words) the basic goals and objectives of engineering.
2. How would you evaluate a system to determine whether engineering objectives have been met? What figures of merit would you use?
3. When in the product-life-cycle time frame is it appropriate to define system requirements? What factors are included?
4. What resources are required to accomplish engineering functions? How should these resources be organized?
5. Refer to Figure 2-1. What is the significance of the corrective-action feedback loop?
6. If you were given the job of organizing an engineering effort, how would you proceed? What are some of the constraints associated with organizational development?
7. What is meant by: Organizational span of control? Delegation of authority? Chain of command?
8. How does decision making relate to the engineering process?
9. Refer to Figure 2-2. What is the most important step in the engineering analysis process?
10. What is meant by a model? How is it employed in engineering analyses?

11. What are some of the benefits of models? Discuss some of the cautions that should be exercised in using models.
12. Refer to Figure 2-2. How do the requirements in the engineering analysis process relate to the basic engineering requisites identified in Figure 1-1?
13. What information does a decision tree provide?
14. What is the purpose of sensitivity analysis? What educational benefits are derived by accomplishing sensitivity analyses?
15. Define risk and uncertainty. What is the difference between the two?
16. Define engineering management. What are the major functions?
17. What are the challenges in engineering management? Name a few.
18. What is the purpose of planning? How does engineering participate in the planning process?
19. The aspect of planning is a dynamic process. True or false? Why?
20. What functions does management perform to ensure that program activities are producing effective results?
21. Engineering management is responsible to whom?

part two

The Product Life Cycle

Part Two covers the product life cycle, commencing with the definition of a need and extending through phase-out and material disposal and/or reclamation. It deals with the process of bringing a product/system into being and then supporting it throughout its planned life cycle. This process is applicable to any product or system, whether small or large, although the specific activities discussed must be tailored to meet each individual need.

Engineering plays a major role in each phase of the life cycle. Specific tasks include planning, marketing, research, design, manufacturing, test and evaluation, and support of the product in the field. Understanding the engineer's role in the life cycle is necessary to understanding the total engineering spectrum. This part is represented by the second ring in the illustrated conceptual outline.

Conceptual Outline

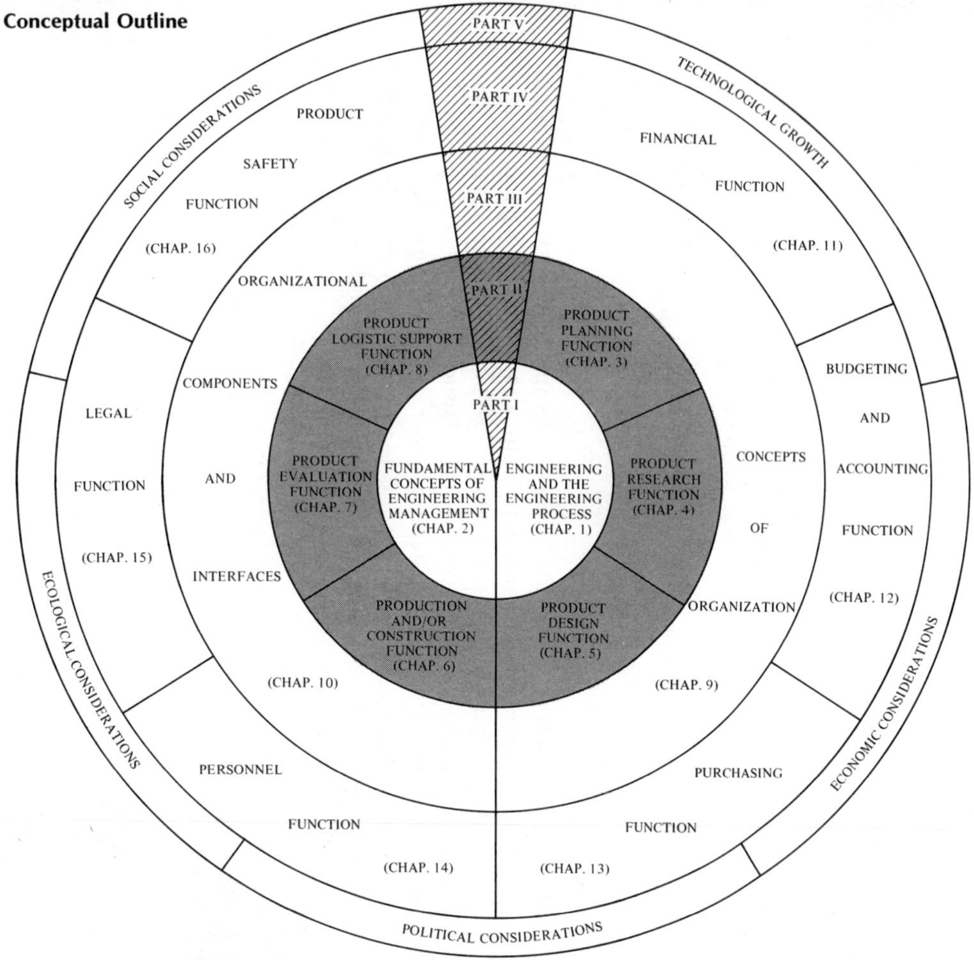

PART I	ENGINEERING IN MODERN TIMES
PART II	THE PRODUCT LIFE CYCLE
PART III	ORGANIZATION AND THE PRODUCT LIFE CYCLE
PART IV	RELATED FUNCTIONS AND INTERFACES
PART V	ENGINEERING AND THE ENVIRONMENT

3

PRODUCT PLANNING FUNCTION

The product planning function, as described in this text, includes marketing and marketing analysis, the performance of feasibility studies, and advanced planning. The planning function constitutes an initial phase in the product life cycle and is represented by blocks 1 and 2 in Figure 1-4. The planning function involves a series of steps leading to product research and design. These steps, as part of an iterative process in itself, are illustrated in Figure 3-1 and discussed in this chapter. Planning should be *functional* and not *organizational*. The functions depicted in the figure are not necessarily aligned to a specific organization, but may be accomplished by individuals representing many organizational components.[1]

3.1 IDENTIFICATION OF NEED

The aspect of product planning stems from the identification of a need for a product or a process (i.e., a system of some type). Needs may evolve from several sources.

A. Needs Determined from Current Deficiencies

This area primarily relates to those needs which are stimulated as a result of some problem or deficiency occuring *today*. Society as a whole, an industrial firm, a governmental agency, or even an individual may have a

[1]An in-depth coverage of such aspects as marketing and marketing analysis, advanced planning, and so on, is not adequate or possible within the confines of this text. The intent is to provide the engineer with some familiarization of terms and definitions, the process of product planning, and some references (from time to time) for further study.

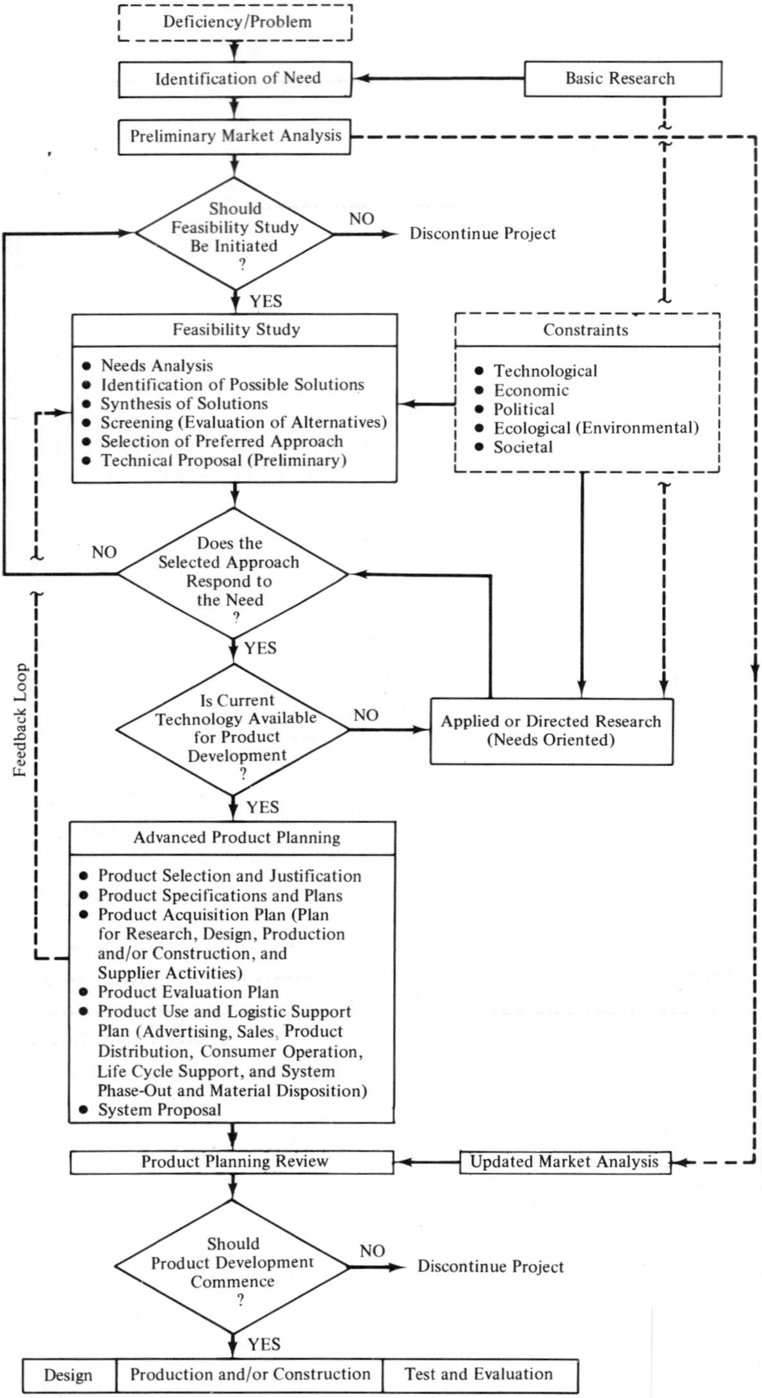

Figure 3-1. Product planning process.

specific problem that requires resolution. Problems may also arise as a consequence of a legislative act or regulatory procedure. In any event, current problems evolve on a day-to-day basis, and resolution is often achievable through the development, production, and use of a new product or process.

B. Needs Determined from Anticipated Deficiencies

Through the prediction of societal, political, ecological, technological, and/or economic trends, anticipated needs can often be determined. For instance, we suspect that economic constraints in the future will likely prohibit the use of certain systems currently in use. As a result, a replacement system that exhibits a lower overall life-cycle cost is needed to accomplish the designated function(s).

Needs in this area should be supported in terms of a description of the nature of the need and the anticipated time frame. As an event in the future involves the aspects of risk and uncertainty, a probability factor should be assigned to define the likelihood that the need will actually exist when predicted.

3.2 PRELIMINARY MARKET ANALYSIS

Once a need has been identified, it is necessary to determine whether the technology currently exists to satisfy the need, and if not, when it will become available. In this instance, "technology" refers to the technical methodology and/or techniques required for achieving a practical purpose. This is basically accomplished through technology review (knowledge of what is currently available) and technological forecasting (knowledge of what is likely to be available in the future).

Given the availability of the proper technology, the producer of the prospective system will complete a preliminary analysis of the market potential and the market share. The objective is to view the system over its projected life cycle on the basis of economic considerations. Gross estimates of revenues from sales and costs are made with the intent of assessing whether or not the venture is worthwhile (i.e., should one pursue the project further?). A projection of economic factors is illustrated in Figure 3-2, where costs are projected together with expected revenue from sales, and an estimate is made of the time required before the investment becomes worthwhile (i.e., the investment recovery point). If this point is too far out in time, it may not be practical to proceed further. A decision is made on the basis of the projected life cycle, the market potential and size, consideration of competition and possible obsolescence, and so on.

The data input at this stage is usually quite rough, since the system is neither defined nor are the cost predictions very accurate. However, an initial projection must be made to justify further effort. Chapter 11 includes additional coverage of investment decisions.

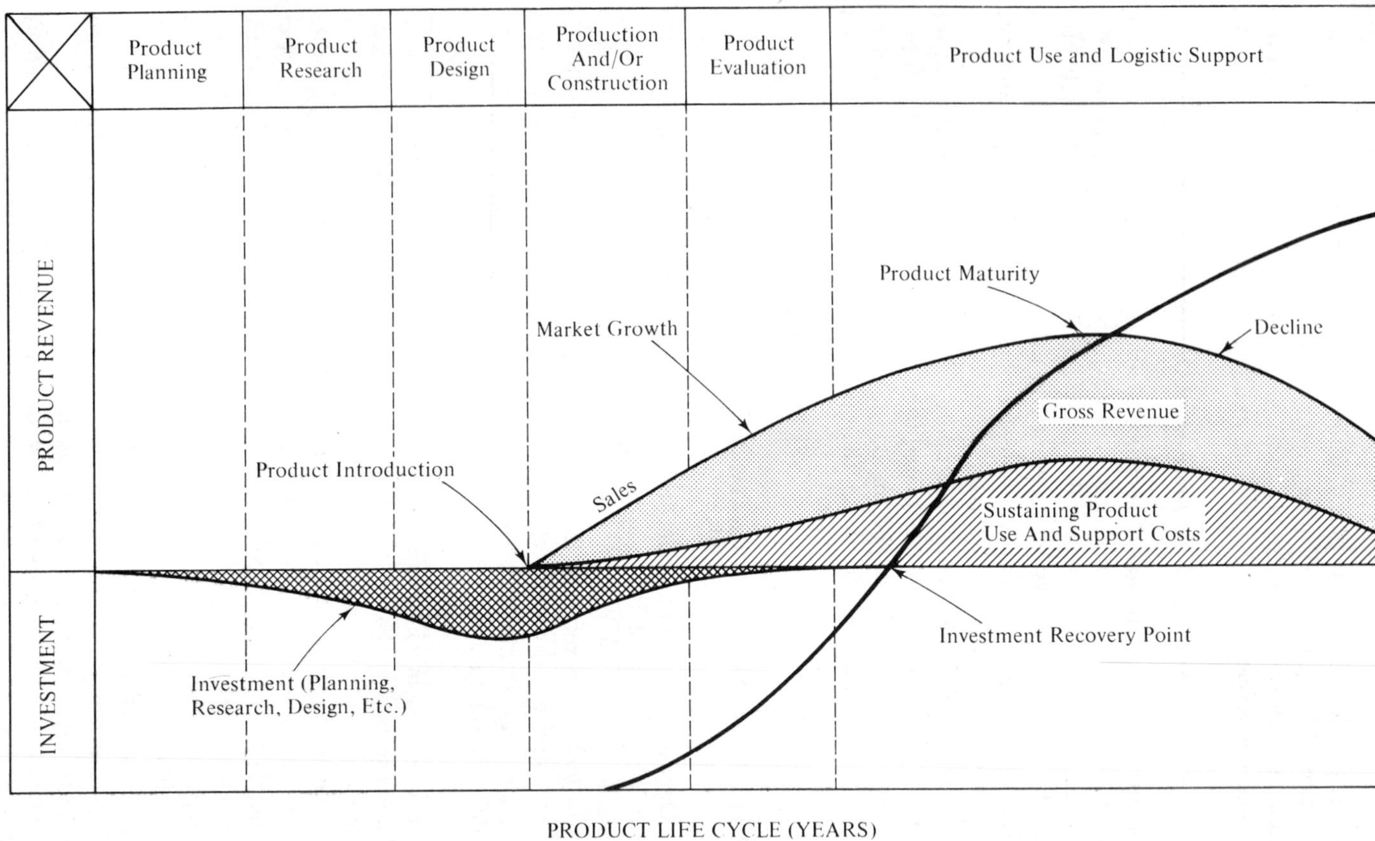

Figure 3-2. Typical product economic factors.

A. Technological Forecasting

Technological forecasting is undertaken in the event that current technology is not available to solve a given problem and it becomes necessary to determine when the appropriate technology will emerge. Specifically, a technological forecast is "a prediction of the future characteristics of useful machines, procedures, or techniques."[2] These characteristics will include performance parameters, effectiveness goals, and the like. Forecasts should cover a complete description of the nature of the technology (e.g., airplane, communication system, building, appliance), the characteristics to be exhibited by the technology output (e.g., performance and effectiveness features), the time frame, and the probability that the technology actually will provide the desired results in a timely manner. This information is evaluated to ensure that the results will, in fact, satisfy an identified future need.

The techniques employed in forecasting may include a variety of qualitative and quantitative methods, such as[3]:

1. *Polling techniques*—the polling of various engineering and management personnel to make estimates on the outlook for future technology. A more precise method that may be classified as being in this category is the *Delphi method* of forecasting. A group of "experts" is assembled and individually asked a series of questions concerning anticipated events of the future. The results from this first series of questions are combined and presented to the entire group. Based on the output, these experts are asked a second series of questions concerning future events and projected dates. The results are again combined and presented, and new estimates are made based on the data presented. This is an iterative process, involving a series of well-structured questions along with the necessary feedback of results, obtaining the best estimates possible through group action.
2. *Analogy techniques*—comparison of the technology being forecasted with similar technologies of the past. Statistical correlation techniques may be used to support these comparisons.
3. *Growth-curve techniques*—the use of growth curves to extend an existing capability in terms of performance, effectiveness, and the like. By modifying current systems, growth is possible in certain areas. An example would be the use of an exponential distribution curve to represent what is likely to occur when certain system modifications are accomplished.

---

[2]This definition and coverage of the subject matter is included in: J. P. Martino, *Technological Forecasting for Decision Making,* American Elsevier Publishing Company, Inc., New York, 1972.

[3]Two sources describing forecasting techniques are: J. P. Martino, *Technological Forecasting for Decision Making,* American Elsevier Publishing Company, Inc., New York, 1972; and P. E. Green and D. S. Tull, *Research for Marketing Decisions,* Prentice-Hall, Inc., Englewood Cliffs, N.J., 1966.

4. *Extrapolation techniques*—the use of historical data and regression analysis to project future events. Historical data points are plotted, and a regression analysis is accomplished to develop a trend curve. This trend curve is then projected into the future.

5. *Analytical modeling techniques*—the use of analytical models and sensitivity analyses to answer the "what if?" questions pertaining to the future. The results aid in determining future events and their probability of occurrence.

Typical forecasts may be presented in various forms, as illustrated in Figure 3-3. These forecasts, employing one or a combination of the techniques described above, are influenced from time to time by breakthroughs in technology, innovations, and/or inventions. At such times, new forecasts should be made, since the listing of future events, the time frame, and the probability factors will probably change.

Engineers play a major role in both technology review and technological forecasting. Relative to the first area, they should already be knowledgeable of what is currently available in their field of expertise (refer to Section 1.1A). In technological forecasting, engineers must initially evaluate the need in terms of technical requirements. Then they must consider alternative approaches to include a range of possibilities, from the modification of an existing capability to the development of a completely new system. Finally, they must recommend a course of action such as "proceed with a feasibility study to determine . . ." or "initiate activity to perform research to accomplish . . ." Engineers may solicit assistance in the application of forecasting techniques, but only they are qualified to evaluate technical requirements.

Regression analysis

B. Market Potential

The *market potential* refers to the total amount of a product or service that the market is *able* to absorb during a prescribed period of time under *optimum* conditions.[4] It is not a sales forecast but represents the market *opportunity*. In other words, given the need and *no* competition, the producer is likely to sell x number of systems. The market may include the domestic market (e.g., private and industrial sectors), the government market, and the foreign market.

Estimates of the market potential are usually made for different product classifications, geographical areas, consumer types, and by application. The engineer may be effectively employed to provide assistance in determining product uses and applications. These estimates are used to establish sales projections.

---

[4]This basic definition is presented in: P. E. Green and D. S. Tull, *Research for Marketing Decisions,* Prentice-Hall, Inc., Englewood Cliffs, N.J., 1966.

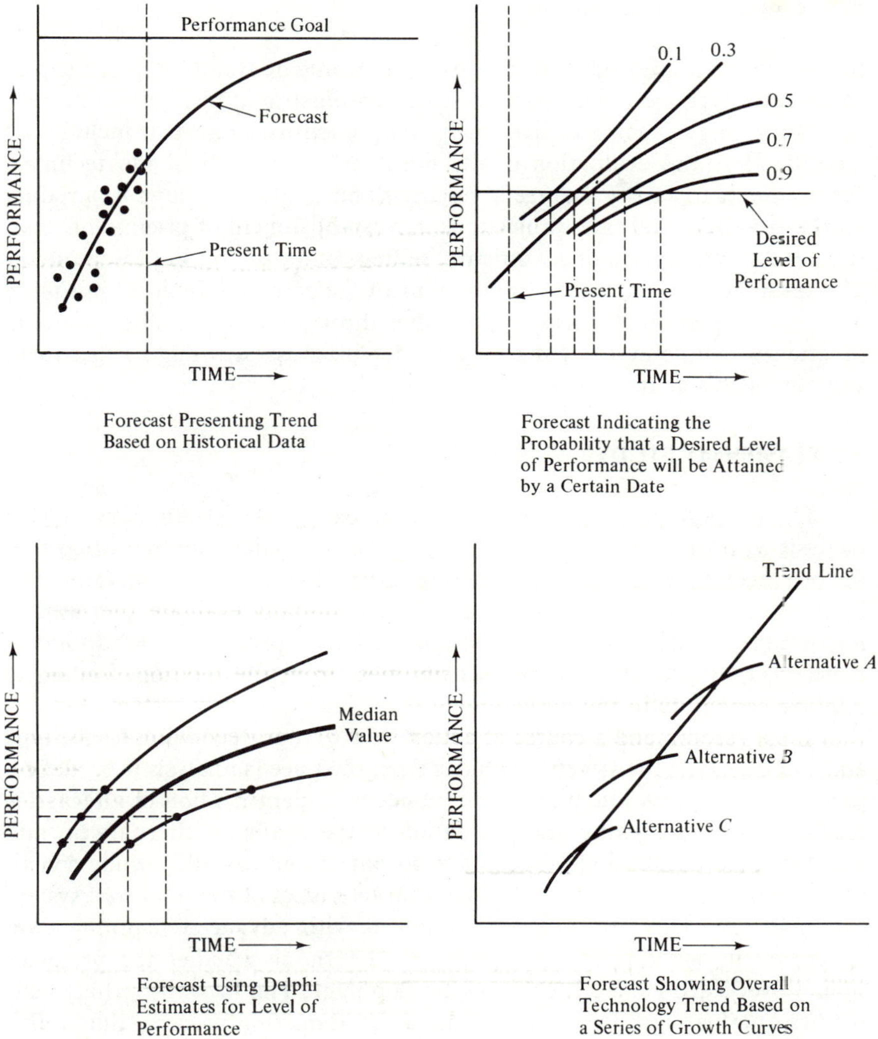

Figure 3-3. Typical technological forecasting approaches.

C. Market Share or Penetration

The *market share* constitutes the ratio of the anticipated sales for an individual firm to the total projected sales based on the market potential. The prospective producer needs to know the estimated percentage share of the market that can be expected. Sales analyses and sales forecasts are accomplished, based on product pricing information, to determine approximate revenue by year. Sales forecasts may include both short- and long-range forecasts, depending on the type of information desired and the time frame. These estimates, to cover both pessimistic and optimistic values, are used in projecting the total system outlook from an economic standpoint. Anticipated revenue over the life cycle is presented as illustrated in Figure 3-2.

Marketing as such is an extremely comprehensive subject. It includes an identification and evaluation of the need, an assessment of the technical requirements to satisfy the need, research and analysis to support product market potential and sales projections, the establishment of pricing information, and the functions of advertising, selling, and distributing the product. Marketing functions are intuitive in many of the activities depicted in Figure 3-1. The integration of marketing activities throughout the product planning function will become evident as the reader progresses through subsequent sections of this chapter.

3.3 FEASIBILITY STUDY[5]

The purpose of a *feasibility study* is to extend the preliminary market analysis with the intent or arriving at a preferred system configuration that the producer is willing to propose in response to an identified need. Thus far, both technical and planning data have been of a cursory nature, and assuming that the results have been encouraging from the producer's standpoint, it is necessary to proceed with another but more in-depth iteration, leading to a better system definition.

The feasibility study, varying in scope of coverage depending on the type and complexity of the system, includes a detailed needs analysis (i.e., definition of system operation and support concepts), identification of alternative configurations, screening and evaluation of the available alternatives, and selection of a preferred approach. The output of the feasibility study constitutes a proposal covering the *technical* characteristics of the preferred system configuration. This information, combined with advanced planning data (discussed in Section 3.4), is reviewed to determine whether the producer should proceed further with system development. The feasibility study constitutes another significant step in the decision-making process, thus influencing future product activities, particularly in the case of large-scale systems.

A. Detailed Needs Analysis

The *needs analysis* in this instance refers to the definition of system operational and maintenance support requirements.[6]

---

[5]A feasibility study, depending on system requirements and on organizational preferences, may be included as an inherent part of the marketing analysis effort or as part of the conceptual design effort in Chapter 5. There is no intent here to respond to individual organizational patterns, and it is realized that some overlap may exist from project to project. However, it is necessary to address these issues at this early phase in the life cycle.

[6]System operational and maintenance requirements are defined further in: B. Blanchard, *Logistics Engineering and Management,* Prentice-Hall, Inc., Englewood Cliffs, N.J., 1974.

1. ***System operational requirements.*** The technical parameters for the development of a system stem from the *operational concept*. One might ask the following questions: What are the anticipated quantities of equipment in the field and where is it to be deployed? How will the equipment be utilized? What is the expected operational environment at each location? The answer to these and comparable questions leads to the definition of the maintenance support concept for the system and the identification of design criteria.[7] Specifically, the operational concept includes the following information:

a. *Mission definition*—identification of the prime mission of the system along with alternative or secondary missions. What is the system to accomplish? How will the system accomplish its objectives? The mission may be defined through one or a set of scenarios.

b. *Performance parameters*—definition of the operating characteristics or functions of the system (e.g., size, weight, speed of performance, accuracy, output rate, capacity). What are the critical system performance parameters?

c. *Operational deployment*—identification of the quantity of equipment, personnel, facilities, etc., and the expected geographical location to include mobility requirements. How much equipment and associated software is deployed and where is it to be located? When are they required? When does the system become fully operational?

d. *Operational life cycle* [*horizon*]—anticipated time that the system will be in operational use. What is the total inventory profile throughout the system life cycle?

e. *Utilization requirements*—anticipated usage of the system and its elements (e.g., hours of operation per day, on-off sequences, operational cycles per month). How is the system to be used in the field?

f. *Effectiveness factors*—system requirements specified as figures of merit for cost/system effectiveness, operational availability, dependability, logistic support effectiveness, mean time between maintenance (MTBM), failure rate (λ), maintenance downtime (MDT), facility utilization (in per cent), personnel efficiency, and so on. Given that the system will perform, how effective or efficient it is? Refer to Appendix A for some definitions of measures of effectiveness.

g. *Environment*—definition of the environment in which the system is expected to operate (e.g., temperature, humidity, arctic, tropics, mountainous or flat terrain, airborne, ground, shipboard). This should include a range of values as applicable and cover transportation, handling, and storage modes. How will the system be handled in transit, and what will the system be subjected to during operational use and for how long? A complete environmental profile should be developed.

---

[7]Design criteria refers to design standards, constraints, guidelines, and the like. This constitutes an input to the system design process.

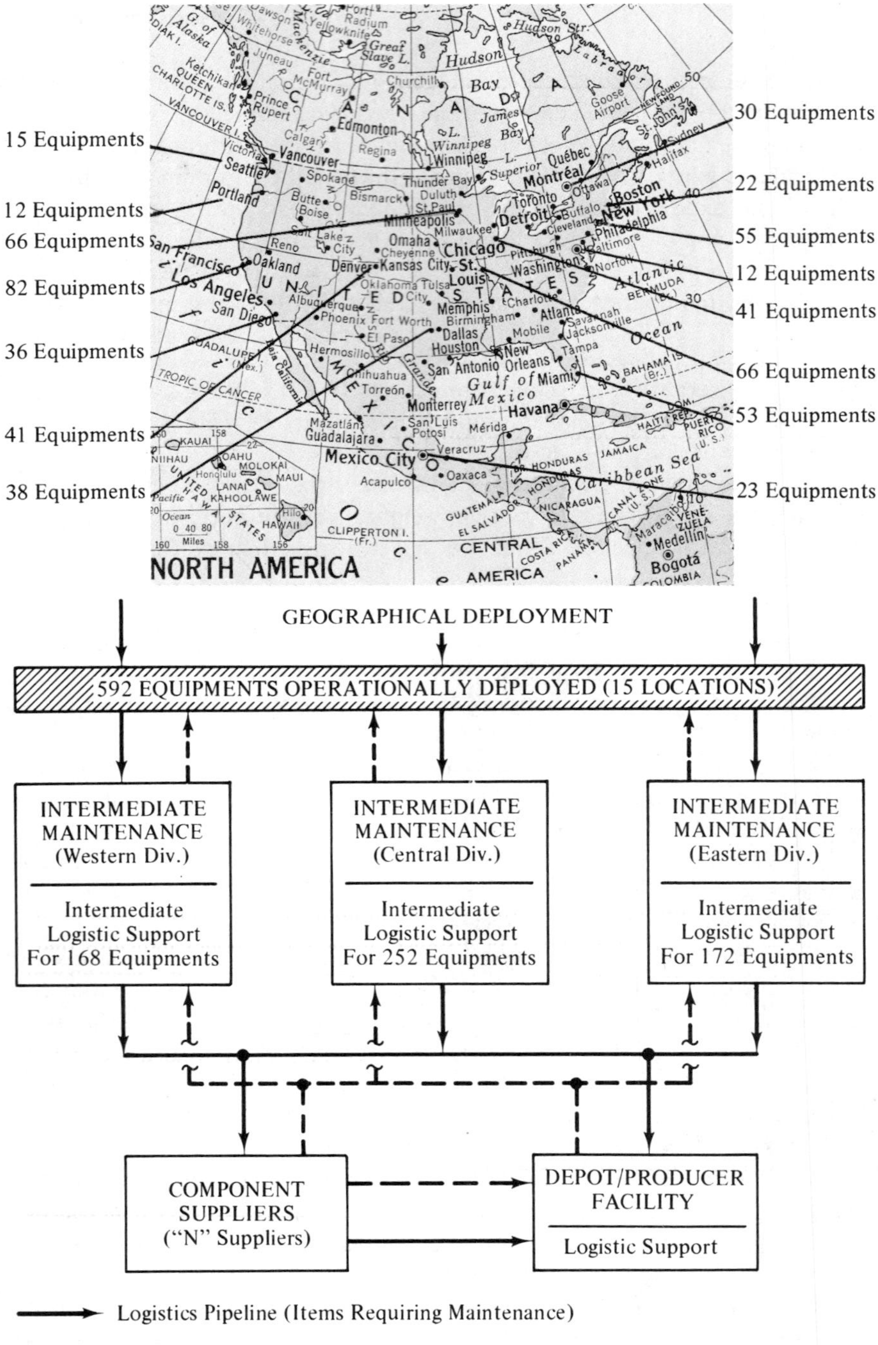

Figure 3-4. Operational/maintenance concept (*example*).

In essence, the producer (management and engineering) must project how the system will be deployed and utilized in the field by the consumer. Engineering must design the system to do the job under the conditions stated, and management must ensure that this is accomplished effectively, efficiently, and in a timely manner.

When reviewing operational requirements in total, many of the specific factors identified above are based on the projected deployment of equipment. This not only influences system operation on a day-to-day basis, since each organizational entity operating the system usually functions somewhat differently, but the equipment site locations intuitively specify environmental conditions, transportation and handling methods, and storage requirements. Thus, one of the first steps in defining operational requirements is to determine consumer deployment locations as illustrated in Figure 3-4. The example shown illustrates that 592 pieces of equipment are located at 15 operational sites. Consumer operations at each site are estimated, and requirements are established based on the demand for the equipment. This, in turn, determines expected utilization and effectiveness factors to which the system must be responsive. The figure also includes the first step in illustrating the level of support anticipated.

2. System maintenance concept. The *maintenance concept* responds to the question: How does the producer envision that the system will be supported throughout its life cycle? It delineates levels of maintenance support, repair policies, effectiveness measures (e.g., maintenance manpower, time, and cost constraints), and serves several purposes.

a. It provides a baseline for the establishment of supportability requirements (e.g., reliability, maintainability, and human factors characteristics) in system/equipment design. For instance, if the repair policy indicates that no scheduled or detailed unscheduled maintenance is to be accomplished at the consumer's operational site, the equipment design should not incorporate diagnostic test provisions, special packaging considerations for the benefit of the consumer, and so on. On the other hand, if it is decided that a certain amount of maintenance should be performed on site, then built-in self-test provisions, accessibility, readily removable functional packages, and related characteristics should be incorporated in the design. The maintenance concept aids in establishing design criteria related to system support.

b. It provides the basis for the establishment of requirements for total logistic support. Given an assumed design configuration of the prime mission equipment (i.e., that part of the total system which is *directly* related to the mission or operational requirement, such as an airplane, train, electronic equipment, or nuclear power plant), it is then necessary to consider how it should be supported.[8] The maintenance concept, supplemented by logistic

[8]This does not mean to imply that the design is by any means fixed. The engineer must start with some baseline assumption, then synthesize it, evaluate it, modify it, synthesize and evaluate the modified configuration, and so on. Logistic support requirements can be determined for each configuration being evaluated.

support analyses, leads to the identification of maintenance tasks, task frequencies and times, maintenance personnel quantities and skill levels, training needs, test and support equipment, supply support (e.g., spare/repair parts), facilities, and data. These support requirements are evaluated and integrated with the prime mission equipment and associated software to form the total system.

The maintenance concept evolves from the deployment profile illustrated in Figure 3-4. Levels of maintenance include *organizational* maintenance, *intermediate* maintenance, and *depot* maintenance. These levels are defined as follows:

a. *Organizational maintenance*—organizational maintenance is performed at the operational site (e.g., the 15 locations identified in Figure 3-4). Generally, it includes tasks performed by the using organization on its own equipment. Organizational-level personnel are usually involved with the operation and use of equipment and have minimum time available for detailed system maintenance. Maintenance at this level normally is limited to periodic checks of equipment performance, visual inspections, cleaning of equipment, some servicing, external adjustments, and the removal and replacement of some components. Personnel assigned to this level generally do not repair the removed components, but forward them to the intermediate level. From the maintenance standpoint, the least skilled personnel are assigned to this function. The design of equipment must take this fact into consideration (e.g., design for simplicity).

b. *Intermediate maintenance*—intermediate maintenance tasks are performed by mobile, semimobile, and/or fixed specialized organizations and installations. At this level, end items may be repaired by the removal and replacement of major modules, assemblies, or piece parts. Scheduled maintenance requiring equipment disassembly may also be accomplished. Available maintenance personnel are usually more skilled and better equipped than those at the organizational level and are responsible for performing more detailed maintenance.

Mobile or semimobile units are often assigned to provide close support to deployed operational equipments. These units may constitute vans, trucks, or portable shelters containing some test and support equipment and spares. The mission is to provide on-site maintenance (beyond that accomplished by organizational-level personnel) to facilitate the return of the system to its full operational status on an expedited basis. A mobile unit may be used to support more than one operational site. A good example is the maintenance vehicle that is deployed from the airport hangar to an airplane parked at a commercial airline terminal gate and needing extended maintenance.

Fixed installations (permanent shops) are generally established to support both the organizational level tasks and the mobile or semimobile units. Maintenance tasks that cannot be performed by the lower levels, because of limited personnel skills and test equipment, are performed here.

High personnel skills, additional test and support equipment, more spares, and better facilities often enable equipment repair to the module and piece-part level. Fixed shops are usually located within specified geographical areas. The three intermediate facilities supporting the equipment illustrated in Figure 3-4 are examples. Rapid maintenance turnaround times are not as imperative here as at the lower levels of maintenance.

c. *Depot maintenance*—the depot level constitutes the highest type of maintenance, and supports the accomplishment of tasks above and beyond the capabilities available at the intermediate level. Physically, the depot may be a specialized repair facility supporting a number of systems/equipment in the inventory or may be the equipment producer's plant. Depot facilities are fixed and mobility is not a problem. Complex and bulky equipment, large quantities of spares, environmental control provisions, and so on, can be provided if required. The high-volume potential in depot facilities fosters the use of assembly-line techniques, which, in turn, permit the use of relatively unskilled labor for a large portion of the workload with a concentration of highly skilled specialists in such key areas as fault diagnosis and quality control.

The depot level of maintenance includes the complete overhauling, rebuilding, and calibration of equipment as well as the performance of highly complex maintenance actions. In addition, the depot serves as a maintenance facility, and may be a special capability established for a specific need or the producer's main plant. The depot facilities are generally remotely located, to support specific geographical area needs or designated product lines.

The above definitions basically cover the maintenance aspects of system support as reflected by the solid lines in Figure 3-4 (i.e., the flow of items from operational sites to the intermediate facilities and to the depot if the requirement exists). The reverse flow, indicated by the dashed lines, represents supply support or the flow of spares and repair parts to support maintenance operations.

Within the scope of the identified levels of maintenance support, the producer should attempt to define a basic repair policy.[9] The repair policy may vary from discarding the entire system when a failure occurs (i.e., the system is replaced as an entity, and no maintenance is accomplished on the faulty item) to complete repair of the system, down to removal and replacement of a component part. There are many combinations of repair policies involving mixed approaches, an example of which is illustrated in Figure 3-5.

---

[9]A repair policy in this instance is an indication of what anticipated maintenance activities are likely to be accomplished at each level of maintenance. In addition, basic logistic support resources are identified along with the effectiveness constraints required to meet system requirements. Repair policies are defined in depth in: B. Blanchard, *Logistics Engineering and Management,* Prentice-Hall, Inc., Englewood Cliffs, N.J., 1974, Chapter 4.

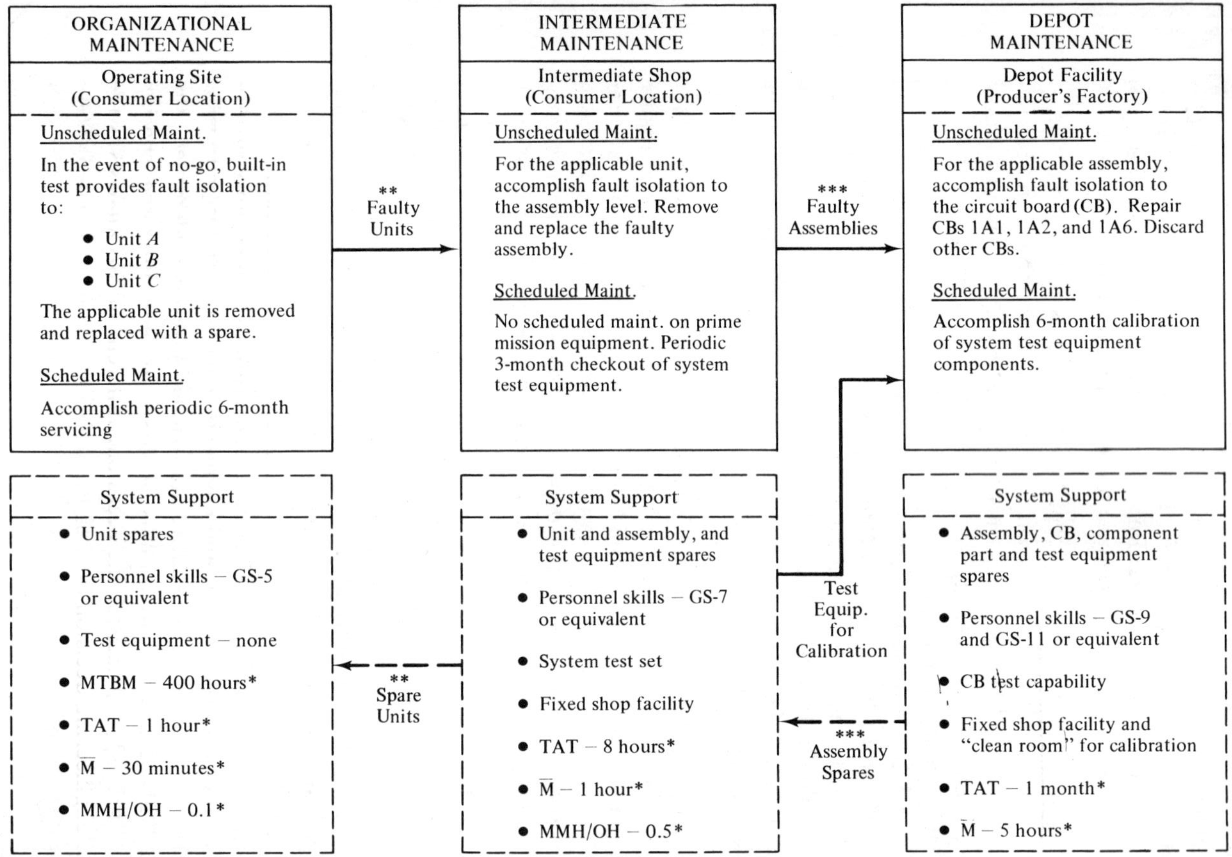

*Refer to Appendix A for appropriate definitions
**Pipeline – 2 weeks
**Pipeline – 1 month

Figure 3-5. System maintenance concept flow (*repair policy*).

Repair policies at this state in the life cycle are by no means fixed. However, assumptions must be made in order to provide an input for advanced product planning and for system design. These assumptions may be based on past experience for comparable systems and should be evaluated in terms of total cost effectiveness. For instance, it may appear to be too costly to discard or throw away the entire system if a failure occurs as a result of the high cost of replacement spares, and it may be too costly to accomplish repair down to the Nth item because of the cost of test and support equipment, personnel, facilities, and other required resources. The best approach may represent a policy somewhere in the middle, as illustrated in Figure 3-5. In any event, a policy must be assumed (i.e., the most likely approach envisioned at this time) with the intent of updating it as system development progresses.

B. Preliminary Systems Analysis

The definition of system operational and maintenance requirements is a starting point in the engineering development process. The responsible engineering activity must know how the system is to be utilized and supported if the output is to effectively meet the need. The operational and maintenance support requirements specified for the system provide goals for the development of prime mission equipment, software, and the elements of logistic support. These goals may be stated both qualitatively and quantitatively with quantitative factors presented in the form of discrete values and/or a range of values. Sometimes, the specified quantitative goals are in the form of minimum or maximum constraints (i.e., the range shall be at least 200 miles, the size shall not be greater than 2 feet by 4 feet, the operational availability shall be at least 98 per cent, the life-cycle cost on a unit-item basis shall not exceed $1,500.00, etc.). The specification of these and comparable factors is based on the engineer's evaluation and interpretation of the need.

Given a set of guidelines for system development, the engineer (through a review of current and predicted future technology) investigates and determines different technical means by which the system requirements can be met. At this point, the various alternatives are defined in rather gross terms. However, decisions must be made concerning the feasibility of proceeding further with the project, and engineers must do their best with what is available. Once feasible alternative approaches have been identified, the engineer will perform a preliminary analysis to determine the specific approach most appropriate for the occasion. Inherent in this analysis is the aspect of engineering decision making discussed in Section 2.3. The process illustrated in Figure 2-2 is followed, and a recommended course of action evolves as a result.

C. Results of Feasibility Study (Technical Proposal)

The results of the feasibility study are generally presented in the form of a technical proposal to management. The information prepared should include the following:

1. A definition of the functional configuration (e.g., functional block diagram) and physical characteristics of the system.
2. A description of the system operational and maintenance support requirements identified in Section 3.3A as applied to the technical approach recommended from the preliminary systems analysis defined in Section 3.3B.

This proposal serves as the engineering technical baseline to support management decisions made throughout the product planning process.

3.4 ADVANCED PRODUCT PLANNING

A. Product System Evaluation, Selection, and Justification

The combined results of the preliminary market analysis (Section 3.2) and the feasibility study (Section 3.3) aid in justifying the product or system in terms of need and technical approach. However, the results of these efforts must be reviewed further from the business standpoint.

As illustrated in Figure 3-2, the selected product or system must be evaluated in terms of life-cycle cost. Life-cycle cost includes all future costs associated with planning, research, and development, investment (production and/or construction), operation and support, and system phase-out.[10] These costs are estimated and projected by year. Subsequently, a product pricing scheme must be established, with the objective of maximizing the return on investment. Pricing decisions raise such questions as:

1. What quantity of equipment will the producer sell to the consumer at a given price?
2. What effect will the established price have on the market share? This factor must consider the reactions of possible competition in price setting.
3. If the price is varied, what effect does the variation have on the system design, performance, effectiveness, and logistic support factors?

Pricing policies can be quite sophisticated and yet are treated very lightly in many instances. The establishment of a product price must consider the producer business objectives, the competition, the producer-consumer relationship in terms of system transition and support, warranties, liabilities, risks, and many other factors. This area is discussed further in Section 3.4F, where system proposals are covered.

With the establishment of pricing data, an estimate of the market share, and a knowledge of the time of need, revenues can be projected by year and compared with the anticipated costs in Figure 3-2. The revenue figures are

---

[10]Refer to Appendix A and Chapter 12 for a definition and typical breakout of total life-cycle cost.

EVALUATION CRITERION FACTORS	WEIGHTING	SCORE*
1. Technology Availability a. Now (3 Points) b. From 1 to 2 Years in the Future (2 Points) c. From 2 to 5 Years in the Future (1 Point) d. More than 5 Years in the Future (– 1 Point)	15	45
2. System Technical Characteristics a. Performance (1) Exceeds Requirements (2 Points) (2) Meets Requirements (1 Point) (3) Does Not Meet Minimum Requirements (– 10 to – 50 Points)	15	15
b. Effectiveness-Operational Availability (1) Greater than 95% (3 Points) (2) Between 90% and 95% (2 Points) (3) Between 80% and 90% (1 Point) (4) Less than 80% (– 2 Points)	10	20
3. Marketing Information a. Market Potential (1) Greater than 500 Equipments (4 Points) (2) Between 300 and 500 Equipments (2 Points) (3) Between 200 and 300 Equipments (– 1 Point) (4) Less than 200 Equipments (– 2 Points)	12	24
b. Market Share (1) Greater than 75% (4 Points) (2) From 25% to 75% (2 Points) (3) Less than 25% (– 2 Points)	8	16
4. Financial Data a. Return on Investment (1) Greater than 25% (3 Points) (2) Between 20% and 25% (2 Points) (3) Between 15% and 20% (1 Point) (4) Between 10% and 15% (– 1 Point) (5) Less than 10% (– 2 Points)	20	40
b. Estimated Annual Sales (1) Greater than $1,000,000 (2 Points) (2) Between $100,000 and $1,000,000 (1 Point) (3) Less than $100,000 (– 1 Point)	10	20
c. Investment Recovery Point (Refer to Figure 3-2) (1) Less than 3 Years (2 Points) (2) Between 3 and 5 Years (1 Point) (3) Greater than 5 Years (– 1 Point)	10	10
Total	100	190

*Maximum Possible Score – 285 Points

Figure 3-6. System evaluation (*typical*).

based on the inventory profile or the quantity of equipment delivered for consumer use by year (assuming that payments are made on a corresponding basis). The results can be used to determine the return on investment for the project.[11]

When evaluating a project representing a particular product configuration, one should select evaluation criterion factors that cover all major considerations necessary to justify the selection decision. A typical listing of such factors is presented in Figure 3-6. Referring to the figure, evaluation factors

[11]Common financial measures, to include return on investment, are discussed further in Chapter 11.

are listed, and weighting values are established for each major category. The weighting values (adding up to 100) indicate the level of importance considered valid by the evaluator.[12] Points are indicated for each evaluation category and are assigned against the system configuration in proportion to the degree of compliance with the criterion in question. The point values are multiplied by the weighting values to determine the score. Referring to Figure 3-6, the configuration being evaluated received a total score of 190 of a possible 285 points, based on the assumptions made.

The product configuration proposed as a result of the feasibility study is evaluated using the above approach or some other comparable method. To ensure that this configuration is the best possible solution from both a *technical* and *business* standpoint, alternatives are considered and evaluated in a consistent manner.[13] The evaluation process follows the logic flow illustrated in Figure 2-2, and the alternatives are presented in the form of a decision-tree matrix shown in Figure 2-3.[14] If the proposed configuration is indeed justified on the basis of both technical and business criteria, a product specification and advanced planning information are prepared as shown in Figure 3-1. On the other hand, if one of the alternatives receives a significantly better evaluation score, the results are reviewed and incorporated into the feasibility study proposal (refer to the feedback loop in Figure 3-1). Some illustrations of the evaluation process are presented in Appendix B.

B. Product Specifications and Plans

With the initial product (or system) configuration fairly well established, advanced planning continues through the development of formal specifications and planning documentation. Specifications basically cover the technical design requirements for the product, and planning documentation includes all management-related activity. The combination of specifications and plans along with the associated cost data constitute the proposal for all subsequent research, design, production/construction, evaluation and test, product use, and logistic support activity. Once these specifications and plans have been reviewed and accepted, they will be considered as the basis

---

[12] Both evaluation factors and weighting values will vary with the application. Emphasis in the evaluation will be different in each occurrence, and no two evaluators will necessarily agree with the evaluation factors or the weighting values. In an attempt to reduce bias in the evaluation, the Delphi technique briefly discussed in Section 3.2A may be employed to develop the desired information.

[13] These alternatives may include some of those initially considered during the feasibility study or may involve relatively small variations of the proposed configuration.

[14] Decision trees in the evaluation of alternatives are covered in: H. G. Thuesen, W. J. Fabrycky, and G. J. Thuesen, *Engineering Economy,* 4th ed., Prentice-Hall, Inc., Englewood Cliffs, N.J., 1974; and D. W. Karger and R. G. Murdick, *Managing Engineering and Research,* Industrial Press, Inc., New York, 1969.

for all future program engineering and management decisions.

The scope and depth of such documentation depends on the nature, size, and complexity of the product. Also, the extent of new development (where extra controls are required) versus an off-the-shelf capability will dictate the amount of documentation necessary. For small and relatively simple items, the technical specification and program planning requirements may be included in a single document. On the other hand, for large-scale systems a significant assemblage of documents may be appropriate. In either case, the extent of documentation must be tailored to the need as dictated by the management and technical controls required to get the job done.

In dealing with large systems, there are many elements that must be covered by specification (dictating the technical requirements) and by planning documentation (indicating the "how," "when," and "where" information pertaining to program implementation and control). Some of the system components may require an extensive amount of research and development effort; other items are procured directly from current supplier inventories. For new items, some are developed within the major producer's facility while others are developed by suppliers located in various parts of the world. The definition of system and program requirements and the subsequent control of program activities are quite difficult unless there is an established method of relating all requirements in an organized manner.

As an initial step in defining program documentation requirements, management should prepare a *documentation tree*, showing the hierarchy of specifications and plans. There should be no doubt as to what document takes precedence over the others.[15] An example of a documentation structure is presented in Figure 3-7. Both specifications and plans are included, with the *system specification* covering all design requirements and the *program management plan* including all planning, organizing, directing and controlling activities.

1. Specifications.[16] Specifications may be classified as follows:

a. *System specification*—states the technical, operational, and support requirements for a product or system as an entity. Further, it allocates these requirements to functional areas and defines the various functional-area

[15] In the past, product/system requirements have included the reference to numerous standards and component specifications, and in many instances conflicts have arisen causing both design and management problems. Through the proper ordering and preparation of all program documentation, it is hoped that such conflicts can be avoided in the future.

[16] Specification types and formats will actually vary from one situation to the next. However, examples of some standard definitions are presented in MIL-STD-490, "Specification Practices," U.S. Department of Defense, Government Printing office, Washington, D.C., Oct. 30, 1968.

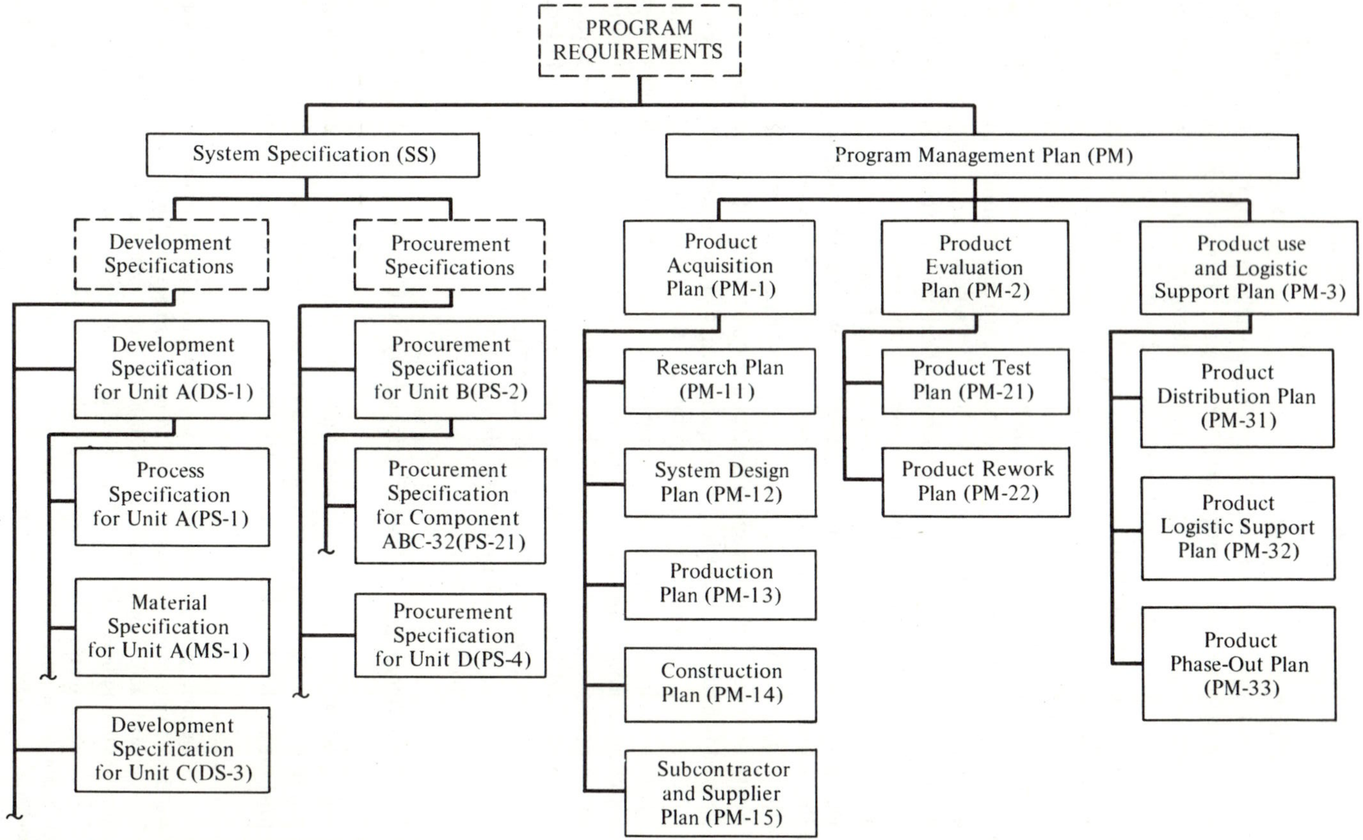

Figure 3-7. Program documentation tree.

SYSTEM *XYZ* SPECIFICATION — SPECIFICATION NO: SS123

1. Scope
2. Applicable Documents
3. Requirements
   - 3.1 System Definition
     - 3.1.1 General Description
     - 3.1.2 Operational Requirements
       - a. Deployment and Utilization
       - b. Mission Definition
     - 3.1.3 Maintenance Concept
     - 3.1.4 System Functional Diagrams
     - 3.1.5 Interface Requirements
   - 3.2 System Characteristics
     - 3.2.1 Performance Characteristics
     - 3.2.2 Physical Characteristics
     - 3.2.3 Effectiveness Characteristics
     - 3.2.4 Design Characteristics
       - a. Reliability
       - b. Maintainability
       - c. Human Factors
       - d. Transportability
       - e. Producibility
       - f. Standardization
       - g. Other
     - 3.2.5 Environmental Characteristics
   - 3.3 Design and Construction
     - 3.3.1 Materials, Processes, and Parts
     - 3.3.2 Workmanship
     - 3.3.3 Electromagnetic Interference
     - 3.3.4 Safety
     - 3.3.5 Interchangeability
     - 3.3.6 Nomenclature
   - 3.4 System Documentation
   - 3.5 Logistic Support
     - 3.5.1 Test and Support Equipment
     - 3.5.2 Supply Support
     - 3.5.3 Personnel and Training
     - 3.5.4 Technical Data
     - 3.5.5 Facilities
     - 3.5.6 Maintenance Planning
   - 3.6 Precedence
4. Quality Assurance Provisions
   - 4.1 Requirements
   - 4.2 Test Preparation
   - 4.3 Test Conductance
     - 4.3.1 Environmental Qualification
     - 4.3.2 Reliability Tests
     - 4.3.3 Maintainability Demonstration
     - 4.3.4 Logistic Support Compatibility
   - 4.4 Test Reporting
   - 4.5 Retesting
   - 4.6 Quality Conformance Inspection
5. Preparation for Delivery
6. Equipment Phase-Out
7. Changes and Revisions
8. Notes

Figure 3-8. Sample system specification format.

interfaces. The information derived from the feasibility study discussed in Section 3.3 is included. Figure 3-8 presents a sample outline of the specification content.

b. *Development specification*—states the technical requirements for any item below the top system level where research and development (to include design) is required. Each development specification must include the performance, effectiveness, and support characteristics that are required in evolving from preliminary design to detail design and production.

c. *Procurement specifications*—states the technical requirements below the system level for any item that is currently in the inventory and procured off-the-shelf.

d. *Process specification*—covers a service (e.g., heat treating, plating, welding, sanding, packing, marking) that is performed on any product or material.

e. *Material specification*—covers the raw material, mixtures (e.g., paints, chemical agents), and/or semifabricated material (e.g., electrical cable, piping) that are used in the fabrication of a product.

The preparation of specifications is an engineering function. The system specification is prepared initially at program inception and later updated to reflect the output of the conceptual and preliminary design activities discussed further in Chapter 5. Development and procurement specifications are based on the results of make-or-buy decisions and are generally prepared during preliminary design.[17] Process and material specifications are more oriented to production/construction operations and are prepared during detailed design. Although these various classes of specifications are directed toward different objectives and phases in the total product development process, it is possible in most instances to identify the need for coverage during the product planning function. Thus, the documentation tree may include reference to a specification prior to its actual preparation.

2. Planning documentation. The top-level program management plan (identified in Figure 3-7) forms the basis for all lower-level plans. Individual supporting plans will be prepared to cover the various phases of product life-cycle activity and will include organizational structure and responsibilities, program functions and tasks, schedules, processes and procedures, and cost projections. These subordinate plans are discussed in Section 3.4C, D, and E. All plans should (if at all possible) be prepared by those responsible to

[17]The results of make-or-buy decisions determine whether an item is to be manufactured within the producer's facility or purchased from an outside source. Economic factors and schedule requirements, combined with the availability of reliable sources of supply, are prime considerations in the decision-making process. Usually, make-or-buy decisions result from a committee effort including representation from engineering, purchasing, manufacturing, quality control, and other related areas.

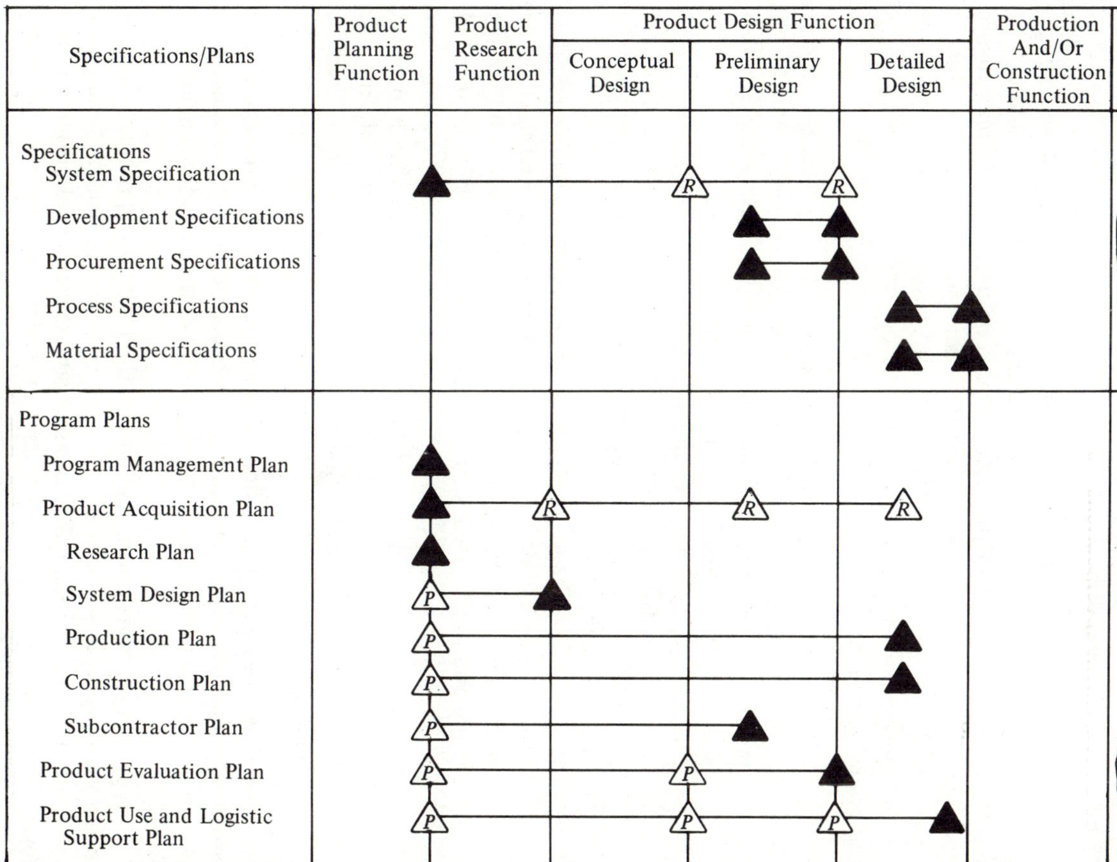

Note:

The product life cycle phases are presented here as in Figure 1-4. Not all functions may be applicable in all instances, and some of the functions may overlap for certain programs (e.g., some conceptual design is accomplished as part of the planning function). Thus, this figure is oriented to the specifications and plans depicted by program phase and not in terms of calendar time sequence.

Symbols:

▲ Formal Document

(R) Document Revision

(P) Preliminary Document

Figure 3-9. Program specifications and plans.

execute the work involved, and must be consistent with available resources and based on a realistic schedule. In addition, there should be provisions for contingency plans for coping with the unexpected.

Figure 3-9 is presented to illustrate the relative timing of specifications and plans in terms of the major product-life-cycle functions. The figure is primarily oriented to a large system in which all the functions identified in Figure 1-4 are applicable. When this is not the case, the requirements illustrated in Figure 3-9 must be tailored accordingly.

C. Product Acquisition Plan

Acquisition as discussed herein refers to the process of acquiring a product or system commencing with the identification of a need and extending through the delivery of the product for consumer use. It may involve research, design, production/construction, and some test and evaluation activities. The plan for acquisition generally covers *producer* operations (versus consumer activities), and will vary in scope and content depending on the following conditions:

1. The product/system need indicates that new development effort is required which involves technology not currently available. In this instance, research, system design, production and/or construction, and some test and evaluation are required.
2. The product/system need indicates that new development effort is required involving current state-of-the-art technology. In this case, research is not required, but activities in the other acquisition areas are necessary.
3. The product/system need indicates that an existing item currently in the inventory will satisfy the requirement if certain modifications are incorporated. Activities in this instance will include some design, production and/or construction, and some test and evaluation.
4. The product system need indicates that an existing capability currently in the inventory and being employed to satisfy a different purpose will also fulfill the new requirement. Here, only the production and/or construction of an additional quantity of items is required.

The above four options are illustrated by the alternative life cycle paths depicted in Figure 1-4, and the requirements relative to acquisition activities stem from the discussions presented earlier in this chapter (particularly in Section 3.4A).

Depending on the appropriate option, the product acquisition plan may include individual lower-level plans covering research, system design, and so on, as indicated in Figure 3-7. In any event, the material presented must cover definition of tasks, scheduling of tasks, organization of tasks, cost/schedule/performance measurement and control, and provisions for program evaluation and corrective action.

1. Definition of tasks. Detailed tasks are identified to cover all acquisition functions. A task is made up of one or more events and associated activities. There are starting events, ending or terminating events, interface events, and so on, which are indicated as discrete points in time that represent a specific identifiable output or accomplishment. Activities constitute the ongoing work needed to complete an event. An initial facet in planning involves the preparation of detailed task descriptions, task input requirements, the procedure for task accomplishment, and task controls (i.e., the provisions necessary to determine task status and when a task has been completed).

2. Scheduling of tasks. Once the tasks have been defined, each must be related in terms of program time (i.e., when must the task be accomplished to produce the desired output?). Task scheduling may be accomplished using one or a combination of techniques, a few of which are identified below.

a. *Simple bar chart*—a presentation of program activities in terms of sequences and the time span of effort. Specific milestones and man-loading requirements are not covered. Figure 3-10 illustrates a partial bar chart.

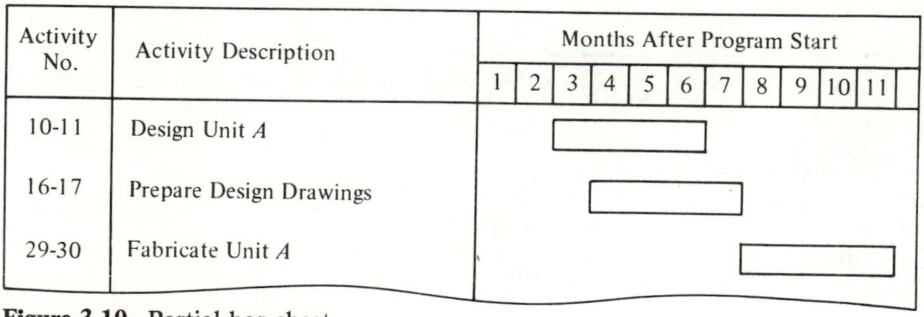

Figure 3-10. Partial bar chart.

b. *Milestone chart*—a presentation of program events (e.g., specific identifiable outputs) in terms of required start and completion times. Events are usually tied to a particular calendar date, as shown in Figure 3-11.

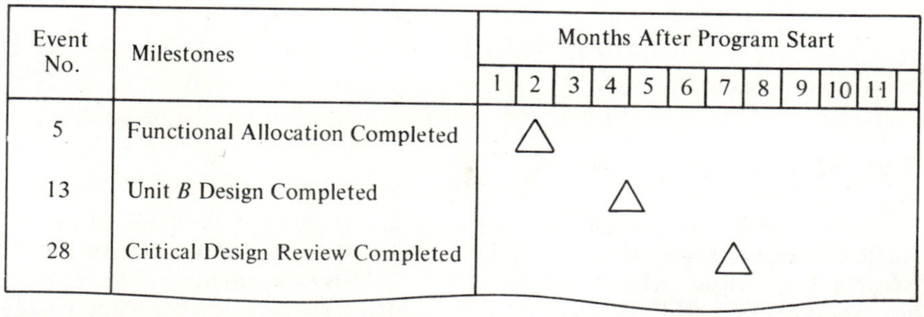

Figure 3-11. Partial milestone chart.

c. *Program evaluation and review technique* [*PERT*]—a scheduling technique that combines events and activities into a program network structure.[18] PERT is ideally suited to early planning, where precise time data are not readily available, and is also effective in showing the interrela-

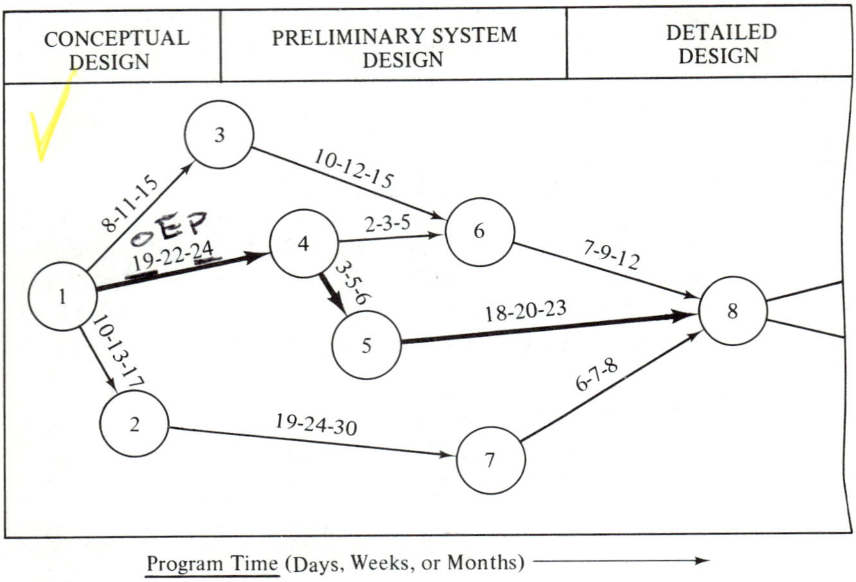

SYMBOLS AND NOTES:

x — Events are usually indicated as circles and are considered as checkpoints to show what has been accomplished to date.

2-3-4 → Activity lines represent the work needed to complete an event. Work can only start on the next activity after the preceding event is completed. The numbers on the activity lines indicate the time required in days, weeks, or months. The first number reflects an optimistic time estimate, the second number indicates the expected time, and the third number constitutes a pessimistic time estimate.

→ Critical path. (Events 1, 4, 5 and 8)

Figure 3-12. Partial PERT network.

---

[18]PERT is described in a number of references, which include: D. W. Karger and R. G. Murdick, *Managing Engineering and Research,* Industrial Press, Inc., New York, 1969; W. J. Fabrycky, P. M. Ghare, and P. E. Torgersen, *Industrial Operations Research,* Prentice-Hall, Inc., Englewood Cliffs, N.J., 1972; and C. S. George, *Management in Industry,* Prentice-Hall, Inc., Englewood Cliffs, N.J., 1964.

tionships of combined activities. PERT introduces the aspects of probability by which better management decisions can be made. A partial example of PERT scheduling is illustrated in Figure 3-12.

In applying PERT to a project, one must identify all interdependent activities and events for each applicable phase of the project. Events are related to program milestone dates that are based on management objectives. Managers and programmers work with engineering organizations to define these objectives and identify tasks and subtasks. When this is accomplished to the necessary level of detail, networks are developed, starting with a summary network and working down to detailed networks covering specific segments of a program.[19] The development of PERT networks is a team effort.

When actually constructing networks, one starts with an end objective (e.g., event 8 in Figure 3-12) and works backward in developing the network until event 1 is identified. Each event is labeled, coded, and checked in terms of program time frame. Activities are then identified and checked to ensure that they are properly sequenced. Some activities can be performed on a concurrent basis, and others must be accomplished in series. For each completed network, there is *one beginning event* and *one ending event,* and all activities must lead to the ending event.

The next step in developing a PERT network is to estimate activity times and to relate these times in terms of probability of occurrence. An example of the calculations that support a typical PERT network is presented in Figure 3-13 and described below.

The following steps are appropriate:

(1) Column 1 — List each event, starting from the last event and working downward to the start (i.e., from event 8 to event 1 in Figure 3-12).

(2) Column 2 — List all previous events that have been indicated prior to the event listed in column 1.

(3) Columns 3-5 — Determine the optimistic (t_a), the most likely (t_b), and the pessimistic (t_c) times in weeks or months for each activity. Optimistic time means that there is very little chance that the activity can be completed before this time, while pessimistic time means that there is little likelihood that the activity will take longer. The most likely time (t_b) is located at the highest probability point or the peak of the distribu-

[19]The level of detail and depth of PERT network development (i.e., the number of activities and events included) are based on the criticality of tasks and the extent to which program monitoring and control is desired. Milestones which are critical in meeting the objectives of the program should be included along with activities that require extensive interaction for successful completion. The author has had experience in dealing with PERT networks including from 10 to 700 events. The number of events will of course vary with each project.

1	2	3	4	5	6	7	8	9	10	11	12
Event Number	*Previous Events*	t_a	t_b	t_c	t_e	σ^2	*TE*	*TL*	*TS*	*TC*	*P* (%)
8	7	6	7	8	7.0	0.108	—	—		46.0	26.6
	6	7	9	12	9.2	0.693	—	—			
	5	18	20	23	20.2	0.693	46.8	46.8	0.0		
7	2	19	24	30	24.2	3.349	37.7	39.8	2.1		
6	4	2	3	5	3.2	0.250	—	—			
	3	10	12	15	12.2	0.693	23.4	37.6	14.2		
5	4	3	5	6	4.8	0.250	26.6	26.6	0.0		
4	1	19	22	24	21.8	0.693	21.8	21.8	0.0		
3	1	8	11	15	11.2	1.369	11.2	25.4	14.2		
2	1	10	13	17	13.5	1.369	13.5	15.6	2.1		

Note: Assume that time values for t_a, t_b, etc., are in weeks

Figure 3-13. Example of PERT calculations.

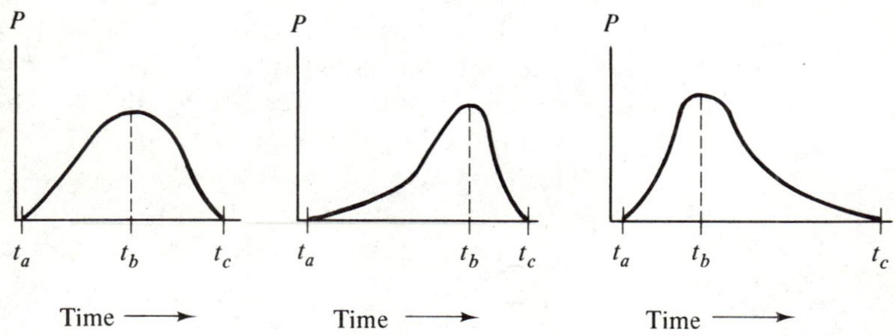

tion curve. These times may be predicted by someone who is experienced in estimating. The time estimates may follow different distribution curves, where p represents the probability factor.

(4) Column 6

Calculate the expected or mean time, t_e, from

$$t_e = \frac{t_a + 4t_b + t_c}{6} \tag{3-1}$$

(5) Column 7

In any statistical distribution, one may wish to determine the various probability factors for different activity times. Thus, it is necessary to compute the variance (σ^2) associated with each mean value. The square root of the variance or the standard deviation is a measure of the dispersion of values within a distribution and is useful in determining the percentage of the total population sample that falls within a specified band of values.[20] The variance is calculated from equation (3-2):

$$\sigma^2 = \left(\frac{t_c - t_a}{6}\right)^2 \tag{3-2}$$

(6) Column 8

The earliest expected time, TE, is the sum of all times, t_e, for each activity and the cumulative total of the expected times through the preceding event, remaining on the same path throughout the network. When several activities lead to an event, the highest time value (t_e) will be used. For example, path 1, 4, 5, 8 in Figure 3-12 totals 46.8.

[20] A comprehensive definition and description of the various aspects of probability distributions, density functions, standard deviation and variance factors, and so on, are not intended to be included herein. If questions arise, the student is advised to review some of the basic fundamentals of statistics.

(7) Column 9 — The latest allowable time, TL, is calculated by starting with the latest time for the last event which equals TE (i.e., where TE equals 46.8). Then, one works backward, subtracting the value in column 6 for each activity, staying in the same path.

(8) Column 10 — The slack time, TS, is the difference between the latest allowable time (TL) and the earliest expected time (TE):

$$TS = TL - TE \tag{3-3}$$

(9) Columns 11-12 — TC refers to the required scheduled time for the network based on the need. Assuming that TC is 46.0, it is necessary to determine the probability (*P*) of meeting the need. This probability factor is determined as follows:

$$A = \frac{TC - TL}{\sqrt{\Sigma \text{ variances}}} \tag{3-4}$$

where *A* is the area under the normal distribution curve and the variance is the sum of the individual variances applicable to the critical-path activities in Figure 3-12 (path 1, 4, 5, 8):

$$A = \frac{46.0 - 46.8}{\sqrt{0.693 + 0.250 + 0.693}} = -0.625$$

Referring to normal distribution tables, the calculated value of -0.625 represents an area of approximately 0.266; that is, the probability of meeting the scheduled time of 46 weeks is 26.6 per cent. If the scheduled time is 45 weeks, then the probability of success would be 0.079. On the other hand, if the scheduled time is 50 weeks, the probability of success is 0.9938.

When evaluating the resultant probability value (column 12 of Figure 3-13), management must decide on the range of factors allowable in terms of risk. If the probability factor is too low, additional resources may be applied to the project to reduce the activity schedule times and improve the probability of success. On the other hand, if the probability factor is too high, this may indicate that excess resources are being applied, some of which should be diverted to other projects. Management must assess the situation and establish a goal.

The application of PERT scheduling is appropriate for both small-and large-scale projects and is of particular value in one-of-a-kind projects or for those program phases where repetitive tasks are not predominant. For highly repetitive manufacturing runs, other scheduling and control techniques are more appropriate.

In summary, the use of PERT scheduling technique is quite popular and offers many advantages:

(1) PERT is readily adaptable to advanced planning and essentially forces the detailed definition of tasks, task sequences, and task interrelationships. All levels of management and engineering are required to think through and evaluate a project carefully.

(2) PERT enables management and engineering to predict with some degree of certainty the probable time that it will take to achieve an objective.

(3) PERT enables the rapid assessment of progress and the detection of problems and delays.

(4) PERT includes a great many data from which performance standards can be established.

The implementation of PERT in a comprehensive and timely manner is possible since the technique is particularly adaptable to computer methods.

d. *Critical path method* [*CPM*]—a network scheduling technique with activities and events analogous to PERT except that emphasis is placed on *critical* activities or those requiring the greatest amount of time for completion. A network similar to the one presented in Figure 3-12 is developed and slack times are estimated as presented in Figure 3-13 (column 10). The critical path is the longest path through the network where slack times are zero, or the path represented by the heavy line in Figure 3-12 (i.e., the path including events 1, 4, 5, and 8). The critical path identifies those activities which may pose problems of both a technical and administrative nature if schedule slippage occurs. In other words, these are the activities that must be closely monitored and controlled throughout the program.

The network paths representing other program activities shown in Figure 3-12 include a slack time, which constitutes a measure of program scheduling flexibility. The slack time is the interval of time where an activity might actually be delayed beyond its earliest scheduled start without necessarily delaying the overall program completion time. Total slack time is available with the objective of possibly effecting a trade-off and the reallocation of resources. Program scheduling improvements may be possible by shifting resources from activities with slack time to activities along the critical path.[21]

---

[21]CPM is discussed further in: W. J. Fabrycky, P. M. Ghare, and P. E. Torgersen, *Industrial Operations Research*, Prentice-Hall, Inc., Englewood Cliffs, N.J., 1972.

e. *PERT/COST*—an extension of PERT and CPM which deals not only with time but includes cost. A cost network can be superimposed upon a PERT network by estimating the total cost and cost slope for each activity line.

When implementing this technique, there is always the time-cost option, which enables management to evaluate alternatives relative to the allocation of resources for activity accomplishment. In many instances time can be saved by applying more resources or, conversely, cost may be reduced by extending the time to complete an activity. The time-cost option can be attained by applying the following general approach:

(1) Determine alternative time and cost estimates for all activities, and select the lowest-cost alternative in each instance.

(2) Calculate the critical path time for the network. If the calculated value is equal to or less than the total time permitted, select the lowest-cost option for each activity, and check to ensure that the total of the incremental activity times does not exceed the allowable overall program completion time.

If the calculated value exceeds the program time permitted, review the activities along the critical path and select the alternative with the lowest cost slope, and reduce the time value to be compatible with the program requirement.

(3) After the critical path has been established in terms of the lowest cost option, review all the network paths with slack time, and shift activities to extend the times and reduce costs wherever possible. Activities with the steepest cost-time slopes should be addressed first.

The time-cost option can be illustrated by referring to two paths of the network presented in Figure 3-12. Data associated with each selection path are as follows:

Path 1, 4, 5, 8				*Path 1, 3, 6, 8*			
Event Number	*Previous Event*	t_e	*Estimated Activity Cost ($)*	*Event Number*	*Previous Event*	t_e	*Estimated Activity Cost ($)*
8	5	20.2	3,500	8	6	9.2	1,050
5	4	4.8	940	6	3	12.2	2,450
				3	1	11.2	1,870
4	1	21.8	4,800				
Total		46.8	$9,240	Total		32.6	$5,370

Path 1, 4, 5, 8 is the critical path, and the alternative path 1, 3, 6, 8 has a slack of 14.2. As management is particularly concerned with the critical path (since it is the governing factor in the overall network relative to program success or failure), the shifting of resources from the alternative path to critical-path activities should be considered. For instance, assume that $200 is reallocated from activity 8, 6 to activity 8, 5. As a result, activity 8, 6 will now require 12.2 units of time, and activity 8, 5 will require 18.7 units of time. The total time along the critical path is reduced, and the project is completed in 45.3 units of time. On the other hand, the time along the alternative path is increased to 35.6 units, and the slack is reduced to 9.7 units. The entire network must now be reevaluated to determine the effects of the changes on the critical path. This process may be continued by trading off resources against time until the most feasible approach is attained.

PERT/COST has proved to be a very useful technique in the planning of program events and activities and allows for the necessary program cost and schedule status monitoring and control requirements accomplished throughout system development and acquisition.

f. *Gantt chart*—a scheduling technique used primarily in production and/or construction planning to show activity or job requirements, facility loading, and work status on a day-to-day basis. This technique was designed for, and is most successfully applied to, highly repetitive operations. An example of one basic form of a Gantt chart is illustrated in Figure 3-14. Gantt charts are used for both long-range production/construction planning as a means for simulating future operations and for short-range scheduling on a day-to-day basis. The short-range aspect may take the form of machine-loading control charts, man-loading control charts, and/or job progress control charts.[22]

g. *Line of balance* [*LOB*]—a technique that complements Gantt in determining production status. While the Gantt technique primarily relates information on the effective and efficient utilization of resources (e.g., machine loading, man loading), LOB is more product-oriented. LOB is not directly concerned with the resources expended but is utilized in determining production progress in terms of per cent of task completion. Major bottlenecks in the production process are emphasized.

3. ***Organization of tasks.*** The events and activities discussed above constitute tasks that are organized into work packages. The identified work packages are evaluated from the standpoint of task type, complexity, and the

[22]Gantt charts are more thoroughly covered in: J. H. Green, *Operations Planning and Control,* Richard D. Irwin, Inc., Homewood, Ill., 1967; and J. W. Gavett, *Production and Operations Management,* Harcourt Brace Jovanovich, Inc., New York, 1968.

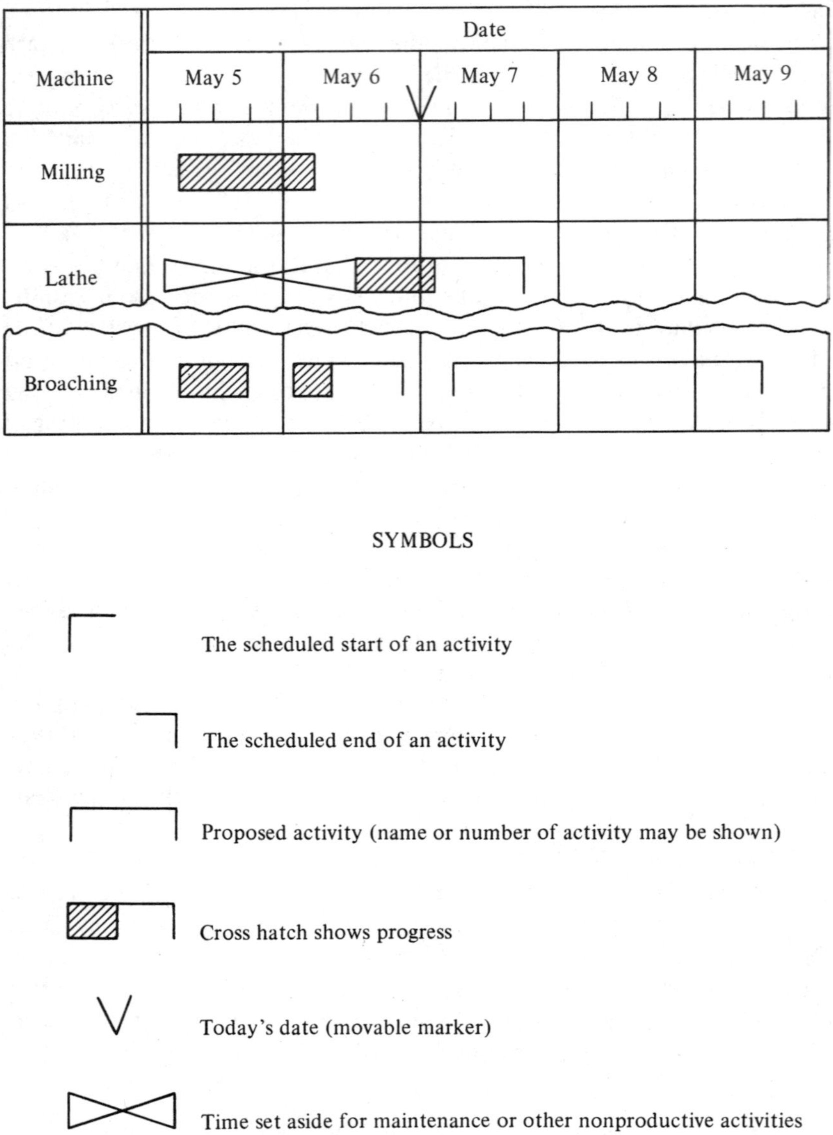

Figure 3-14. Gantt scheduling chart for a machine.

required completion schedule. In addition, estimated costs are established for each activity (as covered in the discussion on PERT/COST) and are then related to work packages. Figure 3-15 illustrates the work package concept as derived from an overall program network.[23] The individual work packages are ultimately combined and integrated with the program *work breakdown structure* (WBS). The WBS links objectives and tasks with resources and is an excellent management tool for program planning and budgeting. A partial sample WBS is presented in Figure 3-16.[24] The WBS is

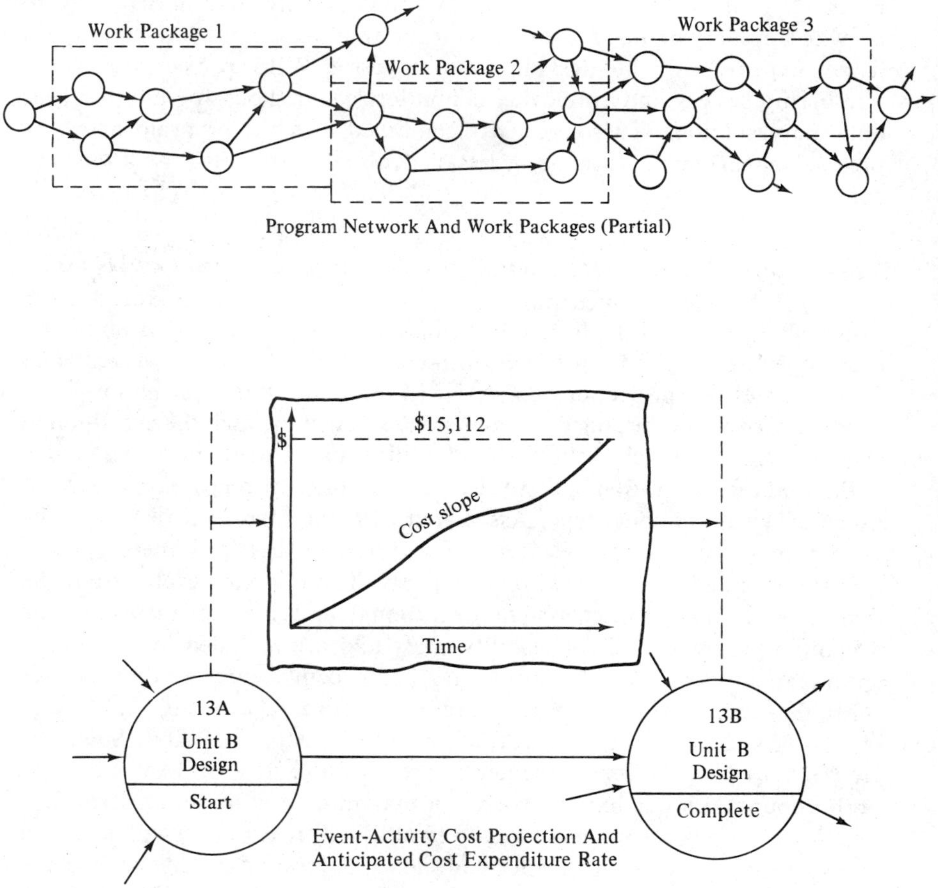

Figure 3-15. Organization of tasks into work packages.

---

[23]The development of program networks and work packages is described in Paul Mali, *Managing by Objectives,* Interscience Division, John Wiley & Sons, Inc., New York, 1972.

[24]In developing a WBS, *all* program tasks must be covered. In addition, *all* elements of the system (e.g., prime mission equipment, software, elements of logistic support) must be identified in the various categories of the WBS.

a logical separation of work-related units. Work packages are identified against each block (see block 3A1100, Figure 3-16, for an example of work package identification by block). These packages and WBS blocks are then related to organizational groups, branches, departments, projects, suppliers, etc.[25] Cost estimating is accomplished for each work package and identified by WBS block. The WBS is structured and coded in such a manner that program costs may be initially targeted and then collected against each block.[26] Costs may be accumulated both vertically and horizontally to provide summary figures for various categories of work. This can be accomplished expeditiously by designing an effective WBS project coding or numbering system and employing computerized methods for input-output data manipulations. These cost data combined with the program networks provide management with the necessary tools for program evaluation and control.

4. Cost/schedule/performance/effectiveness measurement and control. Thus far in the product planning process, we have introduced both *program* objectives in the form of scheduling and cost projections and *technical* objectives in the form of performance and effectiveness requirements. Schedule and cost projections are derived from marketing information (e.g., product need dates, pricing data) and are developed through the definition of specific tasks, task scheduling, the identification of work packages and a WBS, and the integration of these elements of work into a program organizational structure of some type. Associated with the identified tasks are the output requirements in the form of a product that will meet specific performance and effectiveness requirements. These factors evolve from the definition of the product/system operational and maintenance concepts accomplished as part of a feasibility study and are included in the system specification. As a result, initial target objectives are established for schedule, cost, performance (e.g., range, accuracy, capacity, size), and effectiveness factors (e.g., availability, reliability, supportability). Such objectives may take the form of single discrete values, a range of values with both upper and lower limits, or a set of maximum and minimum criteria.

With these requirements initially established, it is now essential that a management information system be developed to report data that enables the

---

[25]The organizational aspect of task assignment is discussed further in Part Three.

[26]In terms of cost collection, the WBS and cost breakdown structure used in life-cycle cost analyses (discussed in Chapter 5 and in Appendix A) are quite similar. The WBS serves as a program management tool and is often used for contracting purposes. The cost breakdown structure is an analytical device used in performing system/equipment analysis and evaluation. Although their purposes are quite different, many of the cost input factors are the same. The two efforts should be coordinated to the maximum extent possible to ensure compatibility where applicable.

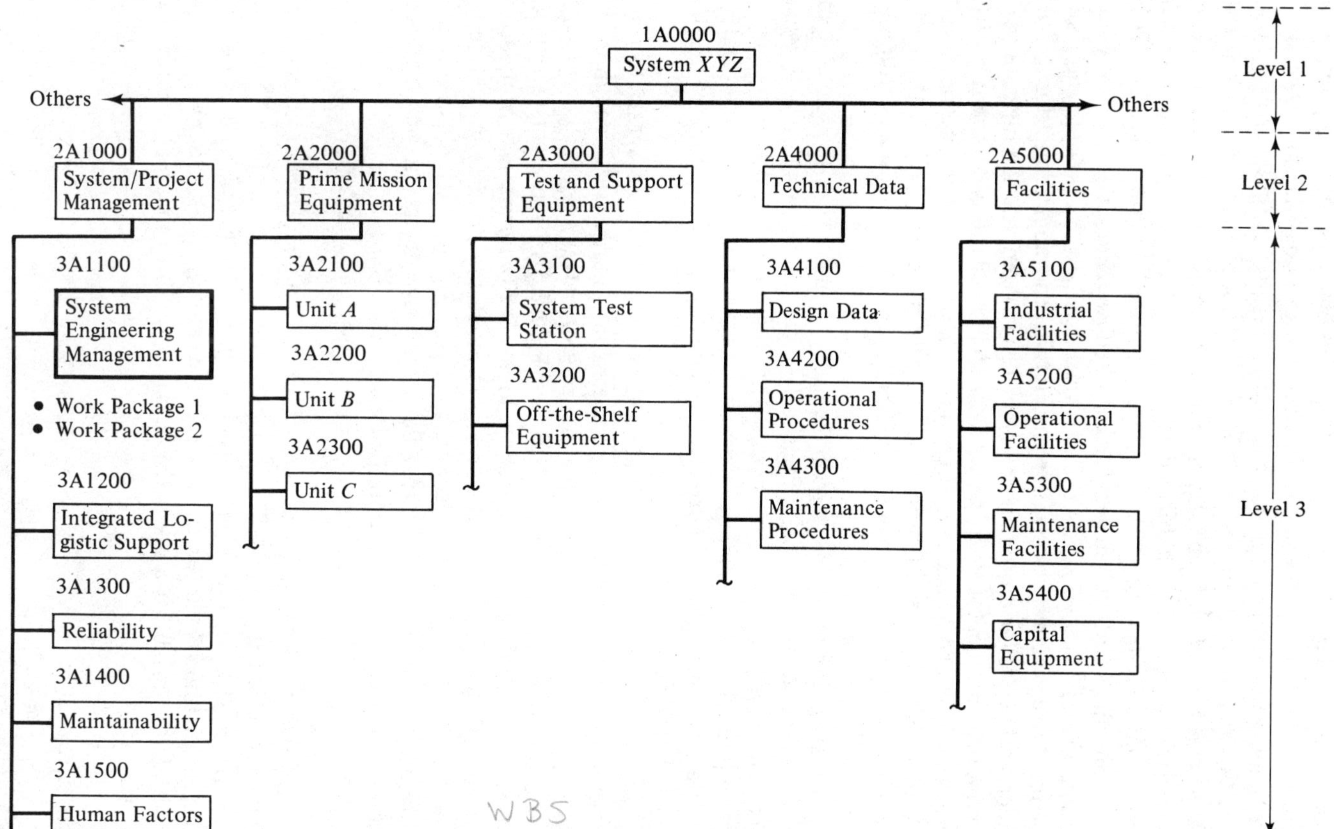

Figure 3-16. Partial work breakdown structure (*sample*).

measurement and control of certain selected factors as the program progresses.[27] This is necessary to provide the assurance that the ongoing system development process is producing the desired results. Program review dates in the product life cycle are established, and data are collected and inputed into the management information system, with the output results being compared with the initially established requirements for compatibility assessment. Input data are taken from accounting records, predictions, analyses, and so on, and will be discussed further in subsequent chapters. The comparison process should answer such questions as: Is the program on schedule? Are program costs within the established budget limitations? Will the currently envisioned product or system design configuration meet all performance requirements? Is the predicted system availability compatible with the established operational requirements? These and other questions will have to be answered on a number of occasions, particularly in the early phases of product development when there are many unknowns and the risks are high.

Product planning at this stage in the life cycle must cover the selection of the management information system to be used in the program and must include a description of its elements. The descriptive information should contain input data needs and the methods for data acquisition, the format(s) for data reporting, and the corrective action process to be followed in the event that problem areas are detected.

Although data reporting formats will likely vary from program to program, a few typical examples are presented in Figures 3-17 through 3-21. Figures 3-17 and 3-18 project task progress (in terms of work completed) against program activities and milestones. Figure 3-19 represents a typical output from PERT time/cost reporting. Figure 3-20 constitutes a cumulative summary of life-cycle cost projections and reliability predictions and compares the results with initial target objectives. Figure 3-21 illustrates the results of an evaluation of three specific system parameters. In each instance problem areas can be easily identified.

5. *Corrective action.* The data from the management information system should readily point out existing problem areas as well as potential areas where problems are likely to occur if program operations continue as originally planned. In order to deal with such contingencies, planning should be initiated to establish a corrective action procedure that will include the following steps:

---

[27]The factors selected will depend on system requirements and will likely vary from program to program. In addition, the management reporting requirements for a given program will differ for each program phase in the life cycle.

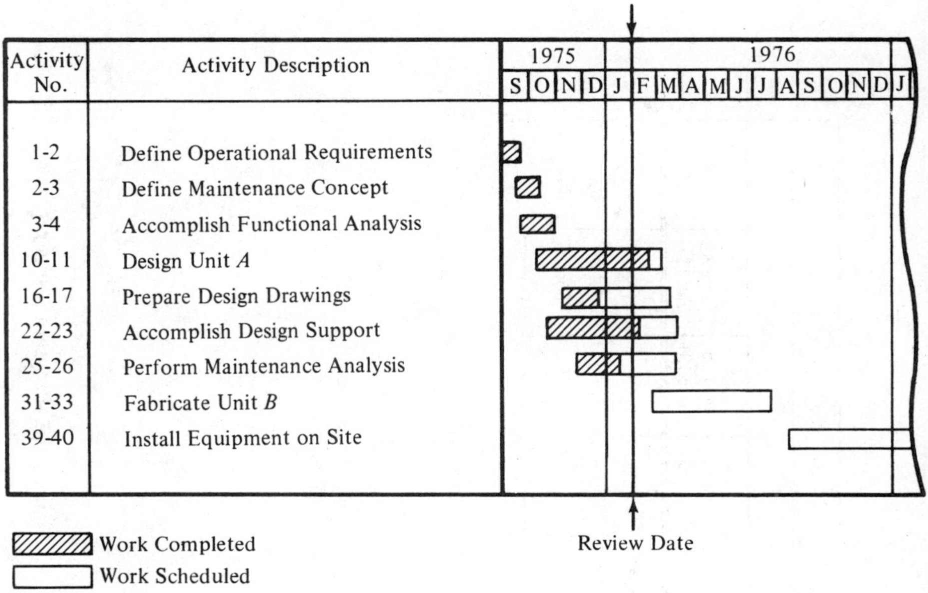

Figure 3-17. Bar chart showing progress against plan.

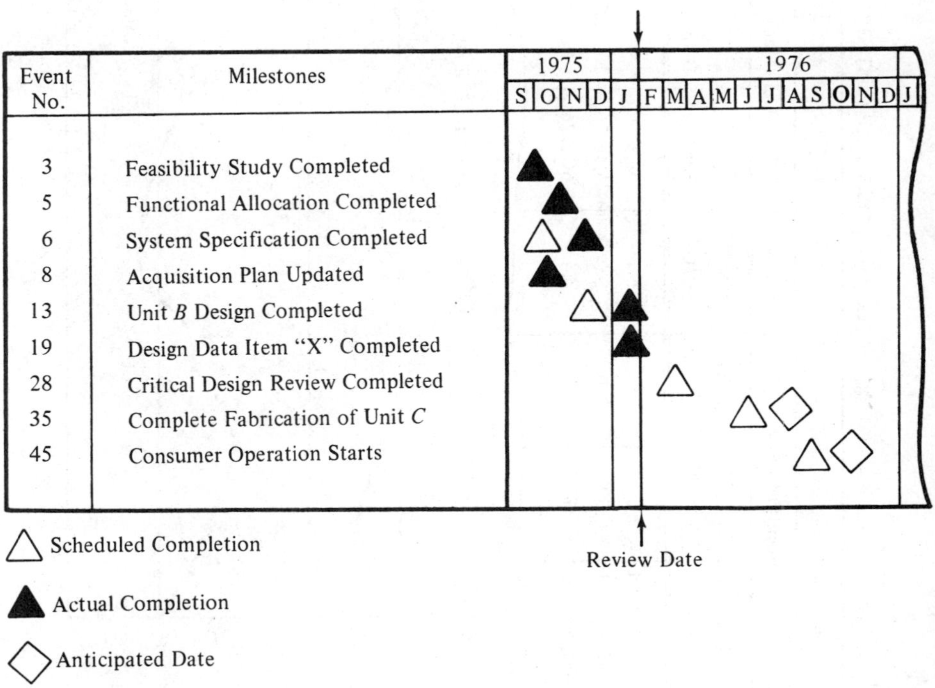

Figure 3-18. Partial milestone chart.

PERT TIME/COST STATUS REPORT

PROJECT: System XYZ				CONTRACT NO: 4B-3015					REPORT DATE: 4/10/76			
Item/Identification				Time Status					Cost Status			
WBS No.	Project No.	Beginning Event No.	Ending Event No.	Exp. Elap. Time (t_e) (Weeks)	Earliest Completion Date (D_E)	Latest Completion Date (D_L)	Slack $D_L - D_E$ (Weeks)	Actual Date Completed	Cost Est. ($)	Actual Cost to Date ($)	Latest Revised Est. ($)	Overrun (Underrun) ($)
3A1100	814763	36	37	5.2	3/4/76	4/11/76	7.8	4/5/76	2000	1950	1950	(50)
2A3100	824823	239	240	3.0	5/15/76	4/30/76	−3.2		3500	3650	4000	500
3A5400	833110	312	313	4.4	6/15/76	8/2/76	9.6		5660	3550	5660	0

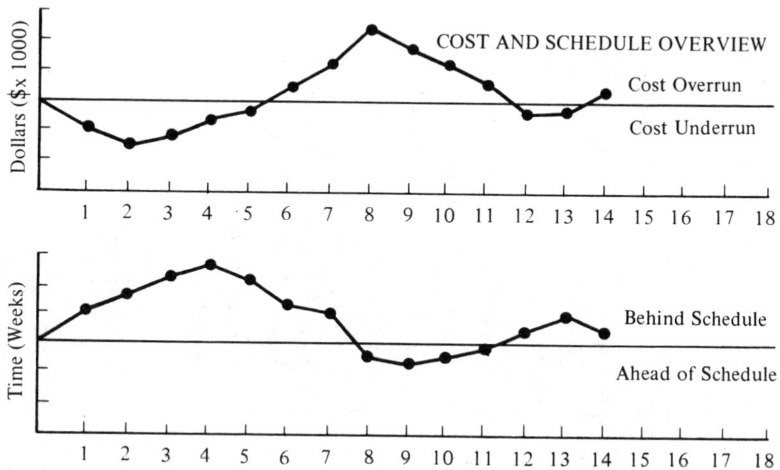

Figure 3-19. Program status reporting.

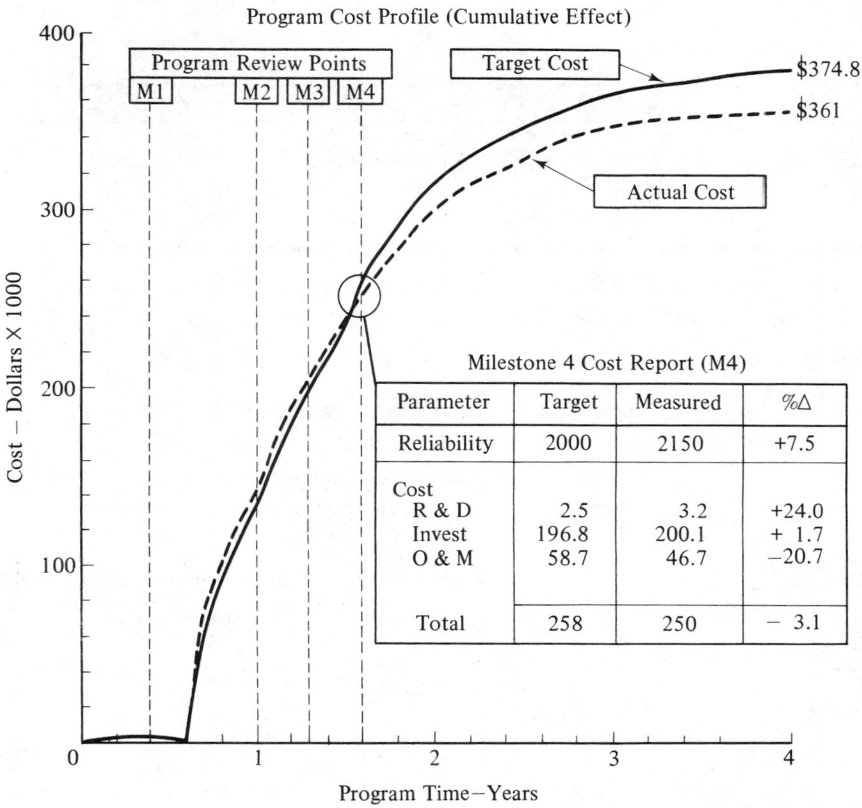

Figure 3-20. Program cost projection.

a. Problems are identified and ranked in order of importance. Ranking should consider the criterion factors listed in Figure 3-6 or some other factors of a comparable nature.

b. Each problem is evaluated on the basis of the ranking, addressing the most critical problems first. Alternative possibilities for corrective action are considered in terms of: (1) effects on program schedule and cost; (2) effects on system performance and effectiveness; and (3) the risks associated with the decision as to whether to take corrective action. The most feasible alternative is selected.

c. Given the decision to take corrective action, planning is accomplished to initiate the steps required to resolve the problem. This may be in the form of a change in management policy, a contractual change, an organizational change, or a system/equipment configuration change.

d. After corrective action has been implemented, some follow-up activity is required to (1) ensure that the incorporated change(s) actually has resolved the problem, and (2) assess other aspects of the program to ensure that additional problems have not been created as a result of the change.

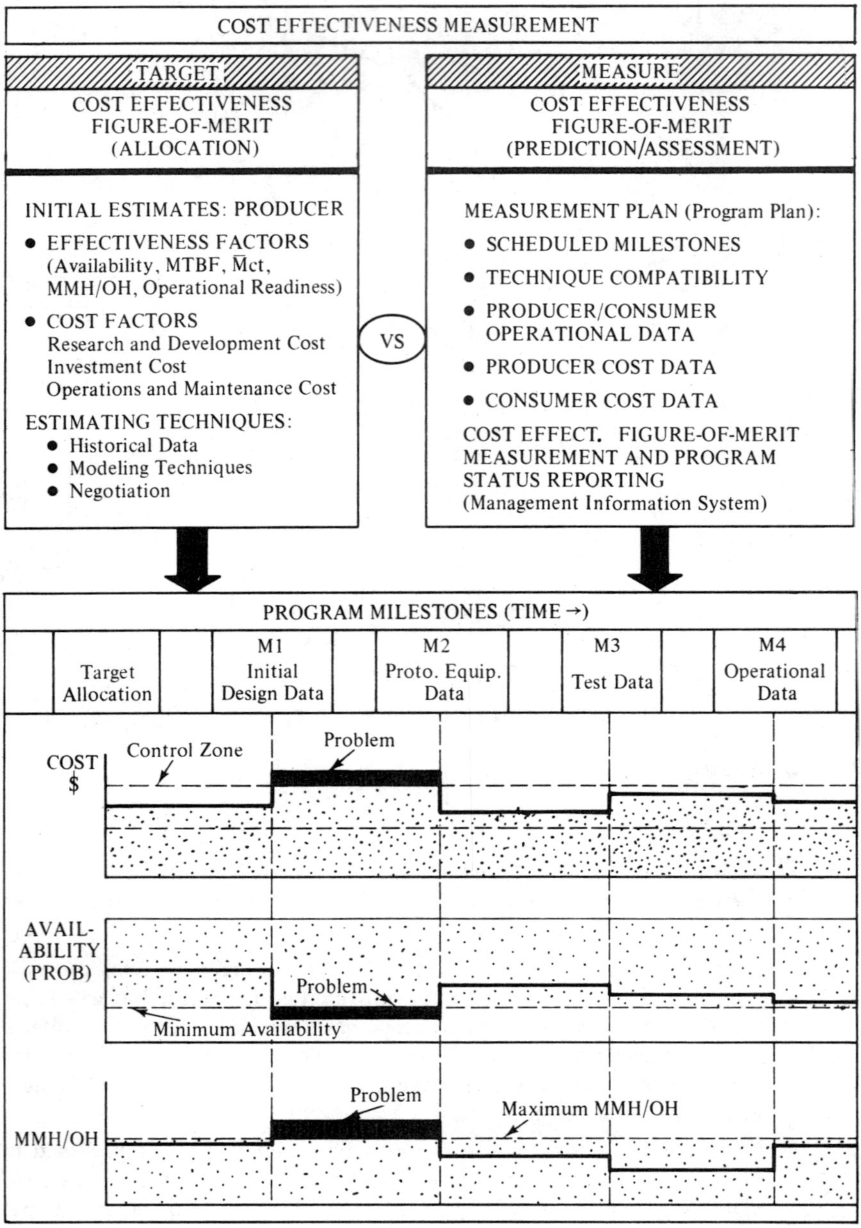

Figure 3-21. Parameter measurement and evaluation.

The corrective action aspect of engineering management is most essential if a program is to be successful.

As Figure 3-7 conveys, the product acquisition plan covers all activities associated with research, system design, production and/or construction, and the various requirements involving subcontracting and supplier activities. Specific content includes (but is not necessarily limited to):

1. *Research plan*
   a. Basic research requirements.
   b. Applied or directed research requirements.
   c. Research methods.
   d. Technology assessment.
   e. Research results and documentation.
2. *System design plan*
   a. Conceptual design.
   b. Preliminary system design.
   c. Detailed equipment design.
   d. Design support.
   e. Engineering model/prototype development.
   f. Transition from design to production.
3. *Production and/or construction plan*
   a. Production and/or construction requirements.
   b. Industrial engineering and operations analysis—plant engineering, manufacturing engineering, methods engineering, production control.
   c. Quality control.
   d. Production operations.
   e. Transition from production to consumer operation and support.

Detailed functions within each of the three areas above are described further in Chapters 4, 5, and 6, respectively. In each instance, tasks are defined, schedules are established, work packages are identified and a WBS is developed, elements of work are assigned to an organization, cost/schedule/performance/effectiveness measurement and control functions are established, and a corrective-action procedure is described.

The product acquisition plan as initially prepared provides extensive coverage of research and early design effort and limited coverage of subsequent detailed design and production/construction activities. As the program progresses, the plan is updated to include additional information as deemed appropriate.

D. Product Evaluation Plan

During the early phases of planning, requirements are defined for the product or system in terms of performance and effectiveness (e.g., range, accuracy, capacity, power, availability, reliability, maintainability, supportability). These requirements, which evolve from the feasibility study, must be designed into the product and incorporated in the production model(s), and the product must be evaluated to ensure that this has been adequately accomplished. Requirements must reflect the need, and when established one must intuitively consider the methods by which these requirements will be verified. Requirements that cannot be verified should not be specified. When a performance and/or effectiveness requirement is written into the specification and program activities are identified to develop equipment such that it will meet the specification, the engineer must establish an approach for the test and verification of that requirement. Thus, it is necessary to develop a preliminary product evaluation plan concurrently with the development of the product acquisition plan.

The preliminary product evaluation plan indicated in Figure 3-9 should include:

1. Test and evaluation requirements.
2. Categories of test and evaluation.
3. Test preparation phase (to include the logistic support resource requirements necessary for test support).
4. Test and evaluation procedures.
5. Data collection, analysis, and corrective action methods.
6. System rework, modifications, and retest procedures.
7. Test and evaluation reporting.

Although much of the information required for detailed test and evaluation planning is not available until after the completion of research and preliminary system design, it is essential that the concepts (i.e., approach) governing test and evaluation be established as an output of the product planning function.

E. Product Use and Logistic Support Plan

Product use and logistic support planning commences with the definition of operational requirements and the maintenance concept for the system. Such planning can be categorized in terms of (1) operational planning or how the system is to be deployed and utilized throughout its life cycle, and (2) logistic support planning or how the system is to be maintained and supported. Since the definition of operational requirements and the maintenance concept is necessary for subsequent product development, early planning in these areas must be accomplished concurrently with the preparation of the product acquisition plan (refer to Figure 3-9). As development progresses, this planning is updated to reflect the operation and support of the final product design configuration as it is to be used by the consumer throughout the life cycle. The ultimate objective is to include the following information:

1. *Product distribution and use planning*

   a. The product marketing and sales strategy for those items in the producer's inventory not yet sold.
   b. The deployment and distribution of the product to designated consumer locations after production/construction (i.e., producer output).
   c. The recommended procedures for product operation.

2. *Logistic support planning*[28]

   a. A detailed maintenance plan which describes proposed echelons or levels of maintenance and the recommended functions (i.e., scheduled and unscheduled maintenance tasks) to be performed at each echelon.[29]
   b. A plan for the acquisition of test and support equipment and/or handling equipment.
   c. A supply support plan to cover the acquisition of spares and repair parts and the maintenance of associated inventories.
   d. A transportation and handling plan to cover the packing, the use of containers, and the shipment of material.
   e. A technical data plan to include system maintenance procedures (i.e., servicing, inspection, calibration, and overhaul instructions).
   f. A facilities plan for the development of new operating and maintenance facilities and/or the occupancy of existing facilities.
   g. A personnel and training plan to cover system operator training, maintenance training, and training equipment (i.e., aids, devices, simulators, etc.)
   h. A system phase-out plan to cover the phase-out of equipment, the reclamation and recycling of items as appropriate, and the ultimate disposal of residual material.

F. Product or System Proposal

A proposal constitutes a bid for undertaking research, engineering design, the production or construction of an item, the testing and evaluation of an item, the support of an item already in existence, or some comparable level of effort. In the context of the material covered in this chapter, a proposal will represent a recommended course of action based on the results of

---

[28]Logistic support planning is covered extensively in: B. Blanchard, *Logistics Engineering and Management*, Prentice-Hall, Inc., Englewood Cliffs, N.J., 1974.

[29]The specific *requirements* for logistic support are defined through the logistic support analysis (LSA) effort, accomplished as part of product or system design. These requirements are then acquired and established for operational support through proper planning.

the technical feasibility study and the advanced product planning discussed earlier and accomplished in accordance with the flow process illustrated in Figure 3-1.

As the figure indicates, a specific need is identified, and a preliminary market analysis is accomplished to determine whether further effort is considered worthwhile by performing a feasibility study. The results of such a study represent the technical characteristics of an item, which are included in the system specification. Given that the defined technical entity is feasible, it is necessary to develop certain planning information (including schedule and cost data) to verify that the item described in the system specification is adequate from a management viewpoint; that is, the proposed item can be produced in a timely manner and at a reasonable cost. The combined output of the system specification and advanced planning data will lead directly to a proposal of some type. Proposals may be classified as follows:

1. *Internal proposals*—a proposal generated within an industrial firm or agency and directed to the management of that firm/agency for approval to accomplish a defined scope of work. This type of proposal is initiated in response to an identified need and generally results in the expenditure of internal resources (e.g., overhead funds) to develop the product desired. It usually involves engineering and middle management proposing something to the next higher level of management, division management proposing an idea to corporate management, and so on.
2. *External proposals*—a proposal generated within a firm or agency (approved by the management of that firm/agency) and directed to an outside consumer, such as a government organization or another firm. The proposal may be directed to a prime contractor or a subcontractor in government work or a private firm involved in some commercial enterprise. This type of proposal may be *solicited* or *unsolicited*, and the accomplishment of the defined scope of work is dependent on external funds. The solicited proposal is prepared in response to an *invitation for bid* or a *request for proposal*, and the solicitation is usually accomplished through a public advertisement or by mailing the invitation for bid to a selected list of qualified firms. On the other hand, the unsolicited proposal is initiated from within the firm in response to a perceived external need.

Although both types of proposals have a great deal in common, the external proposal is usually broader in scope and is more significant in industry since the immediate objective is to win contracts, bring in funding, and make a profit.

1. Proposal strategy. Strategy in this instance refers to the type and amount of effort that should be applied in generating the proposal and in establishing a bidding approach. Proposal preparation often requires the expenditure of a great deal of time and money, and the cost involved is incurred by the firm preparing the proposal. Thus, it is important to thoroughly evaluate (to the greatest extent possible) the probability of success along with associated risks. This is accomplished and supported by the results of the market analysis, the feasibility study, and the planning accomplished up to this point. The prime question is whether "to bid" or "not to bid."

This question is posed at various stages in the process illustrated in Figure 3-1. A no-bid decision may evolve at any time if:

a. The firm feels that the outside competition is really outstanding and that the chances of winning a contract are slim (based on the market analysis).

b. The firm feels technically not qualified to do the job (based on the results of the feasibility study).

c. The firm does not possess the capability to do the job within prescribed schedule and cost goals due to limited resources (based on the results of advanced product planning information).

d. The firm does not wish to take the risks associated with the imposed warranties, liabilities, and the contractual provisions specified in the invitation for bid.

e. The firm, because of prior experience, does not wish to deal with the specific organization responsible for product acquisition.

f. The firm feels that a competing firm has already been selected to receive the impending contract and that response to the invitation for bid is merely a formality (i.e., just going through the motions).

On the other hand, if the process illustrated in Figure 3-1 is pursued and a bid decision is made, management must evaluate the extent of competition that exists. If real competition exists and the firm definitely wants the contract, the effort involved in proposal preparation may be extensive, and management must accordingly assign the appropriate manpower to do the job. In some instances, a firm will spend millions of dollars of its own money in the proposal preparation effort to do the job deemed necessary to win the contract. The results may include going beyond the feasibility-study stage and accomplishing some preliminary design work along with the construction of a representative model or mock-up of the system. A great deal of money is spent, the risks are relatively high, and the losses are extensive if the firm does not win the contract.

Undertakings of this magnitude usually evolve from management decisions made early in the planning process illustrated in Figure 3-1 and are based on the results of a preliminary market analysis supported by the firm's long-range plans and objectives to pursue the type of product being considered. In such instances, a preliminary effort is initiated long before a formal invitation for bid is received, and the tasks required will be guided by a sustaining marketing analysis activity that provides information relative to consumer requirements.[30] In contrast, if the competition is marginal or real competition does not exist at all, the bidding firm's management may wait until the invitation for bid is received and respond to the specified requirements with a minimum expenditure of resources.

In both situations, a bidder, to be successful, must be completely responsive to the invitation for bid, even to the requirements written in fine print, which are not always obvious. Otherwise, the firm stands the chance of being disqualified on the basis of "nonresponsiveness." If the bidding firm feels that it would like to propose options that represent an improved technical approach or one that is more cost-effective, such options should be proposed along with the responsive approach dictated by the configuration conveyed in the invitation for bid.

Thus, proposal strategy is a key factor in the success of a firm. Not only must decisions be made on whether or not to bid, but also decisions involving the start of preposal activity (both of a technical and planning nature) and the amount of resources to be applied must be made. For large systems, the preposal effort may take the form of a project-type organization.

2. Proposal preparation. Proposals are generally prepared by a team of specially selected individuals who represent marketing, engineering and/or research, production/construction (e.g., manufacturing, quality control), purchasing, finance and/or accounting, contracts, and ancillary services (e.g., technical editing, art, printing). The team will analyze the invitation for bid, assess the requirements, and the results will determine the depth of manpower coverage in each area. The team of selected individuals is organized, under the leadership of a proposal manager, with the responsibilities of each member being clearly defined.[31] These individuals may be obtained on a loan basis from other ongoing projects within the firm and in certain instances may even include outside consultants. The proposal manager is responsible for the organization of the proposal, its content, its responsiveness to the invitation for bid, and for completing the job in the allotted time

---

[30]Formal invitations for bid or requests for proposal generally include a required response time on the order of 30 days, 60 days, 90 days, and so on; that is, the bidder must submit his proposal to a designated point not later than x days after receipt of the invitation for bid. The time period allotted is insufficient to accomplish the work required in a highly competitive effort involving a large-scale system. Thus, the firm must decide on when to initiate an early preproposal effort and the amount of resources to be applied.

[31]Selection of the proposal manager should consider an individual who is thoroughly familiar with consumer requirements and who possesses the ability to quickly organize, understand and assign responsibilities, and get the job done! In most instances proposal activities are accomplished in an extremely limited time period, and the effective utilization of the assigned personnel is essential. The proposal manager often is a senior representative of engineering or a member of the marketing staff.

and within the established budget requirements. Other major functions include (but are not necessarily limited to):

a. *Engineering* [*or research*]—responsible for the entire technical content of the proposal. This requires the interpretation of technical requirements and may include the preparation of a preliminary system specification, the establishment of an early design configuration, and so on.

b. *Production*—responsible for all anticipated manufacturing aspects associated with material fabrication, assembly, inspection, test, and related supporting functions. Production also includes all necessary quality-control provisions.

c. *Purchasing*—responsible for all subcontractor or supplier pricing information.

d. *Finance*—responsible for the collection of cost projections that cover each proposed task and for the analysis of these projections in terms of the bid-price goal established by the firm's top management.[32]

e. *Contracts*—responsible for evaluating the imposed consumer contractual requirements, determining producer/consumer risks, and for establishing a satisfactory contractual structure for the bidder. Generally, the invitation for bid will specify the type of contract to be applied (e.g., firm-fixed-price, fixed-price-incentive-fee, cost-plus-fixed-fee, cost-plus incentive-fee), and the contract representative must determine and alert top management of the risks involved.

f. *Marketing*—responsible for the tenor, tone, and quality of the proposal. Marketing also provides considerable support relative to the interpretation of consumer requirements and obtains the necessary ongoing intelligence information pertaining to consumer organization, proposal evaluation procedures and criteria, allocated budget limitations, and so on.

During the proposal effort, each team member is responsible for defining the tasks in his area of responsibility, scheduling these tasks, and establishing the organizational approach proposed for task accomplishment. In addition, cost projections are prepared for each task and identified in terms of work packages and the WBS. The proposal manager reviews each functional area and ensures that the input data are compatible across the board for the overall program. A draft version of the proposal is then submitted to the firm's top management for review and approval. Upon completion of the approved proposal manuscript draft, the material is processed for

---

[32]Through the evaluation of market analysis data (combined with information relating to the firm's product line objectives), top management will generally establish a bid price that will hopefully lead to winning a contract. This is not necessarily the final bid price but is one that serves as a goal for maintaining a competitive posture in the marketplace.

editing, artwork, typing, printing, and ultimate submittal to the consumer.[33]

3. Proposal content. The ultimate content of a solicited proposal will, of course, depend on the requirements included in the invitation for bid. The consumer (i.e., customer) may specify these requirements in a very general form or in a detailed manner. Usually, the former situation is more often the case, and the prospective bidder is forced to contact the consumer in an attempt to obtain additional information. Members of the bidder's proposal team should take advantage of every opportunity possible to discuss system requirements, particularly those of a technical nature, with the consumer prior to proposal submittal. The results may prove to be the difference between winning or losing a contract. On the other hand, unsolicited proposals may take any form or shape depending on the desires of the bidder. Regardless of the situation, however, the proposal in its final form should generally contain the following information[34]:

a. *Proposal title page and outline.* Include a title page and a brief outline of the proposal content. If proprietary information is contained in the proposal, an additional introductory sheet that covers the details of such should also be included.

b. *Introduction.* The bidder should define in a brief, clear, and concise manner the problem at hand (to let the consumer know that the requirements are thoroughly understood) and should summarize his proposed approach to problem resolution. If alternative approaches are considered, these should be covered in terms of advantages and disadvantages, being sure to support any recommendations in this area. Finally, the bidder should briefly describe his capability and experience in the area and his ability and willingness to apply the necessary resources to the proposed effort.

c. *Summary of proposed effort (statement of work).* A definitive statement of work should be included listing (in summary form) each major line-item task, schedule of task completion or delivery date, and cost. This should briefly respond to the question: What is being offered and what are the costs involved? References may be made to supporting data in other sections of the proposal.

---

[33]During the process of manuscript editing and the preparation of art work, a designated representative from each participating functional area should review the results to ensure that the content and meaning of the submitted draft manuscript have not changed and that all figures and illustrations are accurate. The process of translating from a rough manuscript to a finished proposal sometimes results in the inadvertent introduction of errors.

[34]For solicited proposals, the proposal format must be consistent with the requirements specified in the invitation for bid. In the absence of such, the bidder should develop an outline that is compatible with the invitation for bid in such a manner that the consumer can *easily* identify the response to his requirements.

d. *Technical requirements.* This section should cover a definition of the proposed system in terms of all technical requirements. Usually, the extent of definition follows the content of the system specification described in Section 3.4B1 (refer to Figure 3-8).

e. *Management plan.* The bidder's management approach, relative to accomplishing the proposed effort, should be defined herein. Depending on the type of system and the applicable life-cycle phase, the information needed may be derived directly from the plans described in Sections 3.4C through 3.4E. Specifically, this should include (1) a description of the bidder's organization; (2) definition, scheduling, and assignment of tasks (to contain milestone charts, PERT, CPM, and Gantt charts as appropriate); (3) procedures or references to procedures for task accomplishment; (4) cost information for each assigned task, work packages, and the WBS; (5) subcontractor work requirements; and (6) program review, evaluation, and corrective-action provisions. The management plan not only responds to the "what, when, where, and how" questions pertaining to program accomplishments, but must provide the necessary assurance that consumer requirements will be met in entirety.

f. *Contractual and legal provisions.* Any special items or contingencies that pertain to contract type, payment dates, warranties, liabilities, default clauses, penalty and/or incentive clauses, and so on, should be described. The bidder should clearly state his understanding of all contractual requirements.[35]

g. *Supplemental information.* Any supporting data or documentation that may help the bidder in winning a contract should also be included. Such information may contain (1) a statement of the bidder's financial status (assets and liabilities) or a copy of the previous year's annual report; (2) a description of successful work accomplishments on earlier programs of a similar nature; (3) a description of available facilities and resources for the effort proposed; (4) a description of subcontractor and supplier capabilities; and (5) a description (with appropriate resumes) of the personnel scheduled to work on the proposed program, individual background data, and the per cent of time assigned. The objective is to convince the consumer (customer) that the bidder possesses the know-how and can do the job more effectively and efficiently than any competing firm.

The completed proposal should be carefully reviewed by the proposal manager to ensure that all significant areas are covered and that the best interests of the firm are indicated. In the final preparation effort, care must

[35]Refer to Chapter 15 for discussion of contractual and legal provisions. When dealing with government agencies, procurement practices must be in accordance with the Armed Services Procurement Regulations (ASPR). It is recommended that a representative of the firm's contracts department be thoroughly familiar with these regulations and how they are interpreted and applied. It is advisable to have an updated copy of the ASPR on hand.

be taken to check that both text material and illustrations are presented in an optimum manner. Although fancy charts and colored displays will not overcome technical and management deficiencies, a sloppy job (e.g., poor reproduction, mixed type and inconsistent margins, insufficient line drawings) conveys an unprofessional appearance reflecting the firm's attitude in general. This may cause the firm to lose out in a competitive situation even if its technical approach may be superior overall. Thus, it is essential that the final proposal be neat in appearance, contain adequate illustrations and color where appropriate, not be too voluminous, and be bound in such a manner as to facilitate the review process accomplished by the prospective customer.

As a final step, the completed proposal is approved and signed by top management and submitted to the customer via a letter of transmittal. The letter generally contains the sales-pitch data and may cover items of a personalized nature that should be highlighted. A copy of the letter of transmittal should be attached to each copy of the proposal.

*4. **Proposal review, evaluation, and contract negotiation.*** Upon receipt of a proposal from a bidding firm, the consumer/customer proceeds with the review and evaluation process. If the proposal is unsolicited and a single bidder is involved, the consumer may review the proposal and negotiate directly with the bidder resulting in a mutually agreeable statement of work and price. The output may coincide directly with the proposal as originally submitted or may assume a revised posture (e.g., slight variations in scope of work at a reduced cost) and will ultimately be incorporated into a contract of some type. On the other hand, the consumer may take the technical system ideas conveyed by the single bidder (those not proprietary in nature) and initiate solicitations from other firms.

When competitive bidding occurs, the consumer generally establishes an evaluation procedure directed toward selecting the best approach proposed. Initially, each bidder is evaluated in terms of compliance with the requirements (i.e., the invitation for bid in instances where solicited proposals are being reviewed). Noncompliance may result in automatic disqualification, or the consumer may approach the applicable bidder(s) and recommend a proposal revision and/or addition. When two or more bidders meet the basic requirements, the proposals may then be evaluated in terms of criteria such as illustrated in Figure 3-22. The consumer develops a list of items considered relevant in the evaluation and establishes weighting factors to indicate degree of importance. All proposals are reviewed (using individual checklists supporting each criterion factor noted) and ratings are established on the basis of the extent of compliance with the desired objectives. These ratings are multiplied by the weighting factors to provide a score for each item, and the individual scores are then added, with the highest score reflecting the bidder with the best overall approach. In this instance, bidder A seems to be the preferred alternative.

Evaluation Criteria	Weighting Factor (%)	Bidder A		Bidder B		Bidder C	
		Rating *	Score **	Rating *	Score **	Rating *	Score **
A. Technical Characteristics	30						
1. System Performance	10	5	50	6	60	8	80
2. System Effectiveness	4	7	28	5	20	6	24
3. System Design Approach	7	7	49	6	42	8	56
4. Test and Evaluation Approach	2	9	18	7	14	4	8
5. Logistic Support Requirements	6	8	48	5	30	7	42
6. Growth Potential	1	7	7	5	5	9	9
B. Management Plan	25						
1. Organization Structure	6	9	54	8	48	6	36
2. Planning (Scheduling)	8	7	56	7	56	6	48
3. Resources Available	5	6	30	8	40	7	35
4. Management Controls	6	8	48	5	30	6	36
C. Total Cost	35						
1. Acquisition Price	15	4	60	5	75	7	105
2. Utilization Cost	20	7	140	4	80	5	100
D. Supporting Factors	10						
1. Prior Experience	3	9	27	8	24	9	27
2. Past Performance	7	9	63	6	42	4	28
TOTAL	100	X	678	X	566	X	634

* Ratings range from 1 to 10 depending on the evaluation of the proposed system configuration in terms of desired attributes.

** Score is the product of the weighting factor and the rating.

Figure 3-22. Proposal evaluation criteria (*sample*).

Once that a preferred bidder is selected on a technical and management basis, the consumer and bidding firm will proceed toward negotiating a contract. Negotiation may be accomplished quite simply if there is mutual agreement on the type of contract to be implemented. However, contract types differ in terms of assumed risks, and mutual agreement may not initially exist.

For instance, a fixed-price type of contract (e.g., firm-fixed-price, fixed-price-incentive, fixed-price with redetermination) offers the greatest risk to the producer, especially if system performance and design specifications are not reasonably well established. Conversely, cost-reimbursement types of contracts (e.g., cost-plus-incentive-fee, cost-plus-fixed-fee) are more risky to the consumer if performance and design uncertainties exist. Thus, negotiation may not only become an extensive undertaking relative to the contract type established, but may also involve some haggling in the areas of scope and work and the proposed price/cost figures. Further discussion on contracts and negotiation is presented in Chapter 15. [36]

3.5 SUMMARY

This chapter (a) covers the determination of a need for a product; (b) justifies its feasibility from a technical standpoint; (c) discusses and presents techniques for advanced product planning; and (d) includes certain aspects of marketing, the preparation and evaluation of proposals, and contract negotiation. As such, the material content is rather comprehensive; however, addressing each of the steps illustrated in Figure 3-1 is extremely important, and the combined results will determine the nature and extent of subsequent activities in the product life cycle.

Referring to Figure 3-1, the process illustrated represents a series of steps with the necessary feedback provisions required for early decision making. It is iterative in nature and includes options as to whether to proceed further or not to proceed. Assuming that the results of each step in the process are in favor of product development, the ultimate output will constitute a proposal leading into the next logical phase in the product life cycle (i.e., research, design, production and/or construction, etc.). The advanced product planning activities will dictate the particular actions required, and the extent of effort expended in such planning (refer to Section 3.4) must be tailored to suit each individual need.

Finally, it should be noted that the engineer is directly involved in every

---

[36]The reader should undertake further study in the area of contracting since the engineer often serves as an essential component of the negotiation team (i.e., in support of system technical requirements) and should understand the different type of contracts used, together with the associated risks.

step of the process. He participates in the preliminary market analysis to assess technical requirements and proceeds to take the lead in conducting feasibility studies. In advanced product planning, he assists in the definition of tasks (i.e., what has to be done) and assures that performance, schedule, and cost requirements are realistic in terms of attaining technical objectives. When the results of planning indicate the need for a proposal, engineering is responsible for the preparation and review of all technical material and is often involved in contract negotiations from the standpoint of justifying the technical approach, assessing the likelihood of meeting advanced technology objectives, assuring that product performance and effectiveness requirements have been adequately considered, assuring that product safety provisions will be covered, and so on. Engineering is an inherent element of the product planning function.

QUESTIONS AND PROBLEMS

1. What is the first step in the product planning function?
2. What information is obtained from the preliminary market analysis? A feasibility study?
3. Define what is meant by "technology."
4. What is technological forecasting? What purpose(s) does it serve?
5. Describe some of the techniques commonly employed in technological forecasting.
6. Define the difference(s) between market potential and market share.
7. Describe the approach that you would follow in performing a market analysis.
8. After performing a market analysis, what factors would you consider in making a decision as to whether to proceed with additional product development activity or to discontinue the project?
9. Select a system of your choice and define the basic operational requirements for that system.
10. For the system described in problem 9, define the basic maintenance concept (use illustrations where appropriate). What information does the system maintenance concept provide?
11. Why is the definition of system operational requirements and the maintenance concept so important at this stage in the product life cycle?
12. Refer to Figure 3-6. The weighting factors used in system evaluation represent what? How are these values determined?
13. What is the purpose of the system specification? The program management plan? What are the basic differences between the two?
14. Select a system of your choice and develop a system specification.
15. What is the basic difference between a development specification and a procurement specification? A process specification and a material specification?

16. The system acquisition plan generally includes what phases of the product life cycle? What information does the acquisition plan include?
17. List some of the techniques generally used in the scheduling of tasks, and discuss applications in terms of the advantages and disadvantages of each.
18. What are the advantages of PERT? What time information is required for PERT?
19. The following data are available:

Event	*Previous Event*	t_a	t_b	t_c
8	7	20	30	40
	6	15	20	35
	5	8	12	15
7	4	30	35	50
	3	3	7	12
6	3	40	45	65
	2	25	35	50
5	2	55	70	95
4	1	10	20	35
3	1	5	15	25
2	1	10	15	30

(a) Construct a PERT chart from the data above.

(b) Determine the values for standard deviation, TE, TL, TS, TC, and *P*.

(c) What is the critical path? What does this value mean?

20. Refer to the problem example in Figure 3-13. What is the probability of success if the scheduled time is 48 weeks? 44 weeks? What scheduled time would assure a 95 per cent probability of success?
21. When employing PERT/COST, the time-cost option applies. What is meant by the time-cost option? How can it affect the critical path?
22. What is the principal application(s) of the Gantt chart? Name some advantages over other scheduling techniques.
23. What is a WBS (discuss its purpose and application)? How can the activities and events in a PERT network be related to the WBS? How can cost information used in PERT/COST be related to the WBS?
24. Refer to Figure 3-19. There is a trend toward a program cost overrun. How would you determine the cause of the overrun (i.e., where the problem exists)? What would you do about it?
25. Why is the establishment of a good management information system so important?
26. Why is the initiation of a corrective action procedure so important in the management of a program?

27. Why is it necessary to develop a preliminary product evaluation plan concurrently with the development of the product acquisition plan?

28. The product use and logistic support plan stems from what source of data? Why is it essential to consider logistic support concurrently with the development of the product acquisition plan?

29. When are proposals generated? What type of information is usually contained in most proposals? What is meant by "proposal strategy"?

30. When a firm is proposing to design and develop a new system that requires some state-of-the-art advances relative to existing technology, would it be advantageous for the firm to negotiate a fixed-price or a cost-reimbursement type of contract? Why?

31. When a consumer is planning to procure a given quantity of items, would it be most advantageous for the consumer to negotiate a firm-fixed-price, a cost-plus-incentive-fee, or a cost-plus-fixed-fee type of contract? Why?

32. How is the engineer involved in the product planning function?

4

PRODUCT RESEARCH FUNCTION

If a firm or agency is to prosper and grow, it must not only take the steps necessary to keep abreast of its competitors, but it must actually move ahead whenever and wherever possible. The aspect of "moving ahead," in line with the goals and objectives of the firm as defined through advanced product planning (refer to Chapter 3), almost invariably is highly dependent on research.

Research is directed toward making new discoveries, uncovering new principles, and in general results in advancing the state of the art in some technical specialty area. Research may be categorized as *basic* or *applied*.

4.1 BASIC RESEARCH

Basic research, sometimes referred to as *pure* or *fundamental research*, covers those scientific areas of activity accomplished for the sake of gaining knowledge in anticipation of some future need. A specific product application is not known at the time, and the research performed is mainly for the sake of learning nature's laws and studying the physical and natural resources of the universe. The direction of such research is usually not specified in advance but is determined as work progresses and specific accomplishments are identified. The objectives are to establish new concepts and relationships which hopefully will enable the firm to maintain a position on the "leading edge" of technology.

Basic research may be initiated within a firm or agency with the intent of discovering new techniques that will promote the firm's objectives toward the development of some product. The activities required are completely unstructured, and the results may take varying forms, either proving or disproving a certain concept(s). In some instances, the results of research projects have turned out to have very practical applications. An example is the development of nylon by du Pont.[1] Eastman Kodak in the development of film-processing techniques, Xerox in the development of xerography, and many other examples can be cited as the direct result of basic research.

Research in this area is also performed by colleges and universities (probably the largest contributor), where projects are financed by the federal government, industrial firms, or by independent research agencies. In many instances the investment in basic research far exceeds the benefits derived, and the costs are high. On the other hand, the discovery of new phenomena would certainly be inhibited without the pursuit of basic research.

4.2 APPLIED RESEARCH

Applied research, sometimes referred to as *directed research*, constitutes those areas of scientific activity dedicated to advancing the state of the art in one or more technical fields of specialty, with the ultimate objective of solving some problem or correcting a deficiency. Research of this type has a practical purpose and is dependent on the findings of basic research.

Applied research responds to an identified need that has been defined through a request for service from some external organization or agency, a preliminary market analysis, a feasibility study, and the like (refer to Figure 3-1). For instance, there is a need for a product or service with specific performance and effectiveness characteristics, and current technology in the area does not exist. Consequently, research is initiated to establish new methods and techniques that hopefully will lead to the development of a product that will fulfill the desired objectives. Research activity here is more structured, and the output is expected to have direct practical applications.

The pursuit of applied research is more prominent within industrial firms since there are always prime motivators, such as improving the performance of existing products or services, making existing products at a lower overall cost, developing new products, attaining greater customer appeal and a larger share of the market, and so on.[2] In addition, the results

[1]Refer to H. T. Amrine, J. A. Ritchey, and O. S. Hulley, *Manufacturing Organization and Management,* 3rd ed., Prentice-Hall, Englewood Cliffs, N.J., 1975, Chapter 22.

[2]Although applied or directed research has in the past primarily been accomplished within industrial firms, there are trends toward increasing research of this type within colleges and universities and independent research agencies. Colleges and universities are being requested, at an increasing rate, to apply their available expertise in resolving problems at the local, state, and national levels.

may effect improvements within the firm's internal operations. Examples of research in this category include the many activities associated with seeking new sources of energy and reducing energy consumption; establishing better methods of communication (e.g., switching methods, optical techniques, microwave links); and providing new techniques for air, water, and noise pollution control.

Applied research, usually initiated within industrial firms, may be funded by the federal government (e.g., independent research and development funds), other industrial firms, independent research agencies, and/or by the initiating firm itself through the use of overhead funds or through the investment of some percentage of the firm's profit (e.g., retained earnings). Research funded by external sources generally evolves through the submittal of a technical proposal by the firm, approval by the funding source, and the subsequent negotiation of a contract for a specified level of effort and a defined output. The expenditure of resources for applied research can be more readily identified with results since the effort expended is oriented toward practical accomplishments.

4.3 RESEARCH METHODS

The methods for accomplishing research are of course dependent on the type and objectives of the research being pursued. However, the steps often followed are summarized below.

A. Search for Knowledge

Research generally commences with an assessment of current scientific knowledge to determine possible applications of existing theories and concepts to the problem at hand. The engineer or scientist accomplishing the research should perform an exhaustive search of available literature on the subject area and related fields.[3]

B. Formulation of Theories

From the literature search, the researcher identifies those theories and concepts that are appropriate to the situation. He (or she) should soon be able to determine areas where additional effort is necessary for technology advancement and should initiate activities toward this goal. He may pursue tasks leading in a number of different directions at first before arriving at any conclusions. However, with persistence and the necessary initiative, he

---

[3]The literature search will include the review of library material within the firm (e.g., earlier research reports, test reports), published technical journal articles, college/university published theses and dissertations, government reports, and so on.

should ultimately make some headway toward the establishment of new knowledge, which incorporates the principles and concepts necessary to meet his objectives. As he progresses, he should document each step in the process along with the results. Hopefully, the ultimate output will be in the form of new discoveries and inventions, which in turn provide the necessary technology for product development.[4]

C. Verification of Research Results

Prior to the final formalization of results, the researcher should develop methods for testing and verifying any newly established theories. In other words, the results of the research tend to lead toward the establishment of certain conclusions involving physical and natural relationships. Theories alone are not necessarily supportable unless verified through some form of hypothesis testing; that is, assumptions are made and testing is conducted to prove or disprove the validity of these assumptions. Testing may be of an analytical nature or may involve the development and use of physical experimental models or prototype equipment. The intent is to prove that the established theories are indeed valid for certain defined conditions.

The engineer performing research must be clever and perceptive to recognize and define phenomena not previously noticed. In addition, he (or she) must be persistent and patient in the accomplishment of tasks that are exploratory in nature. Much of his activity involves probing and the investigation of new areas, and there is a good deal of trial and error with successful results occurring only after many hours of painstaking effort. This can be very disheartening and yet extremely rewarding when new discoveries are made.

In the past many research projects were completed on an individual basis. However, the scientific and technological challenges in our world today are tremendous, and it has become readily apparent that scientists and engineers could better attain the desired objectives through the pooling of their efforts and knowledge. Thus, research is more often carried on through an integrated team approach rather than by singular efforts.

4.4 RESEARCH APPLICATIONS

Although all types of research can be classified as either *basic* or *applied*, it may be worthwhile to briefly discuss research from the standpoint of a few specific or tailored applications.

---

[4]Resulting discoveries and inventions should be processed through legal channels for patent rights, copyright provisions, and so on. More information in this area is presented in Chapter 15.

A. Product Research

Product research (as the name implies) is directed primarily toward uncovering new ideas for product development and/or product improvement and is basically oriented to system hardware and software. The initiation of product research evolves through the identification of a newly defined need. From the standpoint of the industrial firm, product research may be stimulated by the firm's desire to enter a new market, expand its current product line(s), modify current products for greater consumer appeal, and/or to pursue activities related to associated by-products. Product research is an inherent part of the discussion in previous sections of this text and in the process illustrated in Figure 3-1.

B. Operations Research

Defining operations research is rather difficult since there are almost as many definitions as there are organizations involved in the area. However, it is safe to say that operations research is the systematic application of scientific methods, engineering principles, and technical skills to the study and evaluation of physical configurations and functional operations in industry, government, and military organizations.

Operations research employs analytical techniques such as simulation, linear and dynamic programming, queuing theory, and Monte Carlo analysis to predict future conditions in new systems design and to evaluate ongoing operations and existing design configurations for the purpose of making improvement. Analytical models are generated to reflect real-world situations and are judiciously applied to study and evaluate any aspect of a firm or agency's operation (refer to Section 2.3D). The research required constitutes the development and proper application of the right techniques for resolution of the problem at hand. This necessitates a continuing effort to discover new methods in order to improve the predictions and evaluation assessments made. Operations research methods are used in support of management decision making throughout the product life cycle.[5]

C. Market Research

Market research, often used synonymously with market analysis, is primarily aimed toward learning more about the needs and desires of potential consumers. Some industrial firms have developed products for direct market application without having any idea in advance of consumer accept-

---

[5]Operations research is covered in a number of texts. Several good references are: W. J. Fabrycky, P. M. Ghare, and P. E. Torgersen, *Industrial Operations Research,* Prentice-Hall, Inc., Englewood Cliffs, N.J., 1972; and A. Schuchman, *Scientific Decision Making in Business—Readings in Operations Research for Non-mathematicians,* Holt, Rinehart and Winston, Inc., New York, 1963.

ance of that product(s). This can be quite risky, resulting in high costs if the product is not accepted. On the other hand, a firm may initiate certain technical activities with the intent of learning consumer reaction to the product prior to the investment of resources for detailed product design and production. The objective is to accomplish a needs analysis, a study of market potential and market share, a feasibility study, and the like as necessary to minimize the risk for the producer and to provide greater assurance of consumer acceptance later. Market research is basically accomplished through the steps described in Sections 3.1 and 3.3.

D. Manufacturing Research

Product research is certainly a major activity and contributor to the success of an industrial firm. However, a concurrent consideration should include research activity directed toward the improvement of manufacturing processes and techniques. Manufacturing research deals with the development of tools, fixtures, handling equipment, and methods of product fabrication, assembly, and test with the intent of reducing costs and increasing productivity. Many firms have recognized the importance of manufacturing research and the results have been significant, particularly in the application of numerical control, computer-controlled machines, automatically controlled power tools and handling devices, electrical wire wrap and soldering techniques, chemical milling, photo etching, and others.

E. Materials Research

Research here is directed toward the development and application of new materials and is closely tied to product research and manufacturing research. The discovery of new materials often makes it possible to develop new products and/or manufacturing processes that were impossible earlier. Examples of successes in new material applications include the uses of plastics (in place of wood and metals), composites, fiberglas, epoxies, and others.

4.5 TECHNOLOGY ASSESSMENT

When new technological advances or changes evolve from research, there has been a tendency to proceed with the direct implementation of the results without adequate consideration of the occurrence of side effects on our society in general. In other words, a new technology has been evaluated on the basis of the technology itself without assessing its impact on social, political, economical, and ecological or environmental factors. In many instances the consequences have been costly, and these side-effect impacts have turned out to be quite harmful. For example, the development of a new product manufacturing process in an industrial plant may cause a discharge of waste, resulting in air and/or water pollution which is detrimental to the community and overall geographical area where the plant is located. This, in turn, may create individual health hazards, resulting in a shorter human life span; increase the need for medical services and drugs, and the cost of medical insurance benefits; cause changes in social and political organizations; result in a movement of families out of the community which will reduce the economic growth of the area; and so on. Thus, when evaluating the impact of a new technology on society, the engineer or manager of a firm may wish to address such questions as:

a. What impact does the new technology have on the physical environment in terms of air pollution, water pollution, solid waste, water and power shortage, urban noise, and personal health in general?
b. What impact does the new technology have on the social environment in terms of available human resources (i.e., labor pool), community cooperation and social attitudes, family living styles, local crime and juvenile delinquency, and community population growth?
c. What impact does the new technology have on the political environment of the area in terms of organizations, community planning, and legislation?
d. What impact does the new technology have on the economic environment of the community and surrounding area?

Technology assessment is the process of systematically evaluating a new technology and determining the consequences of a technological change. The evaluation is accomplished in terms of defining all the costs and benefits associated with the change and reviewing various alternatives with the intent of selecting a preferred approach in implementing the change. The selection should be based on a cost/benefit or cost/effectiveness analysis covering both the short-term time frame and all long-term effects possible (both good and bad). Figure 4-1 illustrates the process.[6]

4.6 EVOLUTION FROM RESEARCH TO PRODUCT DEVELOPMENT

The difference between research (whether basic or applied) and development sometimes becomes rather fine. Research deals with the discovery of new facts and fundamental relationships; development is concerned with the most effective and efficient method(s) for applying the facts or principles identified. Development constitutes the process of translating the results of

---

[6]Technology assessment is discussed further in: F. Hetman, *Society and the Assessment of Technology*, Organization for Economic Cooperation and Development Publications Center, Washington, D.C., 1973.

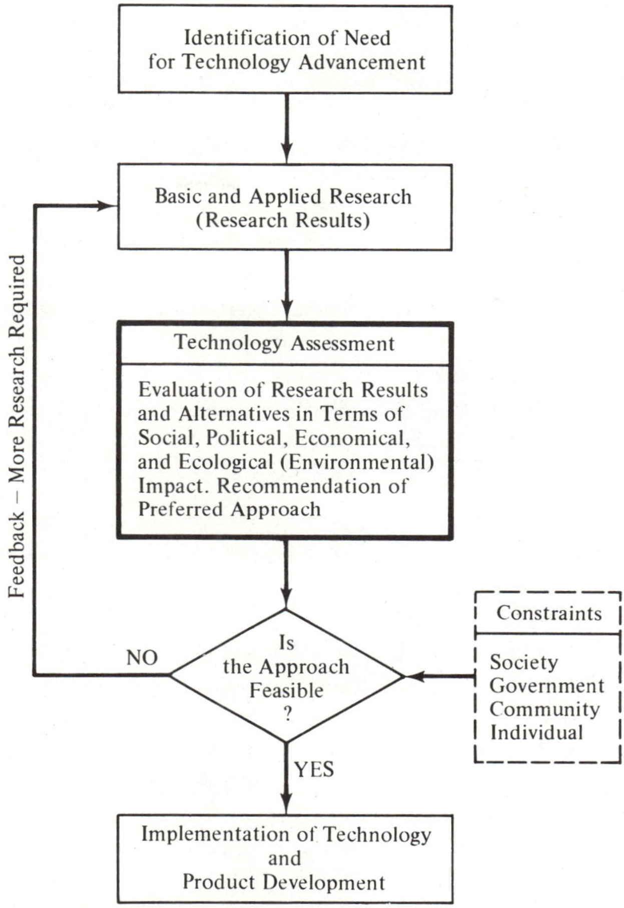

Figure 4-1. Technology assessment loop.

research into a fixed product design configuration of some type and is accomplished through the process discussed in Chapter 5.[7]

In a number of organizations the proper transition of ideas to the product development stage is lacking. Researchers perform the functions of discovering new principles and relationships and then proceed to drop the project(s) when the perceived challenges disappear. For many, carrying on a new idea through product development is too burdensome and inhibits the

[7]In many organizations the functions of research and development are so closely related that certain activities often overlap. Thus, the two functions are sometimes integrated into a single organizational entity and are referred to synonymously. The intent here is not to promote a distinct separation between them but to show the phase relationship.

pursuit of new research possibilities. Thus, a gap often exists between research and product development, and many good ideas are lost in the process. Research must be carried on to the point where intelligent decisions can be made as to whether it is feasible to progress through the steps of product design.

QUESTIONS AND PROBLEMS

1. What is research and why is it important?
2. What are the basic types of research? From your own experience, list some examples of each type indicated.
3. How is research performed (i.e., what basic steps are followed)?
4. What motivates industrial firms to pursue research?
5. How are research projects funded? Discuss some of your own experiences in this area (if any).
6. What is the initial step in performing research?
7. Why is documentation so important in the accomplishment of research?
8. Define: product research, operations research, manufacturing research, and market research.
9. What is the difference between research and development? Give an example.
10. If you were planning to subcontract research, what type of a contract would you be likely to negotiate?
11. Define technology assessment. What is its purpose? Why is it important (if at all)?
12. Refer to Figure 4-1. If the results of a technology assessment are not feasible, what steps would you take?
13. Assume that you are responsible for conducting an assessment of a given technology and establish the evaluation criteria that you would use in the evaluation process. Develop a detailed checklist showing the appropriate weighting factors that you would assign to each item.

5

PRODUCT DESIGN FUNCTION

Product or system design is basically accomplished through the systems engineering process. Systems engineering constitutes the application of scientific and engineering efforts to (1) transform an operational need into a description of system performance parameters and a preferred system configuration through the use of an iterative process of functional analysis, synthesis, optimization, definition, design, test and evaluation; (2) integrate related technical parameters and assure compatability of all physical, functional, and program interfaces in a manner that optimizes the total system definition and design; and (3) integrate reliability, maintainability, logistic support, human factors, safety, security, structural integrity, producibility, and other related specialties into the total engineering effort. The systems engineering process, in its evolving of functional detail and design requirements, has as its goal the achievement of the proper balance among operational (e.g., performance, effectiveness), economic, and logistics factors. The process employs a sequential and iterative methodology to reach cost-effective solutions, and the information development through this process is used to plan and integrate the engineering effort for the system as a whole.

This process of design evolution is illustrated in Figure 5-1 and is tailored to meet the specific need.[1] Actually, regardless of the system type and size, design commences with a feasibility study accomplished for the

[1]Tailoring refers to the application of the proper level of effort to the system being developed. The application of too much or too little effort could be quite costly. Thus, the steps presented in Figure 5-1 should be considered as a thought process, with each step being addressed to the extent and depth necessary to accomplish the job.

purposes of establishing a set of requirements, constraints, and design criteria. Based on the results, functional analyses and allocations are generated to apportion the appropriate system-level requirements down to the subsystem, unit, and lower indenture levels of the system. Trade-off studies and system optimization are performed to evaluate the various alternative approaches that are considered feasible in meeting the identified need with the output reflecting a preferred system configuration.[2] This configuration is then defined through the system specification.

The steps just described represent *conceptual design* and *preliminary system design* and are iterative in nature, as reflected by the feedback loop in block 2 of Figure 5-1. This portion of the systems engineering process may be quite repetitive until the ultimate output constitutes a system configuration that will not only meet performance requirements but one that can be operated and supported in an effective and efficient manner throughout its planned life cycle.

Given a preferred configuration described in the system specification, *detailed system/equipment design* commences with the generation of lower-level supporting specifications (refer to the discussion on specifications in Section 3.4B), functional equipment layouts, and detailed design documentation, as reflected in block 3 of Figure 5-1. The output is reviewed on the basis of compliance with the initially established requirements, and corrective action is implemented where appropriate. The system configuration represented on paper and approved through a series of formal design reviews is then converted into an engineering model or prototype system which can be used for test and evaluation. Test and evaluation at this stage is accomplished to the extent possible to verify that the system requirements have been met prior to entering into the production and/or construction phase.

This process leading from the identification of a need through the development of a physical prototype model of the system (including prime mission equipment, software, and elements of logistic support) is discussed in more detail below.

5.1 CONCEPTUAL DESIGN

Conceptual design constitutes the starting point for a design project and includes a feasibility study directed toward defining a set of useful solutions to the problem being addressed. The output of a feasibility study represents the technical baseline for a proposed system and contains a needs analysis, a definition of system operational requirements, the system maintenance con-

[2]The term *preferred* does not necessarily infer *optimal*, as the operational requirements and maintenance concept may not permit true optimization. *Preferred* does imply the best among a number of alternatives within the given constraints.

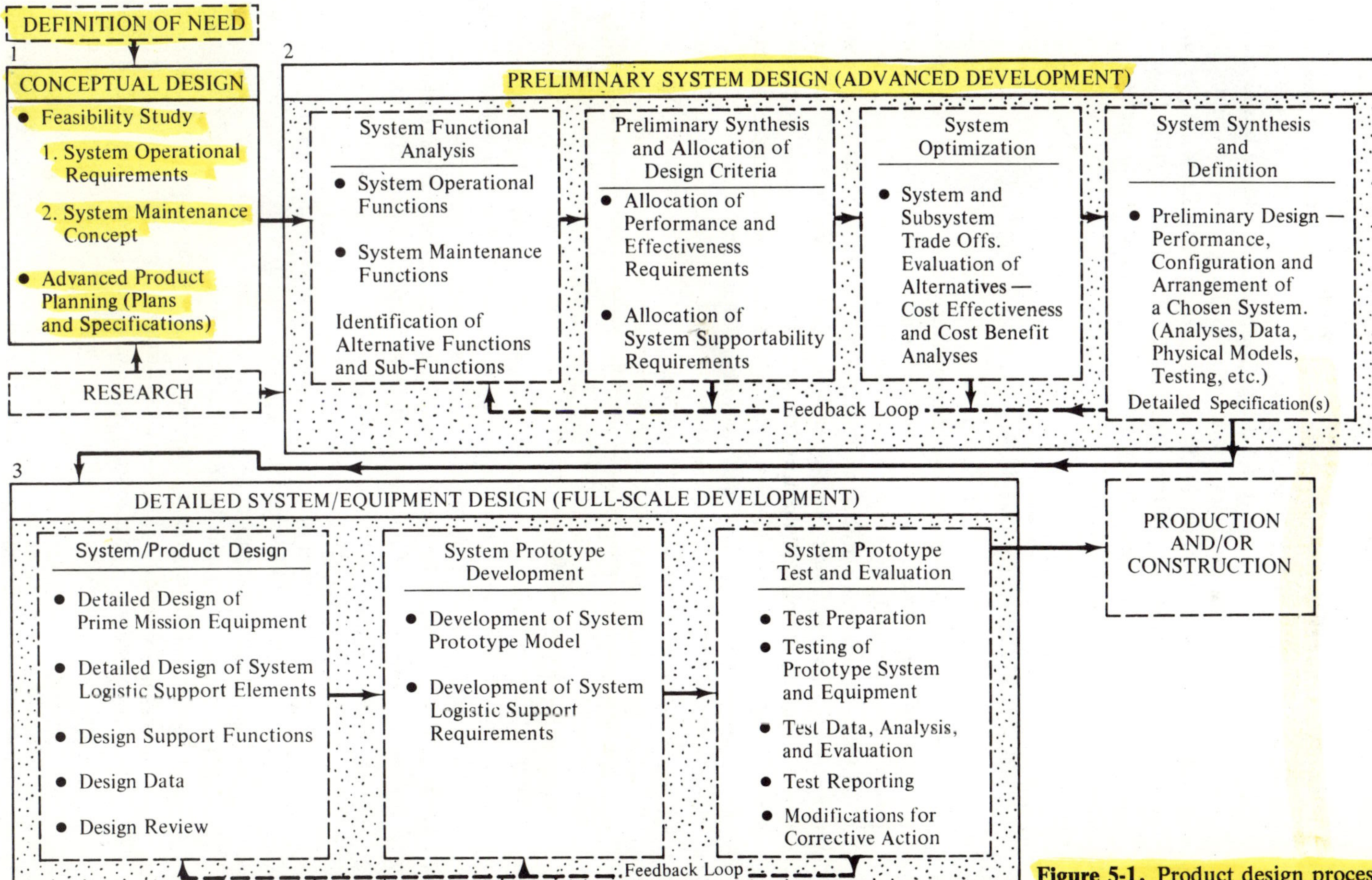

Figure 5-1. Product design process.

cept, and a preliminary systems analysis. The requirements for a feasibility study are covered in detail in Section 3.3.

When a project is initiated, all design activities should be based on the results of a feasibility study of some type. Either a new feasibility study should be initiated to support future design activities if the system requirements are new, or design activities may be based on an existing feasibility study conducted earlier and supplemented by technical management and engineering experience.

5.2 PRELIMINARY SYSTEMS DESIGN

Preliminary system design, sometimes referred to as advanced development, starts with the technical baseline for the system defined in the feasibility study and proceeds to translate the established system-level requirements into detailed qualitative and quantitative design requirements. Preliminary design, illustrated by block 2 of Figure 5-1, includes the process of functional analysis and allocation, the accomplishment of trade-off studies and optimization, system synthesis, and configuration definition in the form of a system specification and supporting specifications as required.

A. Functional Analysis and Allocation

The translation of system operational and maintenance factors into specific qualitative and quantitative design requirements is accomplished through a logical functional analysis and allocation process. Basically, the process provides:

1. The identification and sequencing of operational and maintenance functions necessary to achieve the system requirements.
2. The definition of performance parameters, operational effectiveness, and supportability factors associated with each function. This includes the consideration of system boundaries, constraints, and anticipated use conditions.
3. The identification of performance, operational, and supportability factors with a top-level equipment packaging scheme and the allocation of requirements as appropriate to effect design. The packaging scheme derived may be in the form of subsystems, a series of functional units, or even a category or grouping of physical areas of location where functions are performed. The ultimate output will depend on the type of system (e.g., whether there are multiple quantities of the same equipment or a single facility requiring construction).

These tasks are accomplished early in the system/equipment life cycle and are represented by the first two steps in block 2, Figure 5-1. As such, the functional analysis serves as the starting point for equipment design.

1. System functional analysis. The initial step constitutes the formulation of a functional description of the system and all facets of system development and operation. Functions are identified, and functional flow diagrams are developed for the primary purpose of structuring system requirements in functional terms. Functions represent those top-level performance characteristics and/or actions that must be accounted for! Functions may be classified as independent functions or dependent functions and are presented in a series format, a parallel format, or a combination of both. The functional approach assures:

a. That all facets of system development, operation, and support are adequately covered. This includes design, production/construction, test, deployment, transportation, training, operation, and maintenance. Coverage should include the activities reflected in Figures 3-4 and 3-5.

b. That all elements of the system (e.g., prime equipment, test and support equipment, facilities, personnel) are fully recognized and defined.

c. A means of relating equipment packaging concepts and support requirements to given functions. This identifies the relationship between the need and the resources required to support that need.

Functional flow diagrams are designated as top level, first level, second level, and so on. The top level shows gross operational functions. The first-level and second-level diagrams represent progressive expansions of the

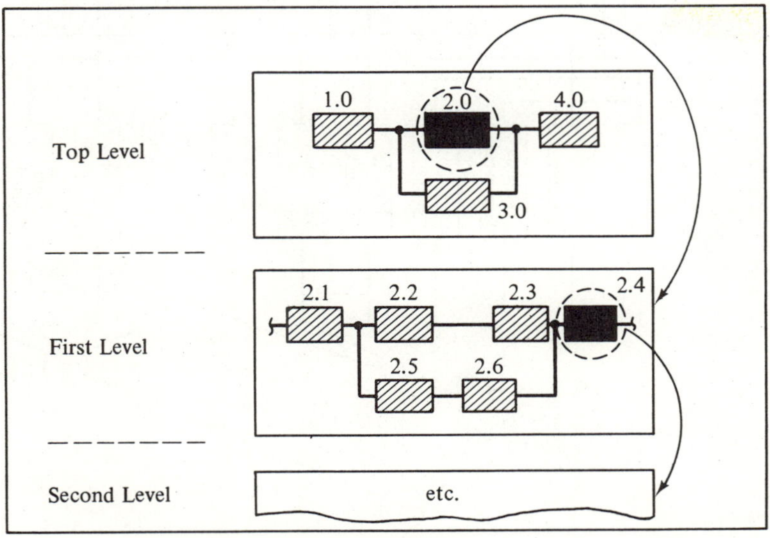

Figure 5-2. Diagram of system indenture levels and traceability.

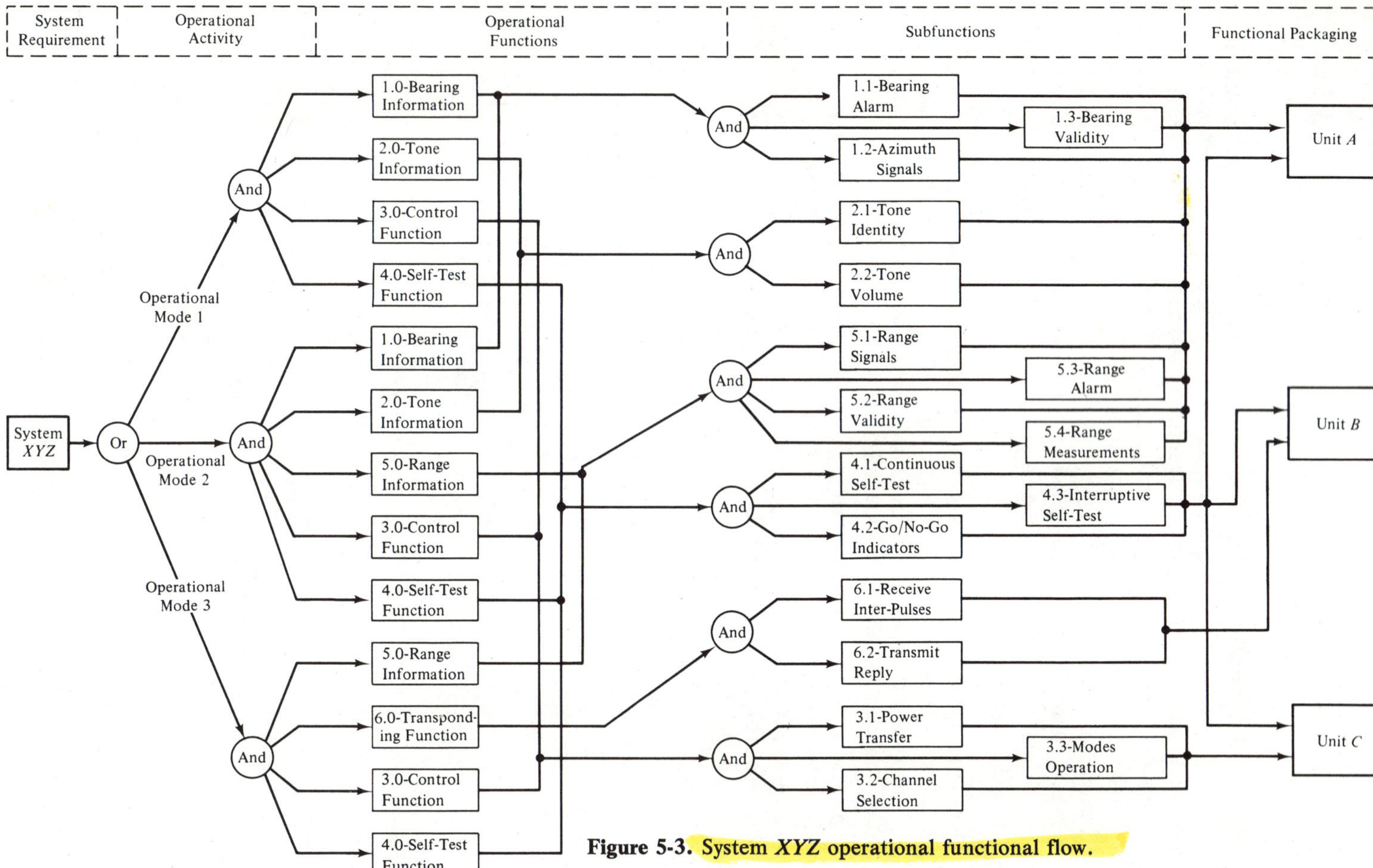

Figure 5-3. System *XYZ* operational functional flow.

individual functions of the preceding level. Functional flow diagrams are prepared down to the level necessary to establish the needs (hardware, software, facilities, personnel) of the system. Functions identified on each diagram are numbered in a manner that preserves the continuity of functions and provides traceability throughout the system to the function origin. The indenture numbering of functions by level is illustrated in Figure 5-2.

a. *Operational functions.* The functional description represents an overall portrayal of the functions that are necessary to describe total system activities. Gross operational activities are defined in terms of mission activities (i.e., system operational requirements described in Section 3.3A1). This may constitute a description of the various modes of system operation and utilization. System functions necessary to support the identified modes of operation are then described. Ultimately, the functional description is completed to the level where initial packaging concepts are visualized.[3]

Figure 5-3 represents a typical system operational functional flow diagram. The system covered will henceforth be noted as system XYZ. The major operational functions of system XYZ are described in Figure 5-3, and the respective blocks are numerically identified for reference purposes. Each major function, in turn, is then analyzed and expanded through the identification of subfunctions, as indicated in Figure 5-4.

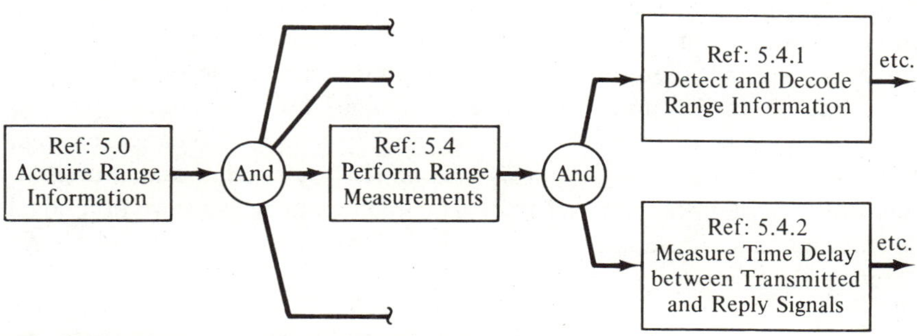

Figure 5-4. Range information functional flow diagram.

Considering the desirability of packaging equipment by function (consistent with system size and weight constraints), the packaging scheme for system XYZ results in three basic units (units A, B, and C). The operational functions incorporated in each of the three units are indicated in Figure 5-3. Developing the design concept further, an analysis of the functions contained

[3]It should be noted that equipment design is by no means formulated at this time. A gross-level configuration is developed for the purpose of allocating requirements. This configuration, which serves as a starting point, may be verified or may change as a result of subsequent analyses.

within each unit will lead to the identification of major assemblies. Given a broad functional packaging scheme, it is then possible to allocate performance parameters and operational effectiveness factors to the unit level and possibly the assembly level. This allocation serves as the basis for subsequent detail design.

b. *Maintenance functions.* Once operational functions are defined, the system description leads to the development of gross maintenance functions. For example, there are specified performance requirements (signal level, tolerances, accuracies, unit dimension, etc.) for each operational function. A check of the applicable function will indicate either a go or no-go decision. A go decision leads to a check of the next operational function. A no-go indication (constituting a symptom of malfunction) provides a starting point for the development of detailed maintenance functional flows and logic troubleshooting diagrams. A gross-level maintenance functional flow is illustrated in Figure 5-5.

Maintenance functions identified at this point will reflect consideration for the effectiveness and supportability factors specified at the system level as well as available logistic resources. In some instances, it may be possible to expand top-level functional flows in a manner similar to the example presented in Figure 5-4. In most cases, however, only an estimate of first-level maintenance functions can be made since the availability of preliminary engineering data is limited. Maintenance functional flows can and should also be prepared for preventive maintenance, transportation and handling functions, support equipment corrective maintenance, servicing, inspections, and so on. In any event, the maintenance functional flow diagrams that are developed are used to update the system maintenance concept discussed in Section 3.3A2 in terms of functions by echelon and a preliminary equipment packaging scheme. The maintenance concept/functional flow development process is iterative and is continued throughout the early system definition process.

The benefits associated with the generation of functional flows are many. First, the process enables the engineer to approach design from a logical and systematic standpoint. The proper sequences and design relationships are readily established. Second, the preparation of functional flows forces the integration of the numerous interfaces that exist in system development and operation. Both internal and external interface problems are quickly identified at an early stage in the life cycle. Sometimes these benefits are difficult to visualize unless one has actually had some experience in functional analysis; however, it has been proved that many design problems which occur later in the life cycle could have been avoided had the engineer(s) followed this approach initially.

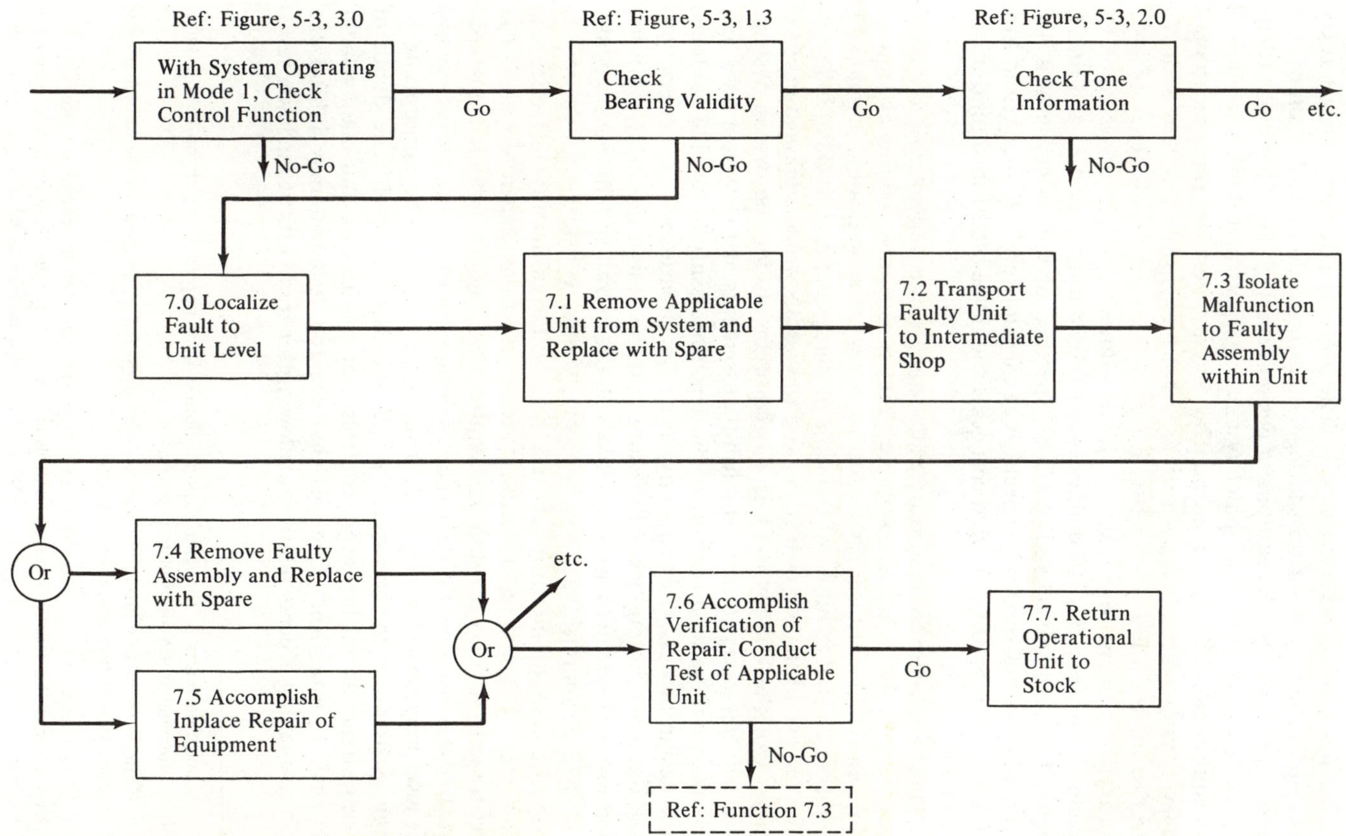

Figure 5-5. Maintenance functional flow diagram.

2. Allocation of requirements.[4] The preceding discussion covered the first step in the process of translating system operational and maintenance requirements into specific equipment design criteria.[5] The functional analysis provides a description of major system functions and defines a preliminary equipment packaging concept. The next step involves the allocation of top-level system factors to the various subelements of the system with the objective of providing some definitive guidelines for the designer of each appropriate element. For instance, if the allowable weight of system XYZ depicted in Figure 5-3 is 750 pounds, what should unit A, unit B, and unit C weigh? If the mission reliability MTBF specified at the system level is 450 hours, what values should be specified at the unit level? If the system life-cycle cost goal is $5,000.00, what cost goals should be established for units A, B, and C?

These and other factors must be allocated to lower indenture levels of the system to provide technical parameters and constraints, functional requirements, and design criteria where needed. Otherwise, individual engineering designers assigned to different elements of the system and working independently will establish their own goals, and the results when combined may not comply with the initially established requirements for the overall system. Thus, it is necessary to first establish requirements at the system level and then allocate these requirements to the depth necessary to provide guidance in the design process. Subsequently, as design progresses, it is necessary to check (on a continuing basis) to ensure that the design results at the lower indenture levels are in turn compatible with the overall system requirements.

When accomplishing allocation, one should consider all appropriate qualitative and quantitative criteria that will significantly influence the design process. These criteria will of course vary from system to system and must directly support operational requirements and the maintenance concept. In some instances the allocation of design criteria to the subsystem and unit level may be sufficient, and in other cases it may be appropriate to assign goals down to the assembly and subassembly level. The depth of coverage depends on the design controls that engineering and management may wish to establish.[6] Controls that are too stringent may inhibit the decision-making process relative to allowing for trade-offs at the lower levels

---

[4]Allocation refers to the distribution, allotment, or appointment of top-level requirements to lower indenture levels of the system.

[5]Criteria refers to a group of characteristics provided as an input to the design process (e.g., factors that dictate equipment design).

[6]If an item is to be procured from an outside supplier, design criteria are established to that level to effect the necessary design controls on the supplier. On the other hand, the level to which criteria are established may be different if the item is to be developed within the major producer's operation. Make-or-buy decisions will influence the depth of coverage.

in order to meet total system requirements. On the other hand, insufficient design controls may not provide the desired results.

The process of allocation stems from the functional packaging scheme identified in Figure 5-3. Although still preliminary in nature since the necessary design data are not available, this scheme (i.e., the identification of units A, B, and C) does form an initial baseline for design guidance. The next step is to attempt to subdivide the units into assemblies, the assemblies into subassemblies, and so on. This may be accomplished through a preliminary equipment breakdown such as the example presented (in a partial form) in Figure 5-6. Some of the parameters normally established at the system level are allocated to the unit level and then to the assembly level. Actually, allocation should consider *all* significant system parameters stated in the form of maximum or minimum requirements (with tolerance bands where appropriate) to include[7]:

a. System effectiveness factors, such as operational availability, readiness, reliability, maintainability, and dependability.
b. System performance parameters, such as range, accuracy, speed of performance, capacity and output rate, power output, weight, size, and volume.
c. Factors covering the system support capability, such as pipeline or supply times between echelons of maintenance, spare/repair parts availability, test and support equipment utilization and availability, facility utilization, personnel effectiveness, transportation and handling rates, and maintenance turnaround times.
d. Life-cycle cost factors, which include research and development cost, investment or production cost, operation and maintenance support cost, and equipment phase-out and disposal cost.

In summary, the purpose of allocation is to provide, as an input, some guidelines to the design engineer to assist him in developing a product that will be compatible with system requirements. These guidelines may be stated both qualitatively and quantitatively and will vary from system to system. At times, it may be impossible to meet the allocated parameters when considering available technology, and trade-offs must be made with other elements of the system. In such instances, the allocations are revised and the best approach is defined representing a modified baseline. In any event, a baseline of some type must be established before proceeding further.

---

[7] Allocation procedures covering reliability, maintainability, logistic support, and cost (representing typical approaches) are discussed in: B. Blanchard, *Logistics Engineering and Management*, Prentice-Hall, Inc., Englewood Cliffs, N.J., 1974.

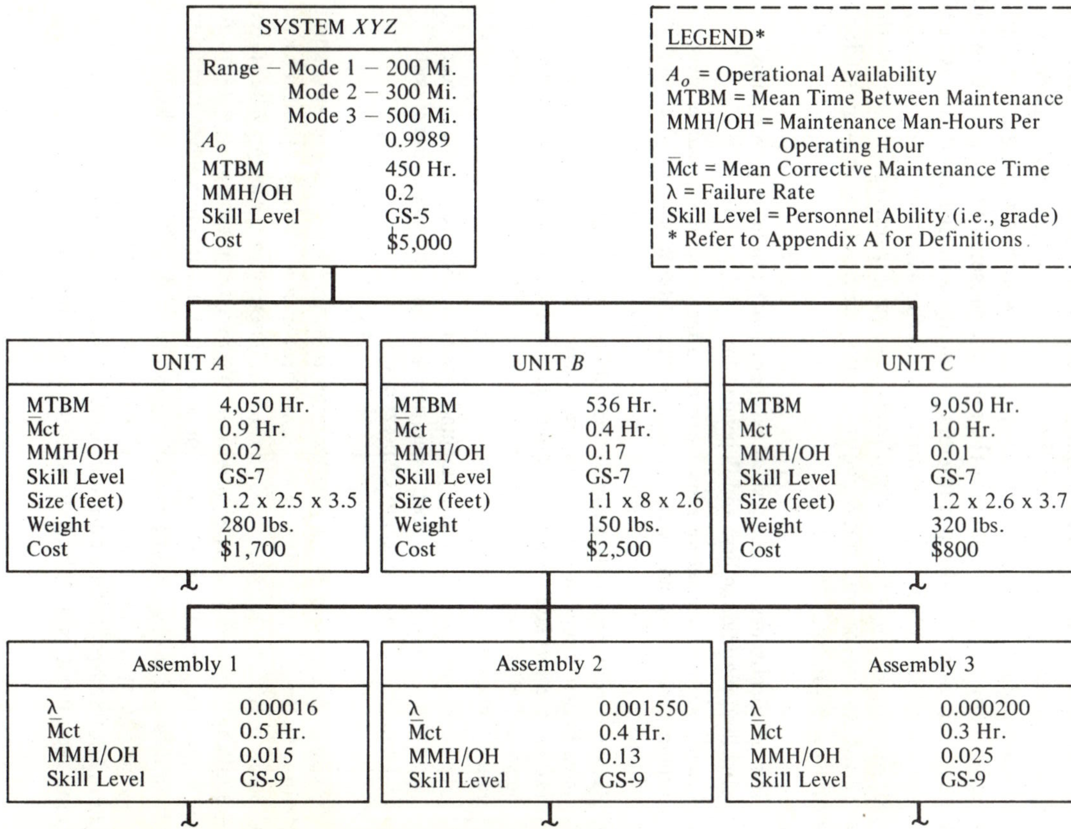

Figure 5-6. System *XYZ* functional requirements (*partial*).

B. Trade-offs and Optimization

The allocation process previously described establishes boundaries and constraints for product/system design (i.e., maximum and/or minimum values to which the design must conform). Within these bounds and constraints, the engineer may envision any number of design configurations that will satisfy the specified requirements. The problem is to select the best approach possible in fulfilling the need.

Figure 5-7 illustrates the steps involves in performing trade-offs and evaluating different alternative approaches. These steps are an integral part of engineering decision making, which is discussed in Section 2.3. Alternatives are identified, evaluation criteria are established, analytical methods are selected, input data are collected, and the various alternatives are evaluated on an equivalent basis. The evaluation itself may involve the accomplishment of specific trade-off studies where two or more individual system parameters are reviewed independently in terms of the effects on each other. For example, Figure 5-8 illustrates a trade-off approach involving MTBM as a function of life-cycle cost. Other trade-offs of this type may deal with different performance parameters in terms of cost, performance as a function of reliability, weight as a function of range, capacity as a function of size and cost, and so on. Ultimately, these individual trade-offs are combined and reviewed in terms of higher-order system parameters such as system effectiveness and life-cycle cost (or equivalent).[8] The results of the overall evaluation will lead to the selection of a preferred approach. If the configuration selected is significantly better than other options, the decision may be clear-cut and the engineer will proceed toward greater system definition. However, in most cases the engineer will perform something of a self-analysis to ensure that he is making the correct decision. The following questions will usually arise:

1. How much better is the selected approach than the next-best alternative? Is there a significant difference between the results of the comparative evaluation?
2. Have all feasible alternatives been considered?
3. What are the areas of risk and uncertainty?

These and other comparable questions should be addressed, and to obtain some of the answers may require one or more iterations of the process illustrated in Figure 5-7. The objective, of course, is to arrive at a decision where the selected approach is clearly the best among the alternatives evaluated, and the associated risk and uncertainty are minimized.

---

[8]Some actual detailed examples of trade-off studies are included in Appendix B.

Reference: Figure 5-6

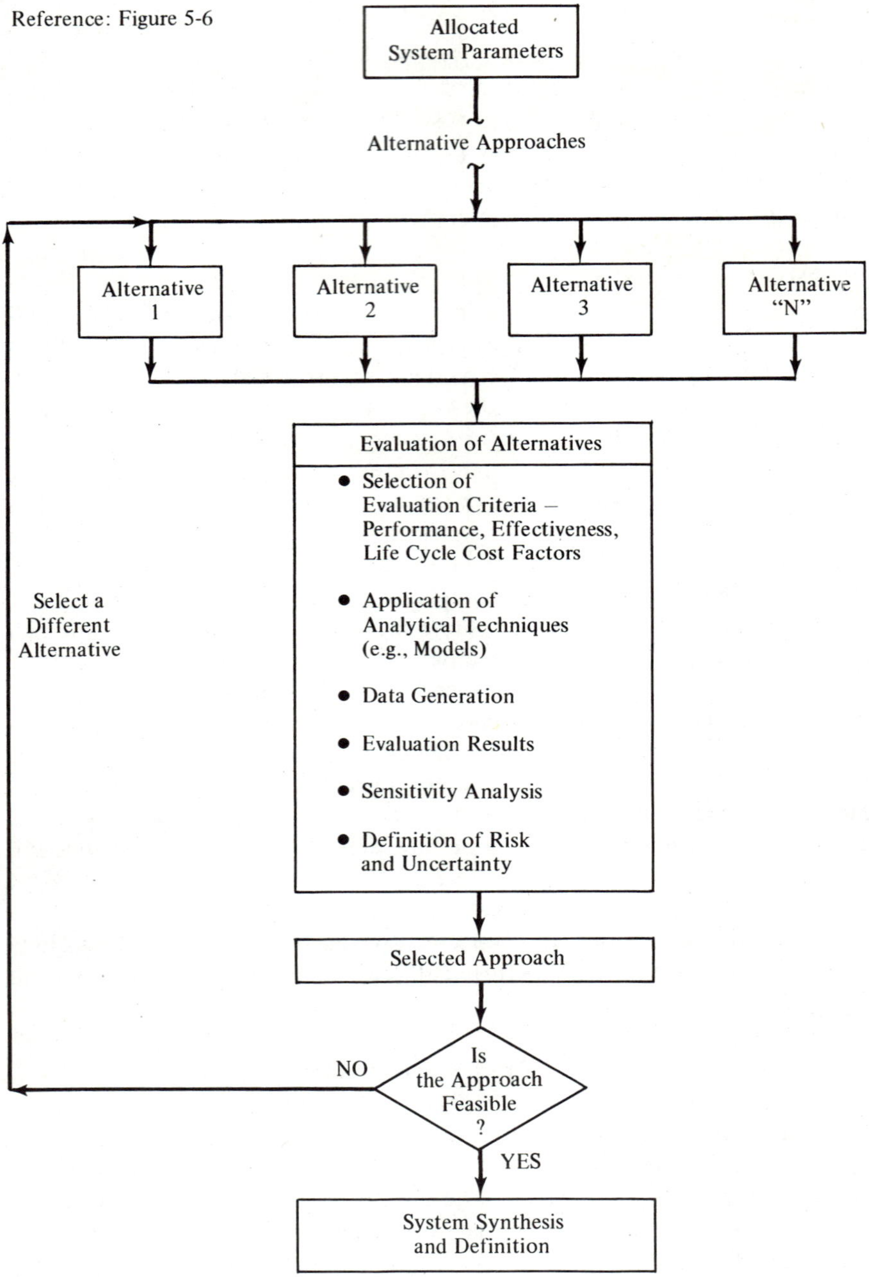

Figure 5-7. Evaluation of alternatives.

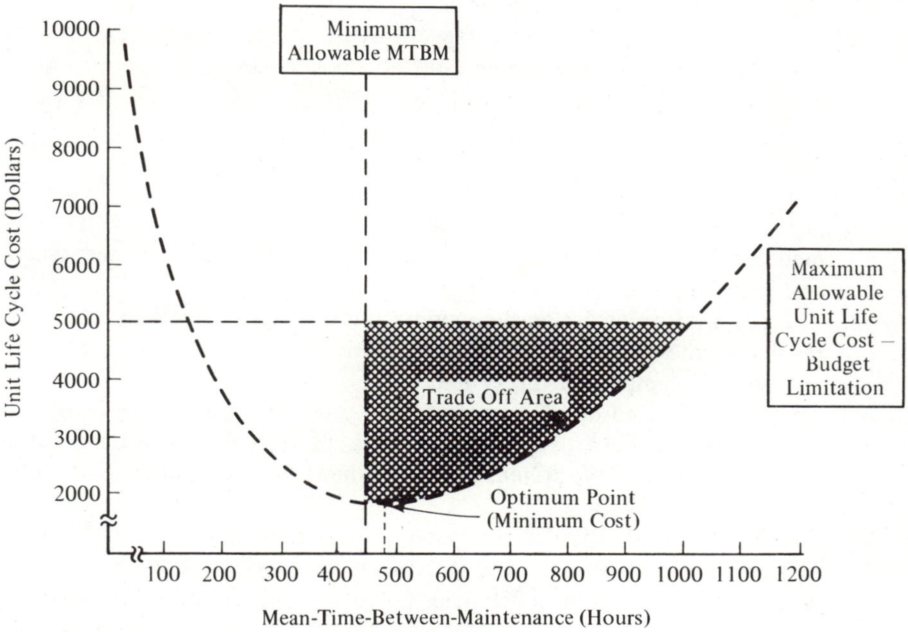

Figure 5-8. MTBM versus cost trade-off.

C. Synthesis and Definition

Synthesis refers to the combination of parts and elements in such a manner so as to form a functional entity. In other words, sufficient trade-offs and preliminary design have been accomplished to confirm and assure the completeness of system performance and design requirements allocated for detailed design. The performance, configuration, and arrangement of a chosen system and its elements are portrayed along with the techniques for their test, operation, and life-cycle support. These portrayals cover intra- and intersystem and item interfaces and provide enough detail forming a definitive baseline that can be presented in the form of a system specification (refer to Figure 3-8). Synthesis may be accomplished by analytical means or through the development and preliminary test of physical models. Thus, the proposed system configuration is synthesized, and the results are included in the system specification (and supporting development and procurement specifications where appropriate,) which is a major output of the preliminary system design phase.

5.3 DETAILED DESIGN

The detailed design phase begins with the concept and configuration derived through preliminary system design; that is, a configuration with performance, effectiveness, logistic support, cost, and other requirements has been described in the system specification. An overall system design configuration has been established, and now it is necessary to convert that configuration to the definition and subsequent realization of hardware, software, and items of support. The process from here on includes:

A. The description of subsystems, units, assemblies, and lower-level components and parts of the prime mission equipment and the elements of logistic support (e.g., test and support equipment, facilities, personnel and training, technical data, spare/repair parts).
B. The preparation of design documentation (e.g., specifications, analysis results, trade-off study reports, predictions, detailed drawings), describing *all* elements of the system.
C. The definition and development of computer software (as applicable).
D. The development of an engineering model, a service test model, and/or a prototype model of the system and its elements for test and evaluation to verify design adequacy.[9]
E. The test and evaluation of the system model that has been developed.
F. The redesign and retest of the system, or an element of the system, as necessary to correct any deficiencies noted through initial system testing.

While the preliminary design stage established the top system-level configuration, now it is necessary to define subsystems, units, assemblies, and on down to the part level. This can be initiated through the development of a detailed system/equipment functional block diagram or *hierarchy tree*, showing all functional indenture levels of the system, which may be an expansion of the breakout used for allocation purposes in Figure 5-6. The system tree should include all elements of prime mission equipment, software, and logistic support.

The system breakout at this stage constitutes a baseline for detailed design. Now each subsystem, unit, assembly, and so on, is looked at as an individual entity. Alternative design concepts are established and evaluated in a manner similar to that accomplished earlier at the system level. The evaluation process generally follows the approach described in Section 2.3.

---

[9]Models in this instance refer to physical models versus the analytical models discussed earlier. For the purposes of discussion, an engineering model represents a working system, or an element of the system, that will exhibit the functional performance characteristics defined in the specification. It does not necessarily represent the system in terms of physical dimensions (i.e., form or fit). A service test model goes one step further and reflects the system in terms of function and physical dimensions. It is like the production configuration except that the same component parts may not be incorporated (i.e., substitute components may be used as long as the system functions properly). The prototype model is a further development and is exactly like the production configuration in all aspects except that it has not been fully qualified in terms of operational and environmental evaluation testing.

When individual design approaches are established at the subassembly or lower level, these approaches are combined and integrated to ensure compatibility at the system level and between the elements of the system. This is an iterative process until a firm design is attained.

Finally, the results of the individual subsystem and lower element designs are projected into the system functional block diagram which forms the basis for translating information into more descriptive design documentation. The resulting documentation, describing the design in detail, is used in the development of system models which are tested to verify design adequacy prior to entering the production and/or construction phase.

A. Initial Design Activities

1. Design objectives. The objectives of design must be consistent with the operational requirements and maintenance concept and will vary depending on the system type and configuration (i.e., whether one is designing a single satellite tracking facility or developing a series of airplanes, cars, radios, etc.). However, in any event the design function must consider the following goals:

a. *Design for functional capability or performance* [*functional design*]—size, weight, volume, range, accuracy, capacity, flow rate, speed of performance, power output, and all the technical characteristics that the system (when operating) must exhibit to accomplish its mission.

b. *Design for reliability*—the characteristic of system design and installation which is concerned with maximizing operational availability and minimizing product/system failures.

c. *Design for maintainability*—the characteristic of system design and installation which is concerned with the ease, economy, safety, and accuracy in the performance of maintenance functions. The objective is to minimize maintenance times, maximize the supportability characteristics in system design (e.g., accessibility, diagnostic provisions, standardization), minimize the logistic support resources required in the performance of maintenance, and minimize maintenance support costs.

d. *Design for human factors*—the characteristic of system design that is directed toward the optimum man-machine interface (i.e., ensuring the compatibility between system physical and functional design features and the "element of man" in the operation, maintenance, and support of the system or equipment). Human factors considers operability and equipment aesthetic features and results in reducing personnel skill levels, minimizing training requirements, and minimizing potential personnel error rates.

e. *Design for safety*—the characteristic of system design that is concerned with the elimination of personnel and equipment hazards due to unsafe conditions.

f. *Design for producibility*—the characteristic of design that allows for the effective and efficient production of one or a multiple quantity of items of a given configuration.

g. *Design for disposability*—the characteristic of design that enables the effective and efficient disposal of system elements when retired or phased out of the inventory. The disposal of material must be consistent with environmental and ecological requirements.

Actually, there are many different considerations in equipment design. The ultimate output must not only meet performance requirements but must do so in a cost-effective manner. Accomplishing such requires an optimum balance between performance features, system availability characteristics, reliability, maintainability, human factors, logistic support, life-cycle cost, and so on. This balance is often difficult to attain since some of the stated objectives are in opposition to others. For instance, incorporating high performance and overly sophisticated functional design techniques will undoubtedly reduce reliability and increase logistic support requirements and life-cycle cost, incorporating too much reliability may require the use of costly system components and thus increase both the system acquisition cost and life-cycle cost, incorporation of too many provisions for maintainability and human factors may reduce reliability, and so on. In essence, the goal is to incorporate only the necessary characteristics to do the job, not too many or too few. In other words, do not overdesign or underdesign.

The proper balance of the various features of design is often attained through the evaluation of alternative configurations using the process described in Chapter 2 on an iterative basis. Although there are many different levels of trade-offs performed (refer to Figure 2-4), the ultimate criterion constitutes some factor of cost effectiveness (which is similar to the standard benefit-cost analysis approach employed for decision-making purposes in many industrial and business applications).[10] Cost effectiveness relates to the measure of a system in terms of mission fulfillment (system effectiveness) and total life-cycle cost and can be expressed in various terms, depending on the specific mission or system parameters that one wishes to evaluate. *True* cost effectiveness is impossible to measure since there are many factors that influence the operation and support of a system that cannot realistically be quantified (e.g., system interaction effects as a result of other systems, political implications, certain environmental factors); thus, it is common to employ specific cost-effectiveness *figures of merit* (FOM), such as

10Although industrial organizations as a whole rely primarily on the benefit-cost analysis approach for top-level management decision making (i.e., considering benefits in terms of revenues and profit along with the cost), the author feels that the cost-effectiveness concept discussed herein is more relevant in making specific design-related decisions. These decisions should then be evaluated in terms of their effects on revenues and profit with the necessary feedback and corrective action provisions incorporated to cover possible changes.

$$\text{C.E. FOM} = \frac{\text{system effectiveness}}{\text{life-cycle cost}} \tag{5-1}$$

$$\text{C.E. FOM} = \frac{\text{availibility}}{\text{life-cycle cost}} \tag{5-2}$$

$$\text{C.E. FOM} = \frac{\text{system capacity}}{\text{life-cycle cost}} \tag{5-3}$$

$$\text{C.E. FOM} = \frac{\text{supply effectiveness}}{\text{life-cycle cost}} \quad \text{(and others)} \tag{5-4}$$

These factors (which may represent a combination of system parameters) are often presented as delta values to allow the comparison of alternatives on the basis of the relative merits of each. Given two or more alternatives evaluated in a *consistent* manner, one can select the best among the alternatives based on these delta values.

The prime ingredients of cost effectiveness are illustrated in Figure 5-9. Although there are different ways of presenting cost effectiveness, this illustration is used for the purpose of showing the many influencing factors and their relationship. Design attributes influence both the system-effectiveness and life-cycle-cost sides of the balance, and the results of design dictate the requirements for logistic support, which in turn affect both areas. For instance, the complexity of prime equipment design influences the personnel skill level and training requirements that can have a significant impact on both system effectiveness and the system operation and maintenance cost. The packaging scheme and standardization in design affect the test and support equipment and the type and quantity of spare/repair parts and supporting inventories, which in turn affect both effectiveness and life-cycle cost. Likewise, the design of the various elements of logistic support (e.g., test and support equipment, facilities) has an impact on prime mission equipment. In essence, there are many system characteristics that interact, and the consequences of these interactions (as well as the affects of changes) must be evaluated and thoroughly understood.

In summary, the basic design objective is to develop a system that will perform its intended function in a cost-effective manner (i.e., do the job effectively at the lowest overall life-cycle cost). To accomplish such requires recognition of the many considerations involved by both engineering and management alike. *All* elements of the system should be addressed on an *integrated* basis, and trade-offs are accomplished to seek a preferred approach in facilitating detailed system design and definition.

2. Establishing the design team. Once the goals and objectives have been defined for detailed design, it is necessary to organize and establish an

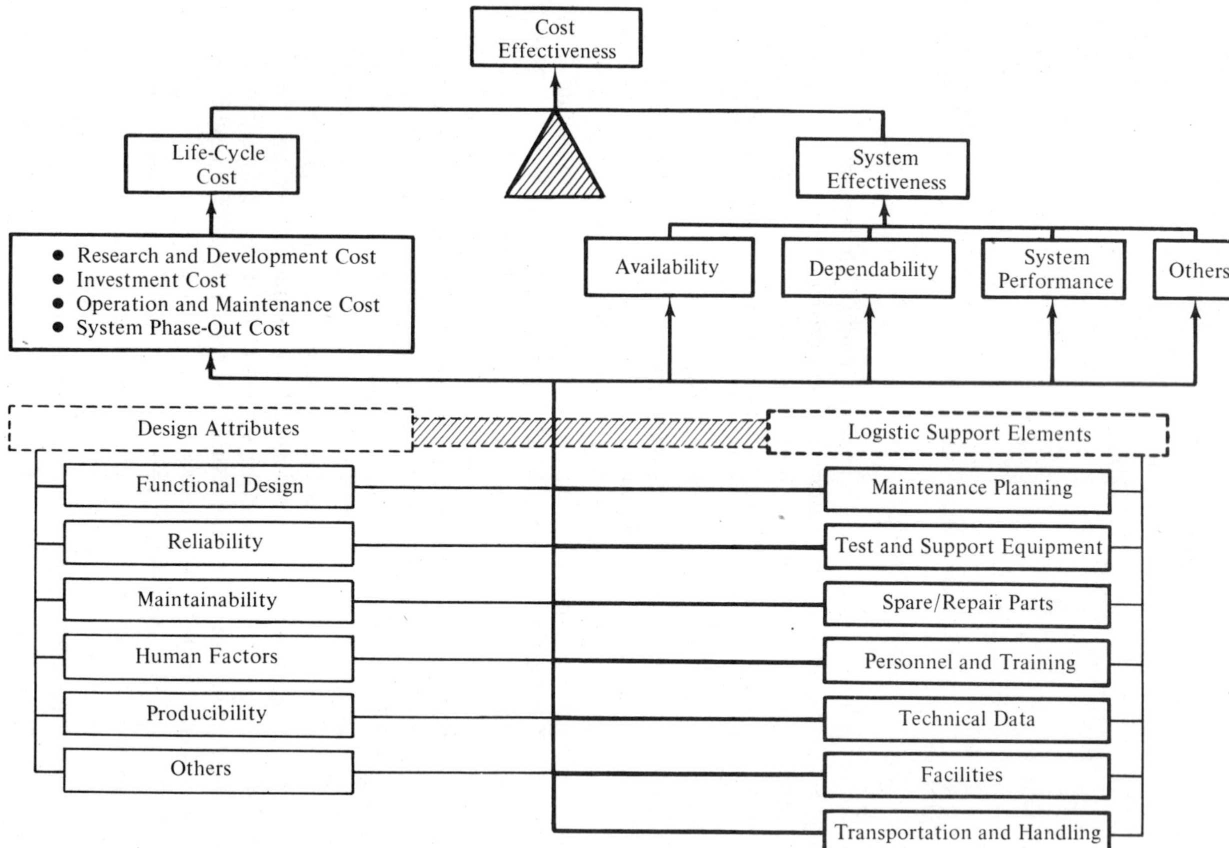

Figure 5-9. The elements of cost effectiveness.

integrated team of management, engineering, and supporting personnel (constituting the proper levels of expertise) for the purpose of performing the transformation functions from a defined "paper" entity in the preliminary system design stage to a prototype system configuration ready for production and/or construction.[11] In essence, a project organization must be established to accomplish detailed design of equipment, software, and elements of logistic support as well as the follow-up design verification functions involving preliminary system test and evaluation. The project organization must be responsive to the system technical requirements covered in the specifications and must accomplish the functions and tasks identified in the product acquisition plan described in Section 3.4C.

As stated earlier (refer to Section 1.4), project organizations will vary with the system objectives. Some projects may involve 5,000 to 10,000 engineers plus supporting personnel, while others may only require a few individuals. Each situation is different, and the outcome will evolve through the proper planning steps described in Chapter 3. In any event, the project team at this stage will probably include a combination of:

a. *Engineering technical expertise*—electrical engineers, mechanical engineers, computer engineers, civil engineers, nuclear engineers, system engineers, reliability and maintainability engineers, and/or others as appropriate to the project needs.

b. *Engineering technical support*—draftsmen, technical publication specialists, component-parts specialists, laboratory technicians, model builders, computer programmers, test technicians, and the like.

c. *Nontechnical support*—marketing, purchasing and procurement, contracts, budgeting and accounting, legal, personnel, and so on.

The project team, consisting of expertise in a number of different areas (especially for large projects), must be properly integrated, and the individuals involved should be highly motivated to accomplish their respective assigned functions in a successful manner. This objective is attained through the proper organization and management of the resources made available. Organizational goals, project organizations, functions and tasks, and the associated management of project resources are covered rather thoroughly in Chapter 10.

3. *Evolution of detailed design.* The evolution of design is illustrated rather basically in Figure 5-10 and amplified in Figure 5-11. The process

---

[11]The design team effort identified here is not necessarily to be considered separate and aside from the system-level design activity required for preliminary design. The overall design effort is continuous from one stage to the next except that the level of emphasis shifts. Additional discussion in terms of project organization is presented in Part III.

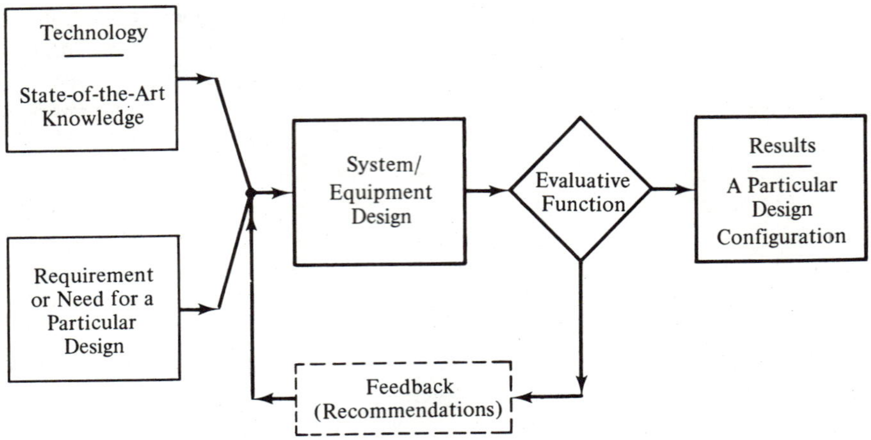

Figure 5-10. Evolution of design.

shown in Figure 5-11 is iterative in working from the system-level definition to an output that can be produced in one or multiple quantities. There are checks and balances in the form of reviews at each stage of design progression, and a feedback loop is provided for corrective action.[12] In this respect, the process is similar in concept to the system analysis, optimization, synthesis, and definition accomplished in the preliminary system design stage except that the requirements now are at a lower indenture level (i.e., subsystems, units, assemblies, subassemblies, etc.).

As detailed design progresses, actual definition is accomplished through documentation in the form of specifications and plans (already discussed), procedures, drawings, material and part lists, reports and analyses, computer programs, and so on. The design configuration may be the "best possible" in the eyes of the designer; however, the results are practically useless unless properly documented, so that others can first understand what is being conveyed and then be able to translate the output into a producible entity.

When addressing the aspect of documentation, the results of design are usually conveyed through a combination of the following:

a. *Design drawings*—assembly drawings, control drawings, logic diagrams, installation drawings, schematics, and so on.

b. *Material and part lists*—parts lists, material lists, long-lead-item lists, bulk-item lists, provisioning lists, and so on.

---

[12]The design reviews reflected in Figure 5-11 represent the day-to-day design progress in working from the system level down to the small component level. These reviews are informal and occur continuously as compared to formal design reviews, which are scheduled at specific times in the life cycle. Formal design reviews are discussed in Section 5.4.

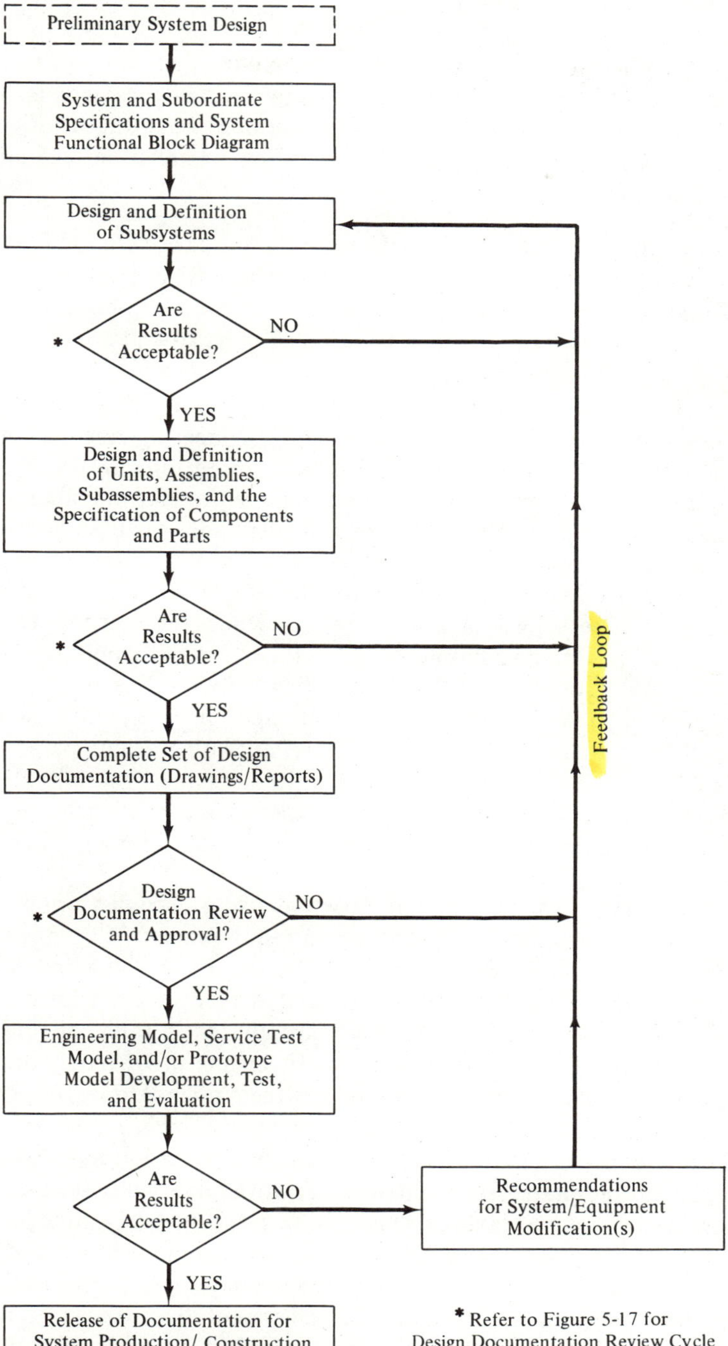

Figure 5-11. Basic design sequence.

c. *Analyses and reports*—trade-off study reports supporting design decisions, reliability and maintainability analyses and predictions, human factors analyses, safety reports, logistic support analyses, configuration identification reports, computer documentation, installation and assembly procedures, and so on.

Design drawings, constituting the primary source of definition, may vary in form and function depending on the design objective; that is, the type of equipment being developed, the extent of development required, whether the design is to be subcontracted, and so on. Some typical types of drawings are specified in Figure 5-12, and examples of these are presented in Figures 5-13 through 5-16.

During the process of detailed design, engineering documentation is rather preliminary at first and then gradually progresses to the depth and extent of definition necessary to enable complete product manufacture. The responsible designer, with the aid of a draftsman, produces a functional diagram of the overall system. The system functions are analyzed and initial packaging concepts are assumed. With the aid of specialists representing various disciplines (i.e., electrical, mechanical, components, reliability, maintainability, etc.) and supplier data, detailed design layouts are prepared for subsystems, units, assemblies, and subassemblies. The results are analysed and evaluated in terms of functional capability, reliability, maintainability, human factors, safety, producibility, and other design parameters to assure compliance with the allocated requirements and the initially established design criteria. This review and evaluation occurs at each stage in the basic design sequence and generally follows the steps presented in Figure 5-17.

Engineering documentation is reviewed using checklist criteria developed from both program and specification requirements. If the review indicates complete compliance with these requirements, the documentation is considered approved on an initial basis and is ready for the formal design review process. Approval at this point usually constitutes signing off or initialing the applicable documentation. If, on the other hand, the design configuration represented is not approved, recommendations for corrective action are prepared and submitted to the responsible designer.

The checklist criteria used in the design documentation review may be of a specific nature, quantitative and/or qualitative, emanating from the specifications and program plans, or may be of a more general nature, constituting characteristics representative of desirable common sense features in design. In any event, appropriate checklists are developed early in the design process to serve as an aid in engineering documentation review. Figure 5-18 constitutes an abbreviated sample checklist which is supported by more in-depth criteria included in Appendix C.

1. Arrangement drawing – shows in any projection or perspective, with or without controlling dimensions, the relationship of major units of the item covered.
2. Assembly drawing – depicts the assembled relationship of: (a) two or more parts; (b) a combination of parts and subassemblies; or (c) a group of assemblies required to form the next higher indenture level of the equipment.
3. Connection diagram – shows the electrical connections of an installation or of its component devices or parts.
4. Construction drawing – delineates the design of buildings, structures, or related construction (including architectural and civil engineering operations).
5. Control drawing – an engineering drawing that discloses configuration and configuration limitations, performance and test requirements, weight and space limitations, access clearances, pipe and cable attachments, support requirements, etc., to the extent necessary that an item can be developed or procured on the commercial market to meet the stated requirements. Control drawings are identified as envelope control (i.e., configuration limitations), specification control, source control, interface control, and installation control.
6. Detail drawing – depicts complete end item requirements for the part(s) delineated on the drawing.
7. Elevation drawing – depicts vertical projections of buildings and structures or profiles of equipment.
8. Engineering drawing – an engineering document that discloses by means of pictorial or textual presentations, or a combination of both, the physical and functional end product requirements of an item.
9. Installation drawing – shows general configuration and complete information necessary to install an item relative to its supporting structure or to associated items.
10. Logic diagram – shows by means of graphic symbols the sequence and function of logic circuitry.
11. Numerical control drawing – depicts complete physical and functional engineering and product requirements of an item to facilitate production by tape control means.
12. Piping diagram – depicts the interconnection of components by piping, tubing or hose, and when desired, the sequential flow of hydraulic fluids or pneumatic air in the system.
13. Running (wire) list – a book-form drawing consisting of tabular data and instructions required to establish wiring connections within or between items.
14. Schematic diagram – shows, by means of graphical symbols, the electrical connections and functions of a specific circuit arrangement.
15. Wiring and cable harness drawing – shows the path of a group of wires laced together in a specified configuration, so formed to simplify installation.

Figure 5-12. Typical engineering drawing classifications.

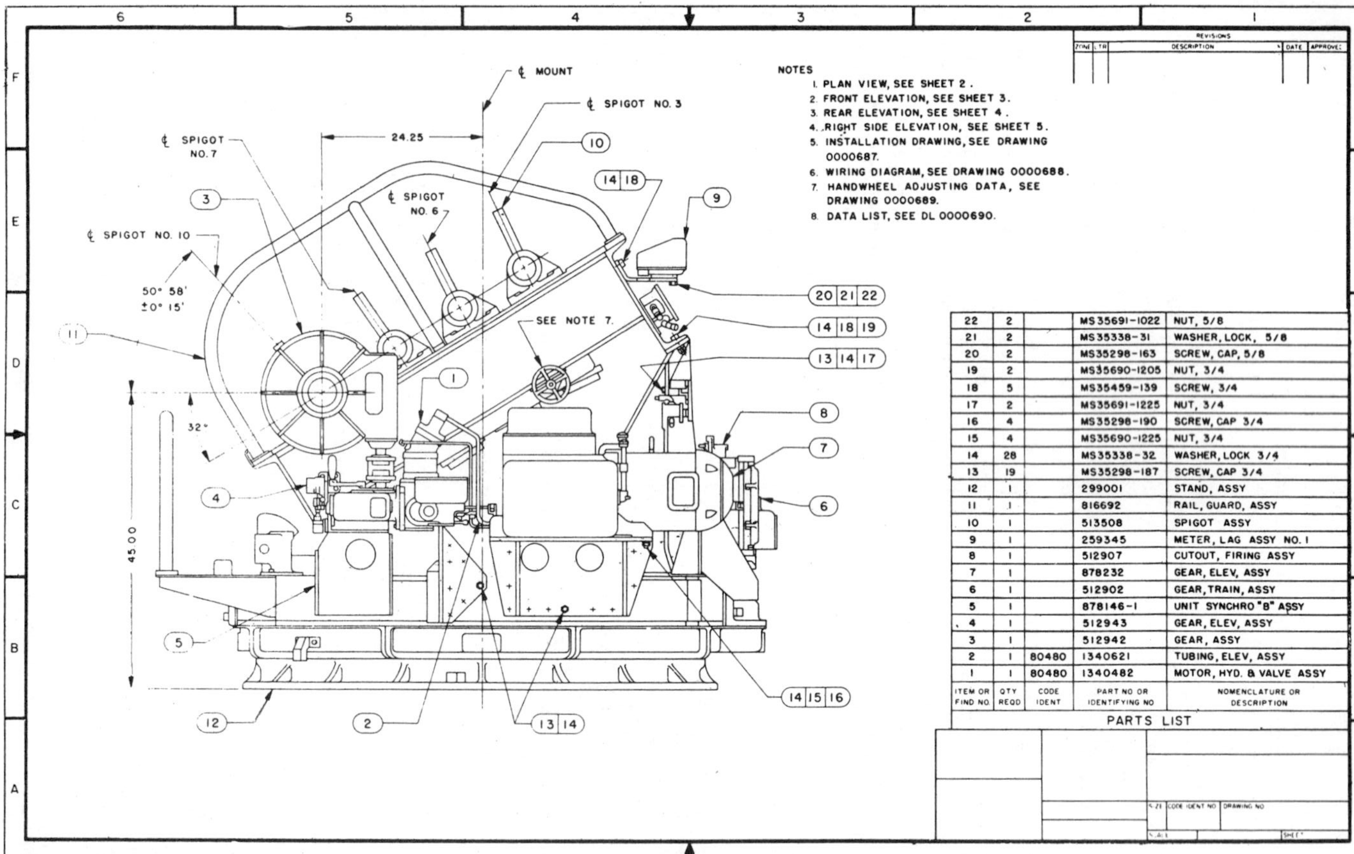

Figure 5-13. Assembly drawing (*sample*). (*Source:* MIL-STD-100A, "Military Standard, Engineering Drawing Practices," Oct. 1, 1967.)

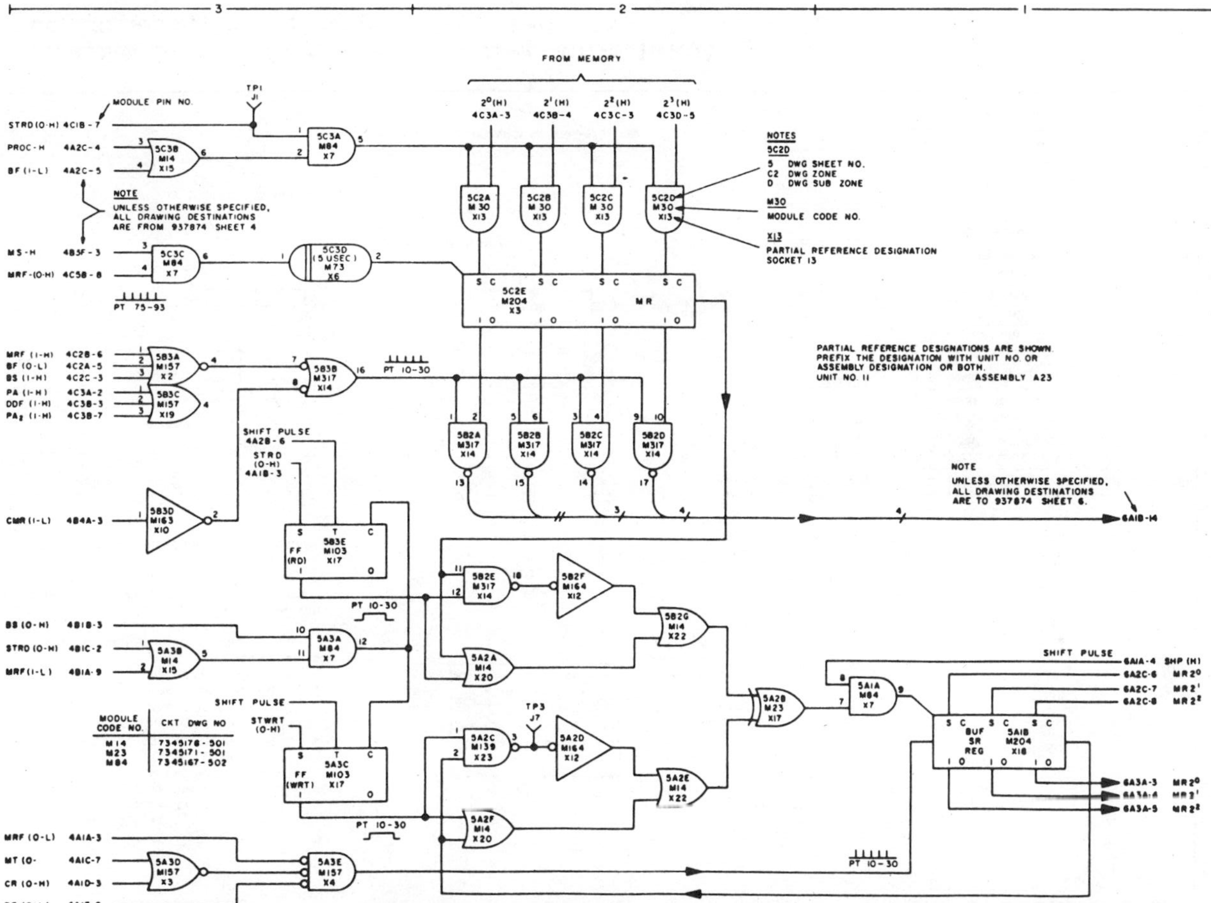

Figure 5-14. Logic diagram (*sample*). (*Source*: MIL-STD-100A, "Military Standard, Engineering Drawing Practices," Oct. 1, 1967.)

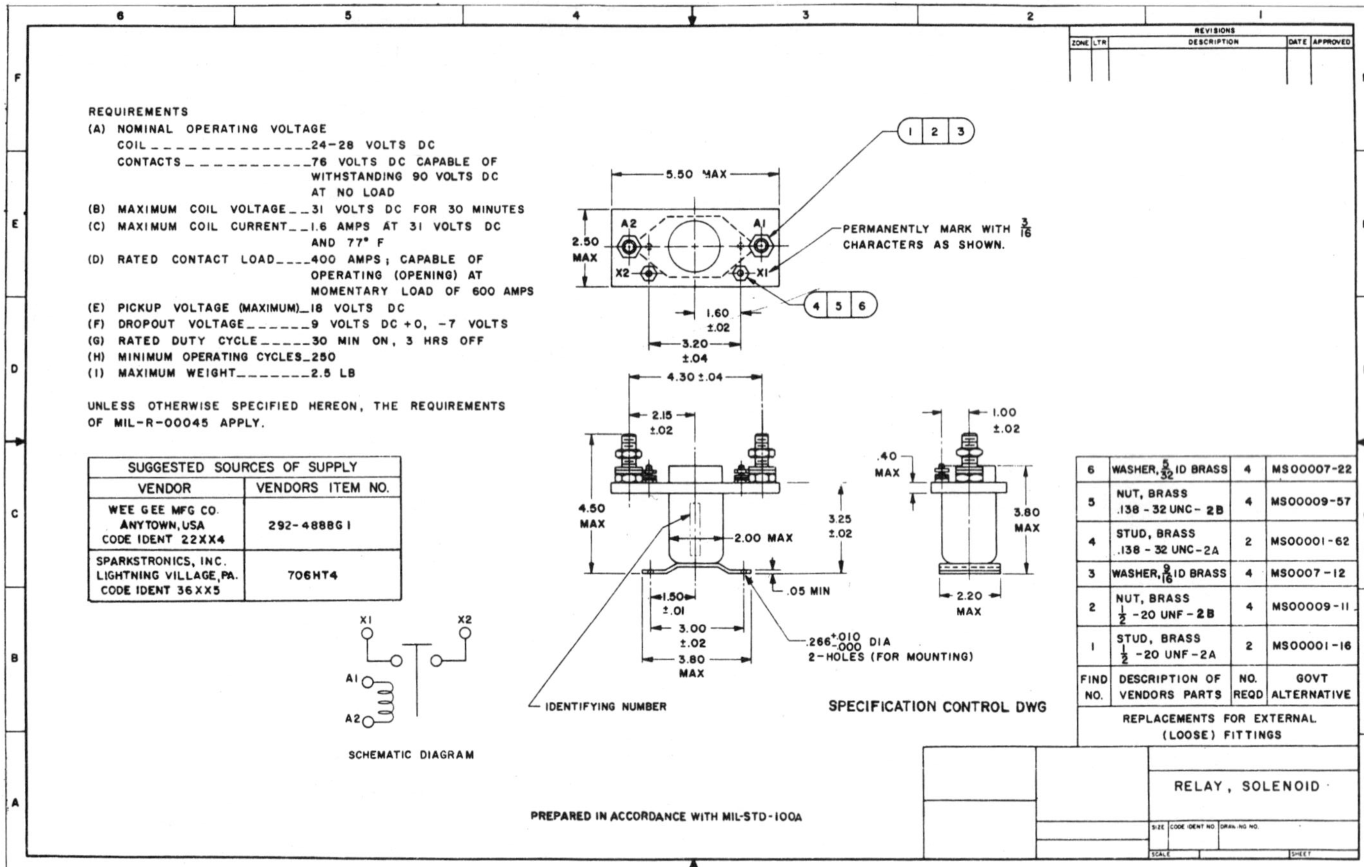

Figure 5-15. Specification control drawing (*sample*). (*Source:* MIL-STD-100A, "Military Standard, Engineering Drawing Practices," Oct. 1, 1967.)

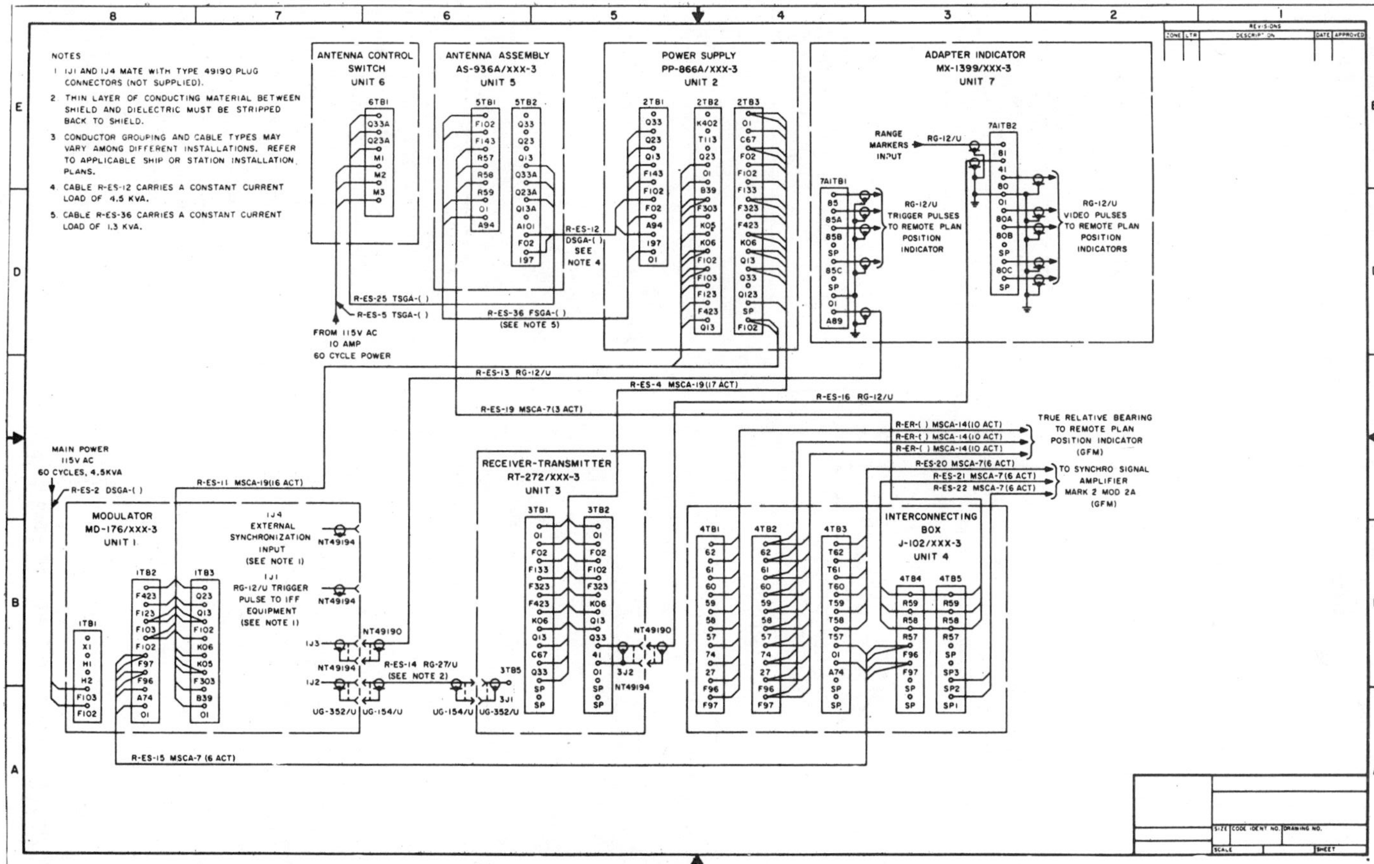

Figure 5-16. Interconnection diagram (*sample*) (*Source:* MIL-STD-100A, "Military Standard, Engineering Drawing Practices," Oct. 1, 1967.)

Reference: Figure 5-11

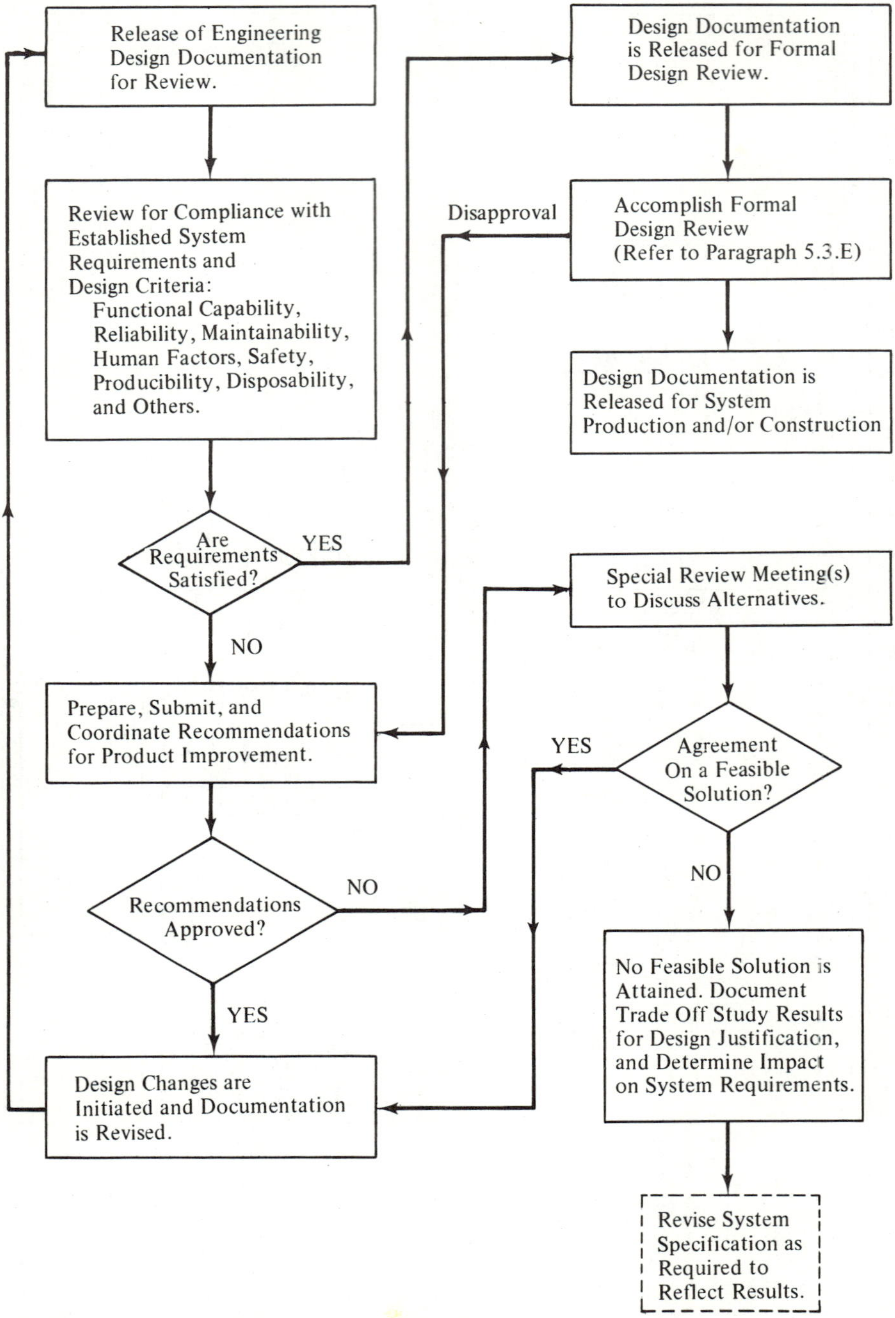

Figure 5-17. Design documentation review cycle.

System Design Review Checklist

General

1. System Operational Requirements Defined
2. Effectiveness Factors Established
3. System Maintenance Concept Defined
4. Functional Analysis and Allocation Accomplished
5. System Trade Off Studies Documented
6. System Specification and Supporting Specifications Completed
7. Product Acquisition Plan Completed
8. Design Documentation Completed
9. Logistic Support Requirements Defined
10. Ecological Requirements Maintained
11. Societal Requirements Maintained
12. Economical Approach Justified

Design Features – Does the Design Reflect Adequate Consideration of:

1. Accessibility
2. Adjustments and Alignments
3. Cables and Connectors
4. Calibration
5. Detailed Packaging and Mounting
6. Disposability
7. Environment
8. Fasteners
9. Handling
10. Human Factors
11. Interchangeability
12. Maintainability
13. Panel Displays and Controls
14. Producibility
15. Reliability
16. Safety
17. Selection of Parts/Materials
18. Servicing and Lubrication
19. Standardization
20. Storage
21. Survivability
22. Test Provisions

When reviewing design (layouts, drawings, part lists, engineering reports, program plans), this checklist may prove beneficial in covering various program functions and design features. The items listed are supported with more detailed criteria in Appendix C. The response to each item listed should be YES.

Figure 5-18. Sample design review checklist.

4. Design aids. During the early stages of the design process, the designer engineer is confronted with the task of defining equipment layout and packaging schemes, performing various types of design analyses, selecting components, incorporating reliability and maintainability provisions, determining man-machine interfaces, and the like. He has to solve many problems and evaluate numerous alternatives in a relatively limited period of time and yet turn out an effective product. With the complexity of products/systems continually increasing, the design engineer often needs all the assistance that he can attain to reach his objectives. This ever-increasing challenge has led to the development of various types of design aids, ranging from design standards documentation to the use of three-dimensional mock-ups or physical models to assist in performing design functions. Many industrial firms and government agencies have developed tools along these lines. Some of the more common categories are noted below.

a. *Design standards documentation*—design standards manuals are developed to cover preferred component parts and supplier data, mathematical and statistical tables, stress/strength tables, engineering drawing practices, and the like.[13]

b. *Design criteria documentation*—individual criteria documentation is often developed to cover areas such as reliability, maintainability, human factors, and safety. Documentation here relates to design features to include accessibility, diagnostic and testing provisions, packaging and mounting techniques, interchangeability, controls and panel display methods, human performance capabilities, reliability and safety characteristics, and so on.[14]

c. *Computer usage in design*—computers are utilized rather extensively in (1) the performance of analytical modeling and design analyses (e.g., system requirements analysis, the evaluation of alternatives, predictions, stress/strength analysis, analysis of design margins, statistical and mathematical analyses, optimization studies; (2) the transmitting of design con-

---

[13]Many industrial firms/agencies generate their own specific standards to suit individual needs and product types. In the absence of such, however, there are available to the engineer numerous documents that cover various areas of design. For instance, there are general handbooks in most engineering fields (e.g., T. Baumeister and L.S. Mark's, *Standard Mechanical Engineers' Handbook,* 7th Ed., McGraw Hill Book Company, New York, 1967) and there are standards covering areas such as engineering drawing, which include: J.H. Earle, *Engineering Design Graphics,* Addison—Wesley Publishing Company, Inc., Reading, Mass., 1969; and MIL-STD-100A, "Military Standard, Engineering Drawing Practices," Government Printing Office, Washington, D.C., Oct. 1967. These and other documents constitute handy references to facilitate the engineering design process.

[14]Specific criteria documents are often generated within firms/agencies from standard texts such as: I. Bazovsky, *Reliability Theory and Practice,* Prentice-Hall, Inc., Englewood Cliffs, N.J., 1961; B. Blanchard and E. Lowery, *Maintainability—Principles and Practices,* McGraw-Hill Book Company, New York, 1969; E. J. McCormick, *Human Factors Engineering,* McGraw-Hill Book Company, New York, 1970; and W. Hammer, *Handbook of System and Product Safety,* Prentice-Hall, Inc., Englewood Cliffs, N.J., 1972.

cepts into drawings, three-dimensional displays, graphic information, and component-parts lists; (3) the storage and retrieval of specific types of design data based on past experience (i.e., the data bank); and (4) in the generation of processes and procedures (e.g., preparation of technical manuals, maintenance instructions, manufacturing procedures/processes).

One of the more significant innovations of recent years is the use of the computer in generating design configurations, drawings, and graphic displays. A designer can draw and evaluate a specific design concept in a few minutes using a light pen, a sensitized cathode-ray tube (CRT), and the full power of a large-scale computer, whereas the use of past conventional methods would have taken hours. The results of the designer's approach are presented in pictorial and/or graphic form on the CRT display. Design changes can be readily incorporated where appropriate. This capability in computer-aided design allows for the evaluation of many design alternatives in a short time frame. Thus, the probability of a better design will evolve with less associated risk and uncertainty, since the designer is able to accomplish a great deal more than what has been possible in the past. In addition, once the designer has established a preferred approach, drawings of standard items can be readily produced and easily replicated when needed. Several illustrations of the computer-aided design concept are presented in Figures 5-19 and 5-20. When using computer-aided design techniques, extreme care must be exercised to avoid design and coding errors and to ensure language compatibility wherever possible. Logic and syntax errors, semantics problems, algorithmic errors, incorrect human inputs, and language variations often inhibit the many benefits that can be derived.

d. *Development of physical scale models or mock-ups*—quite often the design of an item is difficult to visualize in its true perspective when relying on two-dimensional drawings and parts lists. As an aid to the designer, three-dimensional scale models or mock-ups are sometimes developed to provide a realistic simulation of a proposed final system/equipment configuration at an early point in the program prior to the development of final design data or actual prototype hardware.

Models or mock-ups can be developed to any desired scale and to varying depths of detail depending on the level of emphasis required. Mock-ups can be developed for large systems as well as small systems and may be constructed of heavy cardboard, wood, metal, or a combination of materials. Mock-ups can be developed on a relatively inexpensive basis and in a short period of time when employing the right materials and personnel services (industrial design, human factors, and/or model shop personnel are usually well oriented to this area and should be utilized to the greatest extent possible). The uses and values of a mock-up are numerous.

(1) It provides the design engineer with the opportunity of experimenting with different facility layouts, packaging schemes, panel displays, cable runs, and so on, prior to the preparation of final design data.

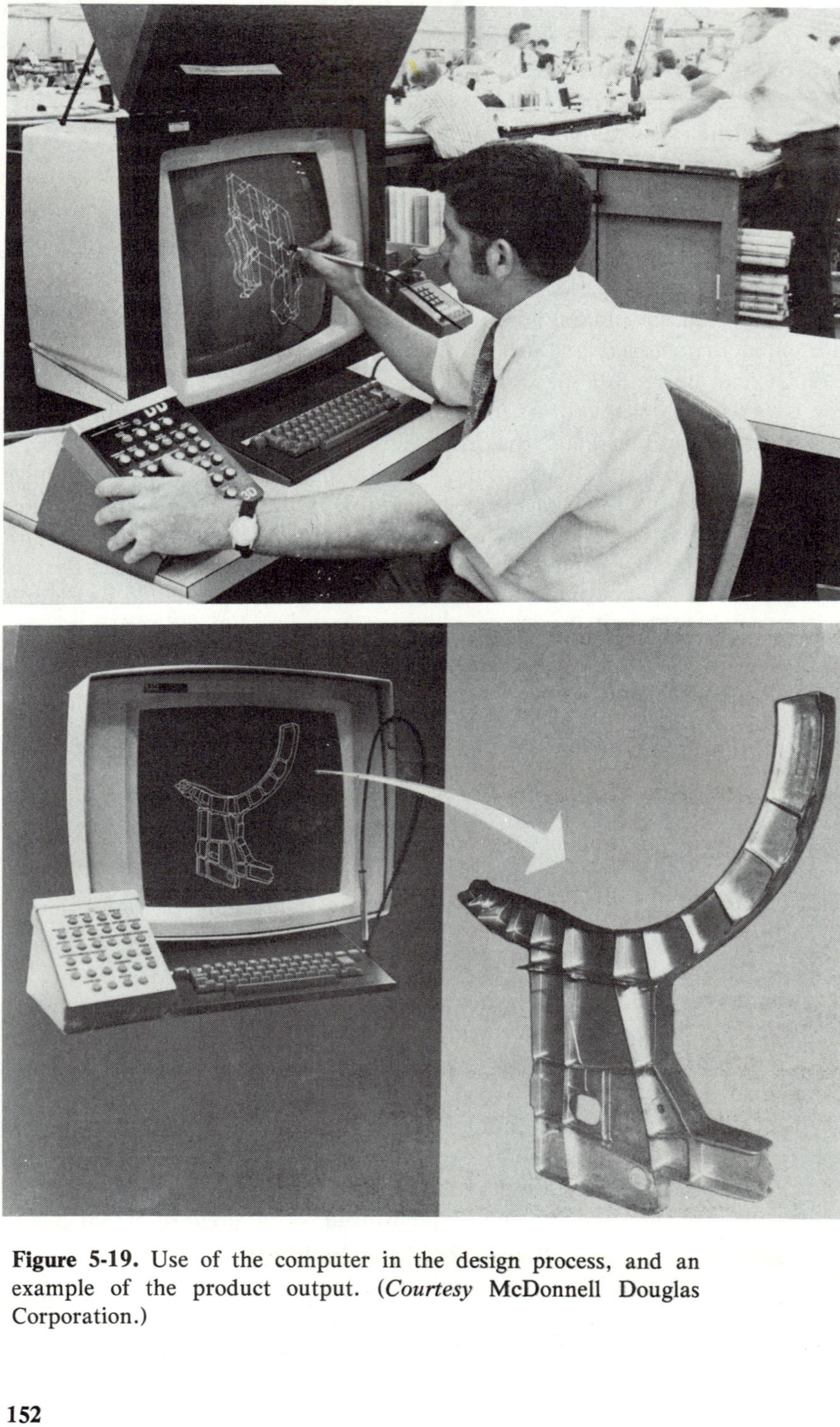

Figure 5-19. Use of the computer in the design process, and an example of the product output. (*Courtesy* McDonnell Douglas Corporation.)

Figure 5-20. Computer-aided design applications. *Top:* Computer-aided large scale integration (LSI) design. The handheld microelectric circuit interconnection mask was generated from logic schematics through computer software for the simulation, routing and test generation, as indicated by the program listing, artwork, and tapes shown. *Bottom:* Fault isolation techniques development. Diagnostic test information is stored on a microfilm cassette and displayed on-line to aid the test and repair of complex large scale integration (LSI) microcircuits. (*Courtesy* Hughes Aircraft Company.)

(2) It provides the reliability/maintainability/human factors engineer with the opportunity to accomplish a more effective review of a proposed design configuration for the incorporation of supportability characteristics. Problem areas readily become evident.

(3) It provides the maintainability/human factors engineer with a tool for use in the accomplishment of predictions and detailed task analyses. It is often possible to simulate operator and maintenance tasks to acquire task sequence and time data.

(4) It provides the design engineer with an excellent tool for conveying his final design approach during a formal design review.

(5) It serves as an excellent marketing tool.

(6) It can be employed to facilitate the training of system operator and maintenance personnel.

(7) It is utilized by production and industrial engineering personnel in developing fabrication and assembly processes and procedures and in the design of factory tooling and associated test fixtures.

(8) At a later stage in the system life cycle, it may serve as a tool for the verification of a modification kit design prior to the preparation of formal data and the development of kit hardware.

In general, the mock-up is extremely beneficial. It has been used effectively in facility design, aircraft design, and in the design of smaller products. Examples of several mock-ups that have actually been employed as design aids are presented in Figures 5-21 and 5-22.

B. Prototype Development

Given that detailed design has progressed to the point where definition can be attained through formalized documentation and the use of design aids, the next logical step constitutes the development of a preproduction prototype model of the system or various significant portions of the system. A prototype represents the production configuration in all aspects of form, fit, and function except that it has not been fully qualified in terms of operational and environmental testing. A prototype may evolve through a series of working hardware configurations such as an engineering model and a service test model, or may evolve directly from a mock-up.

A prototype of the prime equipment, software, and associated elements of logistic support is produced for the purpose of final system/equipment test and evaluation prior to entering into the production and/or construction phase. The intent is to verify design adequacy to the maximum extent considered appropriate at this stage in the life cycle.

C. Engineering Testing Requirements[15]

Thus far, text material has primarily included the specification of requirements, allocation and the establishment of design criteria, design and support activities, analyses and predictions, documentation, and design review. Through the accomplishment of these functions, the designer has been able to establish specific system objectives and later evaluate a given design configuration relative to compliance with these objectives. The evaluation function up to this point has been analytical in nature, providing a certain level of confidence that all qualitative and quantitative requirements have been met. Although the analytical approach fulfills a need, it has its limitations because it does not reflect actual experience with the applicable system hardware and its components. A more realistic evaluation of whether a system or equipment complies with the initially specified requirements is accomplished through the test and demonstration of the prime equipment hardware, software, and its associated logistic support.

1. Categories of test and demonstration. A true test (that which is relevant from the standpoint of assessing total system performance and effectiveness) constitutes the evaluation of a system deployed in an operational

[15]Engineering testing at this stage in the life cycle constitutes an initial part of the overall system-evaluation effort described further in Chapter 7. The categories of test defined here form the basis for subsequent discussion in this area.

Figure 5-21. Full scale aircraft mockup showing structure and equipment installation (*top*) and aircraft crew capsule mating with fuselage. (*bottom*). (*Courtesy* Rockwell International Corporation.)

Figure 5-22. Engineering model of geographical area. (*Courtesy* Martin Marietta Corporation.)

environment and subjected to actual use conditions. For example, an aircraft or a power generating plant should be tested while it is performing its intended mission in actual operational use. User personnel should accomplish operator and maintenance functions with the designated field test and support equipment, technical manual procedures, and so on. In such a situation, actual operational and maintenance experience in a realistic environment can be recorded and subsequently evaluated to reflect a true representation of the system design. A demonstration of this type can best be accomplished by the user during standard operations supported through the employment of normal resources (i.e., product use and logistic support function of the life cycle).

Although idealistically it is desirable to wait until the system is fully operational before accomplishing an evaluation of system performance, effectiveness, and logistic supportability, it is not practical from the standpoint of allowing for possible corrective action. In the event that the evaluation indicates noncompliance (i.e., the system as presently designed will not meet the operational requirements and fulfill its mission), corrective action should be initiated as early as possible in the system life cycle. Accomplishing corrective action after equipment is produced and fully deployed in the field can result in extensive modification programs, which are costly. Thus, it is feasible to establish an overall test program which allows for the evaluation of hardware and its support elements on an evolutionary basis. As is the case for the earlier design evaluation using analytical techniques, an evaluation of hardware is accomplished, commencing with the development of the first engineering model and extending through the test of equipment deployed in the field. This evaluation process includes various types of demonstrations. For the purpose of discussion, tests and demonstrations are classified as follows:

a. *Type 1 testing.* During the early phases of detailed design, breadboards, bench test models, engineering models, and service test models are built with the intent of verifying certain performance and physical design characteristics. These models, representing either an entire system or a designated system component, usually operate functionally (electrically and mechanically) but do not by any means represent production equipment.[16] Tests may involve equipment operational and logistic support actions which are directly comparable to tasks performed in a true operational situation (e.g., measuring a performance parameter, accomplishing a remove-replace action, accomplishing a servicing requirement). Although these tests are not formal demonstrations in an operational environment, information pertinent to actual system characteristics can be derived and used as an input in the overall system evaluation and assessment. Such testing is usually performed in the producer/supplier's facility by engineering technicians using "jury-rigged" test equipment and engineering notes for test procedures. It is during this initial phase of testing that changes to the design can be incorporated on a minimum cost basis.

b. *Type 2 testing.* Formal tests and demonstrations are accomplished during the latter part of detailed design when preproduction prototype equipment is available. Prototype equipment is similar to production equipment (that which will be deployed for operational use) but it is not necessarily fully qualified at this point in time.[17] A test program may constitute a series of individual tests, tailored to the need, to include:

(1) *Performance tests*—tests are accomplished to verify individual system performance characteristics. For instance, tests are designed to determine whether the electric motor will provide the necessary output, whether the

---

[16]Components of design are frequently tested in lieu of testing the entire system. For instance, it may not be feasible to test an entire petroleum distribution system whereas testing a segment of pipeline may be sufficient to verify design adequacy. The objective at this point is to verify design adequacy to the maximum extent practicable in view of the information obtainable via economical means.

[17]Qualified equipment refers to a production configuration that has been verified through the successful completion ot performance tests, environmental qualification tests (e.g., temperature cycling, shock, vibration), reliability qualification, maintainability demonstration, and compatibility tests, Type 2 testing primarily refers to that activity associated with the qualification of a system for operational use.

pipeline will withstand certain fluid pressures, whether the airplane will perform its intended mission successfully, whether the process will provide *x* widgets per given period of time, and so on. Also, it is necessary to verify form, fit, interchangeability, product safety, and other comparable features.

(2) *Environmental qualification*—temperature cycling, shock and vibration, humidity, wind, salt spray, dust and sand, fungus, acoustic noise, pollution emission, explosion proofing, and electromagnetic interference tests are conducted. These factors are oriented to what the·equipment will be subjected to during operation, maintenance, and during transportation and handling functions. In addition, the effects of the item being tested on the overall environment will be noted.

(3) *Structured tests*—tests are conducted to determine material characteristics relative to stress, strain, fatigue, bending, torsion, and general decomposition.

(4) *Reliability qualification*—tests are accomplished on one or more equipment items to determine the mean time between failure (MTBF) and the mean time between maintenance (MTBM). Also, special tests are often designed to measure component life, to evaluate degradation, and to determine modes of failure.

(5) *Maintainability demonstration*—tests are conducted on one or more equipment items to determine the values for mean active maintenance time ($\bar{M}$), mean corrective maintenance time ($\bar{M}ct$), mean preventive maintenance time ($\bar{M}pt$), maintenance man-hours per operating hour (MMH/OH), and so on. In addition, maintenance tasks, task times and sequences, prime equipment-test equipment interfaces, maintenance personnel quantities and skills, maintenance procedures, and maintenance facilities are verified to varying degrees. The elements of logistic support are initially evaluated on an individual basis.

(6) *Support equipment compatibility tests*—tests are often accomplished to verify the compatibility among the prime equipment, test and support equipment, and transportation and handling equipment.

(7) *Personnel test and evaluation*—tests are often accomplished to verify the relationships between man and equipment, the personnel skill levels required, and training needs. Both operator and maintenance tasks are evaluated.

(8) *Technical data verification*—the verification of operational and maintenance procedures is accomplished.

The ideal situation is to plan and schedule these individual tests such that they can be accomplished on an integrated basis as one overall test. Data output from one test may be beneficial as an input to another test. The intent is to provide the proper emphasis, consistent with the need, and eliminate redundancy and excessive cost. Proper test planning is essential.

Another aspect of testing in this category includes production sampling tests when multiple quantities of an equipment item are produced. The tests defined above basically qualify the item; that is, the equipment hardware configuration meets the requirements for production and operational use. However, once a system is initially qualified, some assurance must be provided that all subsequent replicas of that item are equally qualified; thus, in a multiple-quantity situation, samples are selected from the production line and tested.[18]

Production sampling tests may cover certain critical performance characteristics, reliability, or any other designated parameter that may significantly vary from one serial-numbered item to the next, or may vary as a result of a production process. Samples may be selected on the basis of a percentage of the total equipment produced or may tie in with *x* number of pieces of equipment in a given calendar time period. This depends on the peculiarities of the system and the complexities of the production process. From production sampling tests, one can measure system growth (or degradation) throughout the production/construction phase.

Type 2 tests are generally performed in the producer/supplier's facility by personnel at that facility. Test and support equipment, designated for operational use, and preliminary technical manual procedures are employed where possible. User personnel often observe and/or participate in the testing activities. Equipment changes as a result of corrective action are handled through a formalized engineering change procedure. Several examples of type 2 testing are shown in Figure 5-23.

c. *Type 3 testing.* Formal tests and demonstrations, started after initial system qualification and prior to the completion of production, are accomplished at a designated field test site by user personnel.[19] Operational test and support equipment, operational spares, and formal operator and maintenance procedures are used. Testing is generally continuous, accomplished over an extended period of time, and covers the evaluation of a number of system elements scheduled through a series of simulated operational exercises.

This is the first time that all elements of the system (i.e., prime equipment and the elements of logistic support) are operated and evaluated on an integrated basis. The compatibility of the prime equipment with logistic support is verified as well as the compatibility of the various elements of logistic support with each other. Turnaround times and logistics pipeline times, stock levels, personnel effectiveness factors, and other related operational and logistic parameters are measured. In essence, system performance (based on certain use conditions) and operational readiness characteristics

---

[18]These tests are in addition to the normal performance tests that are accomplished on every piece of equipment after fabrication and assembly prior to delivery to the consumer.

[19]The test site may constitute a ship at sea, an aircraft or space vehicle in flight, a facility in the arctic or located in the middle of the desert, or a mobile land vehicle traveling between two points.

Figure 5-23. Examples of two Type 2 testing configurations. F-15 fighter aircraft during a wind tunnel test (*top*) and full scale Apollo docking simulation tests. (*bottom*) (*Courtesy* McDonnell Douglas Corporation.)

(i.e., operational availability, dependability, system effectiveness, etc.) can be determined to a certain extent.[20]

d. *Type 4 testing.* During the product-use phase, formal tests are sometimes conducted to gain further insight in a specific area. It may be desirable to vary the mission profile or the system utilization rate to determine the impact on total system effectiveness, or it might be feasible to evaluate several alternative support policies to see whether system operational availability can be improved. Even though the system is designed and operational in the field, this is actually the first time that we really know its true capability. Hopefully, the system will accomplish its objective in an efficient manner; however, there is still the possibility that improvements can be realized by varying basic operational and maintenance support policies.

Type 4 testing is accomplished at one or more operational sites (in a realistic environment) by user operator and maintenance personnel, and supported through the normal logistics capability. The elements of logistic support as designated through predictions and analyses, earlier testing, and so on, are evaluated in the context of the total system.

Although there are variations from program to program, the four basic types of testing are oriented to the program phases as illustrated in Figure 5-24.

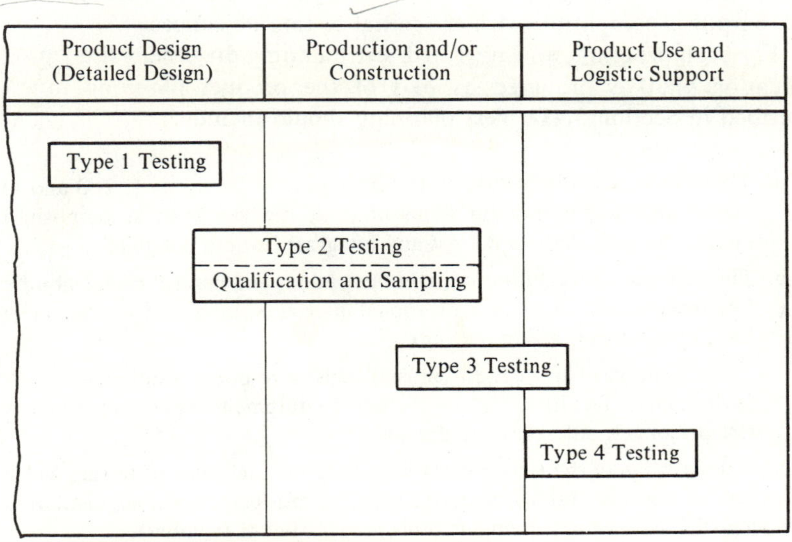

Figure 5-24. Types of tests by program phase.

[20]Type 3 testing does not represent a complete operational situation; however, tests can be designed to provide a close approximation.

2. *Test planning.* Test planning actually commences early in the product planning phase when basic system operational requirements are initially established. If a system requirement is to be specified, there must be a way to evaluate the system later to ensure that the requirement has been met; hence, testing considerations are intuitive at an early point in time (refer to Figure 3-9).

Throughout the various stages of system development, a number of individual equipment tests may be specified. Often there is a tendency to design a test to measure one system characteristic, design another test to measure a different parameter, and so on. Before too long, the amount of testing specified may be overwhelming and prove to be quite costly. Test requirements must be considered on an integrated basis. Where possible, individual tests are reviewed in terms of resource requirements and output results and are scheduled in such a manner as to gain the maximum benefit possible. For instance, maintainability data can be obtained from reliability tests that result in a possible reduction in the amount of maintainability testing required. Support equipment compatibility data and personnel data can be obtained from both reliability and maintainability testing; thus, it might be feasible to schedule reliability qualification testing first and maintainability demonstration second. In some instances, the combining of tests may be feasible as long as the proper characteristics are measured and the data output is compatible with the initial testing objectives.

For each program, an integrated test planning document (i.e., product evaluation plan) is prepared as part of the product planning functions described in Section 3.4D. Test planning should include:

a. The definition and schedule of all test requirements (types 1, 2, 3 and 4). Anticipated test output (in terms of what the test is to accomplish) is defined for each individual test and integrated where possible.
b. The definition of organization, administration, and control responsibilities (organization functions, organizational interfaces, monitoring of test activities, cost control, and reporting).
c. The definition of test conditions and logistic resource requirements (test environment, facilities, test and support equipment, spare/repair parts, test personnel, and test procedures).
d. A description of the test preparation phase for each type of testing (selection of specific test method, training of test personnel, acquisition of logistic resource requirements, and preparation of facilities).
e. A description of the formal test phase (test procedures and test data collection, reduction, and analysis methods).
f. A description of conditions and provisions for a retest phase (methods for conducting additional testing as required due to a reject situation).
g. The identification of test documentation (test reporting requirements).

The basic test plan serves as a valuable reference throughout system design, development, and production. It indicates what is to be accomplished, the requirements for testing, a schedule for the processing of equipment and material for test support, data collection and reporting methods, and so on. It is an integrating device for a number of individual tests in each of the four categories, and a change in any single test requirement will likely affect the total plan.

3. ***Prototype test and evaluation.*** Engineering testing at this stage in the life cycle entails type 1 and a designated amount of type 2 testing and generally includes completing performance tests, initial reliability and maintainability tests, environmental qualification, and prime equipment-logistic support compatibility testing, using the preproduction prototype of the system. Testing is conducted in accordance with the product evaluation plan, and the results are used to determine whether the design configuration meets system/equipment requirements. This does not constitute a complete system evaluation but will hopefully provide an assessment that will enable engineering and management to decide whether or not the system is technically qualified to enter the production/construction phase. Product evaluation is discussed further in Chapter 7.

4. ***System design modifications.*** If the results of engineering testing indicate that system/equipment requirements will not be met, specific problem areas are identified and all approaches for corrective action are evaluated. Each possible solution is reviewed in terms of both technical and cost impact on the system as a whole and on each element of the system (i.e., prime mission equipment, software, and logistic support). The proposed solution, based on the most cost-effective approach, is recommended in a formal engineering change proposal (ECP) and submitted to a review board for final approval. Approved changes are incorporated in engineering design documentation and prototype system hardware (refer to the feedback loops in Figures 5-11 and 5-17).

5.4 FORMAL DESIGN REVIEW

Design is a progression from an abstract notion to something that has form and function, is fixed, and can be reproduced in designated quantities to satisfy a need. The designer produces a model that is used as a template for the replication of additional models (i.e., a qualified prototype). In the course of production, an error made in any one model will result in a single rejection. However, an error in the design, repeated in all subsequent models, may lead to a serious problem (e.g. complete recycling of all equipment for a major modification). Thus, the designer's responsibility is significantly large.

Initially, a requirement or need is specified. From this point, design evolves through a series of stages (i.e., conceptual design, preliminary system design, detailed design and development). In each major stage of the design process, an evaluative function is accomplished to ensure that the design is correct at that point prior to proceeding with the next stage. The evaluative function includes both the informal day-to-day project coordination and data review and the formal design review. Design information is released and reviewed for compliance with the basis system/equipment requirements (i.e., performance, reliability, maintainability, human factors). If the requirements are satisfied, the design is approved as is. If not, recommendations for corrective action are prepared and submitted to the designer for action. If no action is taken as a result of the day-to-day liaison activity, the recommendations are presented and discussed as part of the next formal design review. The informal daily liaison activity is covered earlier. This section is directed toward the formal design review.

The formal design review constitutes a coordinated activity (including a meeting or series of meetings) directed to satisfy the interests of the design engineer, the technical discipline support areas (reliability, maintainability, human factors), logistics, manufacturing, industrial engineering, quality assurance, program management, and so on. The purpose of the design review is to formally and logically cover the proposed design from the total system standpoint in the most effective and economical manner through a combined integrated review effort. The formal design review serves a number of purposes.[21]

A. It provides a formalized check (audit) of the proposed system/equipment design with respect to contractual and specification requirements. Major problem areas are discussed and corrective action is taken.

B. It provides a common baseline for all project personnel. The design engineer is provided the opportunity to explain and justify his design approach, and representatives from the various supporting organizations (e.g., maintainability, logistic support) are provided the opportunity to hear the design engineer's problems. This serves as a tremendous communication medium and creates a better understanding among design and support personnel.

[21]The purposes of a formal design review are also covered in: B. Blanchard and E. Lowery, *Maintainability—Principles and Practices,* McGraw-Hill Book Company, New York, 1969; and B. Blanchard, *Logistic Engineering and Management,* Prentice-Hall, Inc., Englewood Cliffs, N.J., 1974.

C. It provides a means for solving interface problems and promotes the assurance that all system elements will be compatible. For instance, major interface problems between engineering and manufacturing, relative to lack of producibility, are often not detected until after design data are released and production is underway. The results of major problems discovered at that time are quite costly. Another major problem (which seems to be a common occurrence) is the lack of compatibility between the different elements of logistic support with the prime equipment and the elements of logistic support with each other. Such problems are often undetected at an early point in time because of a wide variance of organizational interests and activity, the rush to get the hardware into production, the physical separation of members of the design team, or related reason. A formal design review is intended to prevent the occurrence of these problems.

D. It provides a formalized record of what design decisions were made and the reasons for making them. Analyses, predictions, and trade-off study reports are noted and are available to support design decisions. Compromises to performance, reliability, maintainability, human factors, cost, and logistic support are documented and included in the trade-off study reports.

E. It promotes a greater probability of mature design, as well as the incorporation of the latest techniques (where appropriate). Group review may identify new ideas, possibly resulting in simplified processes and ultimate cost savings.

In summary, the formal design review, when appropriately scheduled and conducted in an effective manner, causes a reduction in the producer's risk relative to meeting contract and specification requirements and often results in improvement of the producer's methods of operation. Also, the consumer often benefits through receipt of a better product.

A. Scheduling of Design Reviews

very formalized

Design reviews are generally scheduled prior to each major evolutionary step in the design process. In some instances, this may entail a single review toward the end of each stage (i.e., conceptual, preliminary system design, detailed design and development). For other projects, where a large system is involved and the amount of new design is extensive, a series of formal reviews may be conducted on designated elements of the system. This may be desirable to allow for the early processing of some items while concentrating on the more complex high-risk items.[22]

---

[22] Items that are procured off the shelf or items for which the design is basic and proved may be processed expeditiously to suit both engineering and production schedules. Complex newly designed items (those pushing the state of the art) will require more in-depth review, followed by modifications and possibly a second review before being released for production.

Although the quantity and type of design reviews scheduled may vary from program to program, four basic types are readily identifiable and common to most programs. They include the conceptual design review (i.e., system requirements review), the system design review, the equipment design review, and the critical design review. The time phasing of these reviews is illustrated in Figure 5-25. In the figure, the conceptual and preliminary system design stages may be considered advanced development, and the detailed design may be considered full-scale development.

1. Conceptual design review. The conceptual design review may be scheduled during the early part of a program (preferably not more than 4 to 8 weeks after program start) when operational requirements and the maintenance concept have been defined. Feasibility studies justifying preliminary design concepts should be reviewed. Logistic support requirements at this point are generally included in the maintenance-concept definition.

2. System design review. System design reviews are generally scheduled during the preliminary system design stage when preliminary system layouts and specifications have been prepared (prior to their formal release). These reviews are oriented to the overall system configuration in lieu of individual equipment items. Supporting data may include functional analyses and allocations, preliminary logistic support analyses, and trade-off study reports. There may be one or more formal reviews scheduled, depending on the size of the system and the extent and complexity of the new design. The purpose of the review is to determine whether the design is compatible with all system requirements and whether the documentation supports the design.

3. Equipment design review. Equipment reviews are scheduled during the detailed design phase when layouts, preliminary mechanical and electrical drawings, functional and logic diagrams, computer data, and component-parts lists are available. In addition, these reviews cover engineering breadboards (hardware), models or mock-ups, and prototypes. Supporting the design are detailed trade-off study reports, reliability analyses and predictions, maintainability analyses and predictions, human factors analyses, and logistic support analyses. The design process at this point has identified specific design constraints, additional or new requirements, and major problem areas. Such reviews are conducted prior to preceeding with finalization of the detailed design.

4. Critical design review. The critical design review is scheduled after detailed design has been completed but prior to the release of firm design data to production. Such a review is conducted to verify the adequacy and producibility of the design. Design is essentially frozen at this point, and manufacturing methods, schedules, and costs are reevaluated for final approval.

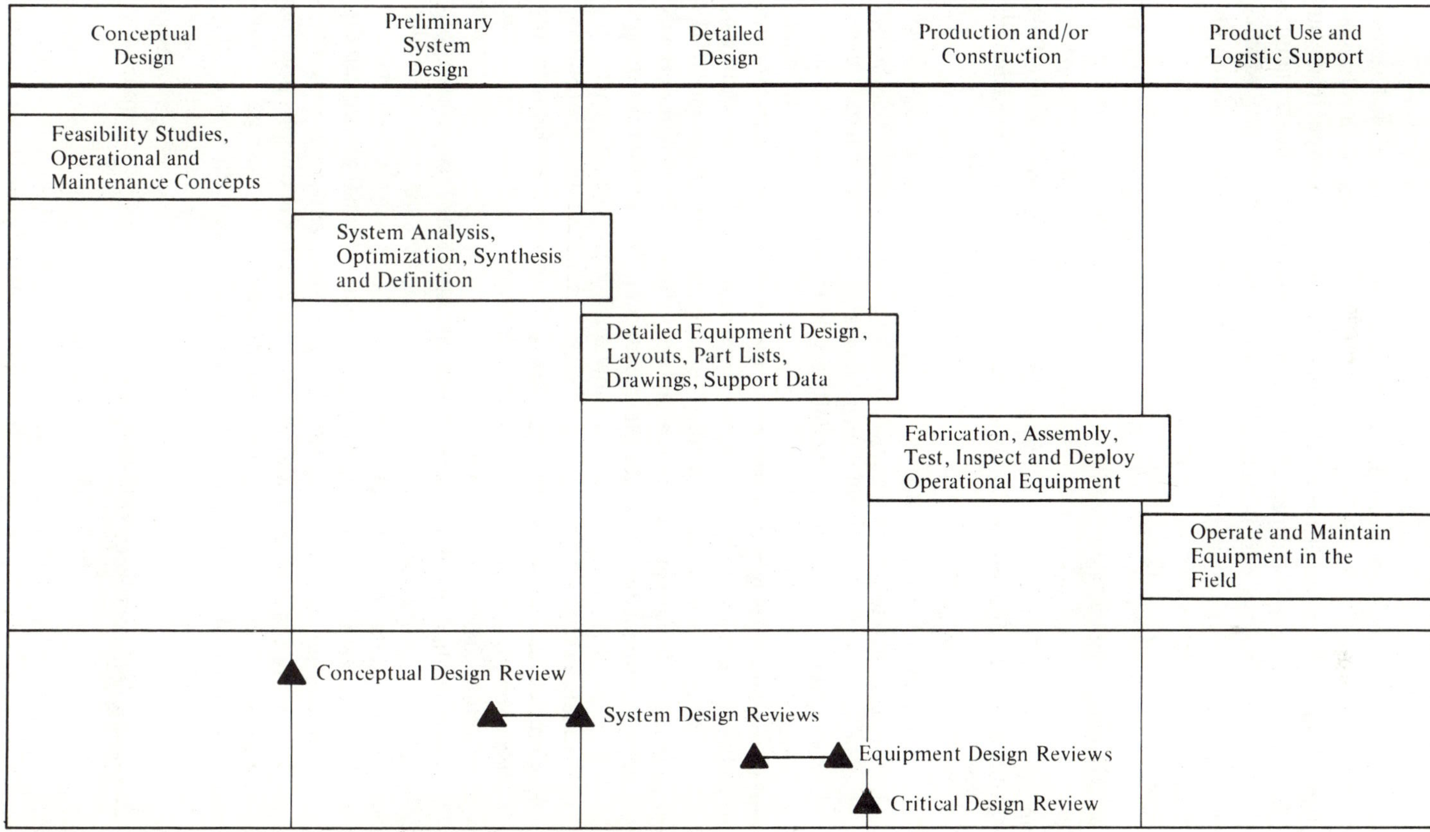

Figure 5-25. Design review schedule in relationship to program phases.

The critical design review covers all design effort accomplished subsequent to the completion of the equipment review. This includes changes resulting from recommendations for corrective action stemming from the equipment design review and prototype testing. Data requirements include manufacturing drawings and material lists, final reliability and maintainability predictions, engineering test reports, a firm logistic support analysis, a production/construction plan, a product evaluation plan, and a product use and logistic support plan.

B. Design Review Requirements

The success of a formal design review is dependent on the depth of planning, organization, and data preparation prior to the review itself. A tremendous amount of coordination is required relative to the establishment of:

1. The item(s) to be reviewed.
2. A selected date for the review.
3. The location or facility where the review is to be conducted.
4. An agenda for the review (including a definition of the basic objectives).
5. A design review board representing the organizational elements and disciplines affected by the review. Basic design functions, reliability, maintainability, human factors, quality control, manufacturing, and logistic support representation are included. Individual organization responsibilities should be identified. Depending on the type of review, the consumer and/or individual equipment suppliers may be included.
6. Equipment (hardware) requirements for the review. Engineering breadboards, prototypes, and/or mock-ups may be required to facilitate the review process.
7. Design data requirements for the review. This may include all applicable specifications, lists, drawings, predictions and analyses, logistics data, and special reports.
8. Funding requirements. Planning is necessary in identifying sources and a means for providing the funds for conducting the review.
9. Reporting requirements and the mechanism for accomplishing the necessary follow-up action(s) stemming from design review recommendations. Responsibilities and action item time limits must be established.

C. Conducting the Design Review

As indicated earlier, the design review involves a number of different discipline areas and covers a wide variety of design data and in some instances hardware. In order to fulfill its objective expeditiously (i.e., review the design to ensure that all system requirements are met in an optimum manner), the design review must be well organized and firmly controlled by the design review board chairman. Design review meetings should be brief and to the point and must not be allowed to drift away from the topics on the agenda. Attendance should be limited to those who have a direct interest and can contribute to the subject matter being presented. Specialists who participate should be authorized to speak and make decisions concerning their area of specialty. Finally, the design review must make provisions for the identification, recording, scheduling, and monitoring of corrective actions. Specific responsibility for follow-up action must be designated by the design review board chairman.

QUESTIONS AND PROBLEMS

1. Define systems engineering.
2. What are the basic steps in product design? Construct an illustrated flow diagram showing these steps.
3. What are the basic differences among conceptual design, preliminary system design, and detailed design? Are these stages of design applicable to all systems?
4. What are the desired results from conceptual design?
5. A functional analysis serves what purpose?
6. Is functional analysis applicable to any system or equipment item? Why?
7. Select a system (or an element of a system) of your choice and develop operational and maintenance functional flow diagrams to the second level.
8. What is the purpose of allocation? How does it affect system design?
9. Select a system (or an element of a system) of your choice and assign top-level requirements. Allocate performance factors, effectiveness factors, reliability and maintainability factors, and cost factors to the second level.
10. What is meant by "design criteria"? How are criteria developed? How are criteria applied to the design process?
11. What are the steps involved in the evaluation of design alternatives? Draw a flow diagram that illustrates these steps and their relationship(s). How would you select a design approach?
12. What is the desired output of preliminary system design?
13. What is logistic support analysis (LSA) and what purpose(s) does it serve?
14. What is meant by life-cycle cost? What aspects of cost are considered? How is life-cycle cost used in the design process?
15. Describe in brief terms how each of the following affects system design: Reliability? Maintainability? Human factors? How are these terms expressed quantitatively?
16. How are supplier requirements identified and specified?

17. Select a system or equipment item of your choice and develop a design review checklist that you can use for evaluation purposes.
18. Define producibility. Why is it important?
19. Define cost effectiveness and system effectiveness. How would you specify and measure each?
20. Design is a team effort. True or false? Why?
21. Describe the checks and balances of the design process as you see them.
22. Why is engineering documentation necessary?
23. Why are drawing standards important?
24. How are design changes initiated and evaluated?
25. Why are design change control and configuration management important?
26. Define computer-aided design. Provide some examples. List some advantages to this approach.
27. What are some of the benefits of physical models or mock-ups?
28. How are system test requirements determined?
29. When is test planning accomplished? What basic factors are considered in test planning?
30. Define the basic categories of testing and their respective applications.
31. What benefits are provided by a formal design review?
32. What are the input requirements for the conceptual design review? System design review? Critical design review?
33. What determines the success of a design review?
34. Define: operational availability, system effectiveness, MTBM, MMH/OH, maintainability, maintenance, MTBF, and MDT. (Refer to Appendix A.)
35. What are the differences among A_o, A_a, and A_i? (Refer to Appendix A.)
36. What is the difference between MTBM and MTBF? (Refer to Appendix A.)

6

PRODUCTION AND/OR CONSTRUCTION FUNCTION

Chapter 5 dealt with system/equipment design and development and an output resulting in a *fixed* system configuration ready for production and/or construction.[1] Production may include the replication of a multiple quantity of identical items or a mixed variety of different item types, whereas construction might involve the fabrication and assembly of a single, one-of-a-kind entity.

When dealing with either production or construction, one must consider the total flow of materials, from the acquisition of raw materials to delivery of the finished product for evaluation and ultimate operational use. The finished product may constitute the prime mission equipment (e.g., a given quantity of airplanes or electrical appliances, a hydroelectric dam, a ground transportation vehicle, a mixed quantity of x and y widgets, and/or elements of logistic support for operational use.[2] The production/construction flow

[1] This does not mean that the proposed design configuration is optimal. However, the design has been validated to the extent necessary to provide a high degree of assurance that the resultant product will fulfill all mission and operational requirements. Design changes from this point on will be limited primarily to the correction of problems related to producibility (or lack thereof).

[2] Logistic support in this instance refers to test and support equipment, transportation and handling equipment, supply support (spares and repair parts, inventories, etc.), personnel and training, facilities, and data intended to support the system while in operational use and throughout its life cycle.

process, regardless of the product type, involves product inventories, material acquisition and control, tooling and test equipment, transportation and handling devices, facilities (manufacturing facilities and warehouses), data, software, and personnel.[3] The extent of this process varies with the requirements from system to system and can become quite complex. In some respects, one might address the production/construction process as a separate system in itself since planning, design and development, test, operations, and logistic support are all key factors in attaining successful results.

The intent of this chapter is to briefly discuss production/construction requirements (established in the plan identified in Figure 3-9 and described in Section 3.4C), design of required capability, production operations, quality control, configuration management and change control, and the transition phase from production to consumer operational use.

6.1 PRODUCTION/CONSTRUCTION REQUIREMENTS

Production requirements stem from the needs analysis and system/equipment operational data (refer to Section 3.3). This includes information pertaining to the type and quantity of products or equipments required, desired product quality, deployment or location where equipment and supporting items are to be delivered, time of delivery, and the general environmental conditions en route.

If an item is to be produced in multiple quantities, the production rate should tie-in with the front end of an inventory profile developed as part of the initially specified system operational requirements, and quantities may be established on a month-to-month basis until such time that the accumulated quantity is equivalent to the demand.[4] Production may be continuous or discontinuous, as illustrated in Figure 6-1.

The production of multiple quantities of an item assumes a flow-shop pattern, where the output variety is limited and each kind of output follows the same basic path and sequence of processing steps. Plant layout, inventory

---

[3]Throughout this text, logistics is addressed in the life-cycle context, beginning with the definition of the system maintenance concept (and the establishment of design requirements) and continuing through system operational use and phase-out. This includes, as an element, the aspects of logistics oriented to the production/construction process often referred to as "industrial" or "business logistics." Logistics from the life-cycle viewpoint is covered in: B. Blanchard, *Logistics Engineering and Management*, Prentice-Hall, Inc., Englewood Cliffs, N.J., 1974. Several good references covering industrial or business logistics are: D. J. Bowersox, *Logistical Management*, Macmillan Publishing Co., Inc., New York, 1974; and J. F. Magee, *Industrial Logistics*, McGraw-Hill Book Company, New York, 1968.

[4]The demand must consider the prime equipment actually employed in the accomplishment of the designated mission plus any spare equipment needed for backup (to cover maintenance and attrition).

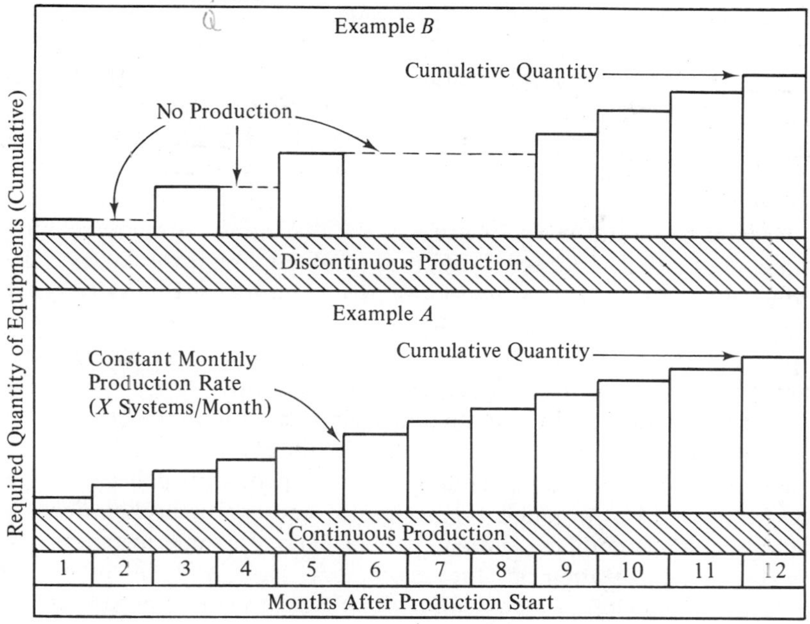

Figure 6-1. Continuous and discontinuous production profiles.

stock points, test and inspection stations, and personnel functions are designed to handle a large volume (i.e., mass production). When production is continuous and at an approximate steady rate (Figure 6-1, example A), the associated logistics resources required are somewhat constant, and production costs are relatively easy to determine. On the other hand, when production is discontinuous (to include production-line start-ups and shutdowns; Figure 6-1, example B), the logistic support resources required are variable and are somewhat greater. Discontinuities result in personnel retraining and problems related to workmanship, quality level, and production realization rates. The probability of items being rejected through test and inspection is high, and the applicable items must be recycled through certain steps in the production process, resulting in added resource requirements and possibly a shift in production schedules. These factors, combined with associated changes in material procurement practices, generally result in higher overall production costs.

Given a large-quantity production requirement, a number of approaches may be considered which will result in the same final output. It may be feasible to establish a faster buildup rate at the beginning of the production cycle or to vary the production rate at other times. In addition, there is the consideration of a *single* versus *multiple* production-line capability, or different combinations of capability in the detailed functions of production (e.g., subassembly, final assembly, inspection and test). Each variation will impact the logistic support resource requirements in the production process to a different extent and must be evaluated in terms of effectiveness and cost.

In contrast to the multiple-quantity situation discussed above, a requirement may constitute the production of a mix of items or the construction of a single item such as a large commercial airline facility. This assumes a job-shop pattern, in which the output varies from item to item, and the activity involves a mix of jobs following different paths through a program network. In the pure sense, no two jobs are exactly alike, and each job requires a unique setup. If a number of different products are involved (i.e., large product mix), the logistic support resources required are usually general-purpose and can be adjusted rather easily to accommodate the variety of output. For the single construction project, the logistics resources will, of course, be peculiar to that particular product output.

In determining specific production/construction requirements, released engineering documentation is reviewed from the standpoint of product explosion and standardization in addition to volume, quality, delivery rate, and other factors.[5] Most products are assembled and can be broken down into assemblies, subassemblies, and component parts. In some instances, there may be standardization at the equipment level and in other cases standardization exists at an assembly or subassembly level, where different product types are required and utilize the same standardized items at a lower indenture level. Thus, in addition to the pure flow-shop and job-shop patterns described above, there may be a mix of each, where some smaller items are mass-produced while larger items of varied configuration are produced on an individual basis. In any event, the ultimate objective in production/construction planning and design is to produce the items necessary to satisfy the need at the lowest total cost and within the constraints of the established system design configuration.

6.2 INDUSTRIAL ENGINEERING AND OPERATIONS ANALYSIS

Industrial engineering in this instance refers to a group of activities responsible for the design and development of a production capability.[6] A production capability is a complex collection of elements (facilities, labor, equipment, and materials) combined and operated to provide a product or service. That is, the production capability is the process of transition from a fixed design configuration (defined by approved design drawings, layouts, material lists, procedures, computer software programs, etc.) to a finished deliverable product. The product may include prime equipment, test and

[5] Production/construction requirements are discussed further in: H. T. Amrine, J. A. Ritchey, and O. S. Hulley, *Manufacturing Organization and Management*, 3rd ed., Prentice-Hall, Inc., Englewood Cliffs, N.J., 1975.

[6] This definition is employed to establish a frame of reference for further discussion. It is not intended that the definition at this point be aligned to any particular organization.

support equipment, training equipment, spare/repair parts, or any other hardware or software item that is part of the system. Industrial engineering activities are concerned with the design and development of this capability and may be subdivided into the following categories.

A. Plant Engineering

Based on the requirements for production, it is necessary to:

1. Determine the capacity and location of both production and storage facilities. Production includes fabrication, assembly, and test operations.
2. Determine utility requirements (e.g., power, telephone, environmental controls).
3. Determine capital equipment needs (e.g., machines, heavy equipment).
4. Accomplish a plant layout of utilities and capital equipment.
5. Design and incorporate material-handling provisions.
6. Establish a maintenance capability for the corrective and preventive maintenance of facilities, utilities, and equipment.

Plant engineering involves the general design, development, construction, operation, and maintenance of production facilities.[7]

B. Manufacturing Engineering

As discussed in earlier chapters, the consumer (or customer) specifies the need for a system. This need is defined in terms of functions, and ultimately a design specification is prepared that forms the basis for equipment design and development. The output, constituting a combination of prototype hardware and various categories of design data, must be converted into process or material specifications. Manufacturing engineering defines the production process through these specifications (refer to Section 3.4B). Specific manufacturing engineering functions are to:

1. Assess the various elements of the system and assist in determining make-or-buy decisions (i.e., whether an item is to be produced within or purchased off the shelf on the outside).[8] This function

[7]Three references in plant engineering include: R. Francis and J. White, *Facility Layout and Location,* Prentice-Hall, Inc., Englewood Cliffs, N.J., 1974; R. Reed, *Plant Location, Layout, and Maintenance,* Richard D. Irwin, Inc., Homewood, Ill.; and J. M. Moore, *Plant Layout and Design,* Macmillan Publishing Co., Inc., New York, 1962.

[8]Generally, a make-or-buy committee is established with representation from design engineering, procurement, manufacturing engineering, logistics, and producer/consumer management. Decisions are made concerning the source of supply and are based on item availability, expected production horizon period, schedule, cost, and related political implications.

must be accomplished early enough in the system design and development cycle to allow for proper production planning.

2. Select the material to be used in the fabrication of each "make" item.
3. Select the basic process for item fabrication. The process should consider the degree of neatness, accuracy, precision, thoroughness, and/or other characteristics that are required and are economically feasible. The process specification should include tolerances and/or upper and lower limits. Variability of materials and processes can be estimated from an empirical frequency distribution. Manufacturing engineering, in conjunction with quality control, establishes the necessary combinations to ensure that the production output is compatible with consumer requirements.
4. Specify, in conjunction with methods engineering, the sequence of manufacturing operations. Develop flow charts and identify functional sequences for manufacturing processes.
5. Identify man-machine operations. Apply human factors criteria in determining a manual task versus a function that can be automated. Decisions in this area must be considered with the results of methods engineering.
6. Select the appropriate tools, test equipment, and handling equipment required for each manufacturing operation. This includes machines, tools, jigs, and fixtures. Numerical control, computer-aided manufacturing, and other techniques are evaluated in terms of need and efficiency.
7. Design special tools, test equipment, jigs, and fixtures. Where feasible, the items of test and support equipment designated for operational use and for support of prime equipment in the field should be used to support manufacturing operations. For instance, if an assembled item requires a test prior to being processed to the next manufacturing station and that test is comparable to a test that is required in the field, the operational test equipment should be used rather than designing a new item specifically for production purposes. This tends to ensure compatibility in system testing, helps to verify the adequacy of operational test equipment, and reduces the cost of production.[9]

C. Methods Engineering

The aspect of analyzing production operations in terms of effectiveness and cost is accomplished through methods engineering. Specifically, this includes the following functions:

[9]Manufacturing processes are covered further in: L. Doyle, C. Keyser, J. Leach, G. Schrader, and M. Singer, *Manufacturing Processes and Materials for Engineers,* 2nd ed., Prentice-Hall, Inc., Englewood Cliffs, N.J., 1969.

1. Establish work methods, time, and cost standards. Select jobs that can be standardized, analyze the jobs in terms of elements, synthesize and evaluate alternative job approaches, select a preferred approach, establish cost and time standards for each job, and apply the established standards for all applicable manufacturing operations. Cost data generated in this area are used in life-cycle cost analyses.
2. Estimate part and manufacturing operation costs.
3. Determine job skill requirements, and estimate manufacturing personnel quantities and labor grades.
4. Design subassembly and assembly operations.
5. Analyze overall production operations in terms of cost and effectiveness, and initiate changes for improvement where appropriate.

D. Production Control Requirements

The remaining aspect of early production design is to establish the procedures and requirements necessary to maintain control of production operations. Primary production control functions are to:

1. Determine the production lot quantities and batch sizes. How many equipment items should be produced in a given lot to maintain the proper control of processes, quality, change implementation, and so on?
2. Determine the economic inventory levels for materials and parts. Establish economic order quantities and procurement cycles.
3. Establish a system for work-order processing and assignment.

As part of the industrial engineering function, one of the initial steps is to develop a process flow chart that covers the overall production operation. This flow is presented in the manner illustrated in Figure 6-2. The intent here is to show all the major functions required to produce the equipment. An expansion of each function can then be accomplished through a work process analysis, illustrated in Figure 6-3.[10] Functions and tasks are identified, analyzed, and the results are employed to determine specific support requirements for the production process. In addition, these results are used to measure the effectiveness of production operations.

When defining production requirements, the industrial engineering activity is charged with the responsibility of seeking an optimum solution. The functions of plant engineering, manufacturing engineering, methods engineering, and production control are accomplished, and the results are considered in the decision-making process. The reader, when analyzing the

---

[10] The work process analysis is discussed further in: R. E. McGarrah, *Production and Logistics Management,* John Wiley & Sons, Inc., New York, 1963.

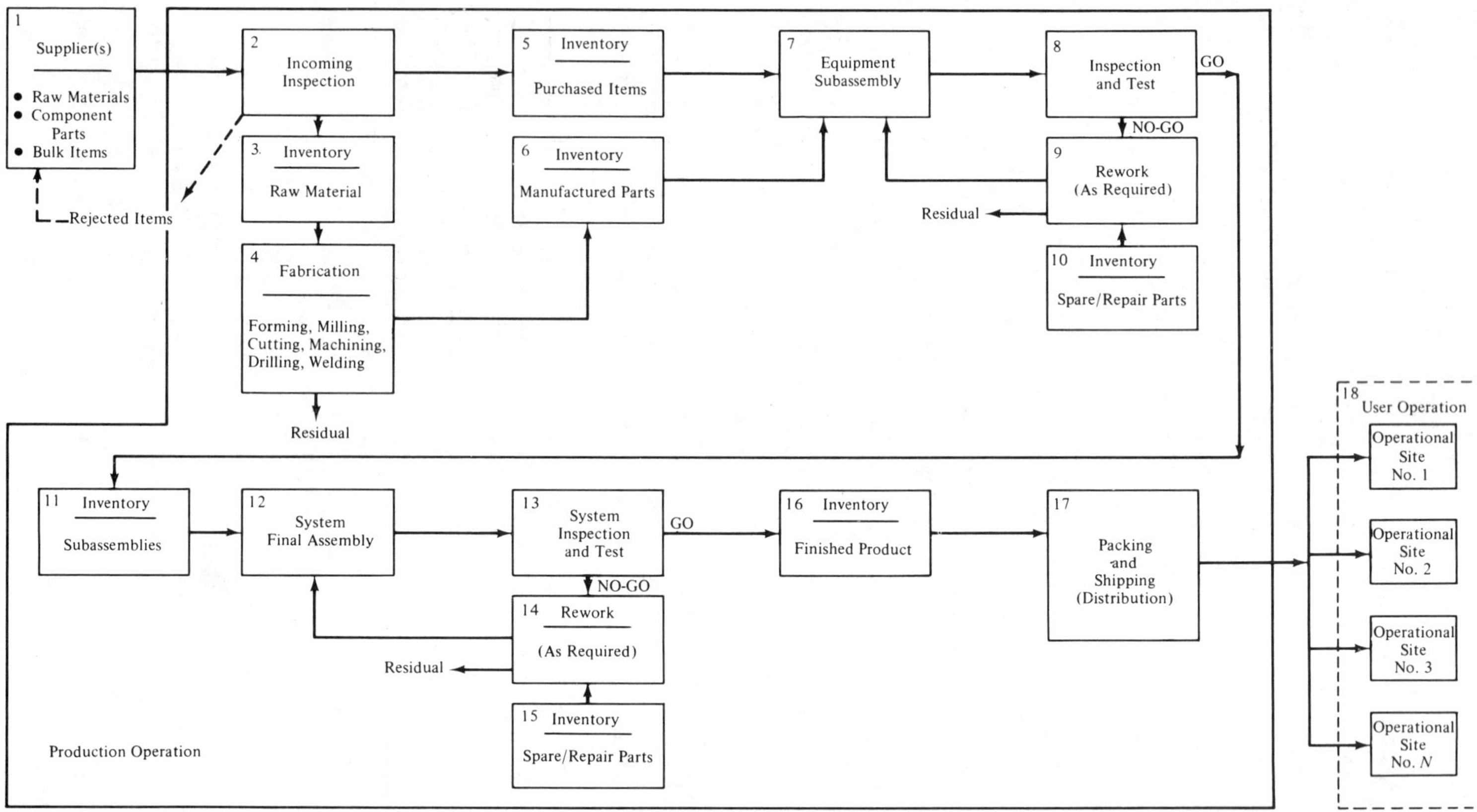

Figure 6-2. Production process flow.

Reference: Figure 6-2, Blocks 7, 8, 11

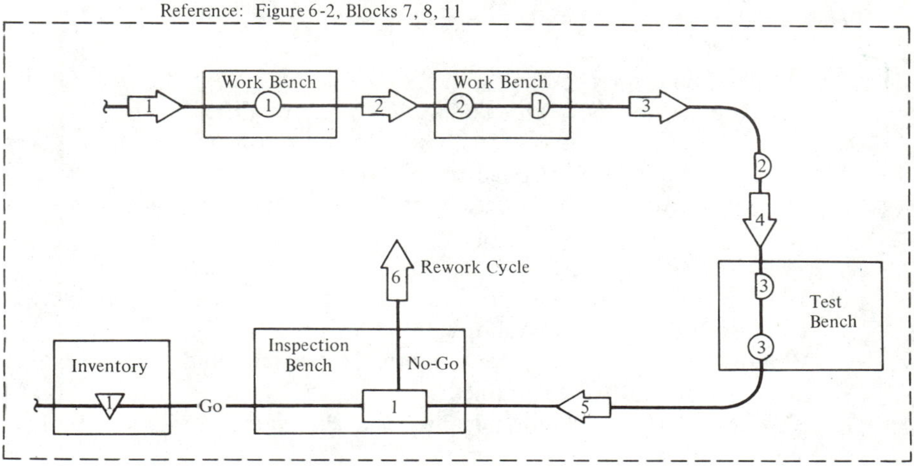

Work Flow Process Chart

Part Name: Chassis *A* Part Number: A12450

Process: Subassembly Department: 312

○ = Operation ⇨ = Transportation D = Delay ▽ = Storage □ = Inspection

Method

Symbol	Description	Distance Moved	Method of Movement	How Accomplished
Transportation 1	Transport to Workbench	10 ft	Pickup and Laydown	Man
Operation 1	Install Parts 1, 2, and 3 on Chassis A12449			Man
Transportation 2	Transport to Station 12	25 ft	Pickup and Laydown	Man
Operation 2	Install Parts 4, 5, and 6 on Chassis A12449			Man
Delay 1	On Workbench			
Transportation 3	Transport to Conveyor	10 ft	Pickup and Laydown	Man
Delay 2	On Conveyor			Power

Figure 6-3. Work process approach.

situation, might consider the production operation as a large system by itself and attempt to evaluate input-output requirements for various internal production configurations. In any event, designing a production capability will lead to many questions, and numerous decisions must be made. For instance:

A. Where should the production facility be located? Should a new facility be acquired or should an existing facility be modified?
B. How many production lines should be established, and how many machines should be assigned to each line?
C. What is the optimum production rate and output on a month-to-month basis?
D. Where should maintenance and rework be accomplished?
E. What preventive maintenance should be accomplished on production facilities, utilities, and capital equipment?
F. Where should inventory stock points be located, and what levels of inventory should be maintained?
G. How should the production facility and associated equipment be laid out for optimum output?
H. What functions should be automated, and what functions should be accomplished through manual means?
I. What modes of transportation should be employed in the accomplishment of transportation functions?

Answers to these and other questions require a rather in-depth operations analysis. These are numerous alternatives to consider, and the input variable relationships are quite complex. In many instances, the analysis effort and the evaluation of alternatives require the development and use of analytical models. Requirements are identified, alternatives are proposed, and the industrial engineer must evaluate each feasible alternative and recommend a preferred solution.[11]

The operational analysis of a production capability is a separate subject in itself, and the reader is advised to review additional literature in order to understand the scope and complexities involved.[12] Typical subject areas that can be addressed through the use of appropriate analytical techniques are:

A. Economic evaluation of production operations.
B. Evaluation of single and multiple production alternatives.
C. Optimum equipment replacement and/or maintenance.
D. Procurement quantities and inventory levels.
E. Evaluation of production sequencing.
F. Evaluation of waiting-line effects.
G. Evaluation of transportation options.
H. Optimum machine assignment.

---

[11]The approach described in Section 2.3, is appropriate here.

[12]An excellent source is W. J. Fabrycky, P. M. Ghare, and P. E. Torgersen, *Industrial Operations Research,* Prentice-Hall, Inc., Englewood Cliffs, N.J., 1972.

I. Optimum acceptable quality-control levels.
J. Evaluation of material purchasing alternatives.
K. Production project evaluation and control.

6.3 QUALITY CONTROL[13]

The quality of a product is the degree to which it satisfies the wants of a specific consumer or user or may be stated in terms of a measure of the degree to which it conforms to specification and workmanship standards. It can be expressed in terms of a given set of attributes required for that product to meet operational requirements. These attributes may include size, weight, shape, durability, hardness, performance, reliability, maintainability, supportability, attractiveness, and the like and should be defined in measurable quantitative terms. Measurement accuracies and tolerances (upper and lower limits) should be defined for each attribute. Quality level is measured by the percentage defective in a given lot or population.

The purpose of quality control is to assure that these attributes are maintained throughout the production cycle. Quality control is a function of management relative to all procedures, inspections, examinations, and tests required during item procurement, production, receipt, storage, and issue that are necessary to provide the user with a product of required quality. Quality control is a production-oriented operation designed to cause a process to manufacture a uniform product within specified limits of per cent defective in accordance with the design requirements.

The conformance to good quality (the adherence of a product to specification requirements) is particularly significant, for the following reasons:

A. It reduces item rejects and the recycling for repair rework, testing, and inspection. The reworking of an item requires additional manpower and machine usage and results in high material waste. Good quality causes a reduction in production costs.
B. It reduces the probability of requiring an extensive amount of maintenance in the field shortly after an item is delivered to the consumer for operational use. An item may pass a series of inspections in the produc-

[13]The subjects of quality and quality control are rather extensive and far beyond the summary discussion presented here. The reader may wish to review the following four references for more in-depth coverage: J. M. Juran, *Quality Control Handbook,* McGraw-Hill Book Company, New York, 1962; A. V. Feignbaum, *Quality Control—Principles, Practices, and Administration,* McGraw-Hill Book Company, New York, 1951; E. L. Grant and R. S. Leavenworth, *Statistical Quality Control,* 4th ed., McGraw-Hill Book Company, New York, 1972; and A. J. Duncan, *Quality Control and Industrial Statistics,* Richard D. Irwin, Inc., Homewood, Ill., 1972.

tion cycle and yet fail shortly after delivery, owing to one or more manufacturing defects. The added maintenance burden will result in a need for additional logistic support resources or cause those resources available to be depleted earlier than expected. Good quality causes a reduction in system operation and maintenance costs.

C. It reduces the probability of product liability suits, particularly if warranty provisions prevail. Manufacturing defects may cause accidents that result in damage to equipment and facilities and/or result in the death of personnel. Good quality reduces the risk to the producer.

D. It causes a favorable impact of future reprocurements and new sales. A producer's reputation stems from the quality of its product.

Items A and B both significantly impact logistic support (in the production facility and in the field, respectively). The initial engineering design may be suitable; however, unless the equipment is producible and the production process is adequate, the product output will not necessarily reflect what was initially intended. Without adequate standards and controls, each individual item of equipment may turn out to be somewhat different, operational effectiveness may fall short of initial objectives, the reliability of equipment may degrade, and so on. Good quality control is necessary to ensure that both prime equipment and logistic support objectives, defined through the design and development process, are realized.

The assurance of good quality is accomplished through (1) the establishment of quality tests, setting up quality standards and acceptance criteria, and interpreting quality data; and (2) the subsequent implementation of a sustaining quality-control effort, constituting a combination of tests and inspections. The development of quality standards and acceptance criteria is accomplished in conjunction with the activities of plant engineering, manufacturing engineering, methods engineering, and production control. Processes are analyzed and quality levels are established. A quality level is chosen by considering the probabilities of accepting bad-quality products as good (i.e., consumer's risk) and of rejecting good-quality products as bad (i.e., producer's risk). Selecting the proper level is dependent on the costs of inspection and the risks.[14] Once quality levels are identified for the attributes to be measured, accuracies and tolerances must be specified, and acceptance criteria are established. Tests and inspections are than planned at different times in the production cycle to ensure proper quality standards at the various stages of production (e.g., after fabrication, subassembly, and/or final

---

[14]The reader would benefit at this point by a review of the basic principles of statistical quality control, particularly the factors dealing with producer's risk (α) and consumer's risk (β). The acceptance of bad products, a function of the established quality level, will directly impact system support.

assembly). These requirements are then included in the applicable process and material specifications developed as an engineering output.

The second aspect of quality control involves the function of inspection and test. Inspection is the process of verifying the quality of an item with respect to a standard. Inspection serves several purposes. The first is to determine the quality of some portion of the items being turned out by a process. The problem is to determine whether to accept or reject a production lot on the basis of an analysis of the selected items. This is often referred to as *acceptance control.* The second purpose is to determine the condition of an operation or process in terms of its quality-producing performance. In this instance, the intent is *process control.* A given inspection may satisfy both purposes.

There are two categories of inspection. The first is *100 per cent inspection or screening*, and the second is *sampling inspection.* For critical components, 100 per cent inspection is often desirable. This includes those instances where measurement accuracies and tolerances of an item are critical or when a given production process is suspect. Each and every item in the production lot is inspected against the process and/or material specification. In sampling inspection, a random sample of items is selected from a production lot (a statistical population), and each item in the sample is inspected in accordance with the requirements of the specification. An accept decision constitutes the acceptance of the entire lot. The frequency of sampling and the sample size are dependent on the type of equipment, the length of the production run, the number of items produced per lot, and the risks involved.

As indicated earlier, the purpose of inspection and test is to ensure that a specified level of product quality is maintained *throughout* the production process. For instance:

A. The size, shape, structure, maintainability, and other parameters of an item may vary in time as a result of gradual changes in material composition and/or manufacturing tolerances. As a result of high costs or long procurement lead times, suppliers may occassionally substitute materials that appear to be completely interchangeable but in actuality are not identical relative to the functional characteristics required. In addition, manufacturing process variations may occur gradually over time and remain unnoticed unless checked periodically. These factors may significantly influence the quality of the product output as well as have a detrimental impact on product support.

B. The equipment reliability (MTBF), demonstrated as part of type 2 testing using the prototype model, should be maintained for all production models. To assure that the required MTBF is attained, a sample number of equipments may be selected from each production lot and subjected to a reliability test. In some instances, a lower MTBF will be evident, which means that reliability degradation is occurring through the production process. This will necessitate an adjustment to input quality standards (e.g., improve component acceptance criteria, tighten manufacturing processes, increase testing) in such a manner as to improve product reliability. On the other hand, the test results may indicate a higher MTBF, which demonstrates equipment maturity and reliability growth.

C. Certain equipment performance parameters (i.e., range, accuracy, flow rate, etc.) are critical to the success of the system mission in the field. Operational test and support equipment has been identified by the logistic support analysis accomplished during detailed design and developed to check the prime equipment in terms of these parameters. A test, accomplished during type 2 testing, initially ensures the compatibility between the prime equipment and its associated test and support equipment. It is required that this compatibility be maintained for all like production items; thus, it may be feasible to select certain equipments at random from different production lots and repeat the test.

Numerous examples and considerations apply to quality control in the production operation. The measure of quality is a function of specified product standards and the established test and inspection requirements. It is not the intent nor is it necessarily feasible to seek quality perfection as the associated production and inspection costs may be prohibitive. On the other hand, poor quality is costly, for the reasons stated above. An optimum acceptable quality level is defined as that point where all product operational and performance characteristics are attained and the total production cost is minimized. The relationship between levels of quality and cost is illustrated in Figure 6-4. Quality is assessed by inspecting the product (to varying degrees) at one or more points in the production process. Accept-reject decisions are made and defective items are recycled for corrective action. The object is to

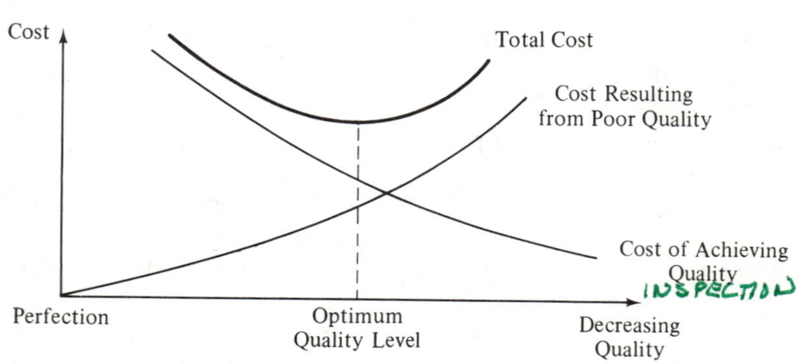

Figure 6-4. Relationship between levels of quality and cost.

establish a production process with the necessary controls that will realize a tolerable defect rate.

6.4 PRODUCTION OPERATIONS

Implementation of the functions defined in previous sections results in the initial design of the production capability, and the engineer's role in this effort is considerable. Materials and services are acquired, and actual production commences in accordance with the prescribed quantities and schedule. Production includes:

A. Fabrication, assembly, inspection, and test of prime equipment.
B. Fabrication, assembly, inspection, and test of logistic support hardware items designated for operational use in the field. This includes:

    1. Test and support equipment.
    2. Transportation and handling equipment.
    3. Spare/repair parts.
    4. Training equipment.
    5. Operational and maintenance facilities.

C. Assembly and inspection of software (e.g., computer program tapes) and operation and maintenance data for use in the field.

D. Provisioning and acquisition of off-the-shelf equipment, associated software, and data (that which is already available in some inventory and is acquired through procurement).

Conventionally, we think of production as being primarily associated with the manufacture of prime equipment only. Although the prime equipment is generally the most significant item, the various elements of logistic support must also be considered in the production process. In some instances, the same producer and physical facility are employed in the production of both prime equipment and logistic support items. In other cases, different production facilities are involved. In any event, the preproduction planning and design effort (i.e., plant engineering, manufacturing engineering, methods engineering, production control, and quality control) must address all elements of the system.

When considering the production operation as a system in itself, we are dealing with the product that is required by the consumer (to include all four categories above, such as prime equipment, spare/repair parts, and software) and the processes, materials, and services necessary to manufacture the product. The combination constitutes the production capability discussed above and illustrated in Figure 6-2. At designated points in the production process, various facets of the operation are measured and evaluated to ensure that a proper level of effectiveness is being maintained. In the design of the production capability, input-output requirements are established for each major function. These requirements may constitute the assembly, inspection, or test of *n* gidgets per unit of calendar time; the fabrication of *x* parts per a given level of cost; the percentage of items reworked; the quantity of man-hours expended per item subassembly; and so on. In addition, the items involved must exhibit a specified level of quality. The combining of these requirements results in a total production capability, and this capability must be assessed on a continuing basis to verify that consumer requirements are being met.

The measurement and evaluation occur through the application of inspection and test functions and work sampling techniques. Inspection and test provisions, including a combination of acceptance control and sampling inspections, verify that product performance and operational characteristics (i.e., proper level of product quality) are being maintained. Work sampling techniques are employed to evaluate production operations.

Work sampling is a method of analyzing and assessing the time (clock time and man-hours) required for the performance of work tasks and is accomplished through a series of random observations which are extended over the production period. These observations may be acquired by establishing timekeeping procedures by job and analyzing the collected time data; observing jobs and monitoring task sequences and time with a stopwatch; taking motion pictures of a job and synchronizing camera speed and film frames with time; and/or by making recordings of certain data and relating the output to probability distributions that are representative of various machine and human activity functions. Each of the available techniques is advantageous in certain situations. For instance, in analyzing an organizational entity that is responsible for a number of functions within the overall production operation, the timekeeping procedure (measuring man-hours expended for a given level of work) may be preferable. When establishing a detailed time standard or measuring the time for mounting one part on another, monitoring and recording time with a stopwatch may be appropriate. On the other hand, in the assembly of an item, particularly if the item is large and the assembly procedure is relatively complex, the motion-picture approach may be the most feasible. The use of motion pictures allows for the evaluation of task sequences, personnel actions and error rates, and task times. The selection of the technique(s) to be employed is obviously dependent on the type of data desired.

The evaluation of production operations is accomplished by using a combination of inspection and test results and work sampling data. The output data are compared with the initially established equipment quality requirements and manufacturing standards. Corrective action is initiated in results) or when changes can be incorporated to improve the effectiveness of production.

Figure 6-5 illustrates the basic production cycle and the feedback corrective action loop. That is, requirements are initially established, a production capability is designed, production commences, the overall process is assessed on a continuing basis to ensure adequacy, and improvements are incorpo-

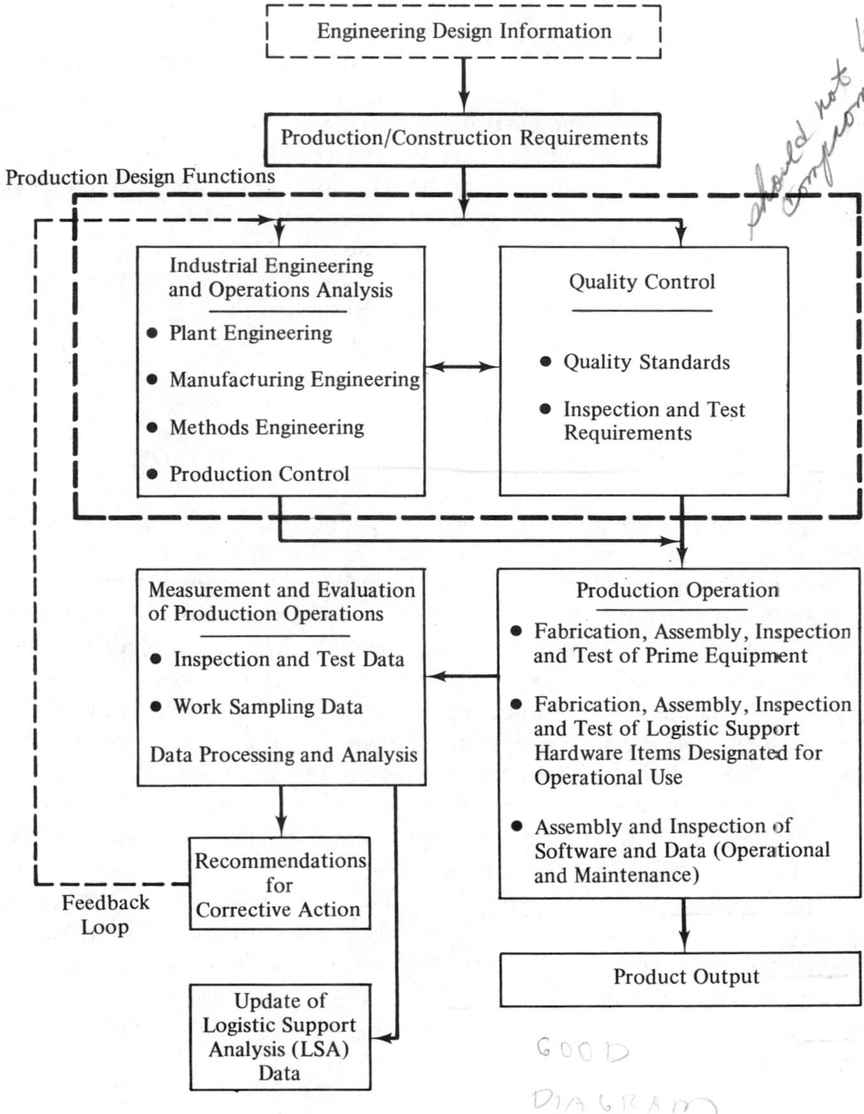

Figure 6-5. Basic production design, operation, and evaluation relationships.

rated where appropriate. It should be noted that measurement and evaluation data covering production operations serve several purposes. The first purpose is to assess the overall production capability discussed earlier. The second purpose is to update logistic support data and resource requirements where appropriate.

Many of the assembly, inspection, and test procedures and quality standards used in production operations are identical to those proposed for operation and support of the system in the field during the product use and logistic support phase. In the performance of maintenance actions, throughout the system life cycle, there are disassembly, reassembly, inspection, and test requirements. Although the environmental conditions differ somewhat, the methods of task accomplishment are directly comparable in terms of task times, sequences, and logistic support needs. Monitoring and collecting data covering these functions provide results that are directly applicable. Thus, it is beneficial to identify and observe applicable functions and gain insight relative to future requirements. This information is then used to update the logistic support analysis where appropriate. Results of the analysis are evaluated (on a continuing basis) to assess the product in terms of design for supportability and producibility.

6.5 CONFIGURATION MANAGEMENT AND CHANGE CONTROL[15]

Quite often in the production process, changes are initiated to correct a deficiency and/or to improve the product. A change may result from the redesign of a prime equipment item, the revision of a production process, or a combination of both. In most instances, a change in one element of a system will have a direct impact on other elements. For instance, a change in the design configuration of prime equipment (e.g., change in size or weight, repackaging, added performance capability) will in all probability affect the design of test and support equipment, the type and quantity of spare/repair parts, technical data, facilities, the production process, and so on. A change in a production process may impact the reliability of the product, which in turn, affects test and support equipment, spare/repair part requirements, and other elements of logistic support.

Each change must be thoroughly evaluated in terms of its impact on other elements of the system prior to a decision as to whether to incorporate the change. Figure 6-6 describes the basic steps in the change procedure. The incorporation of a change after production commences may be quite costly, particularly if the change is accomplished in the latter stages of the production cycle. The change will require the acquisition of additional

---

[15]By change control, the author is referring to the necessary management functions required to ensure that complete compatibility is maintained between *all* elements of a system whenever any single given element is changed for any reason.

materials and services, and a given amount of the already expended product input resources will have been wasted. For equipment that is produced in multiple quantities, the additional resources and resultant waste may be multiplied on the basis of the number of items of equipment that require the change. If a change is incorporated in certain equipment (i.e., equipment item serial numbers 24 and above) and not retrofitted on earlier models (i.e., equipment items with serial numbers 1 through 23), then each of two configurations will require a different type of logistic support capability, which is highly undesirable.

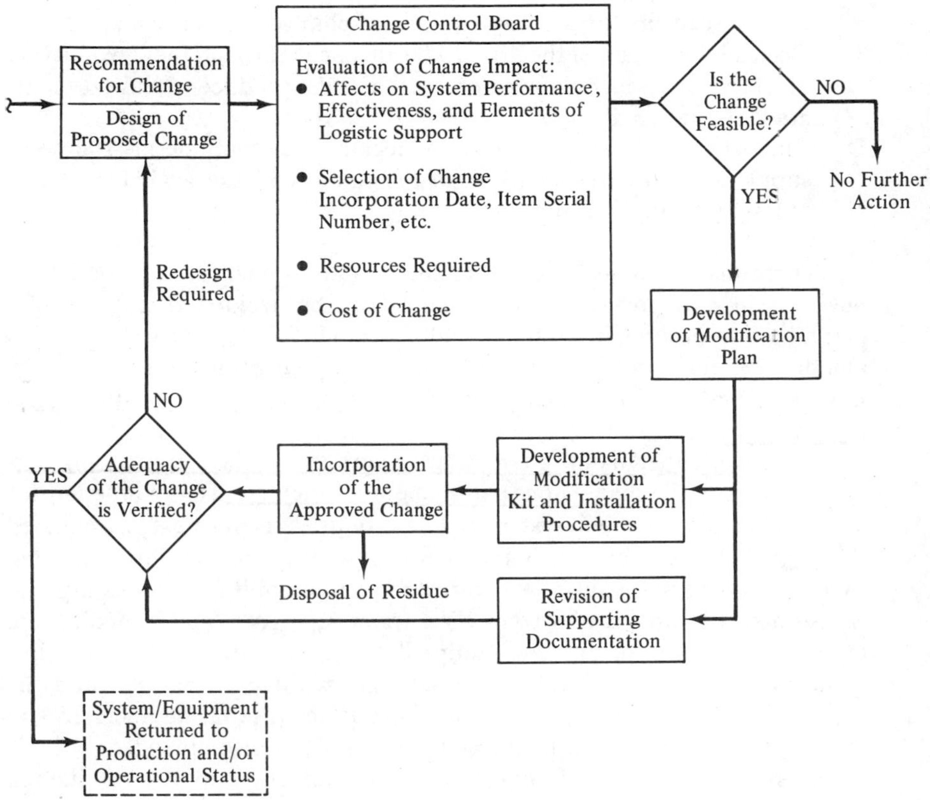

Figure 6-6. System/equipment change cycle.

Past experience with a variety of systems has shown that configuration changes in the production/construction phase are costly. With some systems, many different changes have been initiated at various points. Not always have the required changes been incorporated in all applicable production models, nor have the effects of each change on other elements of the system been considered. This has resulted in the delivery of prime equipment models of different configurations and elements of logistic support that are not compatible with the prime equipment. In such situations, system effectiveness is seriously compromised, the right type of system support is not available, and much waste occurs. The causes related to these occurrences may vary.

A. The system or equipment may enter into the production/construction phase without first establishing a fixed design configuration. In other words, the design has not been formally reviewed, verified analytically, or demonstrated through test.
B. The original need and operational requirements for the system may change, causing a necessary equipment change downstream.
C. The technical state of the art may be advanced through the introduction of a new design technique or process, and the producer decides to initiate a change as a result.
D. The supplier of certain items may decide to discontinue his source of supply and no other suppliers of the same item are available. Thus, redesign is often required.

There may be other reasons for incorporating changes in production; however, in most instances, these causes can be avoided through proper planning and progressing systematically through the steps of system design and development described in earlier chapters. Changes in equipment hardware, software, and manufacturing operations must be controlled and held to a minimum.

If, on the other hand, changes are required for corrective-action purposes, it is essential that change control be implemented. That is, the impact of each change on the total system and its various elements must be evaluated thoroughly. When a change is incorporated in any item, all other affected elements of the system must be modified for compatibility. As an example, it is assumed that unit B of system XYZ (refer to Figure 5-6) is modified to improve performance, and the equipment is to be produced in multiple quantities. The change involves the redesign of three assemblies, and the change is scheduled to be incorporated in equipment serial numbers 25 and above. A retrofit for serial numbers 1 through 24 is recommended.

First, the change will require some modification to the production processes (e.g., fabrication and subassembly methods) and material inventories to cover the future production of equipment serial numbers 25 and above. This, in turn, may affect manufacturing tools, jigs, fixtures, and so on. For the earlier models, modification kits and installation instructions must be developed. The equipment is then pulled out of operation (assuming that the first 24 items of equipment have been delivered to the user) while the change is implemented and verified. The change involves added production costs and operation costs in the field.

As a second consideration, the change will affect to some extent each element of logistic support. Reliability and the frequency of maintenance (anticipated number of maintenance actions) are affected, either upward or downward. Maintainability characteristics will change along with detailed maintenance tasks, task frequencies, and possibly the level of repair decisions made earlier. Test and support equipment must be modified to incorporate test provisions that cover the added performance capability. Spare parts must be changed at the assembly level and below if the assemblies are designated repairable. Personnel training, training data, and operating and maintenance procedures for field use must be revised to reflect the change. Facilities may also require some modification.

Thus, what initially appears to be a simple system modification often has a tremendous impact on the prime equipment, the production capability, and logistic support. The incorporation of such a change must be identified with a specific equipment serial number. Each unit, assembly, and subassembly of that particular equipment must be marked in such a way as to be identified with that equipment. Spares with the same part number as items in the prime equipment must be treated in a like manner. A change incorporated in the system must be traceable through *all* affected elements (i.e., hardware, software, and data).

Configuration change control is particularly significant when the system undergoes a number of changes in the production/construction phase. The results of the single change to system XYZ described above may be multiplied many times. Without the proper controls, there is no guarantee that the production output will provide effective results.

6.6 TRANSITION FROM PRODUCTION TO OPERATIONAL USE

After the system, equipment, and associated elements of logistic support are produced (see block 16, Figure 6-2), the applicable items are distributed or deployed to one or more designated operational field sites. The transition process from production/construction to full-scale operational use is critical since the prime equipment, software, and logistic support must be delivered on a concurrent basis and in a timely manner. The goal is to meet consumer requirements and provide an operational system ready to go at the required point in time.

The designated operational sites and activation dates are specified early in the program through the definition of operational requirements (refer to Chapter 3).[16] The transition from production to consumer operation is covered in the product use and logistic support plan (refer to Figure 3-9)

---

[16]The activation date is considered as the time when the applicable site has initially attained full operational status in accordance with consumer requirements. This date may be different for each site if two or more sites are involved.

prepared prior to the start of the production phase. This plan supplements the operational requirements data and includes:

A. The identification of operational sites or distribution centers, location, and activation date.
B. The identification of items to be delivered at each site, the time and route of delivery, and the mode of transportation used for delivery.
C. The requirements and procedure used to install equipment and make the system completely operational once that the appropriate items have been delivered. This may entail the establishment of a maintenance shop, installation and checkout of test and support equipment, initiation of a supply support capability, and so on.
D. The type and extent (in terms of time duration) of logistic support provided by the producer after the system activation date has passed and the system is considered fully operational. This may entail the immediate transfer of support responsibility from the producer to the consumer, or the provision of a sustaining support capability by the producer throughout the system life cycle.

The period of transition will vary in time, depending on the system type and complexity. If the system is relatively simple to install, operate, and maintain, the transition may require only a few days before the consumer is able to assume full responsibility for system operation and support. On the other hand, if the system is large and complex, the transition may be much longer and accomplished on a gradual basis. An extensive amount of training may be required, producer field service engineers may be assigned to consumer site locations to assist in system operation, and the producer may provide an interim repair capability until the consumer is fully qualified in all aspects of system support.[17] The objective is to accomplish transition at the proper time such that program risks are minimized.[18] Logistic support in this transitional stage addresses two basic areas.

A. The packing and shipping or distribution of the product from the producer's facility to the consumer's operational site, and the inspecting, installing, and checkout of the product at the site. The product may

[17]For some systems the producer will retain full support of the system or equipment throughout the life cycle under a leasing agreement or a repair/rework contract. The reference here pertains to the time until the ultimate support capability (whichever policy is followed) is attained.

[18]In certain instances, it may be preferable to defer the transition of a support capability to the consumer until the consumer is able to provide such a capability without causing a degrading affect on system operation. The proper time phasing of transitioning different aspects of the total logistic support capability (often referred to as "phased logistic support") should be considered from a cost-effectiveness basis and covered in the product use and logistic support plan.

include prime equipment, support equipment, spare/repair parts, and so on. Logistics involves consideration of:

1. Methods of packing and types of containers used for shipping.
2. Mode of transportation used in shipment (e.g., cargo aircraft, truck, railroad, and/or ship).
3. Test and support equipment, personnel, facilities, and technical data required for inspection, installation, and checkout of the product at the operational site. Hopefully, the logistic items assigned to support the prime equipment for the life cycle will be adequate to accomplish these functions.

B. The provision of a logistic support capability for the product/system at each operational site until full activation is attained (i.e., for the duration of the transition period). Although many of the elements of support are identical to those specified for the system life cycle, there are three basic considerations that should be dealt with in arriving at this interim capability.

1. Depending on the type of equipment, established production and quality standards, and equipment maturity acquired through testing during production operations, there may be a greater number of corrective maintenance actions occurring immediately after equipment delivery than the estimated quantity derived from reliability predictions. Reliability predictions, which form the basis for the frequency of maintenance, often assume a constant failure rate. At this early point in the equipment life, the failure rate may not be at the constant state (i.e., it may be higher than initially anticipated). Thus, the logistic support capability must consider the possibility of realizing a burden beyond that which is initially planned. Refer to the early peak in the reliability curve depicted in Figure A-3, Appendix A. Equipment reflecting a representative reliability value at this point will require more maintenance.
2. When considering the introduction of a new system in the inventory, there is a period of time where formal training and equipment familiarization occurs. Until the operator and maintenance personnel are thoroughly familiar with the system, there will be a certain number of operator-induced faults, and the system will not be fully utilized in an effective manner. In addition, the equipment may not be properly maintained, and it is quite probable that total maintenance and support requirements will exceed initial expectations.
3. If the system is large and complex in nature, the consumer may be able to successfully operate the system at the time of activation but is unable to provide full logistic support at all echelons of mainte-

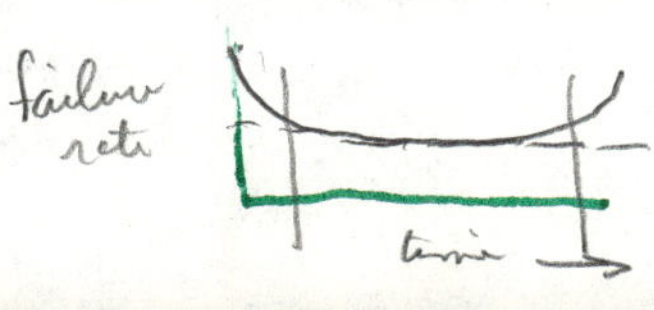

nance. For instance, material procurement lead times may force a delay in establishing a necessary capability in the performance of certain maintenance functions; thus, an interim support capability of some sort must be maintained to ensure successful system operation.

Logistic support requirements for the initial system start-up period, as determined by the logistic support analysis, must address the above considerations. In some cases, however, the support policy may vary somewhat from that recommended for operational use (refer to the operational and maintenance concepts described in Chapter 3). For instance, it may be feasible for a preestablished period of time to accomplish all depot-level maintenance at the producer's manufacturing facility rather than at the consumer's depot installation; it may be desirable for the producer to train and locate experienced field service engineers at the operational sites and intermediate maintenance shops to facilitate consumer personnel training and to assist in the performance of system operating functions; or it may be feasible to provide only higher-level spares (e.g., unit spares in lieu of both unit and assembly spares) at the intermediate shop for a limited period of time and repair the assemblies at the producer's facility until consumer personnel are adequately trained in the maintenance of assemblies. These and other policy diversions may occur to satisfy the needs during the transition from production/construction to consumer operation. However, the ultimate objective is to phase into full operational status defined in the product use and logistic support plan as soon as practicable.

QUESTIONS AND PROBLEMS

1. The initial requirements for production/construction are based on what factors?
2. What considerations are necessary in the design of a production/construction capability? Describe the basic production design functions.
3. If a piece of equipment is to be produced in multiple quantities, what considerations are necessary in the design of a production capability?
4. How does the output of detailed design affect the production/construction capability? Please be explicit.
5. In the production of multiple quantities of a given item, what are the advantages of continuous production over discontinuous production?
6. What is the basic difference between a job-shop operation and a flow-shop operation?
7. How are production standards established and maintained?

8. How would you measure and assess the effectiveness of a production capability?
9. Define logistic support in the context of the production/construction phase. What are the elements of logistic support?
10. What are the differences in producing assembly part 231, as installed in the prime equipment, and assembly part 231, designated as a spare part? Comment on manufacturing standards, quality standards, and inspection processes.
11. Can test and support equipment designated for operational use in the field be employed in the production process? If so, how, and what benefits are derived?
12. In what ways are functions performed in the production process similar to operational and maintenance functions that are anticipated for the field?
13. How does quality control affect system operation and logistic support in the field? Give some specific examples.
14. How does the evaluation of production functions relate to the logistic support analysis?
15. Can the production operation affect the inherent characteristics of reliability and maintainability in the equipment? How?
16. How would you determine the level of quality control that should be applied to a given product? Discuss factors that should be considered.
17. Why is configuration control important? What happens to logistic support when the proper level of control is not maintained?
18. You are considering a design change in the prime equipment. What factors should be evaluated in the decision-making process?
19. How would a design change in the prime equipment affect test and support equipment? Spare/repair parts? Facilities? Personnel and training? Technical data? Why is change control important?
20. Why is the transition process from production/construction to consumer operation so important?
21. What are the logistic support considerations in the transition from production/construction to consumer operation? Discuss the elements of support.
22. Is the level of logistic support provided when the system/equipment is initially deployed to the various consumer operational sites the same as or consistent with that recommended for subsequent periods in the life cycle? Why?
23. What is producibility?
24. What is phased logistic support? What is the major consideration involved?
25. What is the main objective of production planning?
26. In the accomplishment of equipment assembly/disassembly tasks as part of production operations (i.e., during the repair-rework cycle), it was noted that the equipment design lacked certain necessary inherent diagnostic and accessibility features. What would you do to correct the situation?

7

PRODUCT EVALUATION FUNCTION

Evaluation refers to the examination and judgment of a system (or an element of a system) in terms of worth, quality of performance, degree of effectiveness, condition, and the like. Evaluation is an ongoing iterative process which commences early in the system life cycle during conceptual design and extends through the product use and logistic support phase until equipment is removed from inventory. The objective is to determine (through a combination of prediction, analysis, and measurement) the *true* characteristics of the system and to ensure that it successfully fulfills its intended mission.

At first, the evaluation is accomplished using analytical means to predict various outcomes which lead to decisions affecting the early system design process. The definition of operational requirements and maintenance concepts, accomplishing functional analyses, performing trade-offs and optimization, and the establishment of a preferred configuration is a continuing design and evaluation effort. Various aspects of the design are reviewed on the basis of the specified requirements, and corrective action is initiated in areas where nonresponsiveness to these requirements and/or product improvement are indicated. Changes at this stage in the life cycle are relatively easy to incorporate, and the costs are generally low.

As design progresses from a paper definition stage to the point where mock-ups and engineering and service test models are developed, the evaluation process through type 1 testing becomes a little more meaningful since actual hardware is available. Further, when prototypes and production models evolve, the effectiveness of the evaluation in type 2 testing assumes even more significance, as the system approaches its intended operational configuration. After system construction (i.e., one-of-a-kind projects) or during the production of multiple quantities of equipment, the evaluation phases into a consumer or user environment (i.e., type 3 testing), and the true characteristics of the system as an entity can be measured for the first time.

The evolution of system evaluation is illustrated in Figure 7-1. The categories of testing, described in Section 5.3C, are identified by program phase, and the effectiveness of the evaluation effort increases when progressing through types 1, 2, 3, and 4 testing. These steps are necessary to ensure on a continuing basis that system requirements are being achieved. Incorporating changes after the system is in operational use can be extremely costly to both the producer and consumer; thus, it is preferable to eliminate (or avoid) potential problems as early as possible in the life cycle through the best means available.[1]

7.1 PLANNING AND REQUIREMENTS FOR SYSTEM EVALUATION[2]

The evaluation effort is designed based on system requirements and is planned early in the life cycle, and the various categories of test build upon each other. The analytical evaluation is discussed rather thoroughly in Sections 2.3 and 5.2, and the early phases of engineering testing are covered in Section 5.3C. The primary emphasis from here on is the evaluation of the system and all its elements (as a whole) included in types 3 and 4 testing. This facet of evaluation may assume several approaches:

A. The evaluation of a one-of-a-kind item, such as a power generating capability, chemical processing plant, or communications facility, may be accomplished after construction is completed. System evaluation is performed on a continuing basis or for a designated period of time, to assess both performance and effectiveness characteristics.

B. The evaluation of a product where multiple quantities of like items are produced may include the selection and placement of a designated sample number of these items at one or more given consumer locations or test sites. These items are utilized in an operational environment by

---

[1]The scope and depth of testing will depend on the type of system, whether new design and development are required, and the risks involved. Comprehensive testing may be required where design is new and many unknowns exist, whereas off-the-shelf, proved commodities will not require much testing except to assure the proper integration with other elements of the system. Not too much or too little testing should be accomplished. Thus, good test planning is essential.

[2]Planning for system evaluation and the content of the product evaluation plan are discussed in Sections 3.4D and 5.32C2.

Figure 7-1. Stages of product evaluation during the life cycle.

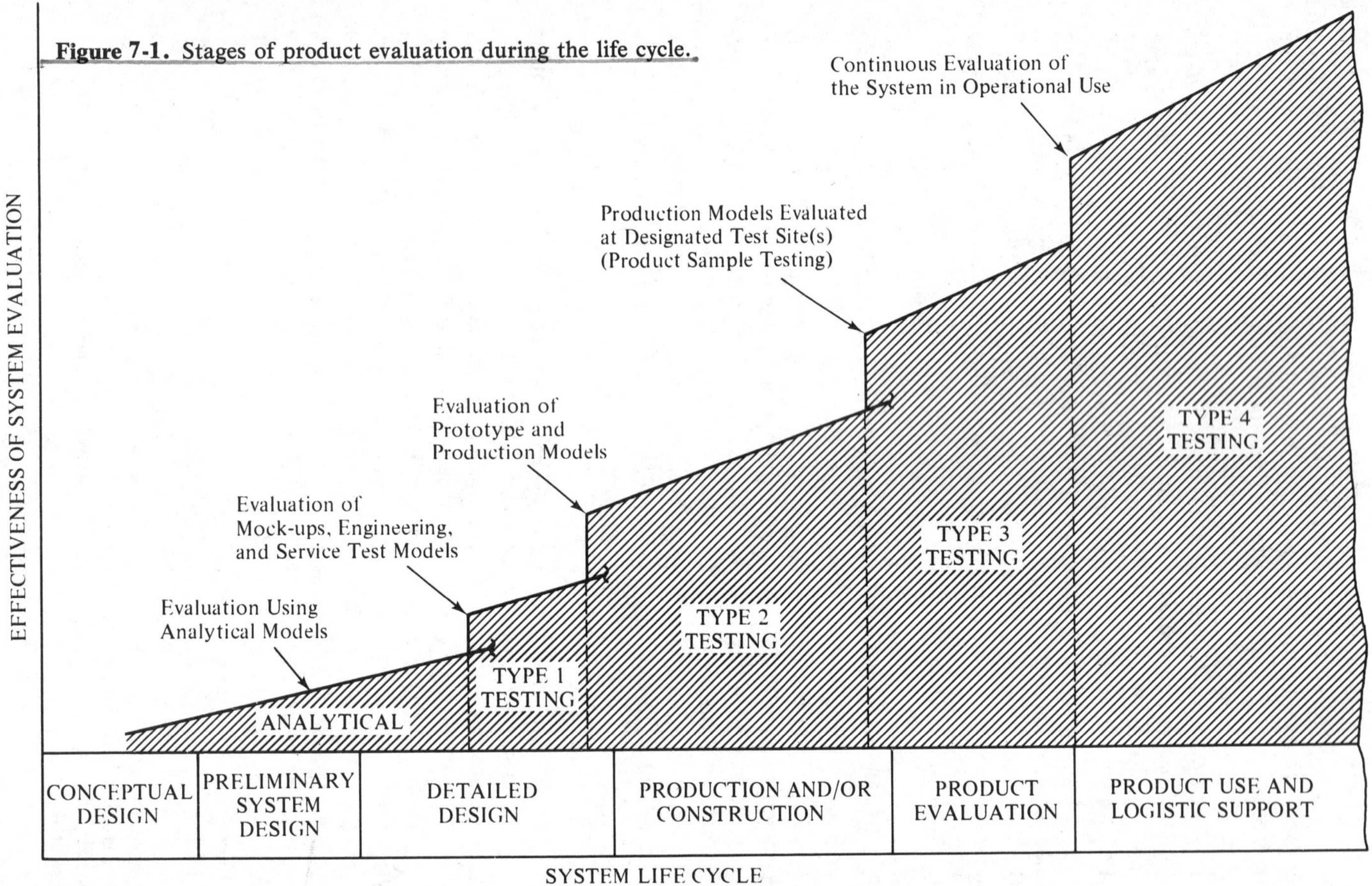

consumer personnel and supported in the same manner as designated for the product use and logistic support phase.[3]

Planning for system evaluation must consider the basic requirements for testing. What is the engineer or manager attempting to accomplish? What are the desired outputs from the evaluation effort? How are these outputs to be assessed? What constitutes satisfactory performance and effectiveness? These and comparable questions must be intuitively addressed in the planning process. The output of the planning process must convey what is to be done, where and when it is to be accomplished, and the methodology of approach to be employed.

7.2 PREPARATION FOR SYSTEM EVALUATION

After initial planning and prior to the start of formal evaluation, a period of time is set aside for preparation. During this period, the proper conditions must be established to ensure effective results. Although there is some variance, depending on the type of evaluation, these conditions or prerequisites include the selection of the item(s) to be tested, establishment of test procedures, selection and training of test personnel, preparation of the test site(s) and facilities, and the acquisition of support equipment and supply support.

A. Selection of Test Item[s]

The equipment configuration used in the test must be representative of the operational item to the maximum extent possible. For type 1 tests, engineering models are used that are not often directly comparable with operational equipment; however, most subsequent testing will be accomplished at such a time that equipment representing the final configuration is available. A prerequisite task involves selecting the test model by serial number, defining the configuration in terms of incorporated versus unincorporated engineering changes (if any), and ensuring that it is available at the time needed.

B. Test and Evaluation Procedures

Fulfillment of test objectives involves the accomplishment of both operator and maintenance tasks. Completion of these tasks should follow formal

[3]A single test site may be selected to evaluate the product in a designated environment; several test sites may be selected to evaluate the product under different environmental conditions (e.g., arctic versus tropical areas, mountainous versus flat terrain); or a variety of geographical locations may be selected to evaluate the product in terms of different markets and consumer conditions. Any one or combination of these methods may be appropriate, depending on the product type and program requirements.

approved procedures, which are generally in the form of technical manuals developed during the latter phases of detailed equipment design. Following approved procedures is necessary to ensure that the equipment is operated and maintained in a proper manner. Deviation from approved procedures may result in the introduction of personnel-induced failures and will distort maintenance frequencies and task times as recorded in the test data. The identification of the procedures to be used in testing should be included in the evaluation plan.

C. Test Facilities and Resources

The necessary facilities, test chambers, capital equipment, environmental controls, special instrumentation, and associated resources (e.g., heat, water, air conditioning, gas, telephone, power, lighting) must be identified and scheduled. In many instances, new design and construction are required, which directly impacts the scheduling and duration of the test preparation period. A detailed description of the test facility and the facility layout should be included in test planning documentation and in the final test report.

D. Test Personnel and Training

Test personnel will include (1) individuals who actually operate and maintain the system/equipment during the test; and (2) supporting engineers, technicians, data recorders, analysts, and test administrators, as appropriate. Individuals assigned to the operation and maintenance of the system should possess backgrounds and skill levels similar to consumer personnel who will normally operate and support the system throughout its life cycle. The required proficiency level(s) is attained through a combination of formal and on-the-job training.

E. Test and Support Equipment

Test and support equipment requirements are initially considered in the maintenance concept and later defined in the logistic support analysis. During the latter phases of detailed design, the necessary support items are acquisitioned and should be available for types 2, 3, and 4 testing. In the event that the proper type of support equipment is not available and alternative items are required, such items must be identified in the product evaluation plan. The use of alternative items generally results in distorted test data (e.g., maintenance times), tends to cause personnel-induced failures in the equipment, and often forces a change in test procedures and facility requirements.

F. Supply Support

Supply support constitutes all materials, data, personnel, and related activities associated with the requirements, provisioning, and acquisition of spare/repair parts and the sustaining maintenance of inventories for support of the system throughout its life cycle. Specifically, this includes:

1. Initial and sustaining requirements for spares, repair parts, and consumables for the prime equipment. Spares are major replacement items and are repairable, while repair parts are nonrepairable, smaller components. Consumables refer to fuel, oil, lubricants, liquid oxygen, nitrogen, and so on.
2. Initial and sustaining requirements for spares, repair parts, material, and consumables for the various elements of logistic support (i.e., test and support equipment, transportation and handling equipment, training equipment, and facilities).
3. Facilities required for the storage of spares, repair parts, and consumables. This involves consideration of space requirements, location, and environmental needs.
4. Personnel requirements for the accomplishment of supply support activities, such as provisioning, cataloging, receipt and issue, inventory management and control, shipment, and disposal of material.
5. Technical data requirements for supply support, which include initial and sustaining provisioning data, catalogs, material stock lists, receipt and issue reports, and material disposition reports.

The types of spare/repair parts needed at each echelon of maintenance are dependent on the maintenance concept and the logistic support analysis. For types 3 and 4 testing, spare/repair parts will generally be required for all echelons since these tests primarily involve an evaluation of the system as an entity and its total logistic support capability. The complete maintenance cycle, supply support provisions (the type and quantity of spares specified at each echelon), pipeline times, turnaround times, and related factors are evaluated. In certain instances, the producer's facility may provide depot-level support. Thus, it is important to establish a realistic supply system (or a close approximation) for the item being tested.

The type and quantity of spare/repair parts and the procedures for spares inventory control throughout the test program should be specified in the evaluation plan. Usage rates, reorder requirements, procurement lead times, and stock-out conditions during the test are recorded and included in the test report.

7.3 SYSTEM TEST AND DEMONSTRATION

With the necessary prerequisites established, the next step is to commence with the formal test and demonstration of the product. This constitutes operating and supporting the system/equipment in a prescribed manner as defined in the product evaluation plan. Throughout this process, data are collected and analyzed, which leads to assessment of system performance and effectiveness characteristics.

7.4 DATA COLLECTION, ANALYSIS, AND SYSTEM EVALUATION

The assessment of performance and effectiveness of a system requires the availability of operational and maintenance histories of the various system elements. Performance and effectiveness parameters are established early in the life cycle with the development of operational requirements and the maintenance concept. These parameters describe the characteristics of the system that are considered paramount in fulfilling the need objectives. Now with the system in operational status, the following questions arise:

A. What is the *true* performance and effectiveness of the system?
B. What is the *true* effectiveness of the logistic support capability?
C. Are the initially specified requirements being met?

Providing answers to these questions requires a formalized data-information feedback capability with the proper output. A data subsystem must be designed, developed, and implemented to achieve a specific set of objectives, and these objectives must relate directly to the preceding questions and/or whatever other questions the engineer or manager needs to answer. The establishment of a data-subsystem capability is basically a two-step function: (1) the identification of requirements and the applications for such; and (2) the design, development, and implementation of a capability that will satisfy the identified requirements.

A. Requirements

The purpose of the data-information feedback subsystem is twofold:

1. It provides ongoing data that are analyzed to evaluate and assess the performance, effectiveness, operations, maintenance, logistic support capability, and so on, for the system in the field. The system engineer or manager needs to know exactly how the system is doing, and needs the answer relatively soon. Thus, certain types of information must be provided at designated times.
2. It provides historical data (covering existing systems in the field) that are applicable in the design and development of new systems/equipment having a similar function and nature. Our engineering growth and potential in the future certainly depends on our ability to capture experiences of the past and subsequently be able to apply the results in terms of what to do and what not to do in the new design.

Supporting the above requires a capability that is both responsive to a repetitive need in an expeditious and timely manner (i.e., the manager's need for assessment information), and one that incorporates the provisions for data storage and retrieval. It is necessary to determine the specific elements of data reporting. These factors are combined to identify total volume requirements for the subsystem and the type, quantity, and frequency of data reports.

The elements of data identified are related to the operational and support requirements for the system. An analysis of these elements will provide certain evaluative and verification type functions covering the characteristics of the system that are to be assessed. A listing of sample applications appropriate in the assessment of system characteristics is presented in Figure 7-2. When determining the reliability of an item (i.e., the probability that an item will operate satisfactorily in a given environment for a specified period of time), the data required will include the system operating time to failure and the time-to-failure distribution, which can be generated from a history of the particular item or a set of identical and independent items. When verifying spare/repair part demand rates, the data required should include a history of all item replacements, system/equipment operating time at replacement, and disposition of the replaced items (whether the item is condemned or repaired and returned to stock). An evaluation of organizational effectiveness will require identification of assigned personnel by quantity and skill level, the tasks accomplished by the organization, and the man-hours and elapsed time expended in task accomplishment. Assessment objectives such as these (comparable to those listed in the figure) serve as the basis for the identification of the specific data factors needed.

B. Design, Development, and Implementation of a Data-Subsystem Capability

With the overall subsystem objectives defined, the next step is to identify the specific data factors that must be acquired and the method for acquisition. A format for data collection must be developed and should include both *success* data and *maintenance* data. Success data constitute information covering system operation and utilization on a day-to-day basis, and the information should be comparable to the factors listed in Figure 7-3. Maintenance data cover each event involving scheduled and unscheduled maintenance. The events are recorded and referenced in system operational information reports, and the factors recorded in each instance are illustrated in Figure 7-4.

The format for data collection may vary considerably, as there is no set method for its accomplishment, and the information desired may be different for each system. However, most of the factors in Figures 7-3 and 7-4 are common for all systems and must be addressed in the design of a new data subsystem. In any event, the following provisions should apply:

1. General Operational and Support Factors
   - (a) Evaluation of mission requirements (operational scenarios).
   - (b) Evaluation of performance factors (capacity, output, range, accuracy, size, weight, volume, mobility, etc.)
   - (c) Verification of system utilization (modes of operation and operating hours).
   - (d) Verification of cost/system effectiveness, operational availability, dependability, reliability, maintainability, human factors, safety.
   - (e) Evaluation of levels and location of maintenance.
   - (f) Evaluation of operation and maintenance function/tasks by level and location.
   - (g) Verification of repair level policies.
   - (h) Verification of frequency distributions for maintenance actions and repair times.
   - (i) Evaluation of software and software interfaces with other system elements.
2. Test and Support Equipment
   - (a) Verification of support equipment type and quantity by maintenance level and location.
   - (b) Verification of support equipment availability.
   - (c) Verification of support equipment utilization (frequency of use, location, percent of time utilized, flexibility of use).
   - (d) Evaluation of maintenance requirements for support equipment (scheduled and unscheduled maintenance, downtime, logistic resource requirements).
3. Supply Support (Spare/Repair Parts)
   - (a) Verification of spare/repair part types and quantities by maintenance level and location.
   - (b) Evaluation of supply responsiveness (is a spare available when needed?).
   - (c) Verification of item replacement rates, condemnation rates, attrition rates.
   - (d) Verification of inventory turnaround and supply pipeline times.
   - (e) Evaluation of maintenance requirements for shelf items.
   - (f) Evaluation of spare/repair part replacement and inventory policies.
   - (g) Identification of shortage risks.
4. Personnel and Training
   - (a) Verification of personnel quantities and skills by maintenance level and location.
   - (b) Verification of elapsed times and man-hour expenditures by personnel skill level.
   - (c) Evaluation of personnel skill mixes.
   - (d) Evaluation of personnel training policies.
   - (e) Verification of training equipment and data requirements.
5. Transportation and Handling
   - (a) Verification of transportation and handling equipment type and quantity by maintenance level and location.
   - (b) Verification of availability and utilization of transportation and handling equipment.
   - (c) Evaluation of delivery response times.
6. Facilities
   - (a) Verification of facility adequacy and utilization (operation, maintenance, and training facilities).
   - (b) Evaluation of logistics resource requirements for support of operation, maintenance, and training facilities.
7. Technical Data
   - (a) Verification of adequacy of data coverage (level, accuracy, and method of information presentation) in operating and maintenance manuals.
   - (b) Verification of adequacy of field data, collection, analysis, and corrective action subsystem.
8. Consumer Response
   - (a) Evaluation of degree of consumer satisfaction.
   - (b) Verification that consumer needs are met.

Figure 7-2. Data information system applications (*requirements*).

System Operational Information Report

1. Report number, report date, and individual preparing report.
2. System nomenclature, part number, manufacturer, serial number.
3. Description of system operation by date (mission type, profile, and duration).
4. Equipment utilization by date (operating time, cycles of operation, etc.).
5. Description of personnel, transportation and handling equipment, and facilities required for system operation.
6. Recording of maintenance events by date and time (reference maintenance event reports).

Figure 7-3. System success data.

Maintenance Event Report

1. *Administrative data*
   (a) Event report number, report date, and individual preparing report.
   (b) Work order number.
   (c) Work area and time of work (month, day, hour).
   (d) Activity (organization) identification.
2. *System factors*
   (a) Equipment part number and manufacturer.
   (b) Equipment serial number.
   (c) System operating time when event occurred (when discovered).
   (d) Segment of mission when event occurred.
   (e) Description of event (describe symptom of failure for unscheduled actions).
3. *Maintenance factors*
   (a) Maintenance requirement (repair, calibration, servicing, etc.).
   (b) Description of maintenance tasks.
   (c) Maintenance downtime (MDT).
   (d) Active maintenance times
   (e) Maintenance delays (time awaiting spare part, delay for test equipment, work stoppage, awaiting personnel assistance, delay for weather, etc.).
4. *Logistics factors*
   (a) Start and stop times for each maintenance technician by skill level.
   (b) Technical manual or maintenance procedure used (procedure number, paragraph, date, comments on procedure adequacy).
   (c) Test and support equipment used (item nomenclature, part number, manufacturer, serial number, time of item usage, operating time on test equipment when used).
   (d) Description of facilities used.
   (e) Description of replacement parts (type and quantity).
      i. Nomenclature, part number, manufacturer, serial number, and operating time on replaced item. Describe disposition.
      ii. Nomenclature, part number, manufacturer, serial number, and operating time on installed item.
5. *Other information*
   Include any additional data considered appropriate and related to the maintenance event.

Figure 7-4. System maintenance data.

1. The data collection forms should be simple to understand and complete (preferably on single sheets), as the task of recording the data may be accomplished under adverse environmental conditions by a variety of personnel skill levels. If the forms are difficult to understand, they will not be completed properly (if at all), and the needed data will not be available.
2. The factors specified on each form must be clear and concise and not require a lot of interpretation and manipulation to obtain. The right type of data must be collected.
3. The factors specified must have a meaning in terms of application. The usefulness of each factor must be verified.

These considerations are extremely important and cannot be overemphasized. All the analytical methods, prediction techniques, models, and so on, discussed earlier have little meaning without the proper input data. Our ability to evaluate alternatives and predict in the future depends on the availability of good historical data, and the source of such stems from the type of data-information feedback subsystem developed at this stage. The subsystem must not only incorporate the forms for recording the right type of data but must consider the personnel factors (skill levels, motivation, etc.) involved in the data recording process. The individual who must complete the appropriate form(s) must understand the system and the purposes for which the data are being collected. If this individual is not properly motivated to do a thorough job in recording events, the resulting data will, of course, be highly suspect.

Once the appropriate data forms are distributed and completed by the responsible line organizations, a means must be provided for the retrieval, formatting, sorting, and processing of the data for reporting purposes. Field data are collected and sent to a designated centralized facility for analysis and processing. The results are disseminated to engineering and management for decision making and entered into a data bank for retention and possible future use.

C. System Evaluation and Corrective Action

Figure 7-5 illustrates a system evaluation and corrective-action loop. The evaluation aspect responds to the type of subjects listed in Figure 7-2 and can address both the system as an entity or individual segments of the system on an independent basis. Figure 7-6 presents some typical examples of evaluation factors. The evaluation approach and the analytical techniques (i.e., tools) used are basically the same as described earlier for decision making, the only difference being the data input. The evaluation effort can be applied on a continuing basis to provide certain system measures at designated points throughout the life cycle (see A and C of Figure 7-6), or it may constitute a *one-time* investigation.

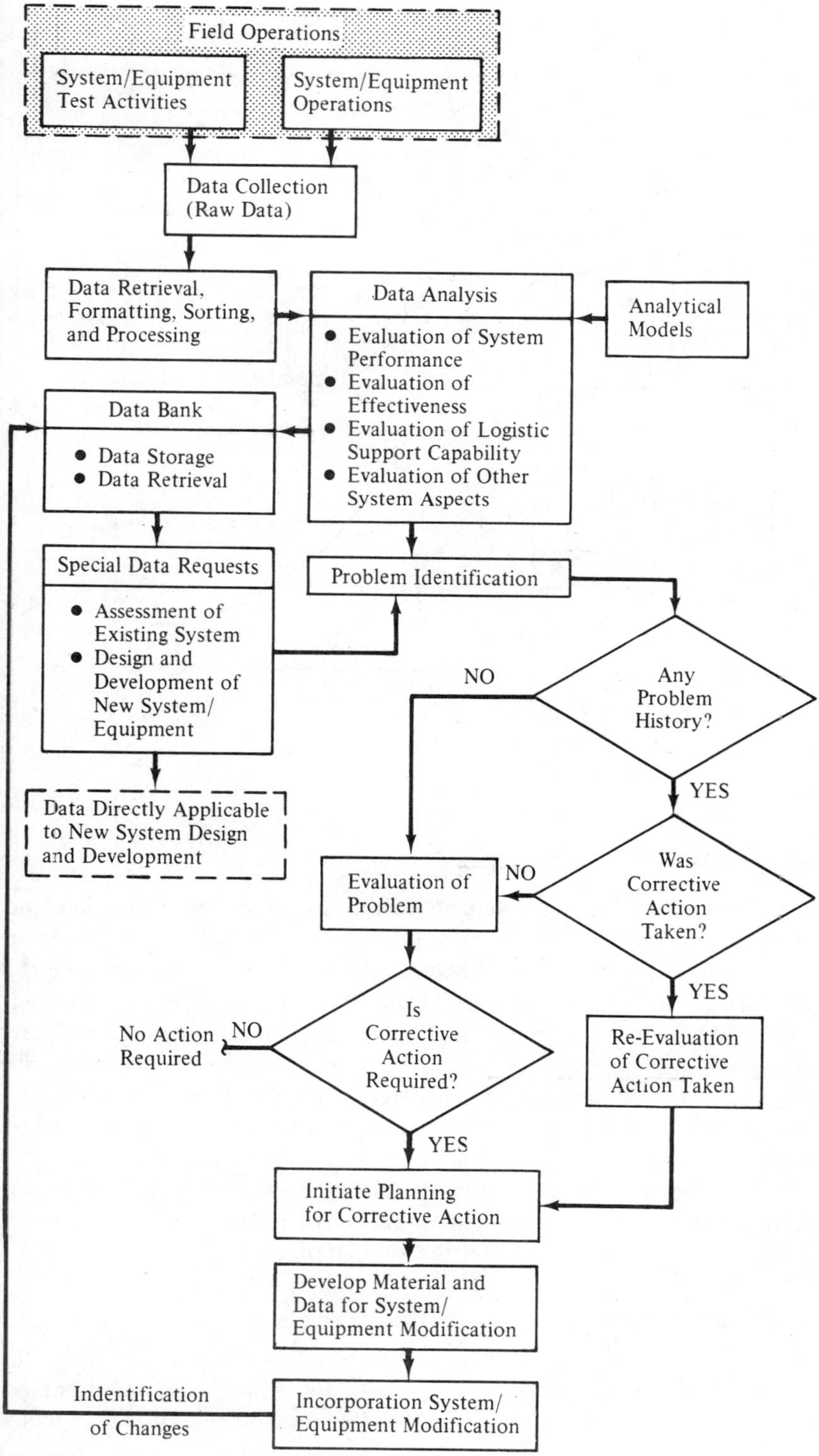

Figure 7-5. System evaluation and corrective action loop.

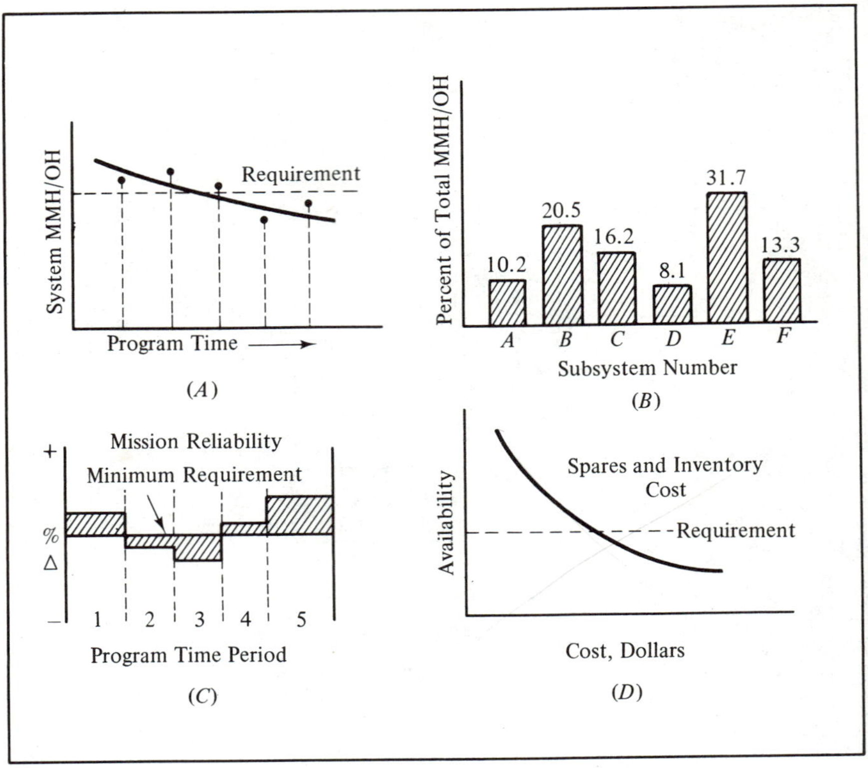

Figure 7-6. Typical evaluation factors.

Problem areas are identified at various stages in the evaluation and are reviewed in terms of the necessity for corrective action. Referring to Figure 7-6(B), subsystem E is a likely candidate for investigation since the MMH/OH for that item constitutes 31.7 per cent of the total system value. In Figure 7-6(C), one may wish to investigate program time period 3 to determine why the mission reliability was so poor at that time. Corrective action may be accomplished in response to a system/equipment deficiency (i.e., the equipment fails to meet the specified requirements) or may be accomplished to improve system performance, effectiveness, and/or logistic support. If corrective action is to be accomplished, the necessary planning and implementation steps are a prerequisite to ensure the complete compatibility of all elements of the system throughout the change process.

7.5 SYSTEM MODIFICATION

When a change occurs in a procedure, the prime equipment, or an element of logistic support, the change in most instances will affect many different elements of the system. Procedural changes will impact personnel and training requirements and necessitate a change in the technical data (equipment operating and/or maintenance instructions). Hardware changes will affect spare/repair parts, test and support equipment, technical data, and training requirements. Each change must be thoroughly evaluated in terms of its impact on other elements of the system prior to a decision on whether or not to incorporate the change. The feasibility of incorporating a change will depend on the extensiveness of the change, its impact on the system's ability to accomplish its mission, the time in the life cycle when the change can be incorporated, and the cost of change implementation. A minimum amount of evaluation and planning are required to make a rational decision on whether the change is feasible.

If a change is to be incorporated, the necessary change control procedures must be implemented. The various components of the system (including all the elements of logistic support) must "track." Otherwise, the probability of inadequate logistic support and unnecessary waste is high. The configuration change control and the change cycle are discussed in Section 6.5. The concepts and concerns in the production/construction stage are equally as appropriate in the product use and logistic support phase. In fact, a change involving the operational system is more critical, since we are dealing with a later stage in the life cycle. Changes will usually be more costly as the life cycle progresses.

7.6 TEST REPORTING

The final effort in the evaluation process constitutes the preparation of a test report. The test report should reference the initial product evaluation planning document and should describe all test conditions, incorporated system modifications during the test (if any), test data, and the results of data analysis. These results may include appropriate recommendations for operation and support of the system as applicable to the product use and logistic support phase.

QUESTIONS AND PROBLEMS

1. Describe the system evaluation process. What is the objective(s)?
2. How is system evaluation accomplished in the following phases: Preliminary system design? Detailed design? Production/construction? Product use and logistic support?
3. Define the basic categories of test and their applications. How do they fit into the total system evaluation process?

4. How does the *effectiveness* of evaluation relate to the system life cycle? Address both the aspects of test benefits and cost.
5. When does the planning for system evaluation commence? What information is included?
6. Select a system of your choice and develop a detailed outline of a product evaluation plan.
7. How are specific system test requirements determined?
8. Why is it important to establish the proper level of logistic support for system evaluation? What would probably happen in the absence of adequate test procedures? What would happen in the event of inadequate training of test personnel (i.e., the consumer)?
9. How are logistic support requirements determined?
10. A good data subsystem serves what purpose(s)?
11. Why is system success data important? What type of information is required?
12. What data are required to measure the following: cost effectiveness, system effectiveness, operational availability, and life-cycle cost?
13. In the event that system evaluation indicates noncompliance with a specified system requirement, what would you do?
14. If during the evaluation the consumer makes a recommendation for product improvement, what would you do?
15. Why is change control so important? How do system changes affect logistic support?
16. What function does the change control board perform?
17. What role does engineering fulfill in the product evaluation function?
18. Briefly describe the producer and consumer functions in system evaluation.
19. What benefits are derived from test reporting?

8

PRODUCT USE AND LOGISTIC SUPPORT FUNCTION

The ultimate objective in accomplishing the activities discussed in Chapters 3 through 7 is to provide the consumer (or user) with a product that will satisfy a designated need for a specified period of time in an effective and efficient manner. Now with the distribution and operational use of the prime equipment and its associated logistic support at hand, the system as an entity is ready for its true test (i.e., to satisfy consumer needs), and the results will be a realistic assessment of the extent to which this objective has been met.

8.1 PRODUCT DISTRIBUTION AND OPERATIONAL USE[1]

Experience has indicated that numerous problems of all types occur when the product or system first becomes fully operational, in spite of the design and development effort described earlier. Many of these problems fall into one or more of the following categories:

[1]Distribution refers to the delivery of goods and services by the producer to the consumer. It includes prime equipment, software, and elements of logistic support. Distribution is accomplished both on an initial basis, when the system is first delivered to the consumer, and on a sustaining basis, relative to the materials and services required for support of the system throughout its life cycle. Depending on the product type, distribution can be quite complex. For readers wishing to delve further into this area, there are a number of texts available which cover the various ramifications of distribution theory in detail.

A. Consumer personnel are not adequately trained to operate and maintain the system and its components. Thus, the product may be utilized improperly, which results in reduced system life, increased logistic support, and possible loss of the system's ability to accomplish its intended mission.

B. One or more of the individual elements of logistic support is incompatible with the prime equipment (e.g., facilities are not adequate in terms of space, utility requirements, and/or environment; the prescribed test and support equipment will not perform the proper functions in verifying prime equipment operation or in the performance of maintenance).

C. The depth and extent of support provided are insufficient (e.g., there are not enough of the required type of spares available, system operating and maintenance instructions are inadequate relative to coverage).

D. The level of support in certain areas is higher than is actually required (e.g., facilities, personnel, and support equipment are not being fully utilized, or there are too many spare parts of a certain type, which results in a higher inventory cost than initially anticipated).

E. The elements of logistic support are not compatible with each other (e.g., the maintenance procedures do not cover the tasks being performed at a given echelon and the support equipment used in task accomplishment).

Because of the inherent problems at this stage in the life cycle, the number of procedural adjustments and/or equipment modifications required to correct the deficiencies has in some instances been extensive and costly. In other words, a major program modification effort has been accomplished to bring the system up to the level of operation required. In other cases, this has not been technically or economically feasible at this late stage, and the system has never been able to fulfill its need objectives; thus, the results of all the previously conducted design and development efforts have been ineffective.

Realization of these rather conventional past problems has caused an increased emphasis in following the system development process discussed previously and the consideration of logistic support early in the life cycle. With the necessary steps accomplished to date, it is hoped that most of the problems falling in one or more of the categories defined above have been eliminated and will not occur when the system is fully operational. Certainly, it is unrealistic to assume that all related problems have been eliminated, but it is not unrealistic to assume that many of the typical discrepancies occurring in the past have been designed out of the system with application of the proper level of effort. If this overall objective is met, the functions remaining will consist of:

A. Operating and supporting the system in the prescribed manner for its planned life cycle.
B. Providing an ongoing assessment of system operation and support in terms of performance, effectiveness, and cost.
C. Initiating any corrective action as necessary for improvement.
D. The phase-out of items from the inventory and the subsequent material disposition actions as required (e.g., processing material for decomposition, reclamation, and/or recycling for other usages).

8.2 ELEMENTS OF LOGISTICS AND LIFE-CYCLE SUPPORT

"Logistics is the art and science of management, engineering, and technical activities concerned with requirements, design, and supplying and maintaining resources to support objectives, plans, and operations."[2] The elements of logistic support include the following:[3]

A. *Personnel and training*—personnel required for the operation and maintenance of the prime equipment, software, and associated test and support equipment throughout the life cycle. Personnel are identified in terms of quantity and skill-level requirements for each operational and maintenance function by echelon and location and may include the consumer, representatives from the producer's organization, or a combination.

Training includes both *initial* training for system/equipment familiarization and *replenishment* training to cover attrition and replacement personnel. Training is designed to upgrade both operator and maintenance personnel to the skill levels defined for the system. Training data and equipment (e.g., simulators, mock-ups, devices) are developed as required to support personnel training operations.

B. *Test and support equipment*—all tools, monitoring and checkout equipment, metrology and calibration equipment, maintenance stands and handling equipment required to support operation and the scheduled and unscheduled maintenance actions associated with the prime equipment, training equipment, facilities, and other items of support equipment.

C. *Supply support*—all repairable spares (subsystems, units, assemblies, modules, etc.), repair parts, consumables, special supplies and related inventories needed to support scheduled and unscheduled maintenance

---

[2]This definition is accepted by the Society of Logistics Engineers (SOLE) and commonly used in many areas.

[3]The elements of logistic support and the determination of logistics resource requirements are also described in: B. Blanchard, *Logistics Engineering and Management*, Prentice-Hall, Inc., Englewood Cliffs, N.J., 1974.

actions associated with the prime equipment, test and support equipment, facilities, and training equipment. Provisioning, material acquisition and processing, and so on, are included. Considerations address each geographical location where spare/repair parts are distributed and stocked, the distances between stockage points, and the methods of material distribution.

D. *Facilities*—physical plant, real estate, portable buildings, housing, intermediate maintenance shops, depots, and so on, required to support (1) production or construction of the system and its components; (2) operational and maintenance functions associated with the prime equipment, test and support equipment, and training equipment throughout the system life cycle; (3) storage for spare/repair parts and data; (4) quarters for operator and maintenance personnel; and (5) training operations. Capital equipment and utilities (heat, air conditioning, power, telephone, etc.) are generally considered as part of facilities.

E. *Technical data*—drawings, microfilm, operating and maintenance instructions, modification instructions, provisioning and facilities information, specifications, inspection and calibration procedures, and computer programs (routines) required to support installation, checkout, operation, and maintenance of the prime equipment and associated test and support equipment.

F. *Transportation and handling*—special provisions, reusable containers, and supplies necessary to support packaging, preservation, storage, handling and/or transportation of prime equipment, test and support equipment, spare/repair parts, personnel, technical data, and facilities.

The requirements for logistic support are analyzed, evaluated, and defined throughout the preliminary system design and detailed equipment design stages (i.e., the completion of logistic support analyses). The provisioning and acquisition of support elements occur during the latter part of detailed equipment design and in the production/construction phase. At this point it is incumbent to verify that those elements of logistics selected will adequately support the prime equipment in the field.

Logistics during this phase of the life cycle may assume different postures, to include the following:

A. The transition of system operation and support from the producer to the consumer, sustaining system operation by the consumer, and the continued support of the system by the consumer throughout its life cycle may be planned. In other words, the consumer assumes immediate and complete responsibility for all system support.

B. The transition of system operation from the producer to the consumer and sustaining consumer operation while the producer retains complete responsibility for life-cycle support may be planned. In other words, the producer is completely responsible for all system support through some type of a contracting arrangement or a leasing agreement with the consumer.

C. The transition of system operation from the producer to the consumer with a split consumer-producer life-cycle support responsibility may be planned. In this case, it may be feasible for the producer to retain the life-cycle support of some items while shifting the support of other items to the consumer.

The support posture assumed is initially defined in the product use and logistics support plan described in Section 3.4E and is dependent on the type and complexity of the system, the goals and objectives of the producer and consumer, program risks and uncertainty, contractual arrangement, liability provisions, and many other factors.

Regardless of the approach selected, the transition process from producer to consumer is extremely important for system acceptance and success.[4] In the area of operation, the producer must ensure that the consumer thoroughly understands how the system is to be utilized (i.e., knowledge of the proper operational procedures). This may be accomplished through a combination of both formal and on-the-job training, accomplished at the producer's facility, at the consumer's operational field site, or both. Training may be conducted initially when the system is first deployed and subsequently on a continuing basis to cover personnel shifts and attrition and may be provided through a combination of live classroom instruction, the use of video tapes or cassettes, computer-based instructions, review of technical data, and so on. In addition, producer engineering personnel may be assigned as field service engineers at the various operational sites to assist the consumer on a day-to-day basis.

For maintenance support (i.e., scheduled and unscheduled maintenance actions), the producer provides the necessary logistics resources, as indicated by the support plan selected. If the consumer is to assume full system support, the producer will usually provide a predesignated level of support for a specified period of time until the system becomes fully operational. This includes maintenance personnel and training, supply support, test and support equipment, and facilities. On the other hand, if the producer assumes the responsibility for life-cycle support, all logistics resources are provided to the consumer on an as-required basis.

Providing the consumer with the necessary system support is essential for ultimate success, particularly if the system is fairly complex in nature. With-

---

[4]Some of the aspects of system transition from production to operational use are discussed in Section 6.6.

out this support, there is a high probability that the system will be utilized incorrectly, causing customer dissatisfaction, personnel-induced faults, increased logistic support requirements, reduced mission capability, and so on. In addition, personnel and equipment safety may be reduced, causing hazardous situations that could result in legal actions taken against the producer. Thus, it becomes incumbent on the producer to minimize these types of problems by following the proper steps in the development process discussed in previous chapters; that is, logistic support must be adequately planned and integrated into the system design process early in the life cycle.

8.3 SYSTEM EVALUATION AND MODIFICATION

As discussed in Chapter 7, system evaluation is an ongoing effort conducted throughout the life cycle. System operation and support assumes a dynamic (rather than static) state as conditions often change, either knowingly or unknowingly.[5] Thus, it is necessary to establish a data collection and analysis capability, which will provide an assessment of system performance and effectiveness on a demand basis. The producer should be interested in the results in terms of how the product is doing in the field, the extent of consumer satisfaction, and an identification of possible areas for improvement (either for this product or for future products with similar characteristics). In addition, the consumer should be interested in determining system performance and effectiveness characteristics (to include cost) and whether the system is fulfilling its intended function in an effective and efficient manner.

The data collection, analysis, and system evaluation effort during the product use and logistic support phase should follow the format described in Section 7.4. The desired information is presented in Figure 7-2. Likewise, in the event that the evaluation effort indicates the need for system modification, the process illustrated in Figure 7-5 should be followed. Recommended modifications must be reviewed thoroughly in terms of effects on system performance, effectiveness, and cost prior to incorporation. If modifications are to be incorporated, kits must be designed and installed, and the system should be tested to ensure proper operation. The affected elements of logistic support (e.g., test and support equipment, technical data) must also be modified as necessary for compatibility purposes.

---

[5]Changes may occur due to *internal* factors such as organizational changes, slight variations in equipment operating procedures, and so on, or due to *external* factors constituting economy shifts, the results of political pressures, public opinion, and the like. Changes may occur suddenly or gradually over a long period of time and, in the case of the latter situation, often are unnoticed.

8.4 SYSTEM PHASE-OUT AND MATERIAL DISPOSITION

System phase-out and the disposal of material no longer required in the operational inventory is a subject not too well covered in the literature. It is common to address the design and development of a system and the operation of that system; however, quite often the phase-out and subsequent disposition of that system is not adequately considered until the time arrives to do something about it.

Realistically, system phase-out and the disposal of material is a definite part of the life cycle. In fact, nonrepairables are removed from the inventory as equipment failures occur in the field. The phase-out and disposition of material should be considered in the early conceptual and design stages of a project. The planning for phase-out and disposal should be responsive to the following questions:

A. What should be done to equipment phased out of the inventory, and where should it be accomplished?
B. At what rate should obsolete items be phased out of the inventory?
C. How is disposability considered in the system/equipment design process and what measures are used?
D. How much of the equipment can be salvaged and used for other purposes? Can the material be processed, converted, and used in the manufacture of another product?
E. Is the method of disposal consistent with ecological and environmental requirements? What are the effects on the environment? Refer to Part V.
F. What logistic support requirements are necessary to accomplish phase-out disposal functions?

System/equipment phase-out and material disposal is another significant stage in the life cycle. There are functions to be accomplished involving transportation and handling, disassembly, decomposition, processing, and so on. Associated with the completion of these functions are the requirements for logistic support (i.e., transportation and handling equipment, personnel, facilities, and technical procedures). These requirements are identified through a functional analysis and the logistic support analysis as applied to this stage of the life cycle. Depending on the type of system and the quantity of equipment in the inventory, the logistic support requirements can be significant; thus, the system program manager must address system phase-out and material disposal in the same level of detail as with any other stage in the life cycle. This area is initially covered in the early system design stage (through the establishment of design disposability criteria) and is subsequently included in the formal product use and logistic support plan.

QUESTIONS AND PROBLEMS

1. What major problems would likely occur in the product use and logistic support phase if the proper steps in the system development process had not been followed? How can these problems be minimized?
2. What factors must be considered in the transition from production/construction to operational use?
3. Why is it important to ensure that the consumer possesses adequate knowledge in the operation and the support of the system prior to delivering the system for operational use?
4. Define logistics. What are the elements of logistic support? How are logistic support requirements for a system initially determined?
5. How are the geographical locations where equipment is to be deployed and the quantity of equipment designated by location determined?
6. How is the maintenance support plan for the system derived?
7. How are operation/maintenance personnel requirements and training requirements (for the system) related?
8. What is included in supply support? Test and support equipment? Facilities? Technical data?
9. What factors are considered in determining whether a system (or an element of a system) should be supported by the producer throughout its life cycle or by the consumer?
10. Why is it necessary to continue with the system evaluation effort throughout the product use and logistic support phase?
11. Why is the producer interested in the results of system evaluation that are derived during the product use and logistic support phase? How could these results be related to marketing activities?
12. What are the consumer interests in system evaluation?
13. How would you verify the adequacy of the following: spare and repair parts, test and support equipment, personnel quantities and skill levels, operating and maintenance procedures, and facilities?
14. What factors should be considered prior to the initiation of system modifications? Why is change control so important?
15. Why is system/equipment phase-out and material disposition significant? What factors influence decisions concerning the methods used in material disposition?
16. How would you consider equipment disposability in the design process? How would you determine the cost of disposal?
17. What (if any) is the significance of logistic support in the system/equipment phase-out and material disposition stage?
18. What role does engineering play in the product use and logistic support phase?

part three

Organization and the Product Life Cycle

Part Three covers basic organizational concepts (i.e., purpose of organization, principles of organization, structuring the organization, etc.) and the organization of engineering functions around the product life cycle. Line and project-staff organizational structures are discussed, together with responsibilities and interfaces. Material in this part is represented by the third circle in the illustrated conceptual outline.

Conceptual Outline

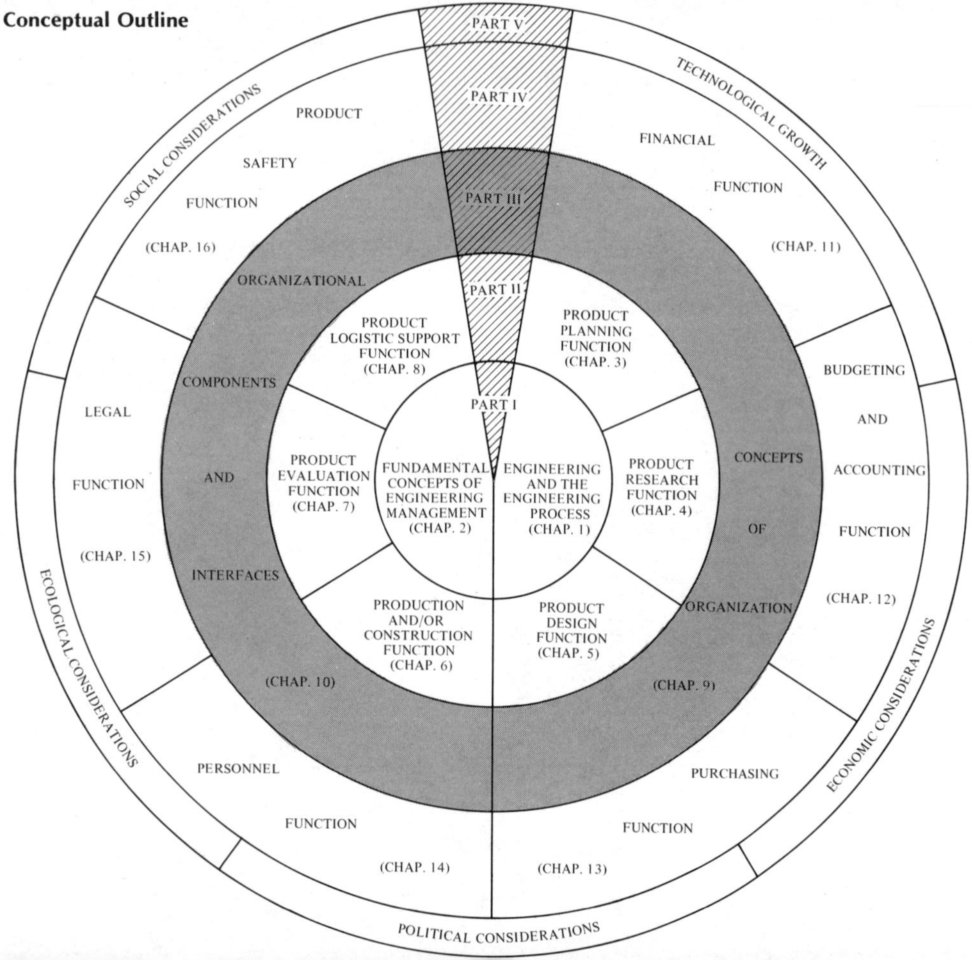

PART I	ENGINEERING IN MODERN TIMES
PART II	THE PRODUCT LIFE CYCLE
PART III	ORGANIZATION AND THE PRODUCT LIFE CYCLE
PART IV	RELATED FUNCTIONS AND INTERFACES
PART V	ENGINEERING AND THE ENVIRONMENT

9

CONCEPTS OF ORGANIZATION

Formal organizations are contrived social entities designed to fulfill specific purposes. Organizations constitute a group of individuals of varying levels of expertise combined into a social structure of some type to accomplish one or more functions. Organizational structures will vary with the functions to be performed, and the results will depend on the goals and objectives established (both on a group and on an individual basis), the resources available, the communications and working relationships of the individual participants, motivation, and many other factors. Some of the more significant aspects of organization are discussed in this chapter.[1]

9.1 PURPOSE OF ORGANIZATION

The purpose of organizing is to achieve the most effective utilization of human, material, and monetary resources by the establishment of decision-making and communications processes designed to accomplish specific objectives. This concept is illustrated in a simple form in Figure 9-1.[2] An

[1]The level of discussion of organization concepts in this chapter is rather cursory in nature and is intended to provide the reader with an *overview* of some key points. More in-depth coverage is provided in numerous texts on organization and management theory and related subject areas. A few recommended references are included in the bibliography in Appendix D.

[2]Figure 9-1, although oriented to the discussion on organization, directly supplements the information flow presented in Figure 2-1.

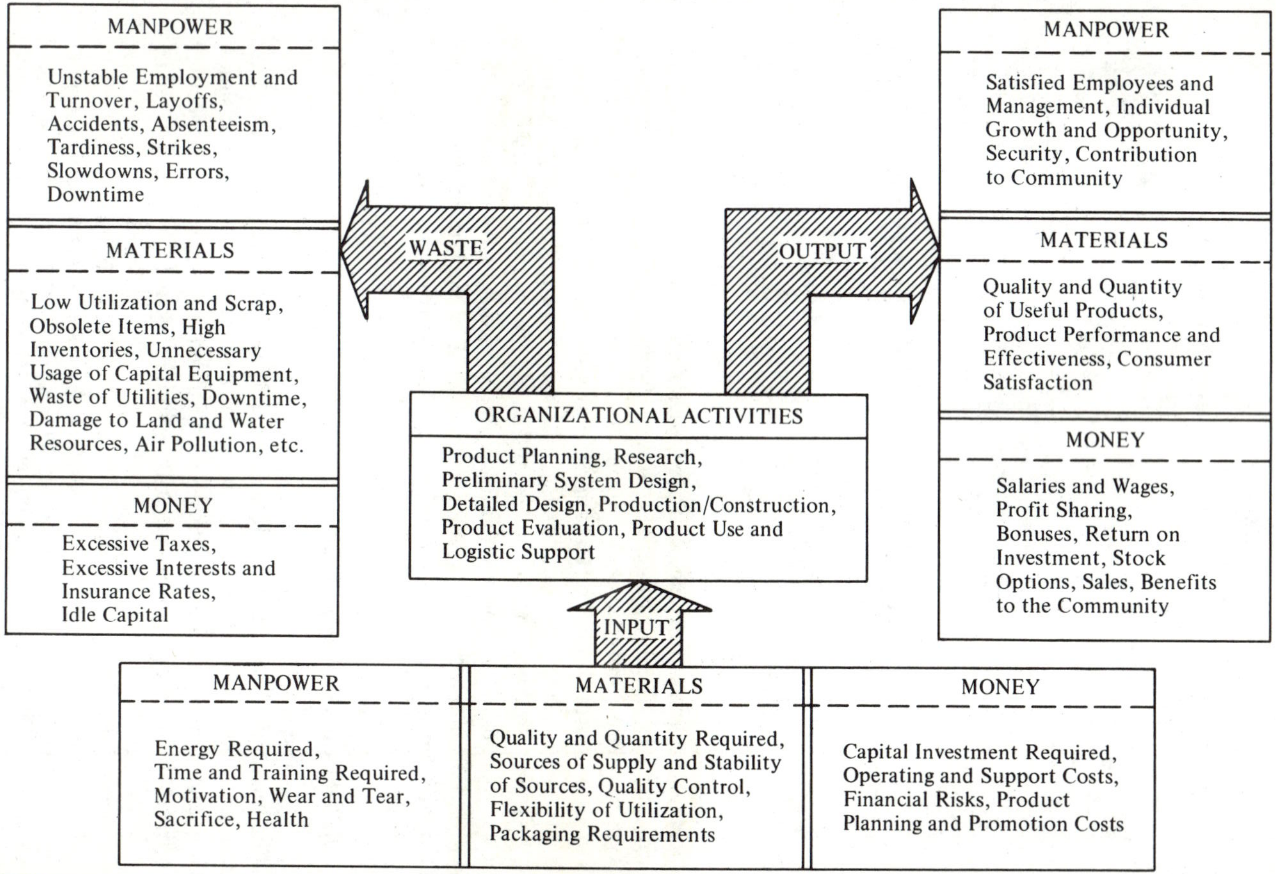

Figure 9-1. The organization in terms of inputs and outputs.

organization requirement is identified to meet one or a set of objectives, and human, material, and monetary resources are necessary to fulfill this goal. The utilization of these input resources to accomplish organizational activities (e.g., product planning, research, design) is then measured in terms of productive output and waste. The purpose, of course, is to maximize the productive output and minimize waste.

9.2 ESTABLISHING GOALS AND OBJECTIVES[3]

Before a definitive organization structure can be established, the applicable firm/agency's top management must first develop the *overall* goals and objectives for that firm. For instance:

A. What technology should the firm or agency attempt to develop?
B. What product lines should be pursued?
C. What is the desired market position? Productivity? Profitability and growth? Product leadership?

Overall goals may be stated both qualitatively and quantitatively and may include financial objectives, marketing objectives, research and engineering objectives, and production objectives. A few characteristic examples are noted below:

A. Achieve a 15 per cent return on investment for the next fiscal year.
B. Reduce the yearly operating cost by 12 per cent.
C. Complete the design and development of a new prototype in 15 months within a cost of $560,000.
D. Develop three new product lines within the forthcoming fiscal year, with a forecasted sales of not less than $1,750,000.
E. Develop and implement a complete logistic support capability for product X within 6 months.
F. Increase the production yield of product Y by 10 per cent in 2 months.

The establishment of goals and objectives is somewhat of an iterative process. Initially, goals and objectives must be identified for the overall firm or agency at the top level of management. Based on the results it is possible to determine what activities need to be accomplished, which in turn leads to

[3]The author believes that "management by objectives" is an excellent approach in the attainment of organizational effectiveness. Two suggested references covering this area are: G. S. Odiorne, *Management by Objectives*, Pitman Publishing Corporation, New York, 1965; and P. Mali, *Managing by Objectives—An Operating Guide to Faster and More Profitable Results*, John Wiley & Sons, Inc., New York, 1972.

the grouping of these activities and the definition of an organizational structure.[4] The established organization is then able to review the top-level goals and objectives in terms of its ability to respond and the methods of approach envisioned. This response may constitute the breaking down of top-level goals into specific lower-level goals for each significant element of the organization and even down to the individual within the organization. Goals often vary with the level of organization. The lower level goals are frequently project oriented, greater in number, and the target period may be shorter. The level of activity (i.e., an organizational group or an individual) directly affected by a specific goal should be included in the goal-setting process, since the participative approach will definitely enhance the motivation required in meeting the goal later. Thus, the establishment of goals and objectives becomes a top down-bottom up process.

Goals and objectives may be stated in rather general terms or in a specific manner. Actually, to be meaningful, they should be as specific as possible and related in quantitative measures where appropriate, to allow for a realistic assessment of organizational status and trends. In many instances, multiple goals are established, and occasionally some of these goals may appear to be somewhat conflicting. When such occurs, a priority system must be established to determine a level of importance for each individual goal. This may constitute a ranking of some type based on highest payoff. In cases of conflicting goals, top management must determine what takes precedence.

The process of establishing goals and objectives often is quite dynamic, since changes may occur as a result of economic and social conditions, technology advances, political influences, and so on. Goal modifications may be induced anywhere in the organizational chain. Top management may change goals because of some political pressure being exerted on the firm, or the individual engineer may force a goal change due to some advancement and/or constraint in technology. In any event, although goals and objectives are initially established as a basis for organization progress toward the fulfillment of a specific need, they must be reviewed on a periodic basis to ensure that they are realistic from the standpoint of both internal and external conditions.

9.3 STRUCTURING THE ORGANIZATION

Once overall goals and objectives have been initially established, it becomes necessary to define the scope of work to be performed. This definition process is basically accomplished through the product planning function

[4]Unfortunately, in real life, organizations are often established and staffed before the identification of overall goals and objectives. This usually leads to redundancies, the hiring of unqualified personnel, incompatibilities relative to individual goals and those of top management (i.e., conflicts in goals), and the expenditure of unnecessary resources, resulting in waste.

discussed in Chapter 3. The results are described in appropriate specifications and plans (refer to Section 3.4B). The next step is to develop an organization structure that will respond to the requirements, and in accomplishing such, one must keep in mind the basic principles of organization and the various organizational concepts.

A. Organization Principles

A great deal has been written about the principles of organization, all of which is intended to assist in the development of an organization structure able to carry out its assigned goals and objectives. The purpose here is to review a few of the more common principles applicable to organization design.[5]

1. Span of management control. Span of management control usually refers to the number of individuals that report to any single supervisor or executive. In other words, the number of people reporting *directly* to a manager is known as his span of management. Although it is a common premise to assume that the ideal span is somewhere between four and eight, actual experience indicates that the number will vary with individuals and from one situation to the next. Major influencing factors include the following:

a. *Management control by organization level.* Individuals in top positions within an organization and responsible for rather difficult policy level decisions usually have fewer people reporting to them than first-line supervisors responsible for carrying out routine procedures. The accomplishment of routine operations should require less individual supervision on a person-to-person basis, and a greater number of subordinates is possible with the same degree of control. Thus, the span of management in numbers will usually vary with the level of organization.

b. *Management control by nature and complexity of function.* Although the level of organization may be the same, the span of management will vary with the complexity of job function. Engineering design decisions generally are based on an analysis of some type which requires more effort and personal supervision than production decisions based on go/no-go criteria and which are repetitive in nature. In this instance, fewer people will report to the engineering manager than to the production manager, even though both may be at the same level of organization. As one deals with the overall system life cycle discussed in Part Two, the span of management control will vary from one program phase to the next and for different categories of functions/tasks within any given phase.

c. *Individual abilities.* The inherent capabilities of individuals will vary in terms of current skills, adaptability to changes, aptitude relative to learn-

[5]Secton 2.2 identifies some of the principles applicable to organization design.

ing new functions, and motivation. In some situations subordinates are able to handle more work and will require less supervision than in other instances. This condition may exist for any given position and is best addressed through the proper definition of job requirements and the subsequent hiring and training of personnel. The span of management control may vary significantly with the individuals involved; thus, the proper matching of abilities and requirements is necessary to promote organizational effectiveness (i.e., the right man for the job).

d. *Willingness of supervision to delegate authority and responsibility.* In some cases managers or supervisors find it very difficult to delegate authority (and associated responsibilities) and will check on every action taken by each subordinate. This will definitely limit the span of control and inhibit organizational growth.[6]

When applying the principle dealing with span of management control to organization design, one can readily see that any number of levels could be established. As the span becomes narrower, more levels of management will be required to accomplish the same amount of work. This results in (a) increased costs due to the number of managers required; (b) ineffective communications since information must travel through more levels, which usually creates distortions or inaccuracies; and (c) less motivation within the organization, since responsibilities are fractionated into many segments, resulting in less challenge to each individual concerned. Conversely, if the span is too broad where a manager or supervisor has too many people reporting to him and he loses control of everything, the organization will also suffer, since the individual elements will tend to stray from its overall purpose. Hence, the intent is to minimize the levels of management and yet maintain the necessary controls to effectively accomplish the organization's overall objective.

2. *Delegation of responsibility and authority.* Quite often a manager is willing to delegate the *responsibility* for a particular activity to a subordinate but will retain the *authority* to control and direct the means by which the activity is accomplished. In this situation the individual assigned the responsibility is powerless when it comes to the full utilization of available resources, and when things go wrong he is often unable to initiate the necessary corrective action without first receiving the proper authority from his superior. This sometimes takes time and causes delays in task accomplishment

[6] Hopefully, supervisors can be trained to delegate more authority and responsibility to their respective subordinates. In some instances, this is forced by assigning more and more work to an individual supervisor until he cannot perform and has to delegate work to a lower level in order to get the job done.

which may be costly. In addition, the responsible individual may see ways in which improvements can be made but will not take action accordingly, owing to a lack of authority. In any event, the individual becomes discouraged, loses motivation, and his productivity ultimately falls off.

This principle can also have a reverse effect, since many individuals are continually seeking more authority to create a feeling of greater self-importance but are unwilling to accept the responsibility that goes along with it. A person desires the authority to control activities but may hesitate to accept the responsibility, particularly when things go wrong. In such instances he or she may be tempted to point a finger at someone else to convey that he was not responsible. In other words, he will not hesitate to accept the glory if all goes well but will quickly shift the blame when problems occur.

Obviously, it can be concluded that responsibility and authority must go together if effective results are to occur. It is essential that each person in the organization know and thoroughly understand the limits of his or her responsibility and authority. Otherwise, he may go too far or not far enough in attempting to accomplish his functions. This, in turn, can result in a trial-and-error situation, with the individual continually probing to determine an approach that will meet with approval. Such probing sometimes leads to opposite results, causing unhappiness and criticism on the part of his supervisor, leading to disciplinary action and personal humiliation which could have been avoided.

3. ***Centralization and decentralization of responsibility and authority.*** When dealing with organizational patterns in general, one should recognize the various aspects of centralization versus decentralization, which relate closely with span of management control and delegation. Centralization implies a centralized organization where decision making occurs at the *highest* possible level within the firm or agency. Proponents of this concept maintain that better decisions can be reached by the most competent individuals, which theoretically represent those in top management. Decentralization, on the other hand, promotes the delegation of responsibility and authority as far down in the accomplishment (i.e., that level at which the action is taking place) to make decisions related to this effort.

Actually, either concept may be appropriate, depending on the size of the firm/agency and its objectives. Also, certain functions within an organization may favor a given approach. For instance, it may be preferable to apply the concepts of centralization to such functions as finance, contracts, and pricing, while implementing the decentralization approach in the areas of research and engineering, manufacturing, and product support.

Concerning the areas of emphasis in this text, the author feels that there are more advantages associated with decentralization, for the following reasons[7]:

a. Decentralization (versus centralization) of responsibility/authority and decision making results in a more rapid response to changes at the working level, as a result of environmental, economic, social, political, and technological influences.

b. Decisions can be made by the individuals who possess the greatest knowledge of local conditions.

c. Decision making at the lower levels in the organization tends to facilitate the personal development of the individuals involved, and hence results in a stronger overall organization.

d. The decentralization concept promotes a greater feeling of satisfaction throughout the organization because of the participative aspects in decision making. Individual motivation and productivity is generally higher since the persons involved feel as though they are contributing something.

One inherent danger in the decentralized approach is that decisions are often made by lower-level managers or supervisors without the knowledge or consideration of the effects of those decisions on other areas. As discussed in earlier sections of this text, the results of certain decisions in one area may significantly impact a dozen other areas. Thus, it is imperative that the decision maker be aware of the cause-and-effect relationships associated with his actions. He must be able to adequately assess the consequences of his decisions prior to making those decisions. This can be facilitated through training, good communications, and the use of the tools available to him (refer to the discussion on engineering decision making in Section 2.3). The checks and balances of decision making as applied to projects are of course covered through periodic program/design reviews, described in Part Two. In spite of this concern, however, it is still felt that decentralization is preferred in most instances.

B. Organization Patterns

The establishment of organization patterns, or the development of an organization structure, generally evolves from the following:

1. The identification of work to be performed (e.g., research, preliminary system design, detailed design of unit A) and the breakdown of this work into specific tasks are accomplished initially through product planning. These tasks are organized into work packages as

[7]The advantages of decentralization are also listed in: D. W. Karger and R. G. Murdick, *Managing Engineering and Research*, 2nd ed., Industrial Press, Inc., New York, 1969, p. 41.

illustrated in Figure 3-15 and described in Section 3.4C. The defined work packages are evaluated from the standpoint of task type, complexity, and the required completion schedule.

2. The individual work packages are then grouped on the basis of similarity and homogenous characteristics to determine positions, and the positions, in turn, are arranged within an organization structure. This arrangement should be compatible with the goals and objectives discussed in Section 9.2 and the principles of organization covered in the previous section. Additional factors should include the projected available resources required for staffing and support, past history (i.e., successes and failures), and the geographical divisions of activity that will directly influence the results.

The manager responsible for the organization may develop a completely new structure, restructure an ongoing organization, or take over a current capability as is. In each situation, he or she is involved in the *continuous* process of evaluating current status and objectives and the subsequent implementation of the organization plan. This process is continuous, since real-world conditions are constantly changing, which directly influences consumer needs and the marketplace. Thus, organization design is dynamic, and established organization structures must be flexible in adapting to new situations.

Organization patterns in general constitute both line and staff functions. The *line* organization represents chains of supervisor-subordinate relationships and involves functions that are *directly* concerned with achieving the prime objectives of programs assigned to the total organization. The *staff* approach refers to functions and individuals that are not in the line-function chain of command and whose efforts do not bear prime responsibility for accomplishing the major objectives of the organization. Staff functions support the line organization in an advisory and/or supporting capacity. Staff functions may appear in the form of single individuals providing ongoing advice to a manager in specific areas or as functional groups of personnel accomplishing designated program tasks. Staff functions may support any number of different line organizations concurrently depending on the needs. Although staff functions are not officially in the direct chain of command in a line organization, considerable influence on line functions can be exerted, depending on the individual relationships of staff personnel and the line manager. More discussion of line and staff functions will follow.

1. ***Functional organization.*** The primary building block for all organization patterns is the approach, which constitutes the grouping of functional specialties or disciplines into separately identifiable entities. In other words, the intent is to perform similar work within one organizational component.

In the *pure* functional structure, all engineering work would be the responsibility of one executive, all manufacturing effort would be the responsibility of another executive, and so on.

For small firms or agencies this approach is often desirable, since it is easier to manage a homogeneous group of personnel with similar backgrounds and interests, and duplication of effort is minimized. On the other hand, for large operations this is not necessarily preferable, owing to the large, impractical centralization of responsibility. Thus, for large, multi-product firms, the pure approach becomes a somewhat modified configuration organized predominantly along functional lines, although some cross-disciplinary activity is introduced.

Although there are variations from one situation to the next, Figure 9-2 presents a functionally oriented organization structure, including activities that are representative for many large companies.[8] The activities presented should be visualized in the context of the product life cycle (covered in Part Two) and vice versa. During the early phases of the life cycle, emphasis is of course placed on marketing analysis, advertising and sales, advanced product planning, research, and preliminary system design. The majority of tasks to be accomplished will be in these areas, and organization staffing (i.e., quantity of groups, people, etc.) will follow accordingly. As the product life cycle progresses, emphasis will shift from preliminary system design to detailed design, production/construction, evaluation, and product support, requiring a comparable shift in organizational content. Thus, the organization structure will not only vary due to external factors such as changing consumer needs and the marketplace, but will vary as a function of the life cycle of the product(s) that it supports.[9]

The majority of engineering functions shown in Figure 9-2 fall under the cognizance of the Vice President for Research and Engineering. However, engineering functions per se are also accomplished in (a) advanced product planning and customer liaison; (b) industrial engineering, manufacturing engineering, and plant engineering; (c) quality control; (d) product support (logistics); and (e) management services.[10] Many of the other areas noted in

---

[8]It should be noted that the organization structure may vary as a function of product types and quantities developed and produced, type of work performed (i.e., research, product design and development, production/construction, etc.), and the number of facilities and geographical locations utilized. Also, variations will occur between profit and nonprofit operations.

[9]The functional organization presented in this context must deal with *all* in-house products. Thus, the various products of the firm are initially addressed on an individual basis and then combined to determine total organization requirements.

[10]These other components may not in actuality be considered as part of engineering in the organizational sense. Nevertheless, engineering expertise is required in the development of system operational requirements and the maintenance concept, technical marketing and planning, industrial and manufacturing engineering, design of test and support equipment, design of training equipment, design of computer support as part of management services, and so on. Much of the discussion in Part Two supports this fact.

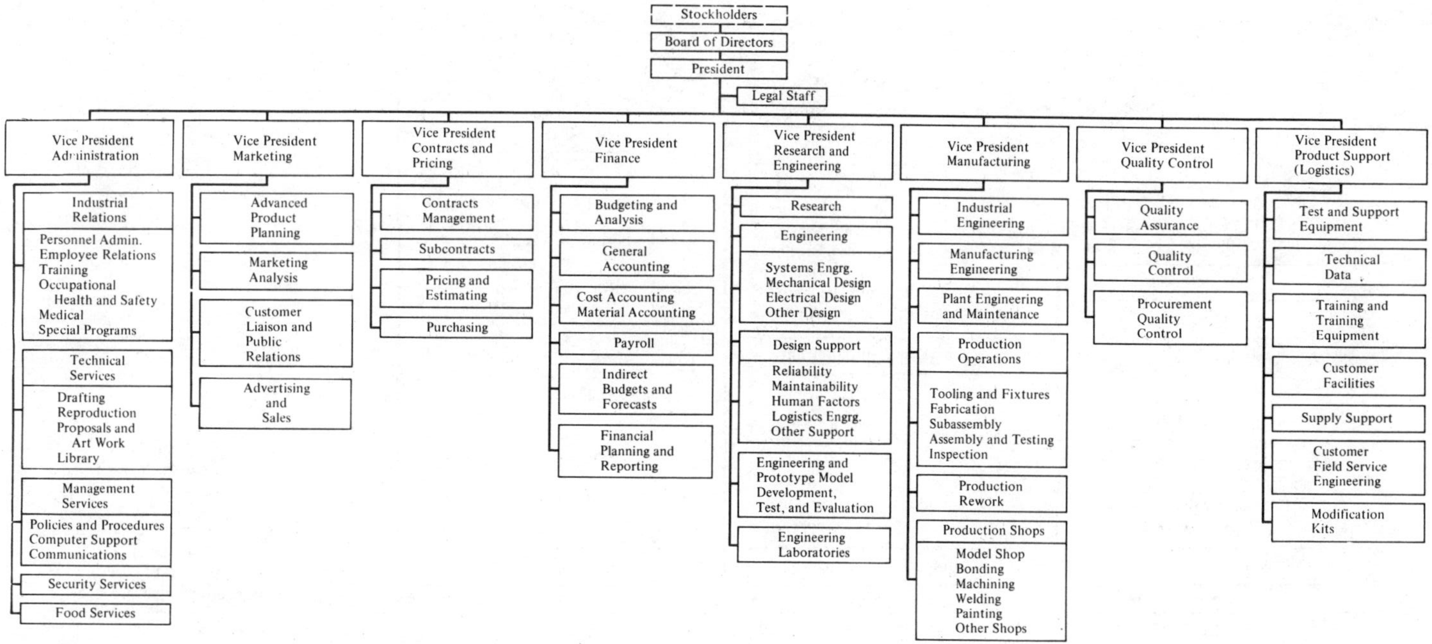

Figure 9-2. Functional organization (*example*).

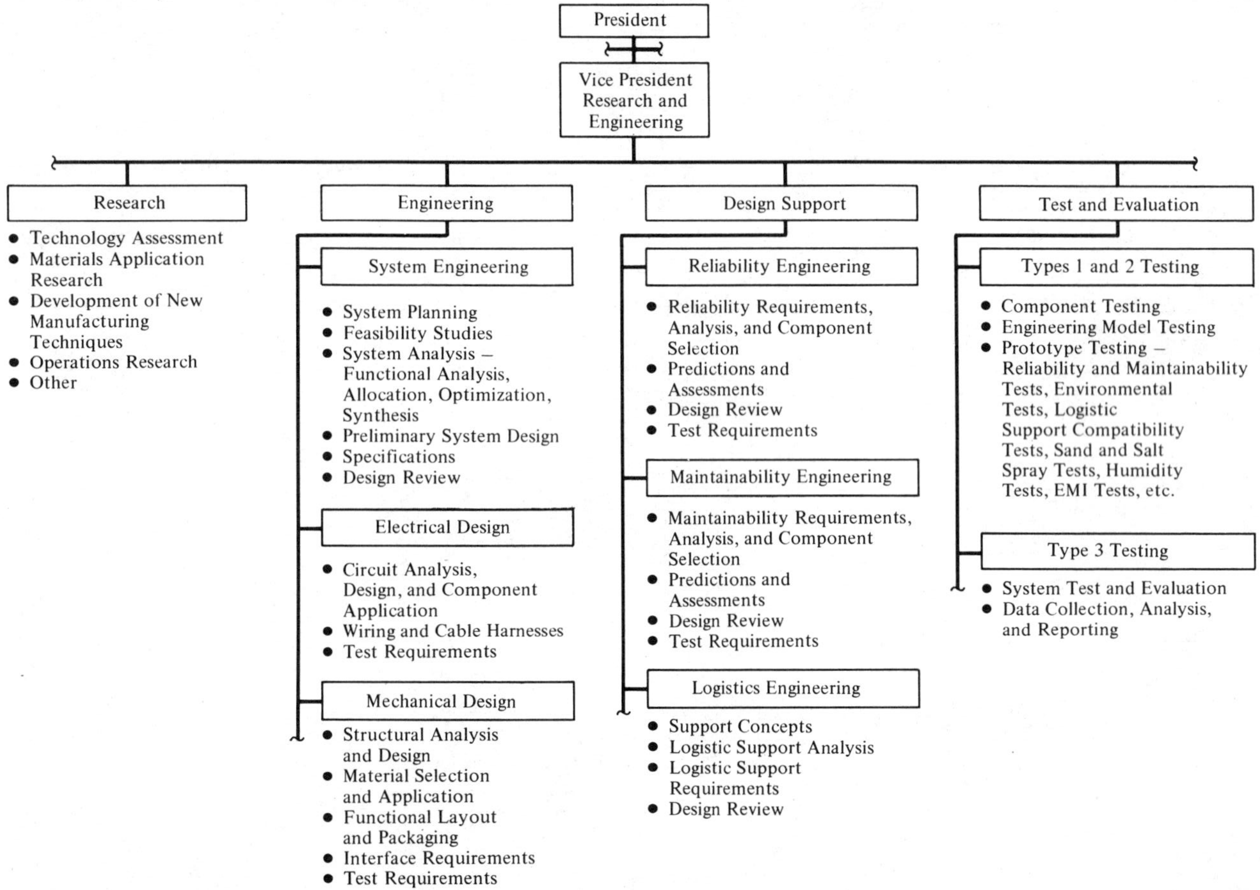

Figure 9-3. Organization activities (*partial*).

the figure either directly or indirectly support engineering. This is not to infer that engineering is "everything to all functions"; however, engineering expertise is required to some extent in all phases of the life cycle, and these requirements seem to be somewhat continuous, if not increasing, in view of technology advances and increasing product complexity.

With the organization established as depicted in Figure 9-2, it is necessary to break the structure down into components and to identify work packages and tasks for each component. An abbreviated illustration of the work package identification by organizational component is presented in Figure 9-3. When identifying work packages for each program, the results should be checked to ensure compatibility with the applicable work breakdown structures (see Figure 3-16). It is essential that all required elements of work be accounted for in the WBS and that an organization component be assigned the responsibility for work completion.

2. Project organization. A project organization is one that is solely responsive to the planning, design, development, production, evaluation, and support of a *single* system or product. It is time-limited, directly oriented to the life cycle of that system, and the commitment of the varied skills and resources required is purely for the purpose of accomplishing system tasks. The project will have its own engineering function, its own support function, and so on.

The manning requirement for a project usually starts with a slow and steady growth, then increases rapidly to maximum size, and gradually declines until the system is eventually phased out of the inventory. As work progresses and the system passes through the various phases of the life cycle, the organization will change significantly in size and structure. A simplified project structure is illustrated in Figure 9-4.

Firms or agencies responsible for the development and support of large systems will generally utilize the project organization approach. The project director has all functional responsibilities for the success or failure of his or her project. All project participants work directly in a line capacity for the project manager. When two or more projects are initiated, each project director, being responsible for achieving his own individual objectives, will compete for the best engineers, facilities and resources, and for priority of funds and supporting services. The competition, although perhaps beneficial from the project standpoint, may have detrimental affects on the overall goals and objectives of the firm if caution is not observed. Top management must continually review *all* project activities for the good of the firm as a whole.

3. Product-line organization. Large firms are often organized on a divisional basis such that each division is responsible for a designated prod-

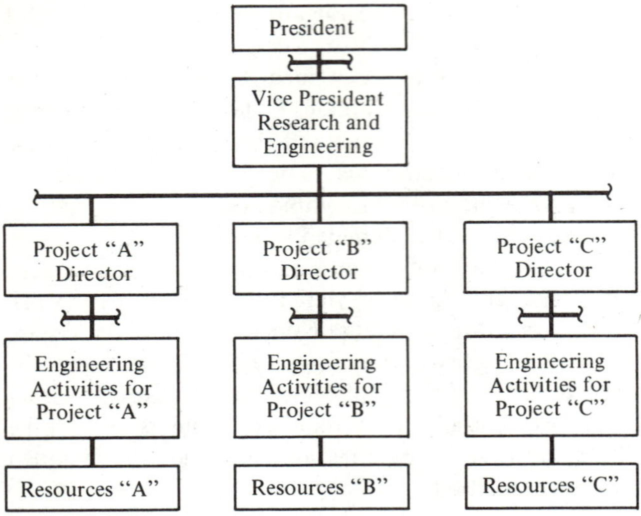

Figure 9-4. Simplified pure project organization.

uct line, and frequently these divisions are separated geographically. Within each division, the product-line responsibilities may be subdivided into either the pure project approach depicted in Figure 9-4, the functional or staff approach illustrated in Figure 9-2, or a combination of both. Figure 9-5 represents a basic product-line organization with project subunits. As in the case of project organizations, each product-line director is solely responsible for the development and support of all products within his or her domain. Each has his own marketing function, research function, engineering function, manufacturing function, and so on.

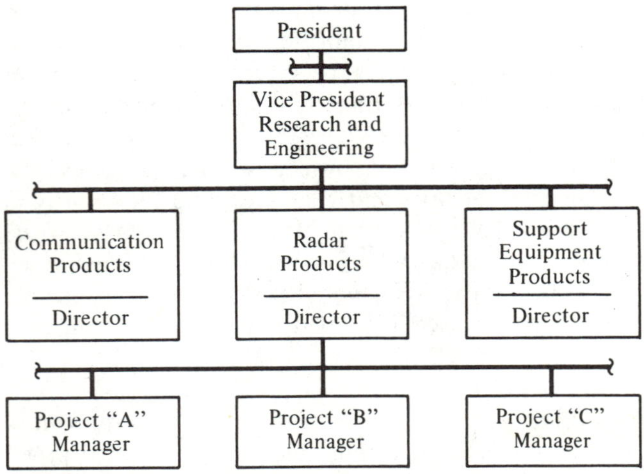

Figure 9-5. Product line organization with project sub-units.

4. Project-functional staff organization. The project-functional staff configuration represents a mix of the pure project organization presented in Figure 9-4 and the functional organization illustrated in Figure 9-2. For large single operations where the scope of work and funding are adequate, the configuration in Figure 9-4 may be appropriate. On the other end of the spectrum, where there are many different in-house product lines and each individually cannot justify a separate project organization, the concept in Figure 9-2 may be preferred. Quite often, however, the actual situation encountered constitutes an approach somewhere in between; that is, a firm or agency may be dealing with a variety of products involving activities differing in magnitude, levels of funding, duration of the life cycle, and so on. In other words, when combining the individual requirements, it may be appropriate to perform some functions through a project structure while others are better accomplished by specialized staff groups.

The mixed project-functional staff configuration imposes an activity matrix illustrated in Figure 9-6, where the firm or agency accomplishes its objectives through both project and functional staff groups. This matrix is more typically represented in Figure 9-7, which illustrates a project-staff organization. Project activities are oriented to individual product types, and the personnel assigned are totally committed to that project only. Their attention is undivided, motivation is high and directed to the project (each

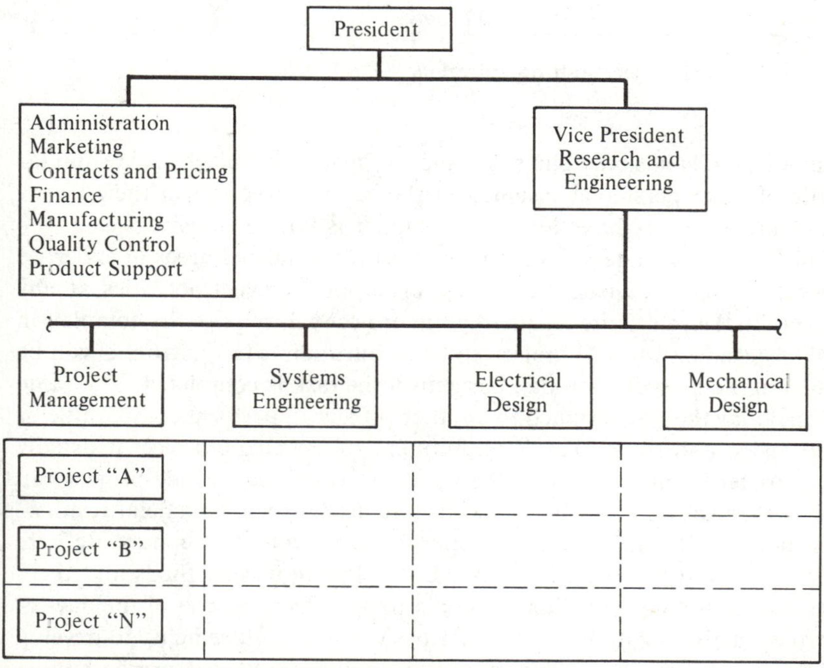

Figure 9-6. A matrix organization concept.

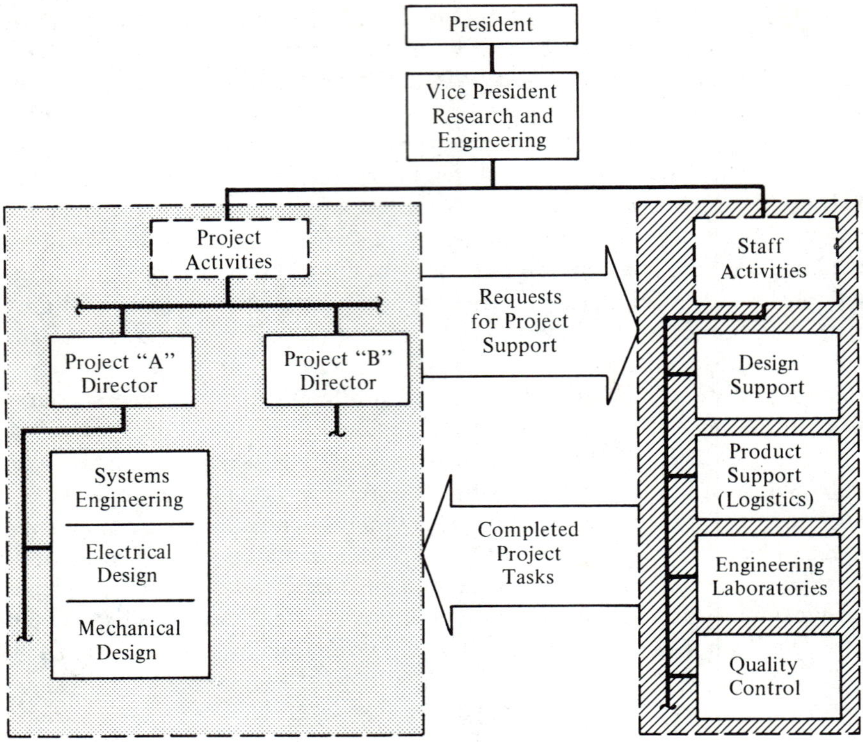

Figure 9-7. Typical project-staff organization.

individual usually has only one boss, and his goals and objectives are similar to those of other personnel assigned to the same project), and their knowledge of project procedures and the consumer is fairly extensive.

Staff groups include service personnel with specialized areas of expertise utilized to advise, counsel, assist, and/or support project activities at any organizational level. Quite frequently the project will require the completion of highly specialized tasks of relatively short duration. Staff personnel will be assigned to the project on a loan basis until the task is completed. The same personnel may then be assigned to another project to perform comparable or related tasks, and so on. Total commitment to any single project is usually for a short term only. However, the experiences gained on one project can readily be transferred to other projects, which often has beneficial results.

In addition to direct project support, staff groups also have definite responsibilities in the areas of research, developing new methods and techniques for improving operations, and planning. The pressures of the day-to-day project activities are not present. However, the requirements to develop new technologies for future business and growth are always present; thus, staff groups must continually maintain a capability current with the state of the art in the areas where top management has specified an interest.

In summary, the project-functional staff organization serves a definite need in responding to a variety of requirements and in allowing for the proper mix (in terms of emphasis) of project activities and those of a more highly specialized nature. This mix, which will be somewhat different in each instance, provides the mechanism for creating the needed communications between the generalist (the project engineer) and the specialist (the staff engineer) in the application of engineering principles to product development and support.[11]

5. Committee activities. Besides the various combinations of project-functional staff organizations, the use of special committees will often occur for one reason or another. Committees, sometimes known as "ad hoc committees," may be established by top management on a temporary basis to address special problem areas or to perform certain designated tasks, which normally do not fall within the purview of either the project or staff groups already in being.

Special committees are often advantageous in that they permit an interchange of ideas and a judicious deliberation on problem areas considered too difficult, too broad, or too important for any one individual or established organizational group. On the other hand, committees sometimes lose sight of their objectives, tend to duplicate the activities of other organizational elements, and may be unproductive. Top management must ensure that established committees have a definite objective and that continual progress is being made toward meeting this objective. When the objective has been fulfilled, the committee should be dissolved.

The challenges in developing and establishing a given organization structure are many, and the goals and objectives of the firm or agency must be met in an effective and economical manner. This requires the correct identification of functions to be performed and the proper balance of resources to accomplish the job. The necessary resources will constitute manpower, material, and money. The organizational elements discussed herein deal with all three commodities. The interactions of these commodities are discussed in greater detail in the sections to follow.

---

[11]The individual engineer assigned to a project has the prime responsibility of applying his knowledge in accomplishing the objectives of that project; that is, developing, producing, evaluating, and supporting the product. His assignments and responsibilities often shift as the product evolves through the various phases of its life cycle. In this sense, the project engineer broadens his areas of expertise and by necessity becomes more of a generalist. On the other hand, the staff engineer usually continues to perform tasks in his own specialized area and is not oriented to all phases of the product life cycle. There is a definite need to integrate these somewhat diverse areas of expertise and to utilize the best of both worlds effectively.

9.4 CHARACTERISTICS OF THE ORGANIZATION

A. Leadership Role

In many instances managers rise from ranks of the organization as a result of being the best engineer, the best scientist, the best technician, and so on. The promotion of an individual into a management position often occurs rather suddenly and does not necessarily result because the individual has demonstrated the proper leadership characteristics for the position. Thus, the individual may find himself charged with the responsibility of directing a small group without being adequately prepared for the task.

In response to this charge, the new manager must quickly acquire the administrative and managerial skills to go along with the technical skills that he or she has already demonstrated. This can be accomplished to some extent through a better understanding of the basic management styles and the desired characteristics of leadership.[12]

1. Managerial styles. Although there are many individual variations, the two managerial styles most often discussed are the "autocratic" approach and the "democratic" approach. The autocratic concept is basically dictatorial in nature, with interests primarily centered around the leader, and is restrictive in that decisions are usually unilateral (i.e., decisions are made from the top down without input from those who are required to carry out the decisions). On the other hand, the democratic concept is participative, non-pressure-oriented, and organizational interests are group-centered. The general theme is that individuals in the group have some say in matters that directly affect them.

Although both styles of management are prevalent for certain situations, the democratic leadership style has been accepted as being more effective from the motivational standpoint.[13] In general, people work harder, are more cooperative, and are more willing to accept changes if they feel that they have some influence in the results. Democratic leadership implies an organization environment where employees have a chance to grow and develop their skills, where formal supervision is considerate and the application of dictates is not arbitrary, and where individual opinions are solicited and respected. As compared to the autocratic approach, management is committed to the recognition of employees as dignified human beings and not merely as factors in a production scheme.

---

[12] These factors are not to be considered as a substitute for actual experience; however, by creating an awareness of the principles involved, it is hoped that the newly appointed manager will be able to do his (or her) job more effectively.

[13] W. G. Scott, *Organization Theory—A Behavioral Analysis for Management,* Richard D. Irwin, Inc., Homewood, Ill., 1967, p. 337.

When applying the democratic style of leadership in a pure sense, some caution must be exercised relative to the extent that democratic principles are followed throughout the entire organization. Obviously, an organization where an individual in the ranks has "equal say" with the boss is not feasible. Some modifications must be made to include the necessary authoritative provisions to allow for positive direction from the top down such that the organization can accomplish its functions with a minimum amount of conflict.

In essence, a manager must maintain authority and provide the necessary direction and control to effectively accomplish the organization's goals and objectives. However, in doing so, he or she can introduce some practices which directly support the democratic style. Of significance are those of *participation* and *communication.* Participation refers to the involvement of persons within the organization relative to:

a. Sharing in the establishment of personal goals and objectives in support of those specified for the organization as a whole. The individual engineer should be able to sit down with his (or her) boss and jointly determine his goals, where he fits into the overall organization, and how he can best contribute to the cause.

b. Sharing in the evaluation and decision-making activities pertaining to alternative ways of meeting both organization and personal objectives. The individual engineer should be able to make recommendations concerning the practices and procedures designated for work accomplishment.

The second aspect, communications, is very important in that the organization is actually a social entity, and there is a need for both formal and informal interaction on a continuing basis. Formal communications may be accomplished both vertically (up and down) and horizontally (across lines of authority) in the organization structure and are usually generated through one or a combination of the following:

a. Periodic group meetings with members of the organization should be held to discuss matters pertaining to management policy interpretations and changes, related activities of importance occurring outside the organization, "grapevine" items, administrative actions, technical and/or organization problems, and so on. These meetings should allow for a two-way *free* (uninhibited) communication between the manager and his subordinates. For large organizations where the daily contact is not possible, meetings of this type should be scheduled on a bimonthly or monthly basis.

b. Individual meetings between the manager and each subordinate member of his organization should be scheduled to review personal goals and objectives and to discuss any items of the subordinate's choosing (not just work-oriented matters selected by the manager).

This area is sometimes neglected because of the pressures of work and time limitations, and direct communications between the manager and his subordinate is lost. Quite often the engineer has something bothering him and is anxious to convey his feelings to his boss. However, his boss is busy and inacessible, and the point of concern becomes a major problem to the engineer which in turn detracts from his assigned work, and the productive output of the organization falls off as a result. In many instances, problems that turn out to be major issues could have been avoided if the communication link had been available earlier. Thus, it is recommended that the manager make it a point to schedule occasional meetings with his subordinates, regardless of whether there is a "hot" item to discuss. Unstructured meetings of this type can be quite beneficial and often alleviate hidden personal frustrations.

c. Written material, such as reports, memoranda, and letters, serve to convey information of a less urgent nature and should be circulated to appropriate individuals within the organization. Care must be exercised to ensure that written communications are prepared in the proper tone, include the necessary subject-area emphasis where needed, and are clear and factual in content. This is important since it is relatively easy to misinterpret information and create problems as a consequence.

Communications of an informal nature constitute the day-to-day interchanges between employees at the same level, employees and managers within a given organization, employees and managers in different organizations, and so on. In reality, the lines of communication generally follow the pattern illustrated in Figure 9-8. This method of communication is necessary to fulfill social needs and is permissible as long as it does not interfere with the lines of authority and the objectives of any specific organizational component. Many organizations experience difficulties that are attributed to a breakdown in communications, which usually can be diagnosed as a problem in leadership. The ability to communicate well is a leading criterion of managerial success and is a necessary requisite in working with other people.

2. Leadership characteristics. The organization is composed of a group of individuals with varying abilities, different roles and expectations, diverse personal goals, and distinct behavioral patterns. People within the organization are highly dependent on one another and yet are often pushing in different directions because of some of these factors. The challenge for the manager is to integrate these various characteristics into a cohesive force, leading to the accomplishment of organization objectives. The manager must not only ensure that the job is accomplished in a satisfactory manner, but hopefully he will inspire and motivate his subordinates to *excell* in this accomplishment of organization objectives. It is apparent that the manager must create a climate that presents challenges (not threats) for the individuals involved.

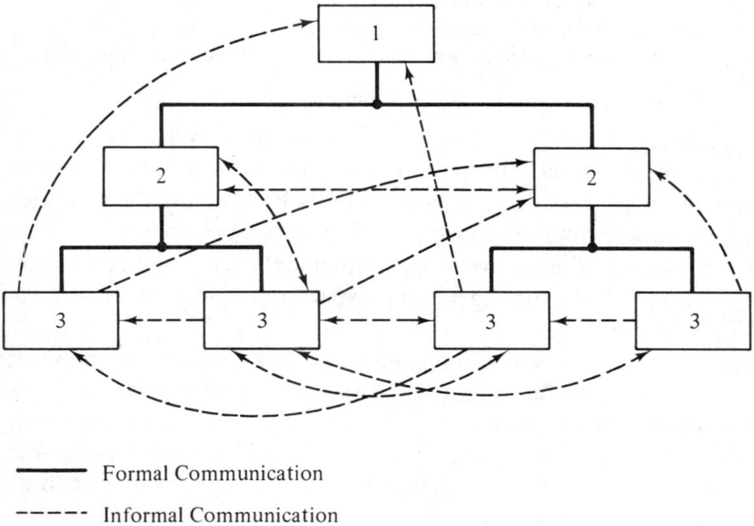

Figure 9-8. Working relationships in a formal organization.

In facilitating this objective, the manager should incorporate certain practices that are responsive to both the organization and to the individual needs of people within the organization. As a start, the democratic style of leadership discussed above tends to promote the necessary environment for the organization. Within this framework, the manager should encourage the participation of individuals in goal setting and decision making and should promote communications. By doing so, he will not only tend to improve the motivation from within, but he will acquire a better understanding of the people in his organization. This understanding should help the manager in the following ways:

a. Recognize the personal characteristics of each individual in the organization in order to better match the individual with the job requirements. A person may excell in one situation while doing only a mediocre job in another situation, even though he has the ability to do both and the overall organization climate remains relatively constant. In essence, the manager should assign employees to the types of work they do most effectively.
b. Inspire each individual to excell in his job by creating an atmosphere of personal interest. An employee will tend to perform better if he knows that the boss is personally interested. This personal interest is developed through involvement at the employee level.
c. Be sensitive to employee problems such that each can be addressed on personal terms. The solution to a problem should, if at all possible, consider the effects on individual employee.

d. Evaluate employees on a personal basis and initiate rewards promptly when warranted. Promotions and merit raises should not be oriented to the organizational hierarchy alone, but should also be directed to the best performers.

Obtaining the desired organization environment not only depends on the intent of the manager in initially establishing practices, but is of course highly dependent on his personal leadership ability in directing the activities of the organization over time. The best planning in the world will have little benefit unless the actions that follow in the implementation phase are directly supportive. As a goal, the manager should strive to exhibit the characteristics listed in Figure 9-9.

In actual practice, it is well recognized that not all managers will meet the criteria specified in the figure; however, the more progressive managers will strive to improve in the direction(s) discussed. Others in management positions will not improve, owing to an attitudinal problem, and the desired organization performance will be compromised accordingly. Some of these inhibiting factors common in many firms/agencies are as follows:

a. The manager will assume an authoritarian style of management stifling participation and communication.

b. The manager may refuse to make decisions when necessary. In many cases, either decisions are not made at all or the "stall" tactic is employed.

c. The manager may place personal interests above those of the organization. Examples are the acquisition of nonessential human and material resources to improve personal status (e.g., "empire building"), taking personal credit when not deserved, emphasizing the deficiencies of others in order to get ahead, and so on.

d. The manager may be resistant to trying something new. Examples include the utilization of available tools that are unfamiliar but will improve job performance, pursuing new technology areas, reorganizing when necessary to improve group performance, and so on.

e. The manager may refuse to disseminate information that could conceivably promote effective job completion and organization growth.

f. The manager may be inconsistent in leadership practices. For instance, decisions may tend to be conflicting, decisions may be reversed, implementation of tasks is not consistent with planning information, and so on.

Reasons for the above inadequacies may stem from personal insecurity on the part of the manager, complacency, a cover-up for administrative or technical incompetence, and/or a combination of any of these factors.

B. Needs of the Individual

Where the discussion above primarily covers the organization environment and the desired characteristics of leadership, it is now incumbent that some attention be given to the needs of the individual employee. As stated

LEADERSHIP CHARACTERISTICS
1. ACCEPTANCE: Earns respect and has the confidence of others. 2. ACCOMPLISHMENT: Effectively uses time in meeting goals and objectives. 3. ACUTENESS: Mentally alert and readily comprehends instructions, explanations, and unusual circumstances. 4. ADMINISTRATION: Organizes his own work and that of his subordinates; delegates responsibility and authority; measures, evaluates, and controls position activities. 5. ANALYSIS AND JUDGMENT: Performs critical evaluation of potential and current problem areas; breaks problem into components; weighs alternatives; relates, and arrives at sound conclusions. 6. ATTITUDE: Enthusiastic; optimistic; and loyal to firm/agency, superior, position, and associates. 7. COMMUNICATION: Promotes communication within and between organization elements. 8. CREATIVENESS: Has inquiring mind; develops original ideas; and initiates new approaches to problems. 9. DECISIVENESS: Makes prompt decisions when necessary. 10. DEPENDABILITY: Meets schedules and deadlines in consistent manner and adheres to firm/agency policies and procedures. 11. DEVELOPING OTHERS: Develops competent successors and replacements. 12. FLEXIBILITY: Adaptable; quickly adjusts to changing conditions; and copes with the unexpected. 13. HUMAN RELATIONS: Is sensitive to and understands personnel interactions; has "feel" for individuals and recognizes their problems; considerate of others; ability to motivate and get people to work together. 14. INITIATIVE: Self-starting; prompt to take hold; seeks and acts on new opportunities; exhibits high degree of energy in work; not easily discouraged; and possesses basic urge to get things done. 15. KNOWLEDGE: Possesses knowledge (breadth and depth) of functional skills needed to fulfill position requirements; uses information and concepts from other related fields of knowledge; and generally understands the "big picture". 16. OBJECTIVITY: Has an open mind and makes decisions without the influence of personal or emotional interests. 17. PLANNING: Looking ahead; developing new programs; preparing plans; and scheduling requirements. 18. QUALITY: Accuracy and thoroughness of work; and maintains high standards consistently. 19. SELF-CONFIDENCE: Self-assurance; inner security; self-reliant; and takes new developments in stride. 20. SELF-CONTROL: Calm and poised, under pressure. 21. SELF-MOTIVATION: Has well planned goals; willingly assumes greater responsibilities; realistically ambitious; and generally is eager for self-improvement. 22. SOCIALNESS: Makes friends easily; works well with others; and has sincere interest in people. 23. VERBAL ABILITY: Articulate; communicative; and is generally understood by persons at all organization levels. 24. VISION: Possesses foresight; sees new trends and opportunities; anticipates future events; and is not bound by tradition or custom.

Figure 9-9. Checklist of leadership characteristics.

earlier, these needs must be thoroughly understood if the manager is to deal successfully with his subordinates.

As a starting point, one should first review A. H. Maslow's theory dealing with the hierarchy of needs.[14] The theory was developed to identify relatively separate and distinct drives that motivate individuals in general. The five basic needs identified are:

1. The physiological needs, such as thirst, hunger, sex, sleep, and activity. These constitute the needs of the body, and unless these needs are basically satisfied, they remain as the prime influencing factor in the individual's behavior.
2. The safety and security needs, which include protection against danger, threat, and deprivation. Having satisfied the bodily needs, the safety and security needs become a dominant goal.
3. The need for love and esteem by others (i.e., the social needs). This includes belonging to groups, giving and receiving friendship, and the like.
4. The need for self-esteem or self-respect and for the respect of others (i.e., ego need). An individual wishes to consider himself strong, able, competent, and basically worthy by his own standards.
5. The need for self-fulfillment or the achieving of one's full potential through self-development, creativity, and self-expression. This relates to man's desire to grow, develop to the point of full potential, and ultimately attain the highest level possible.

The needs are related to each other and are arranged in an order that will stimulate consciousness and activity. As a need becomes satisfied, activity emphasis shifts to the next need category. In other words, a satisfied need is no longer a motivator, and the next need becomes a driving factor.

With this understanding of *basic* needs, one might proceed by asking questions such as: What are the on-the-job needs of scientists? Engineers? Technicians? Manufacturing employees? Actually, these needs will differ somewhat for each profession or major category of individuals as discussed below.

F. Herzberg conducted research, resulting in his "motivation-hygiene" theory, which identifies factors commonly known as "satisfiers" and "dissatisfiers."[15] The theory was developed from research on the job attitudes of

---

[14] A. H. Maslow, "A Theory of Human Motivation," *Psychological Review*, July 1943.

[15] F. Herzberg, "Work and Motivation," *Studies in Personnel Policy Number 216: Behavioral Science, Concepts and Management Application*, National Industrial Conference Board, New York, 1969.

F. Herzberg Research	M.S. Myers Research
1. Satisfiers: • ACHIEVEMENT – personal satisfaction in job completion and problem solving. • RECOGNITION – acknowledgment of an accomplishment (e.g., a job well done). • WORK ITSELF – actual content of the job and its positive/negative effect on the employee. • RESPONSIBILITY – both responsibility and authority in relation to the job. • ADVANCEMENT – promotion on the job. • GROWTH – learning new skills offering greater possibility for advancement. 2. Dissatisfiers: • COMPANY POLICY AND ADMINISTRATION – feelings about the adequacy or inadequacy of company organization and management, policies and procedures, etc. • SUPERVISION – competency or technical ability of supervision. • WORKING CONDITIONS – physical environment associated with the job. • INTERPERSONAL RELATIONS – relations with supervisors, subordinates, and peers. • SALARY – pay and fringe benefits. • STATUS – miscellaneous items such as size of of office, having secretary, private parking place, etc. • JOB SECURITY – tenure, company stability or instability. • PERSONAL LIFE – personal factors which affect the job (e.g., family problems).	Needs • WORK ITSELF (M) • RESPONSIBILITY (D) • COMPANY POLICY AND ADMINISTRATION (D) • PAY (D) • ADVANCEMENT (M) • RECOGNITION (D) • ACHIEVEMENT (M) • COMPETENCE OF SUPERVISION (D) • FRIENDLINESS OF SUPERVISION (D) Note: 1. (M) – Factors which lead primarily toward motivation. (D) – Factors which lead primarily toward dissatisfaction. 2. Description of the above nine factors is self-explanatory and somewhat comparable to the factors in the Herzberg study.

Figure 9-10. Employee motivational factors.

200 engineers and accountants and was based on two questions: "Can you describe, in detail, when you felt exceptionally *good* about your job?" and "Can you describe, in detail, when you felt exceptionally *bad* about your job?" The results were classified in the two categories identified in Figure 9-10.

Referring to the figure, most of the factors listed have some bipolar affects. For instance, advancement certainly is a satisfier when it happens and may be somewhat of a dissatisfier when it does not occur. However, this category weighs much heavier as a satisfier. The item, salary, is definitely a dissatisfier when pay scales and fringe benefits are poor, and is a mild satisfier when the compensation is good. In this case, the predominant classification for salary is that of a dissatisfier. In any event, all the factors listed represent needs of the individual and should be considered by management.

M. Scott Myers conducted research involving 282 subjects (including engineers, scientists, and technicians) interviewed at Texas Instrument, Inc., commencing in 1961.[16] The categories used were motivators and dissatisfiers, and the results for engineers are summarized in Figure 9-10. The items listed and identified with a (M) are clearly motivators and those with a

[16] M. Scott Myers, "Who Are Your Motivated Workers," *Harvard Business Review,* 1970.

(D) are definitely dissatisfiers. For instance, the item, pay, is clearly a dissatisfier if considered inadequate, and advancement is definitely a motivator when it occurs. Once again, there are bipolar affects; however, the categorization does indicate where the greatest impact occurs.

In summary, the needs of the individual will vary somewhat, depending on his or her particular situation. If one need is satisfied, then another need becomes predominant, and so on. In addition, the needs of the individual are also related to the business position of the firm or agency where he is employed. If the firm is in a growth posture, the individual's perceived needs may be somewhat different than if the firm is experiencing a business decline, and the prospect of laying off employees is apparent. Finally, the manager's job is twofold. He must (1) be aware of individual needs in his organization and provide the necessary conditions for motivation, and (2) satisfy the needs of his subordinates on a continuing basis. Human motivation is the key to organizational success.

C. Creativity in the Organization

A successful and progressive engineering organization is highly dependent on the inherent creativity of its individual engineers and on their respective abilities to develop new ideas and to apply the results in a practical manner. Group creativity is a direct result of the manager's ability to stimulate individual creative thinking within the organization and is a function of the leadership style that he (or she) implements and the working environment that prevails.

Creative thinking is an intuitive step-by-step process that is somewhat comparable to the research methods discussed in Section 4.3. That is, one must (1) define the problem, (2) identify new possibilities for problem resolution, (3) select an appropriate approach, and (4) provide the necessary implementation actions in producing the desired results. Although the fulfillment of each step in the process requires some creativity (i.e., the generation and application of new ideas) on the part of the engineer, the first two steps are probably the most challenging!

In many instances, the problem is initially stated in rather general terms, and it becomes necessary to restructure it such that there is a complete understanding of the need, and enough information is available to make it possible for transition to the next step. This structuring or formulation of the problem often requires a number of iterations before the proper level of definition is attained.

The second step, or that of identifying new possibilities for problem resolution, is usually more difficult, requiring intensive concentration on the problem for extended periods of time while alternative considerations are being critically analyzed and evaluated. Creative solutions cannot be forced. Sometimes progress is made almost immediately, whereas at other times an individual will concentrate on the problem for awhile with little or no results, set the problem aside temporarily, and then come back to it from time to time. This process continues until the engineer is inspired by a new thought, and the pieces leading to the solution begin to fall into place. Inspiration often occurs at odd times, both during working and nonworking hours, when driving in a car, during a walk, while laying in bed, and so on.

Creativity on the part of the individual is a function of his technical expertise, individual skills, and attitudinal characteristics (refer to Section 1.1). The highly creative engineer (1) is sensitive to problems in advance, since he is continually questioning and observing; (2) has many ideas, which tend to breed more ideas; (3) introduces a flexibility of thought, which counters stereotyped thinking and the habits of tradition; and (4) fosters originality, which leads to the possibility of offering a greater number of solutions than what otherwise might have been identified.

Individual creativity is directly related to the organization environment to which the engineer is subjected. The manager can do a great deal to promote creativity, or to inhibit creative thinking, depending on his leadership approach. A few points that should be considered in producing positive results are noted below.

1. The manager should strive (within the bounds of the overall organization goals and objectives) to provide engineers a choice of problems to work on within their respective fields of interest. Large firms, in particular, usually are faced with many different types of problems, and the manager should attempt to align problems with employee interests.
2. The manager should allow the engineer adequate time, freedom of thought, and freedom of activity such that new ideas can evolve. This is not to say that engineering functions are not to be accounted for; however, the pressures associated with demanding results too prematurely, monitoring specific personnel actions too closely (e.g., forcing the engineer to "punch the time clock," requiring unnecessary written progress reports), setting the engineer up by comparing his output with that of other less creative engineers performing different tasks, and so on, can be harmful.
3. The manager should provide the necessary resources for the engineer to perform his assigned functions. This includes the supplying of information and materials, the use of facilities, and the provision of supporting personnel, such as draftsmen, technicians, and clerical help. It often becomes quite costly when highly creative engineers are burdened with the performance of noncreative functions.
4. The manager should encourage highly creative engineers to participate in certain seminars and industry association and technical

society meetings so that they can associate with other specialists, learn of new methods and techniques for problem solving, and receive stimulation from other notable people in their field. Quite often an engineer will return from a meeting with many new ideas and is highly motivated to put some of these ideas to work.[17]

5. The manager should be quick to recognize creative work and to reward the responsible engineers accordingly. Rewards should be in the form of both monetary increases and announcements or publicity covering accomplishments. Achievement and the prompt recognition of such are prime motivators.

Creativity will occur in the organization to the extent that the manager understands the benefits of the creative process and personally demonstrates some creativeness in the accomplishment of his own activities. The manager must exhibit the initiative in developing new ideas, and this initiative must be obvious to employees down through the ranks. In addition, the manager must be willing to take th risks associated with new pursuits.

D. Individual Self-Improvement Guidelines

Creativity, as discussed above, can be stimulated if the individual engineer will attempt to cultivate certain practices. Initially, he should pursue a sustaining self-development program with the objective of maintaining currency with the state of the art in his or her technical field of endeavor. Engineers often find that their concepts and ideas are out of date within three to ten years after gradution from college unless they undertake some formalized approach to continuing education. An engineer can hopefully combat this tendency toward obsolescence by taking selected graduate engineering courses at some institution of higher education; participating in appropriate seminars, workshops, and short courses; and/or studying books, journals,

[17]An "industry association" is a bargaining type of organization, which usually represents a number of companies involved in the design and/or manufacture of the same category of products. Examples are the Electronic Industries Association (EIA), representing electrical and electronic equipment producers; and the Aerospace Industries Association (AIA), covering aircraft and space product manufacturers. A "technical society" is an organization that generally represents individual engineers interested in the same technical fields of expertise. Examples include the Institute of Electrical and Electronics Engineers (IEEE), American Society of Civil Engineers (ASCE), American Institute of Industrial Engineers (AIIE), and Society of Logistics Engineers (SOLE). Both industry associations and societies alike sponsor technical meetings, seminars, and workshops throughout the United States and the world for the purposes of promoting technology and the exchange of ideas. Also, individual groups within the association/society organizations often perform documentation reviews and accomplish special studies of general interest. Participating engineers can benefit significantly from these activities, since there is a vehicle for information exchange that is usually not present between individual companies, particularly in competitive situations.

and other publications in his and other closely related technical fields. An up-to-date knowledge base is definitely a prerequisite to the generation of new ideas.

As a second point, individuals often find that certain times of day are more conducive to productivity than other times, particularly when the environment helps to stimulate thought (i.e., when it is quiet and interruptions are kept to a minimum). Thus, if the engineer can arrange his schedule to take advantage of these time periods, he is more likely to be innovative relative to the identification of possible solutions to outstanding problems. Problems are listed in order of importance, and engineering expertise is applied to the most pressing problems first.

Finally, the engineer should maintain a notebook at all times and record ideas that arise as he questions and observes various events. He may be in the middle of any type of activity when a stimulating thought enters his mind, the results of which may lead toward the solution to some problem.

9.5 STAFFING AND DEVELOPING THE ORGANIZATION

A. Requirements

The requirements for staffing an organization initially stem from the results of advanced product planning described in Section 3.4. Tasks are identified from both short- and long-range projections, combined into work packages, and the work packages are grouped into positions. The positions are in turn arranged within the organization structure considered most appropriate for the need (discussed in Section 9.3B). The staffing per se is based on individual position descriptions that are developed by the organization to cover the defined scope of work. A sample position description format is illustrated in Figure 9-11. Position description sheets are prepared to cover each slot in the organization and should clearly identify position title, responsible supervisor, areas of responsibility and job objectives, background requirements, and date of need. Categories of work are evaluated in terms of complexity of task and are assigned by position grade (e.g., manager, section head, supervisor, senior engineer, engineer, technician).[18] Hopefully, the tasks identified for each grade level will represent a challenge to the individual who will ultimately fill the position.

---

[18]The personnel function usually establishes position grades or levels within the organization, prepares general position guidelines, and defines salary ranges for all grades. This is necessary to maintain some degree of consistency within the firm or agency across organization lines and to ensure that the established grades are compatible with those of other firms/agencies pursuing comparable fields. Refer to Chapter 14 for further discussion of the personnel function.

DATE OF NEED: ________

POSITION TITLE	SUPERVISOR
Senior Engineer – Electrical Design	Project Section Head

BROAD FUNCTION

Responsible for the performance of system engineering functions in the design and development of communication products.

FUNCTIONAL OBJECTIVES

1. Develop operational requirements and maintenance concepts for new communication systems/equipments.
2. Interpret and translate system level requirements into functional design requirements.
3. Prepare system and subsystem specifications.
4. Determine requirements and conduct formal design reviews for all system elements.
5. Prepare system test requirements, monitor test functions, and evaluate test results to determine system performance and effectiveness. Make recommendations for corrective action and/or improvement as appropriate.
6. Provide assistance to marketing in product sales activities.

REQUIREMENTS

Degree in electrical engineering plus 5 to 10 years' experience in communications.

Figure 9-11. Position description.

When all positions for a given system requirement are initially determined, the manager responsible for staffing the organization will lay out total position needs by grade category in terms of the program life cycle as shown in Figure 9-12.[19] The engineering position requirements are combined with other personnel projections, and a total system projection is developed as indicated in Figure 9-13. The position requirements are then converted into cost factors for each category by year, and the results are compared with budget requirements and expenditure rates. If the results appear feasible and are approved, the manager will forward the unfilled position descriptions to Personnel in order to proceed with the necessary steps for recruiting and employment.

On the other hand, if the cost projections are out of line, the manager may wish to reevaluate the assigned tasks by personnel category along with the projected manpower distribution. Perhaps certain tasks can be per-

[19]It should be noted that different categories of engineers may be required from one program phase to the next because of the complexity and variation of tasks. For instance, in the conceptual and preliminary system design stages, where a high degree of innovation is necessary, the use of principal and senior engineers is more prevalent than those of a lower grade. Whereas as design becomes more defined, engineers and engineering analysts can be utilized more readily since there are many tasks that are less complex in nature. The reader may wish to refer back to Section 1.3 for a review of the role of the engineer in the product life cycle.

Personnel Category	Concept. Design		Preliminary System Design			Detailed Design					
	Months After Program Start										
	1	2	3	4	5	6	7	8	9	10	11
Engineering											
Manager	—	1	1	1	1	1	2	2	2	2	2
Section Head	—	—	1	3	3	5	6	7	7	7	7
Principal Engineer	2	3	5	7	8	9	10	10	10	9	8
Senior Engineer	3	5	11	12	15	17	22	22	23	23	24
Engineer	1	3	5	6	10	12	24	30	36	47	48
Engineering Analyst	—	—	—	4	6	10	10	15	18	25	30
Technician	—	—	—	—	—	—	—	—	1	2	5
Clerical	1	1	3	3	4	6	7	9	10	11	12
Total	7	13	25	36	47	60	81	95	107	126	136

Figure 9-12. Engineering manpower—Project "A".

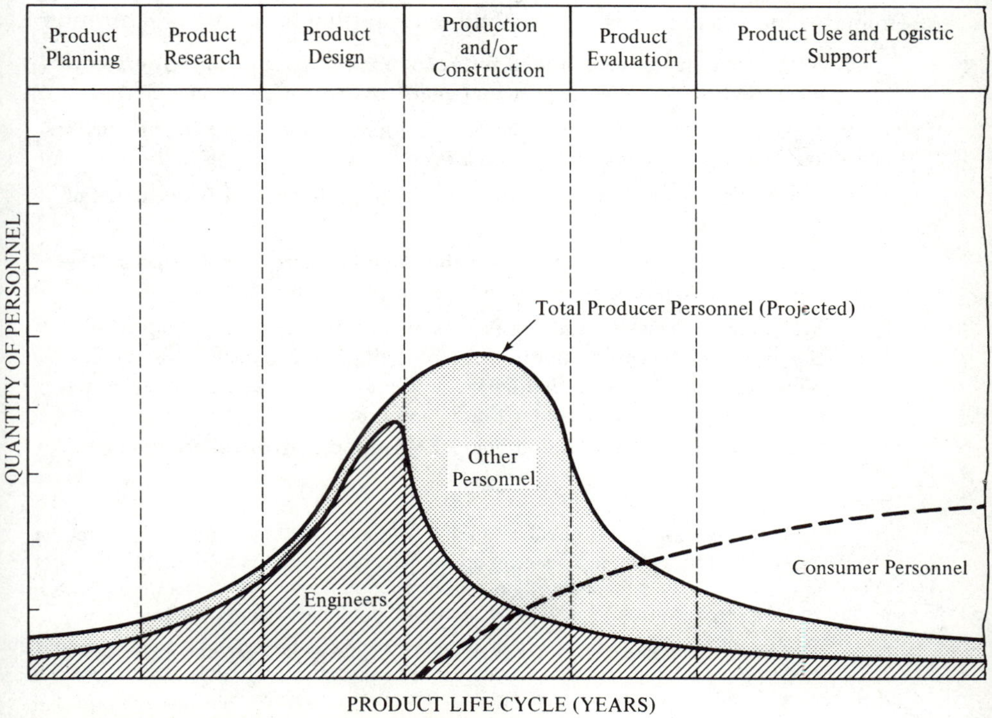

Figure 9-13. Manpower loading for Project "A" (*sample*).

formed by a lower-grade individual through extending the scheduled task span time and starting earlier, which may in turn reduce costs. As a second possibility, the manager may be able to group tasks differently in order to accomplish the work more efficiently. In any event, there are trade-offs that should be conducted to ensure that the most beneficial approach is selected.[20] The end result will be reflected in revised projections (i.e., an update of Figures 9-12 and 9-13), which will serve as the basis for organization staffing.[21]

B. Organization Staffing

Where the specific requirements for personnel are determined by each organization element responsible for accomplishing the work, the follow-on staffing effort becomes a joint activity with the personnel function. Personnel is charged with being responsive to *all* manpower needs throughout the firm/agency and is responsible for the acquisition and processing of qualified individuals while minimizing staffing costs over time. Since there is quite a bit of interaction between the organization groups (where the need exists) and Personnel, one should review the basic functions of each.

1. Responsibilities of the engineering organization. The engineering manager has the following responsibilities in staffing his or her organization:

a. Make forecasts for current and future manpower requirements by employee grade (i.e., number of principal engineers, senior engineers, etc.).
b. Prepare position descriptions for all positions and supply Personnel with staffing requirements and specifications.
c. Review and carefully evaluate all resumes promptly and notify Personnel of any interest.
d. Be available for the interview when the candidate arrives, and conduct the interview to determine areas of mutual interest.
e. Make hiring decisions and notify Personnel of the decision promptly after the interview. Be careful not to overhire in times of expansion. In fact, it is better to staff at a level slightly below the apparent need as organization output seems to be best under these conditions.
f. Keep Personnel up to date on new job openings and/or cancellations.

---

[20] After the responsible manager develops his projections and is satisfied with the results, his supervisor or the firm's top management may impose further cost reductions for one reason or another. This, of course, creates an additional challenge and sometimes forces the manager to modify and improve his overall operation in meeting the ultimate requirements.

[21] This facet of developing manpower requirements and cost projections is an inherent part of the system proposal effort covered in Section 3.4F. Manpower projections indicate staffing requirements for all the individual elements of the proposed organization.

2. Responsibilities of the personnel function.[22] The primary functions of Personnel in the staffing process are to:

a. Develop policies and procedures covering all Personnel actions for the firm/agency as a whole. This should include consideration of employee grades and the salary ranges for each grade, Personnel acquisitions and layoffs, compensation and fringe benefits, employee relations, promotion and transfers, and so on.
b. Locate manpower sources and identify possible candidates that will fulfill the position descriptions and specifications. Make a preliminary screening of candidates, and supply resumes of the applicants to the engineering manager.
c. Invite candidates into the firm/agency for interview, and develop a formalized schedule for each candidate during the visit. If needs exist in two or more groups and there may be a mutual interest in the candidate, Personnel should arrange for several interviews during the candidate's visit.
d. Work with the engineering manager in making a job offer to the candidate or initiate a notification of "no interest" as appropriate.

In staffing an organization, one approach that is often overlooked is the selection and promotion (if warranted) of personnel from within. There is a tendency to immediately look outside for new people when job opportunities evolve since the manager does not want to disturb a smooth-running operation by selecting a key individual from another group in the firm. Also, the manager of the other group may not allow a transfer even if his "star" is selected as a candidate. However, in either case, that particular individual, if not selected, may feel that the opportunities for promotion within the firm are remote and may decide to leave as a result.[23] Thus, if possible, creative individuals from within the firm should receive first consideration when new opportunities prevail. Not only will others feel that growth is possible within the firm, which in turn improves the overall morale, but recruiting costs should be reduced considerably.

C. Personnel Development

Nearly every engineer wants to know the following on an almost-continuing basis: "Where do I currently stand in the eyes of my supervisor?" and "What are my opportunities for growth?" Response to the first question

---

[22]Refer to Chapter 14 for a more comprehensive presentation of Personnel responsibilities.

[23]Unfortunately, this occurs quite often in the industrial world, since individual engineers feel and can in many instances move up much quicker by accepting jobs with other companies rather than by remaining with the same company. Thus, the engineering profession is "tagged" as one that incorporates a large degree of individual mobility, and company personnel recruiting and training costs are high as a result.

is derived through both the *formal performance review,* which is conducted on a regularly scheduled basis (either annually or semiannually) and through *informal communications* with the boss. The formal performance reviews are usually controlled by Personnel in terms of information content, processing requirements, and schedule.[24] These reviews are (1) initially prepared by the engineer's supervisor, (2) reviewed by the supervisor with the engineer, (3) updated to include the engineer's comments of a pro and con nature, and (4) submitted to Personnel for approval and inclusion in the engineer's permanent file. The informal reviews, on the other hand, constitute the supervisor's reaction (on a day-to-day basis) to the individual task accomplishments of the engineer. The engineer seeks approval and recognition and should receive some reinforcement if he is doing a good job. Also, he needs to know as soon as possible when his work is unsatisfactory and improvement is desired. Waiting until the formal performance review is conducted to learn that his work is not satisfactory is a poor practice and is demoralizing to the engineer since he (or she) has assumed, by virtue of not having heard any comments to the contrary, that he has been doing an acceptable job.

The second question pertaining to the opportunities for growth depends on: (1) the climate provided within the organization and the actions of the manager which allow for individual development, and (2) the initiative on the part of the engineer to take advantage of the opportunities provided. The engineering manager should prepare a separate development plan for each person in his organization. The plan, although somewhat tailored to each person's particular needs, should allow for personal development by providing a combination of the following:

1. Formal training designed to familiarize the engineer with the policies and procedures applicable to the overall firm/agency as a whole as well as the detailed operating procedures of his own organization. This type of training should enable the individual to operate more successfully within the framework of the total organization through familiarization with the many interfaces that he will encounter in performing his job.
2. On-the-job training of the engineer in his assigned position. Effort should be directed toward improving the engineer's present performance through day-to-day guidance by his supervisor. The engineer's strengths should be exploited and weaknesses should be corrected to the extent possible. If an individual's work is not improving, consideration should be given to reassigning him to work where he is more highly motivated and which is more in line with his particular capabilities.

---

[24] A sample performance review and evaluation form is included in Chapter 14.

3. Formal technical training designed to upgrade the engineer relative to the application of new methods and techniques in his own field of expertise. This pertains to the necessity to maintain currency and avoid technical obsolescence as discussed in Section 9.4D. The engineering manager, with the assistance of Personnel, should provide opportunities in this area in the form of graduate engineering programs and continuing education short courses, workshops, seminars and individualized instruction.

The engineering manager must realize the need for developing personnel in his organization and should encourage each man to seek a higher level of performance by offering challenging job assignments and training opportunities. In addition, if an individual engineer expects to move up in the organization, he should be quick to take advantage of the opportunities offered. If he does and his job performance improves, he should be recognized accordingly. Personnel development and total organization growth are dependent on both conditions: that is, the manager must provide the opportunities, and the individual engineer(s) must demonstrate the initiative in accepting the challenge.

9.6 MEASUREMENT, EVALUATION, AND CONTROL OF THE ORGANIZATION

The measurement and evaluation of an organization relate back to the purpose and the initially specified goals and objectives for that organization. From these goals, standards are established for both elements of work and cost in terms of the output desired. Activities are initiated, the results are measured and evaluated against the established standards, and corrective action is accomplished as necessary when deviations occur. This process, which is similar to the iterative design and development process discussed in Part Two, is oriented around Figure 9-1, which includes *inputs,* the *processes* associated with ongoing organizational activities, and organization *outputs.* Management's objective is to produce certain desired results (i.e., output) with a minimum expenditure of resources in the form of manpower, materials, and money (i.e., input). This implies a sustaining capability for the measurement, evaluation, and control of organizational functions.

A. Establishing Organization Measures and Standards

The initial and perhaps the most difficult step is that of identifying specific and tangible goals and objectives for the organization and being able to relate these in quantitative measureable terms, where possible. Organization goals are derived from the following:

1. Goals are based on product specifications and plans (refer to Section 3.4B).
2. Goals are established internally by top management for the firm or agency as a whole and allocated down to the organizational level (refer to Section 9.2).

The first category is relatively straightforward since objectives are stated in the form of product performance and effectiveness characteristics, and cost targets are established from marketing information. Organization objectives constitute the performance of certain project tasks in accordance with product specifications, schedules, and allocated costs. In other words, the object is to produce a system that will perform in a specified manner, is effective, and will not exceed a unit cost of x dollars. This objective is tied to the organization through the WBS (illustrated in Figure 3-16), schedules in the form of PERT/COST networks or equivalent, and technical parameters allocated to various elements of the system for which the organization is responsible (refer to Figure 5-6 for an example of requirements allocation).

When discussing the aspect of cost, one should consider both direct costs and indirect costs (often referred to as "overhead").[25] Direct costs cover tasks and materials that are clearly identified as line items in the contract or in the specifications and plans, and are based on man-hour estimates per task and/or actual material acquisition costs. If it is assumed that an individual's pay is proportional to his most effective output, the individual should be assigned to the most challenging job possible within his range of abilities, since efficiency is a key objective. Jobs are evaluated in terms of complexity, and standards (derived from past experience) are used to determine the engineering grades required for task completion. Salary ranges for each grade constitute the bases for estimating direct labor costs, and these costs are spread over the task duration time, as illustrated in Figure 3-15.

Indirect costs, on the other hand, include such supporting costs as the cost of facility space and utilities, top management salaries, employee services, personnel and recruiting expenses, material-handling costs, costs of maintaining certain staff functions, and comparable costs that are intuitive in running an organization but are not directly chargeable against a specific contract line item. These costs are often accumulated and allocated to each organization element on the basis of standards such as a percentage of direct costs, a percentage of occupied space, and so on. In some instances, as the organization increases in activity relative to the volume of project work and revenue to the firm, top management will permit the organization to employ some additional engineers on an indirect-cost basis to accomplish certain functions of benefit to the organization. In any event, organization objectives

---

[25] A more in-depth discussion of cost allocation and accounting is presented in Chapter 12.

in this area may constitute the establishment of a designated level of project work in terms of direct labor output, maintaining a given ratio of indirect to direct cost, generating a backlog of projects equivalent to *y* dollars, or something of a comparable nature.

The second category of goals, as established internally by top management, is more aligned to special situations dealing with the peculiarities of different organization patterns (e.g., staff versus project functions), growth and decline postures of the firm/agency, changes in top management direction, and so on, and might be of a nature similar to those listed in Section 9.2. Organization objectives will relate to the introduction of new products, the modification of an existing product, the development of a new capability, maintaining an operating cost not to exceed *x* per cent of the sales revenue or profit per year, and so on. Other factors of a less significant value might include the quantity and quality of patents per year, the number of reports and publications, or the quantity of citations. A staff function may have one set of goals while a project organization has a completely different set. Likewise, if a company is in a growth condition, established goals may relate more directly to the performance of research, increased product-line volumes, increased market share, and so on, while in a declining condition goals will concentrate on cost reductions.

Obviously, there are many different criterion factors that can be used for the evaluation of organization characteristics. Some can be specified in quantitative terms and others are stated qualitatively, such as: conduct basic research; improve the product to maintain the largest share of the market; incorporate some product diversification; and so on. In any event, goals must be properly applied or tailored to the application, clearly defined, and thoroughly understood at all levels in the organization, particularly since the overall organization goals are a direct function of the performance of each individual in the organization.

B. Evaluation and Control of the Organization

With the establishment of goals as inputs to the organization, there is a need for an assessment of how the organization is progressing (or not progressing) in terms of these goals and whether an effective output will result. The manager of the organization must be aware of his (or her) progress on a continuing basis since he is held directly responsible for the activities of that organization and must be able to react to potential problem areas in an expeditious manner. Top management of the firm or agency should have a general idea of the activities of each organization element to verify that progress is being made toward the overall goals and objectives of the firm. The owners or stockholders of the firm need to know how the firm is doing to ensure the security of their investment. Thus, assessment becomes a matter of significance both internally and externally and stimulates the need for different categories of data at varying frequencies for each level in the total organization hierarchy. Our discussion in this section will primarily focus on the engineering manager and his needs.

The engineering manager must have access to a management information system of some type to provide him with the right kind of data at the appropriate time. When reviewing the total amount of information that a manager requires to effectively do his job, one will soon realize that there is no way to quantify, record, transmit, and project all of the factors that affect the operation of an organization. However, it is possible to select certain types of data presented in the proper format that will provide either direct or indirect indications of organization status. For the majority of organization activities, work is defined in the form of tasks with scheduled starting and completion dates. Also, costs are estimated for task accomplishment. The engineering manager, in attempting to determine status, may wish to utilize a combination of the bar charts, milestone charts, PERT/COST reports, cost projections, and/or performance measurement and evaluation methods discussed in Section 3.4C4. These methods can be readily adapted to a computerized format (creating a way for expeditious handling) and when combined with progress reports, program reviews, design reviews, and so on, should provide the manager with the necessary information on work status in terms of percentage of work completed per dollars expended, milestones attained, and engineering labor costs per task and expenditure rates. Again, the information presented must directly relate to the established goals and objectives.

In reviewing data reflecting organization activities, the engineering manager may detect problems pertaining to delinquencies in task accomplishment, high costs, or both. Each problem (or potential problem area) should be investigated from the standpoint of establishing cause-and-effect relationships. Causes may be pinpointed to either external or internal factors, where internal factors may be attributed to any one or a combination of the following:

1. The organization structure is not compatible with the type of work being accomplished.
2. The adopted managerial style of leadership is not conducive to the desired level of productivity.
3. The personal needs of the individual engineer(s) assigned to the job are not being adequately considered, which results in poor initiative and lack of motivation in work accomplishment.
4. The individual(s) assigned to the job is inadequately trained.
5. Communication between the supervisor and his subordinate is lacking.

When evaluating the individual, one should relate productivity in terms of the standards initially established when activities are first planned, the work packages identified, and organization staffing requirements. One way of addressing productivity is to consider it as a corollary to the efficiency of a machine: that is, efficiency is the ratio of output to input expressed as a percentage. In this context, productivity can be expressed as a ratio of the value of job output to the cost of input, where cost is a function of the engineering man-hours expended and the cost per man-hour. Estimates of the work to be accomplished are made based on standards established from past experience, and job performance results are measured against these standards.

As problems are noted, the engineering manager should strive to isolate the problem on the basis of individual productivity and then attempt to determine the cause or the reason for the deficiency. The cause may be attributed to an individual or to the organization as a whole. In either case, corrective action should be initiated as soon as possible.[26]

C. Summary

This chapter has presented some of the basic concepts of organization in a rather cursory manner. Because of the comprehensiveness of the subject area, an effective summary might well include a review of the following questions:

1. Has all engineering work been identified and assigned to a specific organization element?
2. Has the organization structure been designed to meet the total business objectives of the firm or agency in the most effective and economical manner?
3. Is the organization structure design suited to the type of engineering work at hand?
4. Have qualitative and quantitative goals and objectives been clearly defined for the organization? Are they thoroughly understood by all members of the organization?
5. Have short- and long-range manpower forecasts been developed for the organization?
6. Have organization standards and measures of accountability been clearly established?

---

[26]Corrective action may range from a mere informal conversation with an employee to the reassignment of the job to a different individual or a complete reorganization of the group. The manager should avoid making changes in areas where productivity is high, but must address those areas where the demonstrated productivity is low.

7. Are there provisions for ensuring periodic reviews of organization effectiveness? Are data requirements adequate?
8. Is the span of management control within the organization adequate?
9. Is management willing to delegate responsibility and authority?
10. Have responsibility and authority (i.e., decision making) been designated to the point of action?
11. Have organization redundancies and duplication of effort been eliminated?
12. Have position descriptions been prepared for all organization activities?
13. Have the identified tasks been matched properly with the abilities of the individual personnel assigned to complete the tasks?
14. Does each individual in the organization know who reports to him and to whom he should report?
15. Does the managerial style of leadership promote maximum organization productivity?
16. Are the organization's efforts accomplished both effectively and efficiently?
17. Are the needs of the individual in the organization adequately considered?
18. Is management competent at handling complex problems? Is the manager a skillful communicator? Does the manager support and guide subordinates in work accomplishment? Is the manager an effective leader?
19. Is creativity in the organization encouraged?

In addition to responding to the preceding questions, the reader is advised to review some of the literature covering organization principles and concepts listed in Appendix D.

QUESTIONS AND PROBLEMS

1. What is the purpose of organization?
2. Why is the establishment of organization goals and objectives important? What are they based on?
3. What are your personal goals and objectives?
4. What are some of the influencing factors in defining the span of management control?
5. Why is it important to delegate responsibility and authority to the point of action? Why is it necessary to delegate authority along with responsibility?

6. What are the basic differences between line and staff organization functions?
7. What are the advantages of a functional organization? Project organization? Project-staff organization?
8. Describe the basic differences between the autocratic and the democratic styles of leadership. Which would you select? Why?
9. Name the basic needs of an individual as defined by A. H. Maslow.
10. Refer to Figure 9-10. Are the needs the same for individuals in all job categories? Please explain.
11. How is creativity promoted in an organization?
12. How is organization staffing accomplished? (Identify the steps in staffing.)
13. What is the purpose of position descriptions? What information is included? Prepare a position description that fits your particular objectives.
14. Identify the personnel staffing interfaces between the engineering organization and the personnel function. Draw a flow diagram that shows the functions of each and the interactions between the two.
15. How is personnel development accomplished?
16. What is the purpose of the WBS? How does it tie in with the organization structure?
17. Describe several management techniques that may be appropriately utilized to monitor and control project activities.
18. What is meant by direct cost? Indirect cost?
19. How would you measure and evaluate organization effectiveness?
20. How would you measure and evaluate the productivity of an individual?
21. Refer to Figure 9-9. List the characteristics (in order of importance) that **you** would consider significant in managing an organization. Explain your reasoning.
22. There are no significant differences between objectives for high level managers and those for lower echelon personnel. True or false? Explain.
23. Performance reviews of progress toward accomplishing objectives should be separated from compensation and rewards. True or false? Explain.
24. Clear, well-structured objectives need not be supported by concrete plans for achieving them. True or false? Explain.
25. Management by objectives should facilitate improved communications and a more efficient organization. True or false? Explain.

10

ORGANIZATIONAL COMPONENTS AND INTERFACES

Whereas Chapter 9 introduced some of the fundamental concepts of organization, this chapter deals more with the application of these concepts in terms of the development and operation of various types of organizations, organizational interfaces, and the specific components of the organization related to engineering.

10.1 CATEGORIES OF BUSINESSES, CORPORATIONS, AND AGENCIES

As a start, one should address the various categories of organization on the basis of forms of ownership. Organizations generally fall into either the profit or nonprofit category and include the following.

A. *Profit-oriented noncorporate businesses,* where forms of ownership are basically (1) the individual or sole proprietorship, (2) the general partnership, and (3) the limited partnership. This includes small engineering firms and individual consultants. The vast majority of business *units* are of the noncorporate form, even though the total dollar value of goods produced is relatively small.

B. *Profit-oriented corporations,* where ownership is in the form of individual stockholders. In terms of dollar value of goods produced and the quantity of workers employed, the corporate form of business is the most significant and represents about three fourths of the total national income.

C. *Nonprofit government agencies, research and development organizations, academic institutions, and cooperatives.* This includes federal, state, and local government activities and associated functions, where goals and objectives are not profit-motivated.

The output of each category constitutes goods and services of one type or another; however, the organization goals and objectives, methods of operation, and individual motivation may vary considerably from one situation to the next.

A. Noncorporate Business

1. Individual proprietorship. An individual proprietorship is an enterprise owned by a single person who contributes the original assets to start the business, maintains and controls business operations, reaps full benefits in terms of profit, and is fully liable for all debts associated with the business. This is the oldest and simplest form of ownership and plays a significant role in the present-day economic structure. Individual proprietorships are particularly important in retail trade and service industries and include small engineering firms and private consultants. Engineers in this area provide a variety of services to both large corporations and nonprofit agencies. Some examples are consulting at the local community level in helping to identify and solve problems associated with town/city/county management; consulting at the state and county level in land development and land use; consulting at the federal and state level in the design of bridges, highways, waterways, and structures; performing studies and analyses at the federal and state level in support of budgeting decisions; accomplishing specific tasks for both project and staff functions within a corporation; and so on. The consulting role is significant where tasks are specialized, requiring a high degree of expertise, and where the task span time is relatively short.

The major advantages of the individual proprietorship are as follows:

a. The individual proprietorship is easy to establish, since no legal formalities or filings are required to start a business.

b. The expense in starting a business is minimal (except in instances where capital equipment needs are high), as there are no filing fees, legal fees, or taxes involved in the organization process.

c. This form of ownership is simple and extremely flexible, since business decisions are made by one man (i.e., the owner), and he is not restricted by the limitations of incorporation or a corporate charter.

d. There is a great deal of personal motivation and incentive to succeed. The owner receives *all* the profits.

e. Secrecy can be maintained relative to the operating methods, newly developed techniques, patentable processes, and so on, since stockholder reports and other accountability reports are not necessary.

f. There are no corporate income taxes.

On the other hand, all is not ideal with the individual proprietorship; there are some disadvantages too. The owner is liable for all obligations and debts of the business, and it is often difficult to raise large amounts of capital to buy equipment or hire assistants because of the unstable nature of the business. This instability stems from the fact that if something should happen to the proprietor (i.e., death or disablement), the business may cease to exist. In the event of failure, creditors can take the proprietor's personal property as well as his business property to settle their claims. Thus, the owner of an individual proprietorship enjoys unlimited rights to the control and the profits of the business at the risk of unlimited liability for debts and losses.

2. General partnership. A general partnership is an association of two or more persons to conduct as co-owners a legal business for profit. A general partnership can be established and can operate through a simple oral understanding among the persons concerned, but a written agreement is more prevalent and usually advisable. Such a written agreement may take the form of a filing of articles of partnership in a state or local office of public record. The agreement should specify the name and nature of the business, the initial sources of capital, and the procedures to be followed in distributing residual assets should the business be dissolved for one reason or another. In the absence of a written agreement, all profits and losses are shared equally.

Many small engineering firms (similar to law firms, retail trade organizations, and medical clinics) are based on the general partnership form of ownership. It is quite common to organize a group of engineers who represent different fields of expertise oriented toward the needs of the geographical location and to establish a partnership. If the overall capability and the individual fields of expertise are carefully aligned with the needs, the partnership is likely to flourish, resulting in growth or an excessive backlog of work. If not, the partnership will likely have some difficulty in getting started. In addition, the partnership must consider the advent of changing needs as technology, social, economic, and other influencing factors prevail.

As in the case of individual proprietorships, the general partnership is easy to form, relatively inexpensive in terms of organization cost, and the incentive for success is high. The credit rating in securing additional capital is better than for an individual proprietorship but still limited overall, since the risks are high as compared to a corporate structure.

A major disadvantage is the personal unlimited liability of the individual members of the partnership. Each partner is held responsible for the commitments of the other partners in the organization. In instances where the debts exceed the assets of the partnership, the individual partners are liable to the extent of their personal assets. If the partners also have personal debts, priority of claim is usually based on the "rule of marshalling"; that is, creditors have first claim on personal assets. After personal creditors are satisfied, any remaining personal assets of a partner are utilized to pay partnership debts. Likewise, the organization's creditors have first claim on partnership assets, while remaining assets are available for personal creditors.

Because of this liability, it is incumbent that the partners get along with one another and that there is complete agreement as to who will do what! Major commitments should be reviewed and approved by all partners prior to initiation, since all are affected by the results. Preferably some basic operating procedures are established in writing when the partnership is formed. However, even with such procedures, there can be occasions of misunderstanding, disagreement, or mistrust that could lead to catastrophe for the firm. An awareness of these possibilities and the potential consequences is probably the best means of prevention.

As partnerships grow or personnel changes occur, each "old" partner in the organization has the right to approve any new partners. Likewise, no additional partners should be taken into the partnership without the consent of *all* old partners. This agreement is necessary to minimize the risk of choosing incompetent or dishonest partners.

3. Limited partnership. The limited partnership is an association of one or more *general* partners and one or more *limited* partners. The general partners manage the business and have all the rights and liabilities of partners in the general partnership form of ownership discussed earlier. The limited partners invest their money in the firm and receive a share of the profits, but their liability is limited to the amount of their investment as long as they do not share in the control or management of the firm or attempt to interfere with such. Every limited partnership must have at least one general partner who manages the business and is therefore subject to unlimited liability.

The main advantage of the limited partnership is that it enables the firm to obtain risk capital from limited partners who might otherwise not invest in the venture because of the unlimited liability implications. At the same time, the characteristics of the ease and low cost of formation and the personal incentive to succeed are retained. The basic disadvantage lies with the limited partner, who invests in the firm but has no voice in management.

B. The Corporation

The corporation constitutes a group of people organized to operate a business under a formalized charter granting them, as a body, some of the legal rights and liabilities of an individual. A few of the major characteristics of a corporation are noted below:

1. The owners of the corporation are stockholders who elect a board of directors charged with the responsibility of running the business.

The board of directors then elects officers who make the day-to-day decisions necessary for successful operations. The profits from the operations are paid to the stockholders in the form of dividends as the board of directors dictates.

2. As the corporation is a form of ownership, there are no requirements relative to size. Small corporations may be owned by one or a few stockholders, while large corporations are owned by a wide variety of individuals, other corporations, and so on.
3. The stockholders are not generally liable for the debts of the corporation. Their liability is usually limited to their investment (price that they pay for their stock), and the personal assets of the stockholder cannot be taken by creditors to pay corporation debts.
4. Shares of stock, representing ownership of the corporation, can be freely transferred without the approval of other shareholders and without affecting the corporation as a whole. This infers corporate stability and long life.
5. The earnings of the corporation are subject to federal income tax and, where applicable, state income tax. After these taxes have been paid and a portion of the remaining profits have been distributed to the stockholders as dividends, the stockholders in turn must pay income taxes on these dividends. This constitutes a double taxation of corporate earnings. State taxes will vary from state to state.
6. Accountability and the business posture of the corporation are disclosed periodically through publicly available reports to the government and stockholders.

There are two types of corporations. The *closed corporation* is one that is owned by only a few stockholders, possibly members of a family or a small group. Outsiders are restricted from purchasing the stock. On the other hand, the *open corporation* is one whose stock is held by a wide variety of interests, and shares of stock are being bought and sold on a continuing basis. Trading is carried on by the New York Stock Exchange, the American Stock Exchange, and similar institutions.

1. Formation of the corporation. When forming a corporation, one would intuitively assume that the corporation would incorporate in the state (or country) where it anticipates doing the bulk of its business. This is usually the case for small corporations, whose business operations are primarily limited to the local geographical area. However, this is not necessarily true for large corporations. The general laws of incorporation vary considerably from state to state. Some states have the reputation of being tough on corporations, others are considered to be rather liberal, and others are rated as being fair and equitable. Thus, it may be feasible to incorporate in one state while doing business in another. The decision as to where to incorporate should consider the following:

a. Organization expense—organization taxes (based on x dollars of the corporation's estimated value or on a fixed fee), filing fees, reporting requirements, and so on.

b. Annual corporate income taxes.

c. Incorporation qualification requirements—qualification of corporation directors and officers, rights of stockholders and voting regulations, allowances for corporate charter changes, and so on.

d. Legal precedence or corporate law (i.e., the degree to which past court decisions have favored or gone against other corporations).

Some states will change their respective requirements from time to time to attract business and may later revert to former practices. In any event, when the decision has been made as to where to incorporate, the incorporators will file a notarized certificate of incorporation with the secretary of state and will pay a filing fee and an organization tax. The certificate of incorporation generally includes the name of the corporation, the address of corporate headquarters, a statement covering the corporation's purpose, details about the general nature and administration of the corporation, and the value of the corporation (i.e., amount of capital stock and par value, other assets). When the certificate of incorporation is approved, it becomes the corporate charter.

The incorporation in any given state does not automatically give license for a corporation to freely operate in other states. To other states the corporation is a "foreign corporation" (or "alien corporation" if different countries are involved), and operations of the corporation may be prevented.[1] However, in the interest of the state's economic welfare, foreign corporations are usually admitted with certain provisions. These provisions often constitute the payment of certain taxes and the filing of annual reports.

*2. **Stock and stockholders***.[2] As indicated earlier, stockholders are the actual owners of the corporation, although their share in management is usually limited, as it is delegated to the board of directors.[3] A stockholder is a *holder of record* to the extent of the number of shares of stock that he owns.

---

[1]A corporation that operates in the state in which it is incorporated is known as a domestic corporation. One that operates in a state (or states) other than the one in which it is incorporated is called a foreign corporation. A corporation that is doing business in one country but is incorporated in another is known as an alien corporation.

[2]Stock and the methods of corporate financing are discussed further in Chapter 11.

[3]In open corporations the ownership function and the management function are basically divorced since the stockholders elect a board of directors and then delegate the management tasks to the directors. In small closed corporations, where the majority of stock is owned by a family or a small group, the owners may or may not delegate the management tasks to the directors. In many instances the owners themselves may also be the directors of the corporation, and in such cases the ownership and management functions are not split.

Shares of stock are issued in the form of stock certificates, which in turn may represent one or more shares. A stock certificate is a document that is recorded by the corporation (or its agent) and includes the name of the corporation, the state of incorporation, class of stock, par value, and the name of the owner and the number of shares held. The certificate of incorporation will specify the *par value* to be assigned to each share of a given class of stock. This par value is an arbitrary amount equal to or less than the price per share at which the corporation is willing to sell its stock.[4]

There are many classes of stock; however, the most prevalent categories are preferred and common stock. In either case, the holder of a share of stock is a part owner of the corporation. He does not hold title to the corporation assets, since title is held by the corporation itself; however, he does have a pro rata interest in the assets.

Preferred or *preference stock,* as the name implies, takes precedence over common stock in the event that the corporation is dissolved. Preferred stockholders' claims on income and assets come after those of the corporation's creditors and bondholders but prior to those of common stockholders.[5] The amount paid to the preferred stockholder is usually equivalent to the par value per share of stock or the amount stated on the stock certificate.

Preferred stock also has priority in the payment of dividends. Dividends are generally fixed (e.g., 5 per cent) at the time of issue and are almost invariably cumulative. This means that all current and outstanding back dividends must be paid prior to any distribution to the common shareholders. In the event that the corporation does not distribute dividends in a given year when business is down, the preferred stockholder has the right to his fixed dividend in a subsequent year when business conditions improve. On the other hand, common stockholders may not receive any dividends during an off year.

The advantages in owning preferred stock can be seen from the discussion above. It offers security to the stockholder since the rate of return is usually constant, and there are privileges relative to dividends and the possible liquidation of assets on the part of the corporation. Conversely, a major disadvantage is that the yield is often higher than for bonds, and dividends are not tax deductible.

Common stock, on the other hand, is the simplest form of corporate ownership. Common stockholders are entitled to share in the residual earn-

---

[4]The par value is usually some original value assigned to the stock and is a reference point in the event of liquidation. Also, the par value is fairly close to the price paid by the original purchaser of preferred stock. The actual selling price of a share of stock or market value may be more than or less than the par value, depending on the business posture and the assets behind the stock. Thus, the par value sometimes has little meaning, particularly in transitions involving common stock.

[5]Bonds used as a method of corporate financing are covered in Chapter 11.

ings of the corporation if dividends are paid, and common stockholders collectively have a voice in management through the board of directors they elect. If the corporation profits are high, the common stockholders' return on investment may be high, but if the profits are low, he may receive nothing. Common stock is, of course, more speculative than preferred stock, although an individual can assume different degrees of risk through the careful selection of investments. An investor may prefer to purchase stock of the "blue-chip" corporations, whose dividends are paid on a fairly regular basis and where history has indicated that the applicable corporations are relatively stable. On the other hand, an investor may wish to risk more capital in "playing the stock market" by buying and selling at specific times, depending on hunches, trends, or inside knowledge. In doing such, he may double or triple his investment or he may lose everything. The degree of speculation depends on one's aversion to risk (i.e., the willingness of a person to assume risk versus avoiding risk).

Corporations in general prefer that their ownership lie entirely in the hands of common stockholders without being tied to the special privileges granted preferred stockholders. Usually when a corporation is initially formed, shares of common stock are issued. Later when business declines occur and the corporation is seeking additional sources of financing, it will issue preferred shares of stock since investment in common stock is not attractive from the standpoint of the investor. In most instances the preferred stock issued is convertible into common shares at the stockholder's option and is subject to repurchase at a specified call price at the option of the corporation. The purchaser of convertible shares is assured of a dividend as long as he continues to hold the preferred stock and has the opportunity to participate in the benefits of a corporation's growth by converting to common shares when the dividends and market value of a share have exceeded those values associated with the preferred stock.

3. Control of the corporation. The stockholders of most corporations collectively retain the rights to adopt and amend the bylaws, elect the directors, amend the corporation charter, sell corporate assets, or dissolve the corporation.[6] However, the control of the corporation is generally in the hands of management (i.e., the directors and major officers) rather than the stockholders. In small corporations, if one person or a family owns the majority of stock, there is obviously no problem in control. This person will

[6]The corporate bylaws usually include such information as the date and place of the annual stockholder's meetings; list of corporate officers and their functions; the quantity, term of office, and methods of selecting and appointing directors; the process for amending the bylaws; and the information content that is released to the stockholders for review and/or approval (i.e., annual reports, proxy statements, etc.).

select directors of his choosing, who in turn will appoint corporation officers with the approval of the major stockholder. In some instances, the major stockholder may also fulfill the position of a director and a corporate officer.

For large corporations, if the corporation is doing reasonably well (i.e., paying dividends), most of the stockholders will automatically support management by returning management proxies or voting for what management recommends. If the stockholders are dissatisfied, they will sell their stock rather than attempt to replace management. The replacement of management (except through the regular voting process) requires solicitations from other stockholders, proxy fights, and so on, and is rather expensive for noninsiders, with the costs being borne by the dissident stockholders.

Management controls and functions within the corporate organization structure are discussed further in Section 10.2.

4. Business combinations. Market growth and decline situations often result in business combinations of one type or another. The types of combinations may fall into one of the following categories:

a. *Horizontal integration*—the combination of corporations producing the same type of products (i.e., competitors). If the market declines, two or more corporations may decide to combine efforts.

b. *Vertical integration*—the integration of a corporation and one or more of its suppliers.

c. *Conglomerates*—the combination of corporations, producing unlike products with the objective of diversification.

The methods through which business combinations occur may be either informal (noncontractual) or formal in nature. Informal means may be based on a gentleman's agreement, simple written statements, or the interlocking of directorates (i.e., combining the boards of directors of two or more corporations). Formal methods primarily constitute consolidations and mergers. A *consolidation* takes place when two or more corporations are integrated into a single *new* corporation. None of the original constituent corporations remain as an identifiable entity. On the other hand, a *merger* occurs when one corporation absorbs other corporations by taking over the rights, assets, and liabilities of the corporations that are absorbed. The absorbed corporations lose their identity in the process. In both instances when formal combinations occur, stock transitions are initiated, where old stock is converted into shares of new stock that represents ownership of the new corporation.

The *holding company* is another form of business relating to consolidations and mergers. A holding company is a corporation organized primarily for the purpose of owning stock in other corporations to achieve voting rights and control in those corporations without resulting to an actual merger or consolidation. In other words, the general intent is to hold stock without getting involved in the management and operations of the business. The holding company is known as the parent company, and each subsidiary retains its own corporation identity.

5. Government controls. As indicated earlier, the laws of incorporation created and regulated by each state impose local government controls in the areas of taxation on income, the purchase and sale of assets, and reporting. The federal government also has controls that affect many phases of the ownership and operations of a corporation. When a corporation is engaged in interstate commerce (i.e., doing business across state lines), there are additional taxes and regulations. Also, there are regulations governing ownership functions, such as the exchange of securities and the formation of new business combinations. The following regulatory acts are presented as examples:

a. The Security Exchange Act of 1933 was initiated to promote full disclosure of all facts dealing with the purchase and sale of securities and to ensure that the appropriate information is filed with the Securities Exchange Commission. This was implemented after the depression in the early 1930s when many investors lost millions of dollars in worthless securities.[7]

b. The Security Exchange Act of 1934 provides for the supervision of the stock exchanges and the activities of the security markets. Again, full disclosure relative to trading is necessary.

c. The Sherman Antitrust Act was designed to protect trade and commerce against unlawful restraints of trade and monopolies. The objectives were also to prohibit unfair methods of competition. This act applies primarily to the earlier discussion of business combinations.

In spite of these and many other regulatory requirements now in existence, there are violations occurring on a continuing basis which ultimately result in the initiation and enforcement of additional controls on corporation activities.[8]

---

[7]When a corporation has available cash in excess of current operating needs, it is often invested in government securities, high-grade commercial notes of other firms, and similar credit mechanisms. Securities are usually sold on short notice at a relatively stable market price when cash shortages occur.

[8]As an example, a series of acts were initiated as a supplement to or an amendment of the Sherman Antitrust Act. The reader may wish to review the objectives of the Clayton Act, Robinson-Patman Anti-Price Discrimination Act, Tydings-Miller Act, Federal Trade Commission Act, and the Wheeler-Lea Act.

C. Government Agencies

Another major grouping of organizations includes government agencies. Federal government agencies, such as the Departments of Agriculture, Commerce, Defense, Interior, and Transportation, are quite extensive in organization content and depth and are the sponsoring agents for a large quantity and variety of programs. For instance, the National Science Foundation sponsors a great deal of institutional research, and the Department of Transportation is heavily involved in the program initiation and direction of ground and air transportation systems and transportation vehicle system design. The Department of Interior is responsible for the national parks, forests, waterways, and the design and operation of reservoirs and dams. The Department of Defense and the Army, Navy, and Air Force determine requirements for all national defense projects, which include aircraft, ships, ground vehicles, missiles, electronic systems, and facilities. NASA is responsible for space projects and the development of space vehicles.

At the state level, the state highway department is responsible for the planning and design of highways, bridges, and waterways. The health department deals with air and water pollution and sanitation requirements. Land subdivision and the design of sewers and septic tank systems are also concerns of the health department. State and local agencies also deal with industrial development, urban planning, building construction, and functions associated with county/city/town management.

As one can see, the federal and state governments sponsor and direct numerous programs that require engineering expertise. In many instances, the responsible agencies accomplish product planning, conceptual design, and some preliminary system design activity prior to initiating contracts to businesses and corporations for final system design, detailed design, production/construction, test and evaluation, and product support. In addition, these agencies continually monitor and review the project activities of contractors throughout the duration of the applicable project span time. Engineers representing a wide variety of fields are required by government agencies to accomplish the necessary planning and design functions. Also, engineers are required to accomplish follow-up program review and evaluation effort to ensure that all system requirements are met.[9]

D. Nonprofit Research and Development Organizations

The nonprofit category of organizations also includes research and development institutes, which are similar to large partnerships and profit-motivated corporations but are supported by private endowments, govern-

---

[9]The employment of engineers by government agencies is accomplished in compliance with federal and state civil service requirements. These requirements are briefly discussed in Chapter 14.

ment agency funding, university/college sponsorship, or a combination of these. Institutes in this category (which are relatively few in number) are basically independent in terms of activity and are usually chartered as not-for-profit corporations. A few examples of some of the larger institutes are listed below.

1. Battelle Memorial Institute, Columbus, Ohio
2. Institute for Defense Analysis, Washington, D.C.
3. Mellon Institute Research Foundation, Pittsburgh, Pennsylvania
4. Mitre Corporation, Bedford, Massachusetts
5. Research Triangle Institute, Durham, North Carolina
6. Southwest Research Institute, San Antonio, Texas
7. Stanford Research Institute, Menlo Park, California
8. System Development Corporation, Santa Monica, California

Highly qualified and specialized researchers and engineers are actively engaged in a wide variety of complex problems and projects. The orientation is toward accomplishing basic and applied research, the performance of system studies and analyses, the establishment of advanced design concepts, and in some instances the development of prototype equipment.

E. Universities and Colleges

One of the primary missions of universities and colleges is to participate in both basic and applied research to acquire the expertise necessary to support teaching programs. Such research is generally obtained through grants or contracts and is usually conducted by individual professors and supporting graduate assistants on a part-time basis.[10] The result increases scientific and engineering knowledge, which is not only used in the classroom but is made available to the sponsor through research reports. In addition, universities often reserve the right of free publication; therefore, research results are not always the exclusive property of the sponsor but can be made available to the public.

In many universities, the individual expertise gained from sponsored research is made available to businesses, corporations, and agencies by the release of research reports and through private consulting. Depending on the policies of the university, professors may be allowed to do a certain amount of private consulting on their own during the university's academic year (i.e., September to June) when classes are in session. During the summer months or vacation periods, most professors are free to participate in full-time con-

---

[10] In instances when the size of a given research project is large enough and/or the priority of the research is high, professors may be relieved from teaching and other duties and assigned to the research project on a full-time basis.

sulting. In any event, a good deal of engineering work is accomplished by professors from engineering colleges throughout the world operating on the basis of the individual proprietorship discussed in Section 10.1A1.

F. Cooperatives

Cooperatives contain some of the features of partnerships and corporations. The principal intent of the cooperative is to combine a group of individuals into an organization to enable the conduct of business in a manner to eliminate profit to outsiders and thereby provide goods and services to its members at cost. There are shareholders in cooperatives, and they elect directors in a way similar to the profit-motivated corporation. Members buy share and pay fees, and the profits (if any) are distributed to the members periodically. The advantage of a cooperative lies in the buying and selling of products in bulk quantities rather than on an individual basis, thus realizing the benefits of price breaks and certain market controls.

10.2 CORPORATE ORGANIZATION

The corporation represents the largest category of business activity on the basis of quantity of goods produced, number of employees, percentage of national income, and so on. Corporations in general can significantly influence the status of the other categories of business discussed in Section 10.1 and may have a considerable impact on social, political, environmental, and economical conditions at the federal, state, and/or local levels. The extent of impact will vary, depending on the type of product(s) produced, the geographical location of the corporation, the processes utilized and resources required for product manufacture (e.g., real estate, structures, power, water, sanitation and facilities for waste disposal, availability of transportation), the size and stability of the corporation, and so on.

As stated earlier, the corporation is owned by the stockholders and generally controlled by the board of directors and major officers. Corporations vary in size from large corporations such as the General Motors Corporation, the General Electric Company, and the Monsanto Company to the small corporation with a handful of people.[11] The large corporation is pri-

---

[11]To gain some perspective of magnitude, the General Motors Corporation in 1973 included over 30 major divisions, approximately 811,000 employees, and was owned by 1,305,998 common and preferred stockholders with an equity of $12,566,776,627.00 (as indicated in the GMC Annual Report for 1973). In 1974, the Monsanto Company had employment in excess of 60,000 and approximately 98,500 stockholders. These and other large corporations of this type produce many products and conduct widely dispersed activities throughout the world, and their growth is highly dependent on the direction and extent of the research and engineering functions conducted within.

marily discussed in this section, particularly in regard to engineering activities.

A. Corporate Office Function

The corporate office usually includes the board of directors, the president of the corporation (who may also be a director), and his officers. Figure 10-1 illustrates a sample organization structure, with the corporate function included in the shaded area. The board of directors, under the leadership of a chairman, is responsible to the stockholders for the following:

1. The election of corporation officers (president, executive vice president, corporate vice president, secretary, treasurer, etc.).
2. Determining corporation policies.
3. Initiating and implementing activities within the corporation (e.g., making decisions relative to the pursuit of new product lines, negotiating contracts in the name of the corporation, making large investment decisions).
4. Issuing corporation stock and declaring dividends.

The board may constitute from 5 to 20 directors elected by the stockholders for a specified term of office.[12] Directors of large corporations (with the possible exception of the president) are normally not employees of the corporation but include high-level businessmen from outside organizations. In smaller corporations, the makeup of the board often includes a number of directors who are also employees of the corporation.

The directors act as a group in making decisions for the corporation. Board meetings are scheduled in advance, an agenda is prepared and distributed, stockholders may be invited to attend, and the directors review and vote on specific action items. Many of the action items on the agenda will evolve from recommendations submitted by the president or the officers of the corporation as they are usually more familiar with the specific operations of the corporation than the directors. Minutes of the board meetings record the actions taken by the board as a whole, and in certain instances the actions of individual board members are noted when accountability is desired. The results of board meetings are often covered in special reports

[12]The number of directors of the corporation will vary. The laws of incorporation of a state may specify a minimum number so as to protect the stockholders against the actions of a few, who may make decisions that are not in the best interests of the corporation as a whole. The term of office for each director is usually from one to three years, and the terms for the various directors are often staggered for continuity purposes (i.e., to avoid the possibility of a completely new group of directors not familiar with corporation policies from taking over at any given point in time).

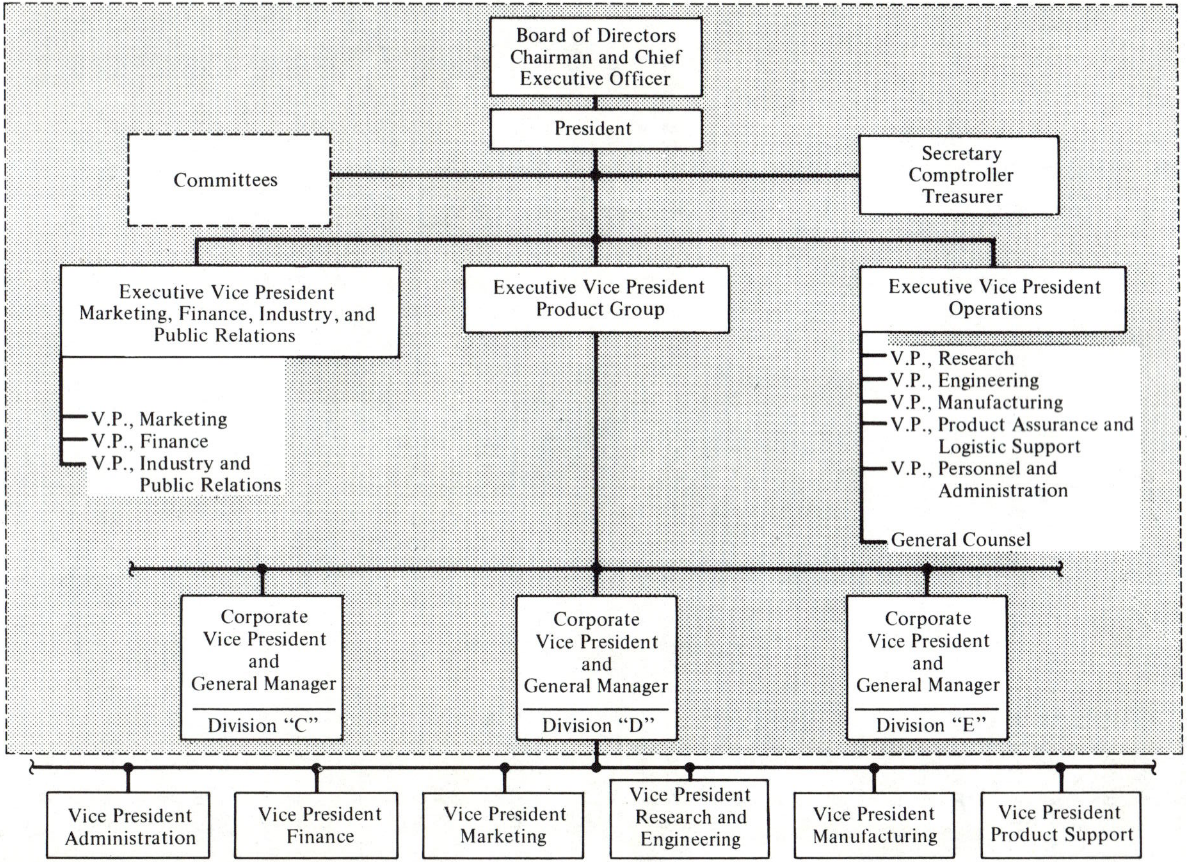

Figure 10-1. The corporate office.

distributed to the stockholders separately or in the corporation's annual report.

The president of the corporation is usually the chief executive officer and is delegated the responsibility and the authority to carry out the policies of the board of directors. He (or she) gives guidance to other executives on his staff and keeps the board of directors informed on the progress of the various phases of the business. The president also becomes heavily involved in public relations at all levels, industry-government relations, and attends many meetings both directly and indirectly related to the operations of the corporation. As a great deal of the president's time is consumed by these highly important activities (which if unattended could hurt the corporation), he has to rely primarily on his executive staff to carry on the day-to-day internal business operations of the corporation.

The president's staff constitutes those executive and operating vice presidents that establish the policies and deal directly with the business of the operating divisions of the corporation. As illustrated in Figure 10-1, certain vice presidents are responsible for policies guiding functional activities *throughout the corporation* in areas such as research, engineering, manufacturing, marketing, product support, and personnel and administration.[13] Other vice presidents not only serve on the corporate staff but function as general managers of the operating divisions. These individuals are located at the physical site of the division but maintain an ongoing close working relationship with other members of the corporate staff, thus promoting the necessary integration of corporate and division activities where appropriate.

In addition, a corporation secretary is resonsible for performing a number of legal duties such as preparing and filing reports required by federal and state regulatory agencies, sending out notices of director and stockholder meetings, sending out stockholder proxy statements, maintaining stockholder lists, and handling stock transfers. The treasurer and comptroller functions are established to evaluate the corporation's financial status and to account for the receipt and disbursement of corporate funds.

Referring to the example illustrated in Figure 10-1, the policies governing engineering and related areas are established through various functionally oriented vice presidents reporting to the Executive Vice President for Operations, and policy direction flows from that level through the general managers of the various divisions to the appropriate operating units within the division organization structure. This flow is illustrated in Figure 10-2.

The policy direction indicated in Figure 10-2 stems from five functional vice presidents. The relationship to engineering ties into the discussion in

---

[13]Policies refer to top-level guidelines applicable to conducting functional activities within the corporation. Policies generally dictate *what* to do and *when* and *where* to do it. On the other hand, procedures are more detailed in nature, directly support the policies, and usually indicate *how* to do something.

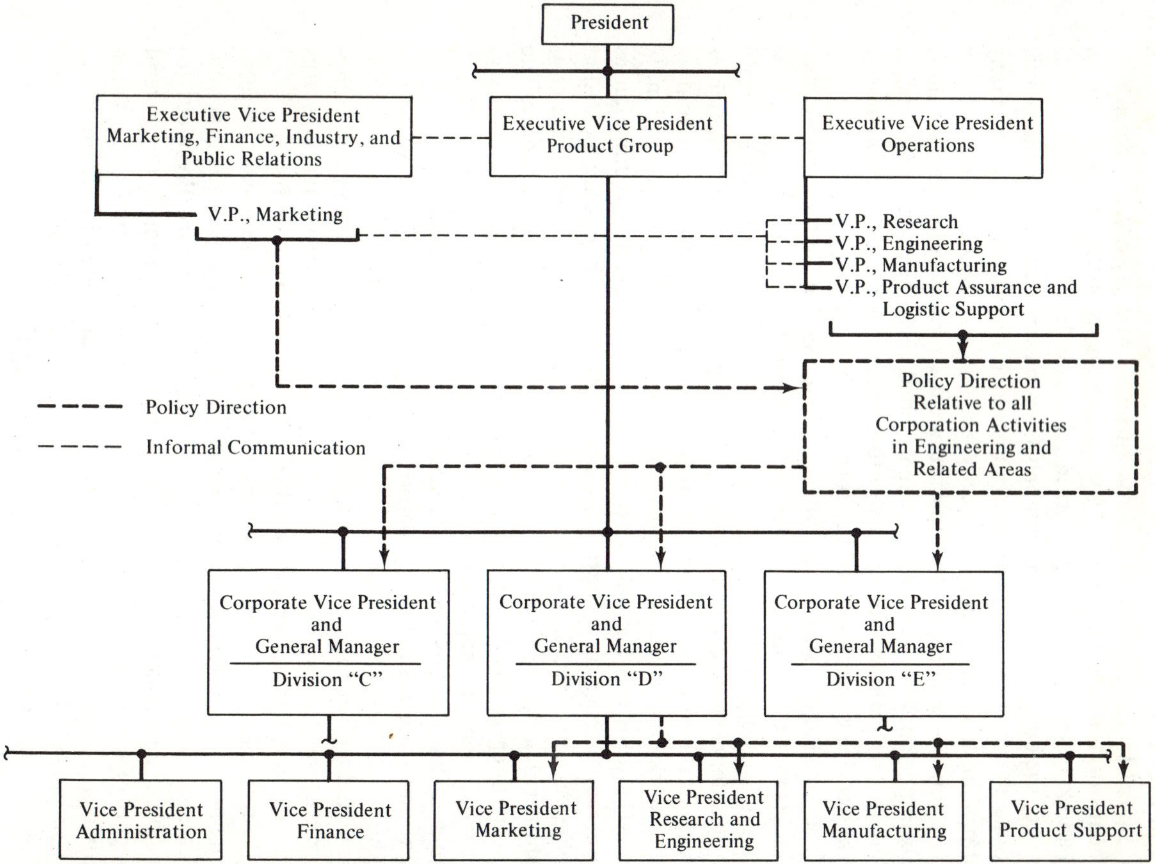

Figure 10-2. Engineering policy flow.

previous sections. For instance, policies pertaining to research and basic product design are established by two vice presidents (working together) and applied as guidance to the division Vice President for Research and Engineering, Division D. Engineering functions pertaining to product production/construction and product support are covered through policies generated by another corporate vice president. Thus, engineering activities in a large corporation are affected by a number of executives who establish policy at the corporate level, and as in any organization structure, these executives must communicate with each other on a continuing basis to ensure that the various policies generated are complete and reflect a consistent approach.

B. Division Responsibilities and Functions

The division organization responsibilities and functions are based on the corporate office structure and the policies established at that level. Policies associated with product line and product mix objectives will influence the organization pattern developed within. The established pattern may assume any one or a combination of the configurations discussed in Section 9.3B (i.e., project, project-staff, product-line, etc.). However, for the purposes of discussion, the basic functional organization illustrated in Figure 10-3 will serve as a primary reference point. There may be any number of divisions within a large corporation. The general manager (or president) of a division, who also serves as a corporate officer, is responsible for all functions within his domain. The vice presidents who report to the general manager are not corporate officers, but are executives of the division and have been delegated the authority and responsibility for large segments of the activity accomplished within the division. In some corporations, personnel in these positions are called directors or chiefs. The hierarchy of titles may constitute vice presidents, directors or chiefs, department managers, section managers, section heads, and supervisors in that order of descendency; or they may assume some comparable set of titles tailored to the size and type of organization structure. The ultimate positions within the division will stem from the policies established for that division by the Corporate Vice President for Personnel and Administration (refer to Figure 10-1).

The engineering activities conducted within a division of a large corporation will in general be tailored to the product life cycle presented in Part Two. A simple transition from these activities to the organization structure discussed above in included in Figure 10-4.

1. Project engineering function. Project engineering refers mainly to those activities necessary in the planning, research, design and development, production/construction, evaluation, and support for a *single* system. If the project is large enough (i.e., a large aircraft program, construction of a large

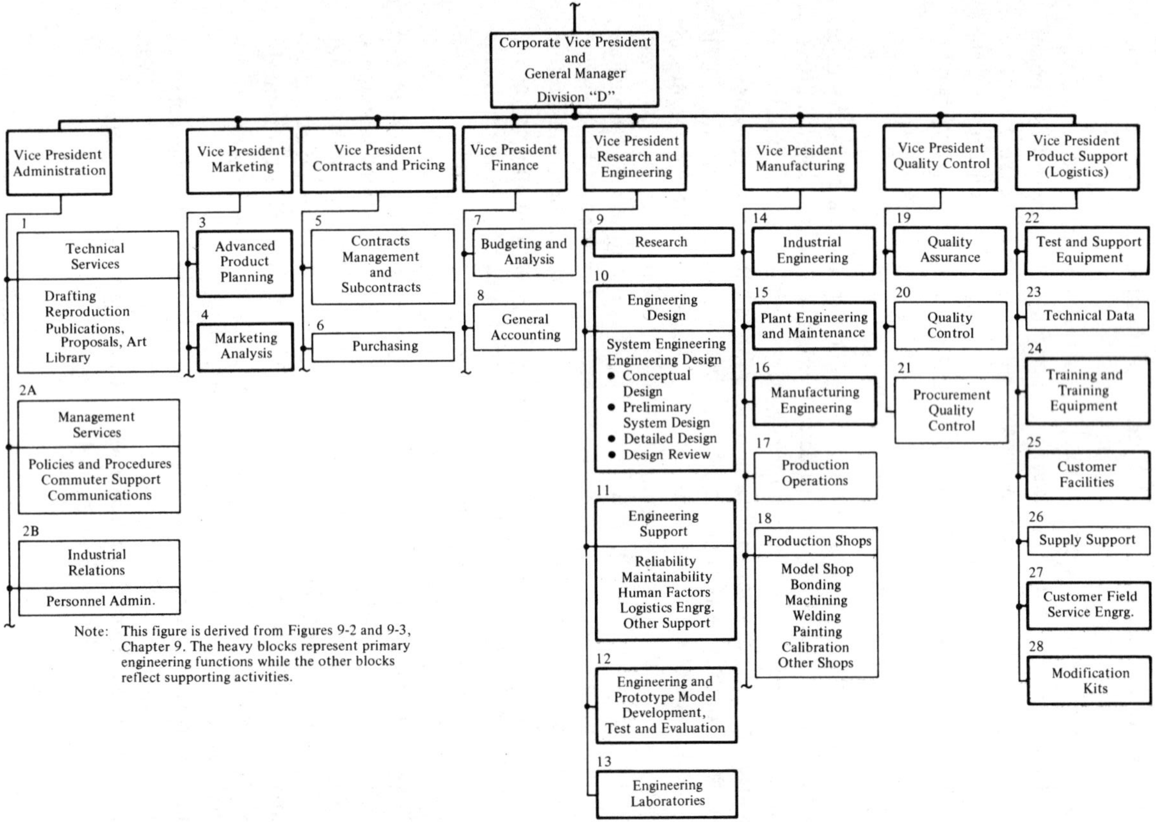

Note: This figure is derived from Figures 9-2 and 9-3, Chapter 9. The heavy blocks represent primary engineering functions while the other blocks reflect supporting activities.

Figure 10-3. Engineering activities within a division of a large corporation.

Figure 10-3 Block No.	Title	Engineering Program Functions/Tasks	Part II Paragraph No.
3	Advanced Product Planning	Product evaluation, selection, and justification; specifications and plans; definition and scheduling of tasks; proposal preparation; and supporting contract negotiation.	3.4
4	Marketing Analysis	Identification of product need (technical requirements); technological forecasting; and feasibility studies.	3.1, 3.2A, 3.3
9	Research	Basic research; applied research; research methods; research applications; and technology assessment.	4.1, 4.2 4.3, 4.4, 4.5
10	Engineering Design	Conceptual design (system operational requirements and maintenance concept); preliminary system design (functional analysis and allocation, trade offs and optimization, synthesis and definition); detailed design (electrical design, mechanical design, design support, design aids, design review, etc.); engineering test and evaluation requirements (all applicable engineering disciplines).	5.1, 5.2 5.3A, 5.3C, 5.4
11	Engineering Support	Reliability, maintainability, human factors, logistics, engineering, value engineering, producibility, safety, and other design support activities. Design criteria, design consultation, predictions and analyses, design review, and test and evaluation requirements.	5.1, 5.2. 5.3A, 5.3C, 5.4
12	Engineering Prototype Development	Engineering model, service test model, and prototype model development; test and evaluation; and system modifications.	5.3B, 5.3C, Chap. 7
13	Engineering Laboratories	Research laboratories (basic and applied research); and engineering laboratories (industrial design, component testing, materials testing, technique/process development, mock-up development, etc.).	Support of activities in Chap. 4 and 5
14	Industrial Engineering	Industrial engineering and operations analysis; methods engineering; and production control requirements.	6.2, 6.2C, 6.2D
15	Plant Engineering	Plant engineering (plant layout and design, utility requirements, etc.); and plant maintenance.	6.2A
16	Manufacturing Engineering	"Make or buy" decisions; material selection; design of manufacturing sequences and processes; determination of man-machine operations; and tool, test equipment, and jig/fixture design.	6.2B
19	Quality Assurance	Requirements and procedures for product quality and for maintenance of the desired quality characteristics throughout production/construction.	6.3
22	Test and Support Equipment	Specification of test and support equipment requirements for operational system support throughout the life cycle; and design of special support equipment items.	8.2
24	Training and Training Equipment	Specification of training and training equipment requirements for consumer personnel; and design of training aids, devices, simulators, etc.	8.2
25	Customer Facilities	Specification of consumer requirements for new facilities in support of system operation and maintenance throughout the life cycle; and design of new facilities.	8.2
27	Customer Field Service	Sustaining producer field service engineers assigned to operational sites to support the consumer in operating and maintaining the system during the life cycle.	8.2
28	Modification Kits	Design of modification kits and installation of system modifications in the field.	8.3

Figure 10-4. Major engineering functions for a large project.

facility, etc.), the functions identified in Figure 10-4, or individual representatives from the responsible organization elements, may be combined under a vice president or a project director as illustrated in Figure 9-4. Project manpower loading assumes the approach conveyed in Figures 9-12 and 9-13, and requirements will include a mixture of expertise, ranging from the systems engineer defining operational requirements and performing optimization studies to the design engineer involved in the detailed design of an electronic circuit or a mechanical device of some type. The expertise required in these areas will usually shift somewhat as the system evolves from the planning stage to research, preliminary system design, detailed design, and so on. Thus, manpower requirements may vary not only in terms of quantity but in areas of individual expertise.

Engineers initially assigned to the project and who are no longer required after their functions have been completed are either (a) reassigned to another task on the same project, (b) transferred directly to another project within the division, (c) transferred to a staff engineering position within the division, (d) transferred to another division within the corporation, or (e) are terminated in employment.[14] On the other hand, when new project tasks evolve and manpower needs increase, the project director will initiate position descriptions and forward them to Personnel for action (refer to the position description example in Figure 9-11). Fulfilling these positions may result from transfers into the project from other sources within the division or corporation, the recruiting and hiring of new personnel, and/or the use of outside consultants.

The number of personnel actions from a manpower-requirement point of view will depend on the nature of the project, the project time frame, and factors such as the methods of contracting and funding. Manpower loading curves will differ from project to project as illustrated in Figure 10-5.[15] The ideal situation is to conduct projects where the man-loading is representative of project A in the figure. The curve presented reflects both growth and stability. On the other hand, there are many projects where the direct manpower requirements are more representative of project B, which poses a few problems, as noted below:

a. The project director may strive to maintain a level of man loading (constituting a composite mix of technical skills) in accordance with the *actual* manpower requirements. Since the curve fluctuates significantly over time, the number of personnel actions in terms of hiring and firing is high (from the standpoint of this particular project). This, in turn, creates instability and a general decline in work quality, as many of the better engineers are

---

[14]Reassignment of personnel within the project organization is usually handled on an informal basis by the project director or his designee. Transfers across organization lines or termination actions are handled by Personnel (refer to Chapter 14) at the request of the project director.

[15]The discussion on project man-loading requirements should consider attrition rates (due to resignations, deaths, and retirements) and effectiveness factors related to motivation, morale, initiative, absenteeism, and general quality of work. Many of these factors directly relate to the discussion on individual needs and managerial styles in Chapter 9.

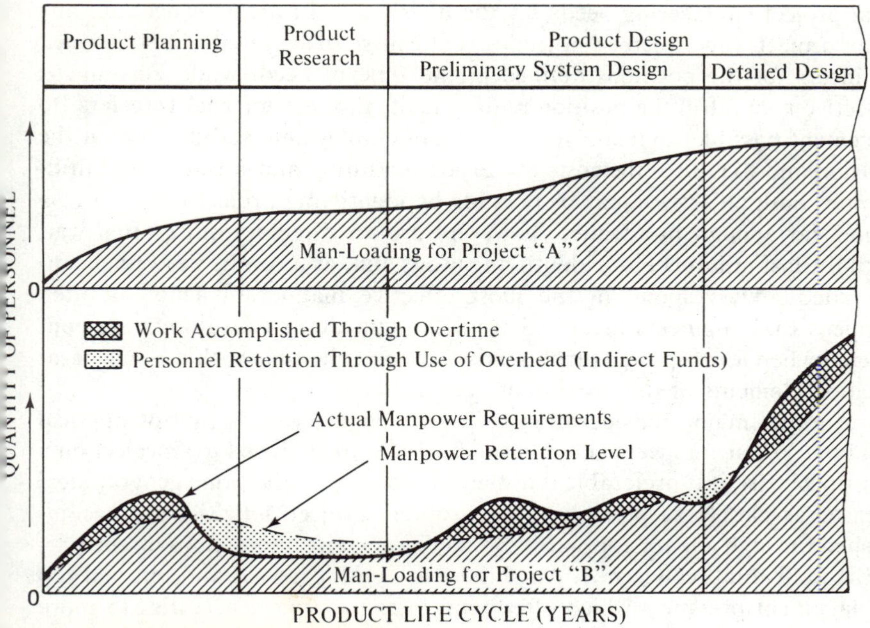

Figure 10-5. Project man-loading requirements.

not likely to be attracted to such a situation. In addition, some of those already assigned to the project, because of the envisioned instability, may actively pursue a transfer or look for another job on the outside.

b. The director may attempt to maintain a level of manpower in accordance with a curve reflecting an average condition represented by the dashed line in Figure 10-5. This indicates both growth and stability; however, the peaks and valleys in terms of actual requirements are covered through project personnel working overtime in some instances and management supporting salaries by using overhead funds and/or by diverting corporation profits in other cases. If the overtime is excessive in terms of both duration and the quantity of hours worked above the average work week, the effectiveness of the employees involved will ultimately decline. After awhile, constant overtime "gets old" and employee fatigue occurs, which reduces the effectiveness of the output. In the other situation, where direct funding will not support all the personnel assigned to the project, management must make the trade-off between continued support and the reduction of project personnel.

The man-loading projections are usually predictable when the proper level of advanced product planning has been accomplished. Each project is reviewed individually and then combined with other projects to determine total project engineering needs for the division.[16] In areas of need on any given project, the director's objective is of course to acquire the best trained engineers for the positions defined at the time of need, while Personnel's objective is to fulfill the position requirements through internal transfers (to the extent possible) to instill some degree of employment stability within the division and to reduce the costs of outside recruiting and hiring. The fulfillment of both objectives may be possible by identifying project needs to Personnel far enough in advance; however, project directors quite often wait until the last minute before identifying needs and, as a result, Personnel cannot adequately respond in the most effective manner.[17] Thus, it often happens that engineers are hired from the outside or consultants are contracted when a manpower surplus exists at the same time within other organization elements of the division or corporation.

Another major consideration of project engineering is that of physical location. When engineers and other individuals are assigned to a project on a full-time basis, it is preferable that they are located in the same geographical area or as closely as possible to each other. Project activities are accomplished through a team approach and there are many interactions on the part of the individuals concerned. Also, there are both external and internal top management pressures which often tend to cause project activities to move along at a rapid pace. For these reasons, it is essential that key project personnel be readily accessible to each other and that both formal and informal communications be maximized to the extent necessary to preclude the "out of sight/out of mind" situation that results in omissions causing problems later. Thus, although the available facilities may not provide all the luxuries and benefits as perceived by different individuals (which may vary considerably), the close proximity of project personnel to each other should be an overriding factor.

In summary, project engineering is oriented toward individual systems. The management of projects includes (a) the establishment of project goals and objectives, (b) the definition of functions/tasks, (c) organizing and staffing to fulfill the requirements, and (d) measuring and evaluating performance on a continuing basis to ensure satisfactory results. These factors of management are discussed rather thoroughly in Chapters 3 and 9.

2. Staff engineering function. Staff engineering pertains to those activities that are conducted within the functional organization illustrated in

---

[16]This review is generally conducted by the division's top management to assess overall status and to provide Personnel with an estimate of *total* needs.

[17]The delinquency on the part of the project director may be self-generated, through lack of internal planning, or caused by an external event which he could not have predicted and cannot control, such as a large contract aware at an unexpected time, a delay in contract negotiation with a customer or supplier, an unexpected budget cut, or a contract cancellation.

Figure 10-3 and which usually support more than one project concurrently. In other words, a division may be engaged in the design and development of a variety of product lines, each representing a relatively small project with only a few individuals assigned on a full-time basis while requiring additional support from other areas of expertise in a part-time capacity. Staff organizations, as illustrated in Figure 9-7, are highly specialized and operate in response to specific requests for project support. In addition, staff groups may perform outside contracted work or may be involved in company-sponsored research. Thus, the combination of assigned project tasks generated from within the division or corporation, contractual work from outside sources, and basic or applied research establishes the requirements for manpower loading and the size of the staff group.

Engineers performing staff functions are often engaged in marketing analysis, research, system analyses and predictions, special investigations, design of small system components, and so on. These engineers may be assigned to an internal short-term project task, research, a special study under separate contract, or a combination of these during any designated period of time. The experience gained in applying knowledge and the techniques developed for one project to other projects is definitely beneficial from the standpoint of promoting the state of the art in technology and enhancing the project results as a whole, and the variety of problems encountered is usually quite challenging for most engineers. However, a major disadvantage is encountered if the total staff effort is not substantial enough to maintain a relatively constant level of employment. If projects are few and gaps exist in the workload, the man-loading curve may be even more cyclic than the one illustrated in Figure 10-5 for project B. Again, management must make the trade-off between a reduction in force or supporting staff personnel on overhead.

When an engineer in a staff organization is assigned to accomplish project work, he is often faced with a situation in which he has two bosses. First, he must be responsive to the project requirements and the project director or his designated manager. Second, he must be responsive to the manager of the functional organization to which he is assigned, as his performance evaluation and prospective promotions and salary increases are often initiated from this source. Hopefully, the goals and objectives of the two managers are compatible since it is beneficial to the project manager to promote a high-quality output from the staff engineer, and it is beneficial for the functional manager to produce a high-quality product if he expects to receive additional requests for work in the future.

Unfortunately, this situation does not always exist, and the lack of communication between the two managers forces dichotomies in objectives and undesirable pressures on the engineer responsible for completing the work. If this continues and the engineer is "pulled" in different directions, he may decide to resign or to initiate a request for transfer, and the project manager may decide to solicit assistance elsewhere. Such actions are not always in the best interests of the division, as the staff organization will lose opportunities for work, which may have detrimental affects on the man-loading requirements. Thus, it is incumbent on the division's top management to ensure harmony between the project and staff functions within.

3. Technical services. Technical services as presented herein represent the miscellaneous supporting activities noted in Figure 10-3, block 1, the combining of which in this manner is peculiar to the organization structure illustrated in the figure. This is not to convey a lack of importance, as each of these functions is a necessary part of the engineering process.

a. *Drafting.* Drafting is the major engineering support activity which produces, processes, and updates all the design layouts and drawings that are necessary in defining the product. The output may constitute any one or a combination of the different types of drawings listed in Figure 5-12 (i.e., assembly drawings, control drawings, detail drawings, installation drawings, blueprints, etc.).

Conventionally, the drafting function has been performed by one or more technicians involved in (1) the translation of basic engineering ideas and concepts into sketches and layouts, (2) the conversion of sketches and layouts to preliminary engineering drawings, and (3) the ultimate detailing of preliminary engineering drawings leading to final drawings which are released for production or construction of the product. The output of this process represents the design engineer's thoughts at various stages as system design progresses, and as such, is highly significant, since drawings are the primary vehicle for communicating these thoughts to others. The engineer may develop all sorts of innovative ideas, but his efforts will be wasted unless he can convey these on paper such that they can be converted into a useful product output. Engineering drawings, combined with other design documentation, serve this purpose.

Drafting utilizes graphical methods and arrangements in presenting engineering details. Knowledge of graphical practices, geometry, engineering and manufacturing processes, dimensions and tolerances, material characteristics, and so on, is required. Drafting is a "language" by itself, used to define physical objects, and is based on the utilization of certain accepted universal standards. Consistency in the use of standards is important if adequate communications are to be maintained.

In the past, drawings at the early stages of system design have been prepared by a layout draftsman working closely with the design engineer on a day-to-day basis.[18] Detailers and senior draftsman work out the particular

---

[18]The particular titles of drafting personnel (whether male or female) may vary from one corporation to the next. As is the case for engineering and management classifications, personnel and administration policies will specify the job position titles.

details in producing a finalized drawing. Drawings are forwarded to a checker for review, and many are returned to the draftsman for corrections. When the drawing is considered ready for release, it will be processed through the review cycle illustrated in Figure 5-17, where engineers from *all appropriate* design and supporting organizations are provided the opportunity for review and approval. This process, particularly for large projects for which hundreds of drawings are required, can be quite extensive, and tight management controls are necessary to determine the status of each drawing in the loop and to ensure that drawings are being produced on schedule. Further, complications can result as design changes are introduced by the engineer while drawings are either in the loop or have been through the final approval cycle and must be revised and recycled for a second review. Thus, a good drawing control system is mandatory if the drafting objectives are to be met in accordance with schedule and cost requirements.[19]

More recently, techniques have been developed where some of the steps in the engineering design process can be simplified to a considerable extent through the use of computerized methods (refer to Section 5.3A4). Computer-aided design techniques will allow a design engineer to visualize his or her ideas, evaluate design alternatives, and to make changes without going through all the drafting steps described above. For large projects, in particular, the use of these improved methods has resulted in both a cost and time savings in terms of design and drafting requirements. A comparative example of such for one large project is related in Figure 10-6.

In determining whether or not to implement computer-aided methods, management must evaluate the benefits derived along with the acquisition and support costs (i.e., life-cycle cost) associated with the necessary computer equipment, terminals, software, and so on. For large projects or in instances where a division of a corporation has *many* small projects, it may be very worthwhile to implement computerized methods. On the other hand, it may be too costly in cases where the overall projected workload is relatively small and sporadic.

As a final point, one should address the aspect of where to physically locate drafting resources within the division. Draftsmen who work directly with the design engineer on a day-to-day basis and/or the computer terminal used by the engineer for design support purposes should be located in the same physical area as the design engineer to provide the accessibility and interaction needed. Drawing checkers who review drawings for completeness

---

[19]Quite often, when the drawing control system is inadequate and bottlenecks occur which create slippages in schedule, individuals may attempt to circumvent the system with the objective of making up time and reducing cost. Experience on many projects has indicated this to be more costly in the end, since drawing errors and design misrepresentations are allowed to remain in the system until detected at some point in the production/construction stage when the initiation of corrective action is even more costly. Performing the job correctly at the appropriate time in the design stage produces more effective results in the long run.

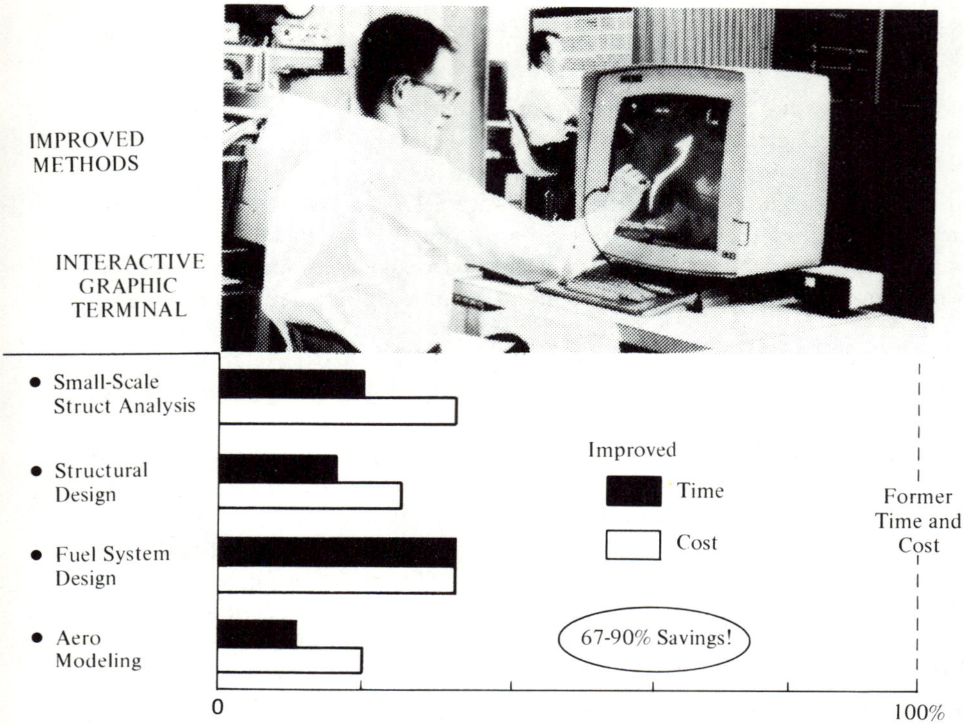

Figure 10-6. The application of improved computer-aided methods in design and documentation. (*Courtesy* Grumman Aerospace Corporation.)

and accuracy and the centralized division drawing file necessary for record keeping and configuration management purposes do not have to be as accessible, since direct interfaces with the design engineer are not as frequent.

b. *Publications and reproduction.* Engineers in performing system planning and design functions are usually burdened with the preparation of proposals, technical reports, manuals, program plans, and specifications.[20] In addition, there are numerous instances when presentations to top management and the customer are required (utilizing flip charts, slides, movies, and/or overhead viewgraphs). Obviously, the engineer must provide the raw technical information required; however, the processing of material can be carried out more efficiently by publications specialists trained to accomplish:

(1) Preparation (from rough engineering data), editing, and production of technical and administrative proposals, plans, reports, and manuals.

[20] In large organizations, technical data (operating and maintenance manuals, installation and service instructions, calibration procedures, etc.) covering the operation and support of the system in the field throughout its life cycle are included as an element of logistic support and are usually produced by a separate function (refer to Figure 10-3, block 23).

(2) Preparation of graphics and charts for presentations.

(3) Processing of artwork and material to be photographed and reproduced.

Support in the preparation of publications, brochures, and presentation material can be of tremendous benefit to engineering organizations, particularly if the publications function includes writers and editors possessing some background in engineering and scientific fields. A certain amount of technical training will certainly prove beneficial relative to ensuring the completeness and content accuracy of the publication. On the other hand, if the publications specialists do not possess some technical knowledge in the subject areas covered, the responsible engineer(s) should review the applicable material prior to production and release.

c. *Technical library.* The division's technical library serves as the primary reference source for permitting researchers and engineers to keep abreast in their respective fields and in fulfilling their objectives for self-development by extending their knowledge. Thus, it is incumbent on management to provide the resources necessary for data acquisition and for the librarian to ensure that the most appropriate material is made available on request. Technical libraries generally include the following categories of data:

(1) Magazines, periodicals, technical society journals, industry association newsletters, pamphlets, and technical conference proceedings.

(2) New textbooks, technical handbooks, dictionaries, and encyclopedias.

(3) Government and commercial specifications and standards, manufacturers' catalogues, and component-parts data.

(4) Research reports, copies of division proposals, and selected project reports.

(5) Indexes to publications and reports.

In some instances the technical library will maintain a separate section for classified and company proprietary material where security is required and tight controls are essential. This material will be contained in locked files and checked out only to those individuals who possess a security clearance and have a "need to know."[21]

---

[21]In dealing with government agencies and defense systems, a security classification (i.e., top secret, secret, or confidential) may be applied to certain designated information. The storage and use of this information will require special facility provisions that will vary with the level of classification. For instance, the handling and storage of secret information will necessitate a facility clearance usually maintained by security guards stationed at each entryway, employee badges or passes for entry into the building, and documentation sign out/sign in procedures with one or more locked safes for storage. To gain access to the information, individuals must be cleared by the responsible government agency (or as otherwise designated) for secret information. Top secret requires more stringent controls, while confidential material is controlled to a lesser degree.

An engineer, when soliciting additional information, may approach the library to borrow a publication or may request the librarian to order a document if it is not already available. If the newly ordered document is to remain on file in the library, the costs will be incurred by the library; however, when an engineer orders the document for retention in his own organization, the library performs the service, and the costs are charged to the engineer's organization.

The librarian usually is faced with the problem of currency as well as space limitations. Available material must be up to date if it is to be useful, and there is the continuing problem of what to purchase within limited budget constraints in order to best meet the need. Also, what items are no longer in demand and can be removed to make room for new material? The library, in an attempt to provide a useful service, should periodically make surveys throughout the division to assess needs and should check on the demand for and frequency of use of documentation already on file. The library must maintain a dynamic posture at all times.

4. Management services. Management services (noted in Figure 10-3, block 2A) can include a wide variety of activities. Of significance here are those dealing with division policies and procedures for engineering operations and the ongoing automated data-processing methods for the handling and transfer of engineering information.

a. *Engineering policies and procedures.* Generally within the administrative structure of a division there is an organization function responsible for the interpretation of corporate policies and for the preparation of a division-wide *Policy and Procedures Manual.* Although individual project and staff functions will vary with product requirements, certain standard procedures for accomplishing work are common throughout the division. For instance, a sample of the subjects often covered by formal procedures is noted below.

(1) Engineering design standards
(2) Engineering data preparation, review, and release
(3) Drafting practices
(4) Configuration management
(5) Preparation of process specifications
(6) Formal design reviews
(7) Engineering change proposals
(8) Quality assurance requirements
(9) Processing of purchase requests
(10) Personnel actions (e.g., recommendations for promotion)
(11) Indepartmental work requests
(12) Engineering notebooks

(13) Patent applications
(14) Capital equipment requests
(15) Planning documentation

In many large organizations the *Policy and Procedures Manual* will cover standard functions involving each facet of the division's operation. The manual should be directly accessible to all engineering personnel, include a copy of each applicable procedure, and indicate the contact point responsible for implementation of the procedure.

When an engineer initially joins a corporation, he (or she) should become thoroughly familiar with all existing policies and procedures. The manual becomes his *guide* in the accomplishment of the many routine tasks that are performed on a day-to-day basis. If new procedures are required and/or changes to existing procedures are considered feasible, the organization responsible for the manual should be contacted for appropriate action. Newly generated procedures or changes thereto should be circulated for review and approval by key management personnel (i.e., those affected by the addition or change) prior to their release and incorporation into the manual.

b. *Data processing, automation, and computer support.* There are a number of functions requiring the creation and processing of information that can be effectively handled through automation. Engineering and management functions adaptable to automation techniques can be classified in the categories noted below.[22]

(1) *Creative functions* are defined as those that result in the development of new concepts and methods. Examples include the development of analytical models for performing operational analyses and evaluating alternative system design configurations, the solving of specific design problems, utilizing computer-aided design techniques and the development of design data, developing new procedures, developing management information systems for project status reporting, design of numerical control techniques for manufacturing or computer-aided manufacturing methods, and development of automated inventory control systems for material handling and warehousing functions.

(2) *Data transfer functions* are primarily those where information flow occurs on a periodic basis. This category addresses repetitive operations and

---

[22]The use of data-processing and computer methods directly supports (a) the management activities discussed in Chapter 3 (i.e., advanced product planning, cost/schedule/performance measurement and control), (b) engineering design functions covered in Chapter 5 (i.e., analyses and trade-offs, predictions, computer-aided design), (c) industrial engineering, numerical control, and operations analysis in Chapter 6, (d) data collection and product evaluation in Chapter 7, and (e) product logistic support in Chapter 8. All phases of the product life cycle include functions that might be ideally suited for automation.

enhances communications within the division, between two or more divisions in the corporation, between the division and the corporate office, and sometimes between the consumer and the producer. Examples in this category are project status reports covering activities and events in a PERT/CPM network, task schedules, completed milestones, and manpower/material expenditures against budget projections; division and organization financial status and measures of effectiveness; the replication of design data such as drawings, procedures, and component specifications and parts lists; fabrication of items using numerical control methods; inventory status; capital equipment utilization data; and the reporting of effectiveness data from the field for system evaluation purposes.

With the ever-increasing growth in job complexity and the need for large amounts of information in a timely manner, both engineers and managers should take advantage whenever possible of the data-processing and computer capabilities provided within the division. Machines can perform many operations faster, more consistently, and more economically than can be accomplished manually.

Data processing refers to the converting, transporting, reproducing, sorting, storing, and retrieving of information on demand. Computer capabilities exist in one form or another in many corporations and often include a centralized computer center with terminals located at designated physical sites throughout the division. The terminals are tied to the centralized computer(s) on a time-sharing basis and are available to engineers and managers who may acquire and process data remotely.[23] Figure 10-7 illustrates a sample division network showing the ties between the computer center and the remote terminals.

As engineers and managers become more aware of the usefulness of the computer and data-processing methods, the demands on usage within the division will increase. This will undoubtedly result in an increased backlog of work in the computer center, particularly if large programs are generated and machine time becomes limited. Large programs, once started, cannot be stopped without losing all the work already completed. Thus, the manager of the computer center is faced with a scheduling problem of when to run large programs, when to schedule small programs, and which programs take priority. Sometimes this is resolved by assigning certain hours of the day to different organizations, and sometimes premium computer charges are

---

[23]Electronic computers are currently available in many sizes and shapes to fit a wide variety of needs (e.g., physical size, speed of data processing, storage capability, language requirements). In some divisions, small computers are also available and physically located within different organizations and are utilized to support limited operations while the terminals and the centralized computer are used for larger operations. The object is to provide the proper mix of hardware/software capability to satisfy the needs—not too much or too little!

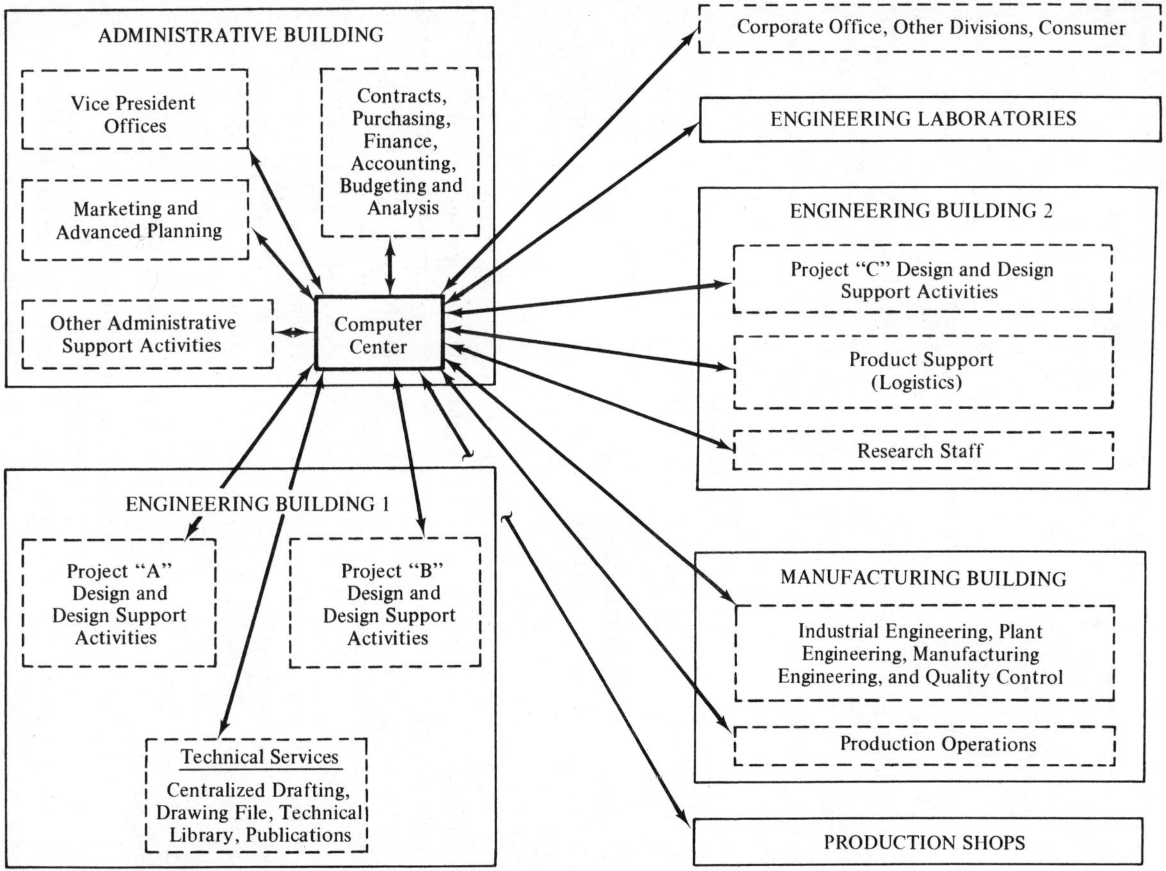

Figure 10-7. Data processing and computer support requirements.

applied to levels of priority, forcing a possible reordering of priorities (i.e., maybe the importance in running a program at a given time is not quite as important as initially perceived).

The engineer or manager in utilizing data processing and computer methods may require the assistance of a programmer to translate engineering ideas into the proper computer language (e.g., Fortran, PLI, Cobol), run the program(s) either remotely or through the computer center, and analyze the results through data interpretation. A *good* programmer knows the computer language, the machine characteristics, data formatting, and quickly recognizes problem cause-and-effect relationships. In addition, the good programmer will carefully document all program routines through the use of computer logic flow diagrams, coding explanations, computer operating instructions, and diagnostic provisions. He or she should verify the program prior to implementation to ensure that the results actually reflect what is needed. On the other hand, inexperienced programmers can introduce all kinds of errors into the process through a lack of understanding the hardware, carelessness in writing the program, mistakes in key punching, and/or misinterpretations of data results. Often these errors are allowed to persist until some "panic" evolves, at which time the blame is quickly attributed to the computer itself and not to the actual cause of the problem.

5. Centralized test laboratories. An inherent part of the research and engineering design functions discussed in Part Two is the experimentation and testing of new concepts. This, in turn, requires the establishment, operation, and maintenance of test facilities. In most medium-sized and large organizations, test facilities are developed and maintained to support different projects throughout the division. The project organization is the customer, and the test facility provides the required services. If testing requirements are too extensive for the available in-house facilities, the project director may solicit support from an outside source.

Testing, which normally supports research (refer to Chapter 4), will include the measurement of basic material properties and the interaction of physical phenomena necessary to establish certain relationships and associated design criteria. Much of the testing accomplished may seem somewhat remote in terms of system/equipment design; however, the results may directly apply to material applications, component selection, process design, and so on.

Testing in support of engineering design constitutes those activities included in the type 1 and type 2 categories of test described in Section 5.3C. Such tests make use of mock-ups, engineering and service test models, prototype equipment, and sample equipment from the production line when multiple quantities of an item are produced. Testing of these items, conducted by the corporation (or producer), is part of the overall system evaluation effort discussed in Chapter 7.

Support of testing activities may require the services of one or more laboratories with trained technicians prepared to assist in the determination of test requirements and the test preparation phase and who will conduct the test, collect data, and help in the analysis and reporting of test results. Some of the laboratory facilities required in support of research and engineering functions are as follows:

a. *Environmental laboratory* for temperature cycling and humidity tests, electromagnetic interference, vibration, shake testing, shock, salt spray, and sand and dust.

b. *Structural test laboratory* for material stress-strain analysis and the determination of fatigue characteristics.

c. *Chemical laboratory* for chemical analysis of materials, fluids, and gases.

d. *Component test laboratory* for electrical and electronic testing of component parameters, tolerances, stresses, and failure modes. Component life testing may also be accomplished in this area.

e. *Wind tunnel* for measuring aerodynamic and wind effects on aircraft configurations, automobiles, buildings, and other structures.

f. *Hydraulic laboratory* for measuring fluid pressures and flow characteristics.

g. *Instrumentation laboratory* for the development of test fixtures and the instrumentation required for specific tests in all areas.

h. *Model shop* for the development of scale models and mock-ups. Other shops are often called upon to support mock-up development in the areas of painting, welding, bonding, and machining.

i. *Calibration laboratory* utilized for the periodic checkout and calibration of all applicable mechanical, electrical, electronic, and other test equipment located throughout the division. The calibration laboratory contains secondary or transfer standards employed to verify the accuracy of the test equipment used by researchers, engineers, and technicians in the laboratories and on the production line. The traceability of test from laboratory equipment to secondary standards and from secondary standards to primary standards is necessary to ensure that all experimentation and testing serves a meaningful purpose.

Laboratory usage is based on the test requirements for each project. The laboratory manager must satisfy project needs, schedule the test facilities accordingly, and hopefully reflect a steady workload and an effective rate of facility utilization. Idle facilities are costly, and extensive testing is also costly. Laboratory facility requirements evolve from the advanced product planning information discussed in Chapter 3 and must be considered from the total division standpoint. Management must continually make the tradeoffs between capital equipment costs and facility workload.

6. Other engineering support activities. Figure 10-3 reflects many additional engineering support activities, such as personnel, industrial relations, contracts management, and purchasing. These areas where major interfaces exist are discussed in Part Four.

10.3 INTERFACE RELATIONSHIPS AMONG BUSINESSES, CORPORATIONS, AND AGENCIES

The interfaces between the different categories of businesses discussed earlier may assume a variety of forms. Formal activities are generated through contracts (or subcontracts) between government agencies and corporations, between small and large corporations, between noncorporate businesses and corporations, and even between individual divisions within the same corporation. These activities, constituting negotiated statements of work along with cost and schedule information, may be accomplished through fixed-price or cost-plus contracts, grants, consulting agreements, and so on. The variety of interactions with organizations and individuals accomplishing engineering and other work is illustrated in Figure 10-8.[24]

The consumer may be a government agency, a corporation, a business, or an individual citizen. As illustrated in Figure 1-3, the consumer initially specifies the need for a product and will ultimately use the product. For the purposes of discussion, the consumer is represented by one organization even though separate segments of that organization may actually deal independently relative to the acquisition of a system and the follow-up operation and support of that system.[25] The producer, on the other hand, attempts to respond to the needs of the consumer by developing and producing a system in accordance with the specified requirements.

When a formal consumer-producer relationship is established between organizations, engineering activities are usually accomplished on both sides. In the requirements definition stage, the consumer assumes a major role while during system design and development, production/construction, and evaluation the producer maintains the predominant role, with the consumer

---

[24]Actually, all formal agreements between organizations are contracts of one type or another. The differentiation noted in Figure 10-8 (i.e., grant versus contract) is intended to comply more with the terms generally used in lieu of the proper legalistic language associated with the agreement structure.

[25]Quite often one segment of the consumer's organization, assigned to represent the *total* organization, will define the requirements for a system while another segment of the consumer's organization will have to operate and maintain the resulting product throughout its life cycle. Hopefully, the procurement facet of the organization will truly reflect the needs of the user element of the organization. Otherwise, the user will likely receive a product that will not meet the intended need.

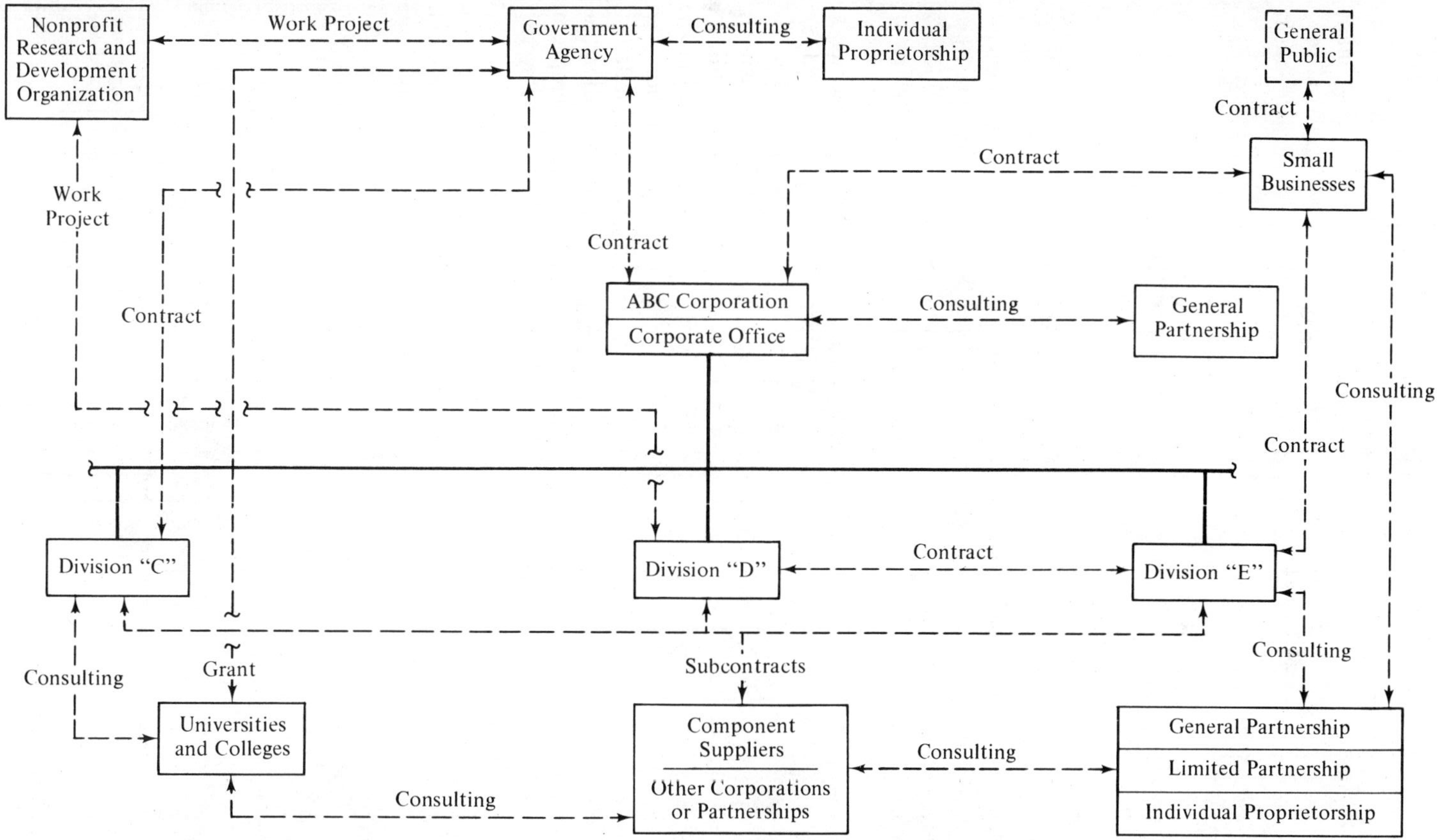

Figure 10-8. Interfaces between businesses, corporations, and agencies.

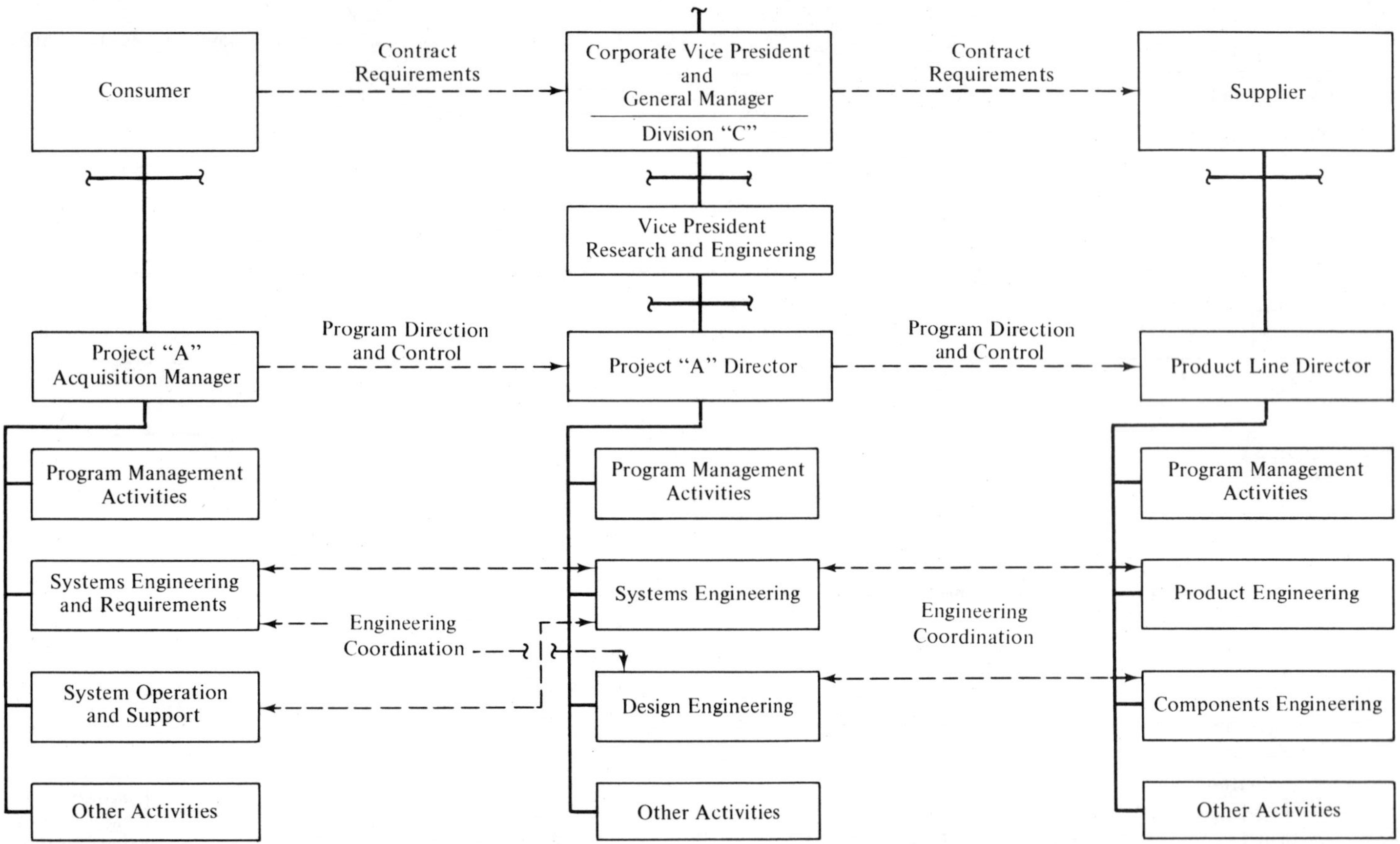

Figure 10-9. Producer-consumer organization relationships.

performing a review and approval function. Finally, as the system phases into operation, the consumer takes over, with the producer providing system support as required. Thus, the consumer-producer relationships are extensive and continue throughout the product life cycle discussed in Part Two.

In the interest of promoting communications and providing an effective product output, the consumer and producer organization structures must be compatible to the extent that comparable functions can be readily identified and the proper coordination at all appropriate levels of organization can be established. In some instances, the producer will strive to set up a parallel organization structure to that of the consumer so as to simplify communications by establishing a one-to-one relationship in terms of the functions being accomplished. This point is illustrated to a certain extent in Figure 10-9. The goal is to identify comparable functions between the consumer and the prime contractor and the prime contractor and its suppliers and to establish vertical program direction and control within each organization while allowing some technical horizontal coordination across organization lines. The manager of the consumer's organization will deal directly with his counterpart in the producer's organization, and each will in turn provide project direction and control within his own organization. Engineers, while responding to project management direction, will communicate with their counterparts in outside organizations on technical matters not of a contractual nature. This technical communication is necessary to ensure that a mutual understanding exists between those defining requirements and those responsible for fulfilling these requirements.

QUESTIONS AND PROBLEMS

1. Describe the basic categories of business.
2. How does the consulting engineer fit into the categories described in problem 1?
3. Briefly define the various forms of ownership.
4. What are the advantages of the individual proprietorship? Name some of the disadvantages.
5. What is the basic difference between a general partnership and a limited partnership? What are the advantages of the partnership over the individual proprietorship?
6. Describe the form of ownership of a corporation. How does ownership relate to corporate management and control?
7. Describe the general steps in the formation of a corporation.
8. What are the differences between common stock and preferred stock?
9. What is meant by "horizontal integration" and "vertical integration" as used in this book? Briefly define each and give examples.

10. How is a corporation internally controlled? Name some of the external factors that control a corporation.
11. How are an organization's engineering requirements determined?
12. Within a corporation, how are engineering activities influenced at the corporate office level?
13. Select a company or business of your choice and draw an organization chart of the operation. Identify all engineering functions, define the quantity and types of engineering expertise required, and draw a man-loading curve.
14. Suppose that the man-loading curve in problem 13 has numerous peaks and valleys. How would you staff the organization to meet the requirements?
15. Assume that you are responsible for developing a new product. Define the system life cycle in a flow-chart format. Identify engineering requirements and describe the organization approach (i.e., draw an organization chart) that you would employ in fulfilling your responsibilities.
16. Assume that you have just been appointed as the corporate vice president of engineering and you have the responsibility to develop engineering policies for the corporation. What areas would you cover for the corporation through policy development? Why? How would you ensure that these policies were being adhered to over time?
17. If you had the choice of accepting a job in a project organization or a staff organization, which would you choose? Why?
18. As a staff engineer, you have been assigned to perform a project task and you have been given conflicting instructions from the project director and the staff group manager (i.e., your two bosses). How would you resolve the problem?
19. How can the use of data-processing and computer techniques facilitate the engineering design process? Identify some applications. When considering the use of computer-aided design, what factors must be evaluated?
20. Assume that you as an engineer have developed a new innovative automated approach to engineering design and that your boss refuses to even consider your idea. What would you do?
21. When designing a new data-processing system, a computer programmer can "make" or "break" you. Why?
22. A technical library can assist an engineer in a number of ways. How?
23. The engineering process described in Part Two includes an ongoing system test and evaluation effort. In determining test requirements, why would it be beneficial to contact centralized test laboratory personnel for support?
24. When first joining a firm, an engineer should review all applicable policies and procedures. Why?
25. In performing engineering test functions, why is calibration so important?
26. If you were assigned the responsibility of directing a new product development effort and you were asked to determine space requirements for engineering design, what factors would you consider? You may wish to develop a simulated layout, locating project personnel and supporting functions.

27. Refer to Figure 10-9. Why is it usually necessary to establish engineering communications across organization lines as illustrated?

28. As a young engineer you are evaluating the alternatives of going into private consulting, joining a small business organization, accepting a job in a large corporation, or working for a government agency. What factors would you consider in the decision-making process? Be prepared to discuss the pros and cons of each alternative.

part four

Related Functions and Interfaces

Given the primary functions of engineering (Part One), the engineering process or product life cycle (Part Two), and the organization aspects of engineering and related activities (Part Three), it is now necessary to briefly address a number of additional functional areas that directly support engineering throughout product development, production/construction, operational use, and logistic support. It is not intended to emphasize the principles of management related to these areas but to address them from the standpoint of major engineering interfaces. This part is represented by the fourth circle in the conceptual outline of the text.

Conceptual Outline

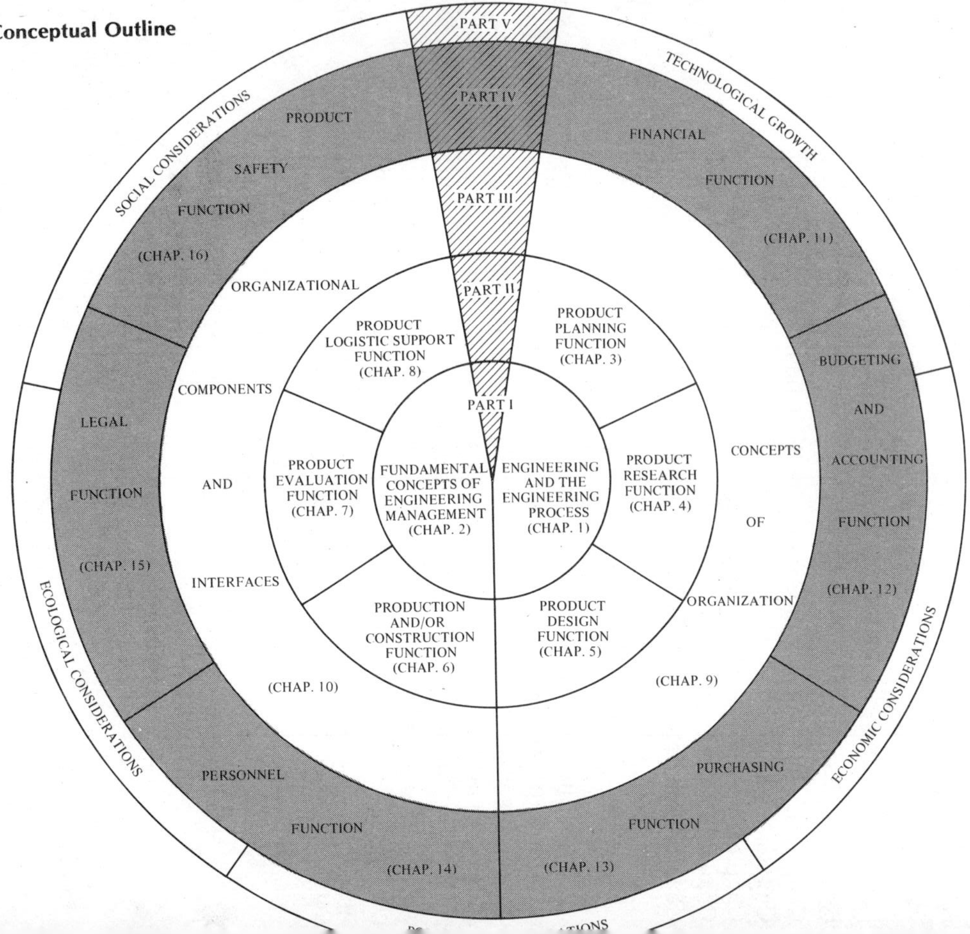

PART I	ENGINEERING IN MODERN TIMES
PART II	THE PRODUCT LIFE CYCLE
PART III	ORGANIZATION AND THE PRODUCT LIFE CYCLE
PART IV	RELATED FUNCTIONS AND INTERFACES
PART V	ENGINEERING AND THE ENVIRONMENT

11

FINANCIAL FUNCTION

The financial function of a business is concerned with maximizing the value of that business to its owners (e.g., the stockholders of a corporation). For a corporation, value is often based on the market price of its common stock, which is a function of the corporation's investment policies, methods of financing, and dividend decisions.

Investment policies dictate the process associated with capital budgeting and expenditures. Proposals to spend money are generated on a continuing basis. Some of these proposals will recommend the expenditure of funds for product development and production in response to an identified consumer need, while others will cover improvements and/or modifications to the corporation's current capability. Proposals of an engineering nature may evolve as a result of the advanced product planning process discussed in Section 3.4F, or may include recommendations for improvements such as the installation and operation of new computer-aided design equipment, an additional test facility, or a new automated inventory control capability.

Proposals must be evaluated along with alternative possibilities (including the alternative of doing nothing) in terms of their expected benefits and associated risks. The proposals are then ranked using some predetermined criteria, and management will make decisions on whether or not to commit capital to the proposed ventures. Investment decisions will then determine the amount and composition of the corporation's assets, which will affect its future business-risk profile as viewed by the investors. Investment decisions are discussed further in Section 11.1.

The second major consideration is that of establishing the best mix of financing. Funds required for operating the business are derived from a combination of direct revenue from sales, loans from banks, sale of stock, sale of securities and bonds, mortgages, and so on. The financial manager is responsible to ensure the proper mix of short- and long-term financing in order to provide the necessary funds for operation at a minimum risk to the corporation. Methods of financing are briefly covered in Section 11.2.

The third major factor in financial management addresses decisions pertaining to dividends. Decisions here affect the percentage of earnings paid to the stockholders in cash, the distribution of additional shares of stock (e.g., stock splits), the rights to purchase additional shares of stock at a lower-than-market-value rate, and the general stability of dividend distributions over time. No matter how high the profits may be for any given time period, the corporation will suffer unless the stockholders receive some of the benefits. Otherwise, the stockholders will probably sell their shares of stock and reinvest in other ventures. Dividend decisions are covered in Section 11.3.

Financial management in the broadest sense deals with investments that require financing. Successful financing is dependent on how well the corporation is doing in terms of its business posture, which in turn is influenced by the stockholders, who expect some benefits from their investment. Thus, investment decisions, methods of financing, and dividend decisions are closely interrelated and must be evaluated on an integrated basis. Figure 11-1 shows some of these relationships.

11.1 INVESTMENT DECISIONS

Section 3.4F briefly covers internal and external engineering proposals with the basic differences relating to how the proposal was generated initially and the ultimate source of funding. When viewing the life cycle of a product, there may be a series of proposals as conveyed in Figure 11-2. Initially, an engineer may come up with an idea and generate a proposal to accomplish a feasibility study. At this point, internal funding is generally required, and management must decide whether or not to invest in the effort associated with proposal preparation. The next stage results in a proposal for product development, which may also require internal funding or funding from an outside sponsor. Again, management must assess the proposed effort, evaluate alternatives, and make a go/no go decision based on the estimated benefits and risks to the corporation.

Proposal generation occurs whenever there are new ideas or whenever the results of work in process indicate the desirability of a follow-on effort. Proposals may cover new product developments, product modifications, changes in the producer's internal design or manufacturing capability, suggestions for cost reduction, and so on. Some proposals reflect effort that is to be funded entirely by corporate financing through bank loans, sale of

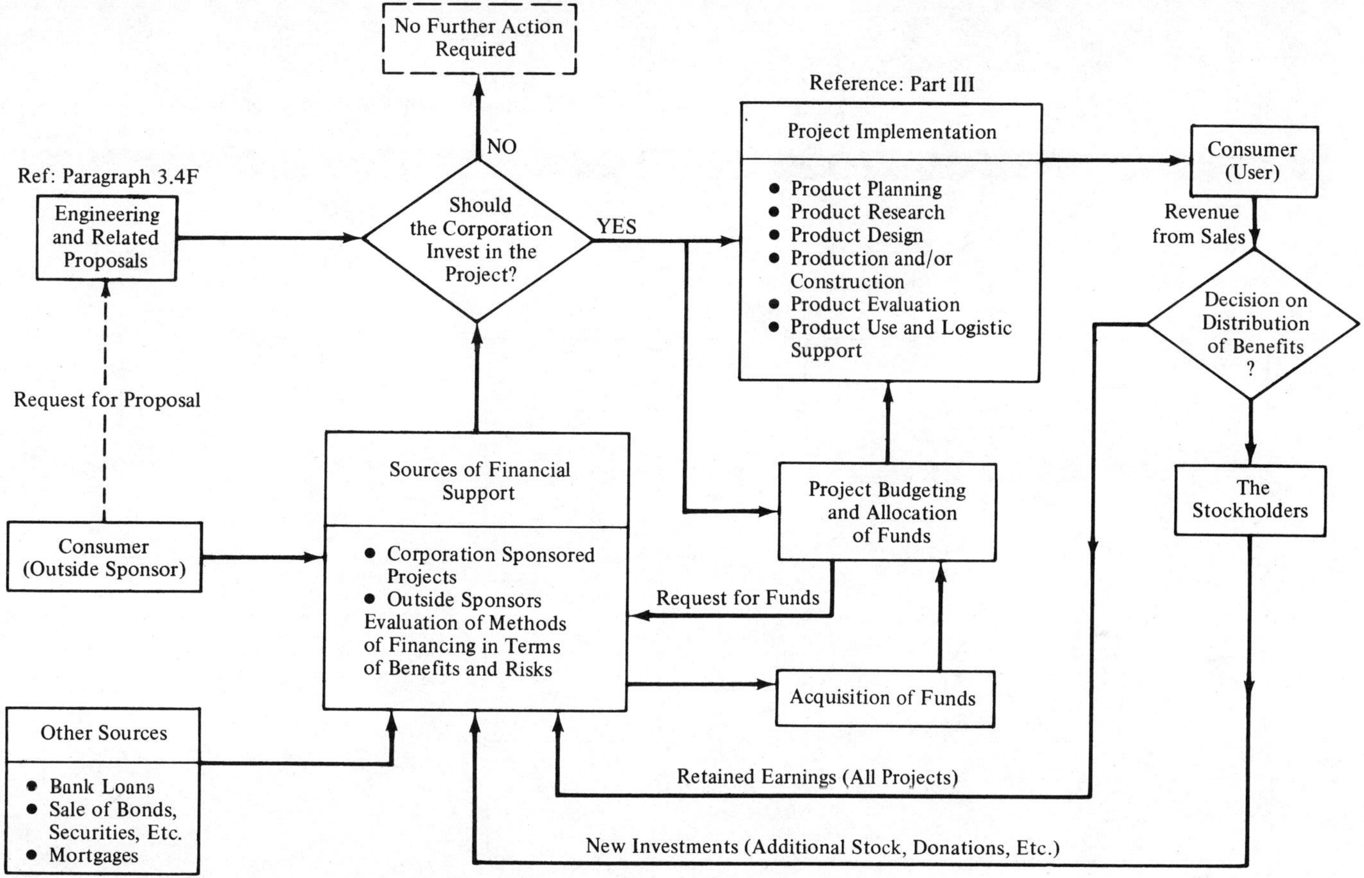

Figure 11-1. Major project financial activity interfaces.

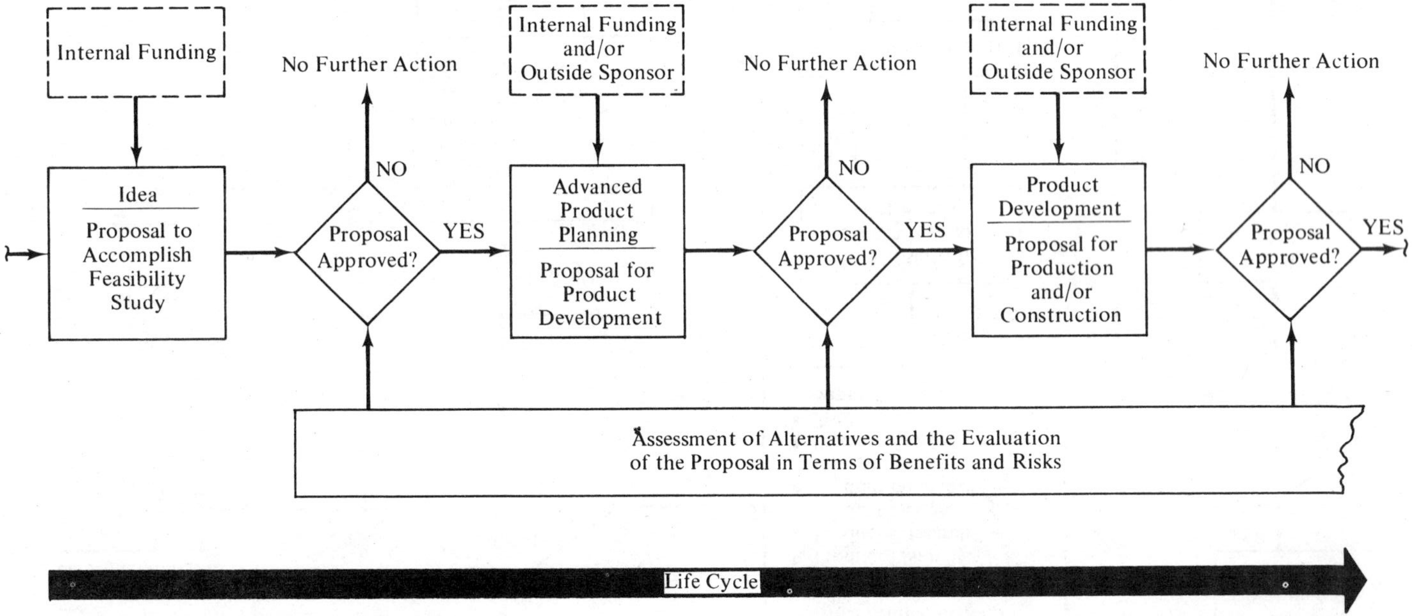

Figure 11-2. The evolution of proposal development.

bonds or securities, or through the use of retained earnings from revenues generated by product sales. Other proposals (e.g., solicited proposals) may cover effort that is to be funded by a government agency, another corporation, a small business, or an outside sponsor. Proposals may be *independent,* where acceptance or rejection decisions do not affect other proposals, or proposals may be *dependent,* relating to each other in such a manner that the acceptance or rejection of any one will influence the decision on the others.

When evaluating proposals at each stage in the life cycle, benefits and costs are estimated and converted into a common basis for comparison.[1] Alternatives are evaluated and rated on an equivalent basis so that the most desirable one can be selected. In reviewing proposals from a specific division of a corporation, management evaluates all engineering proposals and may select as many proposals as considered feasible within the budget structure, proceeding from the highest-ranking proposal on down.

Discussion of proposal evaluation criteria can lead to the identification of many different figures of merit (refer to Figure 3-6). However, of significance in terms of financial considerations are measures such as the internal rate of return, the net present-value factor, and the payback or investment recovery point. The methods for deriving these measures are discussed following a brief introduction to the time value of money concept.

A. Time Value of Money[2]

One should recognize that a dollar today is worth more than a dollar in the future, owing to the fact that money through its investment can earn interest over time. For instance, would you prefer to receive a dollar now or a dollar at sometime in the future? Since the dollar in hand can be invested and accumulate interest each year, your preference should be to receive the dollar now. This relationship between interest and time leads to the concept of "time value of money."

The evaluation of each proposal, representing a potential investment, is viewed from the standpoint of the anticipated receipts and disbursements that will occur over the project life cycle or the operational horizon period selected. Alternatives, such as the investment in other projects or the investment of money in a bank, are evaluated in a like manner. Since receipts and disbursements are related to different points in the life cycle, a common

[1] The criterion measures for evaluating proposals within a corporation and within the government are somewhat different. Although both *benefits* and *costs* are estimated in each instance, benefits to the corporation are in terms of value in the marketplace and to the stockholders, and costs are those accruing only to the corporation. On the other hand, government agencies must identify and measure social benefits and costs; that is, the social effects on particular groups in society.

[2] A good source for further discussion in this area is: W. J. Fabrycky and G. J. Thuesen, *Economic Decision Analysis,* Prentice-Hall, Inc., Englewood Cliffs, N.J., 1974.

point of reference must be assumed such that all alternatives can be compared on an equivalent basis. This point is generally the "present time" or "now," when decisions that have a significant impact on the future are being made; thus, all future receipts and disbursements for each year in the life cycle are discounted to the present value.

The aspect of discounting refers to the application of a selected rate of interest to measure the difference in importance or preference between dollars at the present time and anticipated dollars in the future. Discounting allows for the evaluation of the time-phased profiles of money streams for various alternatives as if they occurred at one point in time (the only fair method of evaluation if one is to make a decision today) rather than spaced over the project life cycle. *Interest* is the rent paid for the use of money or the gain received from an investment, depending on whether you are the borrower or the lender. An *interest rate* is the ratio of this value over time, usually expressed annually. For instance, if you were to invest $1,000 at a simple interest rate of 8 per cent per annum, the interest at the end of 1 year would be

$$I = 1{,}000(0.08)(1) = \$80$$

The principal amount plus the interest received at the end of the year would be $1,080.

Interest rates vary with market conditions and the overall economy. Interest rates applied to individual investments are usually based on the *prime rate,* which is the lowest rate generally charged on business loans to large, well-established, financially sound corporations. This rate is specified by high-volume money-market banks and is fairly uniform throughout the country. [3]

As indicated earlier, discounting is a technique for converting different cash flows occurring over time to an equivalent amount at a common point. For the purposes of illustration, assume that a 5-year project is undertaken, resulting in a series of cash receipts, as illustrated in Figure 11-3. Although receipts (and costs) usually occur continually or at discrete points throughout each year, they are treated either at the beginning or at the end of the applicable time period. In this instance, the receipts are related to the end of each year throughout the project life cycle. The object is to convert these receipts to the present value. For year 1 the sum of all estimated receipts are represented at point A. The question is: What is the value of these receipts in terms of the decision point? This can be determined from the following single-payment present-value expression:

---

[3] Interest rates and their application to loans and investments are discussed in depth in: J. C. Van Horne, *Financial Management and Policy,* 3rd ed., Prentice-Hall, Inc., Englewood Cliffs, N.J., 1974.

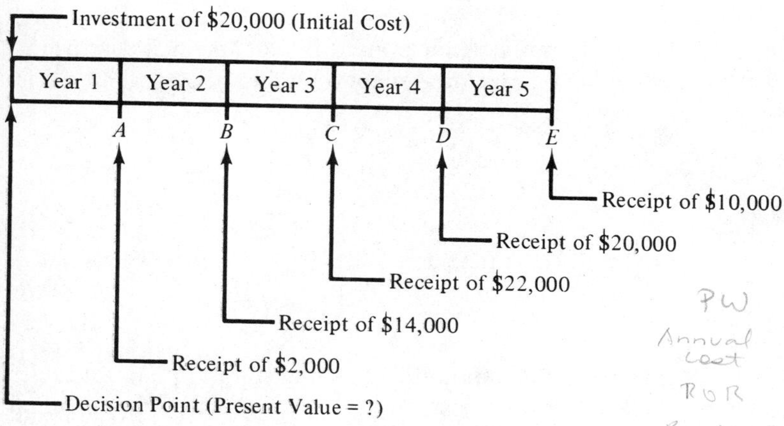

Figure 11-3. Present value of cash transactions.

$$P = F \left[\frac{1}{(1 + i)^n} \right] \tag{11-1}$$

where P is the present value or present principal sum, F the future sum at some interest period hence, i the annual interest rate, and n the interest period.

Assuming an interest rate of 10 per cent, the present value of the receipts at point A is

$$P = 2{,}000 \left[\frac{1}{1 + 0.1} \right] = (2{,}000)\ (0.9091) = \$1{,}818.20$$

For point B, this becomes

$$P = 14{,}000 \left[\frac{1}{(1 + 0.1)^2} \right] = \$11{,}569.60$$

Next, consider the present value of the receipts at point A and point B combined. This is calculated from the expression

$$P = F_A \left[\frac{1}{1 + i} \right] + F_B \left[\frac{1}{(1 + i)^2} \right] \tag{11-2}$$

Assuming the undiscounted estimates of $2,000 and $14,000 at points A and B, respectively, the present value is

$$P = 2{,}000 \left[\frac{1}{1 + 0.1} \right] + 14{,}000 \left[\frac{1}{(1+0.1)^2} \right] = \$13{,}387.80$$

This continues until the receipts for each year in the life cycle are discounted to the present value and totaled. The total present value of receipts for the project represented in Figure 11-3 is $49,565.40. (Note that the undiscounted total value is $68,000.) When the receipts are different for each year in the life cycle, the present-value expression is a continuation of equations (11-1) and (11-2), or

$$P = F_A\left[\frac{1}{1+i}\right] + F_B\left[\frac{1}{(1+i)^2}\right] + F_C\left[\frac{1}{(1+i)^3}\right] + \cdots + F_n\left[\frac{1}{(1+i)^n}\right] \tag{11-3}$$

Present-value calculations can be simplified by using standard interest tables (included in most textbooks on the subject of economic analysis) and multiplying the future sum by the appropriate factor. Figure 11-4 presents an abbreviated example of present-value factors.

In the event that cash flows in each future period are identical, the present value of the series (representing an *annuity*) can be determined by multiplying the value of the cash flow by a compound discount factor. For example, the present value of a future stream of $1,000 receipts for a 4-year period using an 8 per cent discount rate is

$$P = 1{,}000(3.3120) = \$3{,}312.00$$

The compound discount rate is the summation of the individual discount factors for the first four periods indicated under the 8 per cent column in Figure 11-4 (i.e., 0.9259, 0.8573, etc.).

The discussion above pertains to the conversion of anticipated future receipts to the present value. This is necessary for the evaluation of proposals in terms of the present time. However, if one wishes to relate cash flows to some future point in time, equation (11-1) can be transposed to

$$F = P(1+i)^n \tag{11-4}$$

This expression can be used to determine the results of various investments. For instance, assume that $4,000 is invested today in some venture and the annual interest rate is 6 per cent, the compound amount at the end of year 2 is

$$F = 4{,}000(1 + 0.06)^2 = \$4{,}494.40$$

Receipts and disbursements may be treated differently, depending on the problem at hand. In addition, interest rates often change from year to year, which must be considered in the final calculations. In order to get an in-depth feel for the proper treatment of cash flows, the reader should review a textbook on engineering economy. Refer to Appendix D.

n \ i	2%	4%	6%	8%	10%	15%	20%	25%
1	0.9804	0.9615	0.9434	0.9259	0.9091	0.8696	0.8333	0.8000
2	0.9612	0.9246	0.8900	0.8573	0.8264	0.7561	0.6944	0.6400
3	0.9423	0.8890	0.8396	0.7938	0.7413	0.6575	0.5787	0.5120
4	0.9238	0.8548	0.7921	0.7350	0.6830	0.5718	0.4823	0.4096
5	0.9057	0.8219	0.7473	0.6806	0.6209	0.4972	0.4019	0.3277
6	0.8880	0.7903	0.7050	0.6302	0.5645	0.4323	0.3349	0.2621
7	0.8706	0.7599	0.6651	0.5835	0.5130	0.3759	0.2791	0.2097
8	0.8535	0.7307	0.6274	0.5403	0.4665	0.3269	0.2326	0.1678
9	0.8368	0.7026	0.5919	0.5002	0.4241	0.2843	0.1938	0.1342
10	0.8203	0.6756	0.5584	0.4632	0.3855	0.2472	0.1615	0.1074
11	0.8043	0.6496	0.5268	0.4289	0.3505	0.2149	0.1346	0.0859
12	0.7885	0.6246	0.4970	0.3971	0.3186	0.1869	0.1122	0.0687
13	0.7730	0.6006	0.4688	0.3677	0.2897	0.1625	0.0935	0.0550
14	0.7579	0.5775	0.4423	0.3405	0.2633	0.1413	0.0779	0.0440
15	0.7430	0.5553	0.4173	0.3152	0.2394	0.1229	0.0649	0.0352
16	0.7284	0.5339	0.3936	0.2919	0.2176	0.1069	0.0541	0.0281
17	0.7142	0.5134	0.3714	0.2703	0.1978	0.0929	0.0451	0.0225
18	0.7002	0.4936	0.3503	0.2502	0.1799	0.0808	0.0376	0.0180
19	0.6864	0.4746	0.3305	0.2317	0.1635	0.0703	0.0313	0.0144
20	0.6730	0.4564	0.3118	0.2145	0.1486	0.0611	0.0261	0.0115
25	0.6095	0.3751	0.2330	0.1460	0.0923	0.0304	0.0105	0.0038
30	0.5521	0.3083	0.1741	0.0994	0.0573	0.0151	0.0042	0.0012
40	0.4529	0.2083	0.0972	0.0460	0.0221	0.0037	0.0007	0.0001
50	0.3715	0.1407	0.0543	0.0213	0.0085	0.0009	0.0001	....
100	0.1380	0.0198	0.0029	0.0005	0.0001	....	....	....

Figure 11-4. Present-value factors (*typical*).

B. Financial Evaluation Criteria[4]

1. Internal-rate-of-return method. The internal rate of return is a widely used measure of investment worth and is defined as the interest rate that equates the present value of expected receipts with the present value of

[4]An adequate discussion of financial criteria is beyond the scope of this book. Only a cursory presentation is included, to convey an overview of the approach. For a more in-depth level of coverage, the reader is advised to review some of the references on finance and economic analysis listed in Appendix D.

expected disbursements. It can also be stated as the rate of discount that will equate the present value of the net benefits with the cost of the project. Mathematically, this can be expressed as

$$\sum_{t=0}^{n} Q_t \left[\frac{1}{(1+r)^t}\right] = \sum_{t=0}^{n} C_t \left[\frac{1}{(1+r)^t}\right] \tag{11-5}$$

where Q_t is the net cash receipts or benefits during period t, C_t the net cash disbusements or cost during period t, r the rate of return, and n the last period in which cash flow is expected.

The calculations required for solving r are relatively complex and are best accomplished using a computer as an aid. Actually, the solution is obtained through an iterative process of trial and error. To illustrate by using a simplified example, assume that you are evaluating a project that requires an initial investment of $20,000 at time zero with expected cash inflows of $6,000 at the end of each of the next 5 years. The problem can be expressed as

$$20{,}000 = \frac{6{,}000}{1+r} + \frac{6{,}000}{(1+r)^2} + \frac{6{,}000}{(1+r)^3} + \frac{6{,}000}{(1+r)^4} + \frac{6{,}000}{(1+r)^5}$$

With the expected yearly benefits being equal, it is necessary to determine the compound discount factor that, when multiplied by $6,000, equals the initial investment of $20,000 at time zero. Figure 11-5 presents a partial table of compound discount factors, included for convenience. From the figure it can be seen that the rate of return (i.e., the optimum interest rate) is somewhere between 15 and 16 per cent:

Discount Rate (%)	*Compound Discount Factor*	*Cash Flow Each Year*	*Present Value of Stream*
15	3.3522	$6,000	$20,113.20
16	3.2743	6,000	19,645.80

To approximate the actual rate, one can interpolate between 15 and 16 per cent as follows:

Discount Rate (%)	*Present Value*
15	20,113.20
?	20,000
16	19,654.80

$$\text{incremental value} = \frac{113.20}{467.40} = 0.242$$

$$\text{actual rate of return} = 15 + 0.24 = 15.24\%$$

Present Value of One Dollar per Year "n" at Interest Rate "i"

Year	10%	11%	12%	13%	14%	15%	16%	17%	18%	19%	20%
1	0.9091	0.9009	0.8929	0.8850	0.8772	0.8696	0.8621	0.8547	0.8475	0.8403	0.8333
2	1.7355	1.7125	1.6901	1.6681	1.6467	1.6257	1.6052	1.5852	1.5656	1.5465	1.5278
3	2.4868	2.4437	2.4018	2.3612	2.3216	2.2832	2.2459	2.2096	2.1743	2.1399	2.1065
4	3.1699	3.1024	3.0373	2.9745	2.9137	2.8550	2.7982	2.7432	2.6901	2.6386	2.5887
5	3.7908	3.6959	3.6048	3.5172	3.4331	3.3522	3.2743	3.1993	3.1272	3.0576	2.9906
6	4.3553	4.2305	4.1114	3.9976	3.8887	3.7845	3.6847	3.5892	3.4976	3.4098	3.3255
7	4.8684	4.7122	4.5638	4.4226	4.2883	4.1604	4.0386	3.9224	3.8115	3.7057	3.6046
8	5.3349	5.1461	4.9676	4.7988	4.6389	4.4873	4.3436	4.2072	4.0776	3.9544	3.8372
9	5.7590	5.5370	5.3282	5.1317	4.9464	4.7716	4.6065	4.4506	4.3030	4.1633	4.0310
10	6.1446	5.8892	5.6502	5.4262	5.2161	5.0188	4.8332	4.6586	4.4941	4.3389	4.1925
11	6.4951	6.2065	5.9377	5.6869	5.4527	5.2337	5.0286	4.8364	4.6560	4.4865	4.3271
12	6.8137	6.4924	6.1944	5.9176	5.6603	5.4206	5.1971	4.9884	4.7932	4.6105	4.4392
13	7.1034	6.7499	6.4235	6.1218	5.8424	5.5831	5.3423	5.1183	4.9095	4.7147	4.5327
14	7.3667	6.9819	6.6282	6.3025	6.0021	5.7245	5.4675	5.2293	5.0081	4.8023	4.6106
15	7.6061	7.1909	6.8109	6.4624	6.1422	5.8474	5.5755	5.3242	5.0916	4.8759	4.6755
16	7.8237	7.3792	6.9740	6.6039	6.2651	5.9542	5.6685	5.4053	5.1624	4.9377	4.7296
17	8.0215	7.5488	7.1196	6.7291	6.3729	6.0472	5.7487	5.4746	5.2223	4.9897	4.7746
18	8.2014	7.7016	7.2497	6.8399	6.4674	6.1280	5.8178	5.5339	5.2732	5.0333	4.8122
19	8.3649	7.8393	7.3658	6.9380	6.5504	6.1982	5.8775	5.5845	5.3162	5.0700	4.8435
20	8.5136	7.9633	7.4694	7.0248	6.6231	6.2593	5.9288	5.6278	5.3527	5.1009	4.8696
21	8.6487	8.0751	7.5620	7.1016	6.6870	6.3125	5.9731	5.6648	5.3837	5.1268	4.8913
22	8.7715	8.1757	7.6446	7.1695	6.7429	6.3587	6.0113	5.6964	5.4099	5.1486	4.9094
23	8.8832	8.2664	7.7184	7.2297	6.7921	6.3988	6.0442	5.7234	5.4321	5.1668	4.9245
24	8.9847	8.3481	7.7843	7.7829	6.8351	6.4338	6.0726	5.7465	5.4509	5.1822	4.9371
25	9.0770	8.4217	7.8431	7.3300	6.8729	6.4641	6.0971	5.7662	5.4669	5.1951	4.9476

Figure 11-5. Compound present-value factors (*typical*).

The rather basic example above is presented to illustrate the approach used. The complexity in calculating the rate of return becomes apparent when introducing different values of receipts and disbursements for each year of the designated project life cycle. In such instances, all cash flows are treated individually, using specific discount rates obtained from an interest table, and the results are then combined and compared with the objectives expressed in equation (11-5).

When evaluating investment proposals, the estimated internal rates of return for all possible alternatives are compared, and the proposals are ranked in order of priority, the one with the highest rate of return being first. Figure 11-6 illustrates such a ranking. If the internal rate of return is higher than the minimum acceptable rate of return, the proposal will be accepted. The minimum acceptable rate of return is the return that investors expect the corporation to earn on projects. This is not to be considered the same as the cost of capital which is a composite rate that represents the cost of providing money from external sources through the sale of securities and bonds, the sale of stock, and by borrowing from banks. Normally, the minimum acceptable rate of return is higher to compensate for sudden changes in the composite cost of capital rate and the inherent risks associated with most projects.

2. Net-present-value method. The net-present-value method is similar to the internal rate of return in that the discounted cash-flow approach to

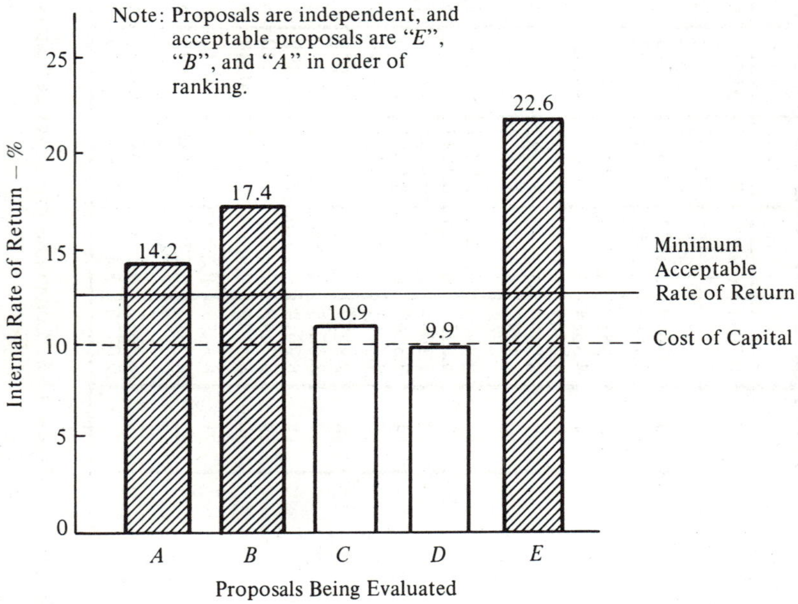

Figure 11-6. Proposal ranking (*rate of return method*).

capital budgeting is utilized. Net present value is the difference between the present value of benefits and the present value of costs and can be expressed as

$$\text{NPV} = \sum_{t=0}^{n} Q_t \left[\frac{1}{(1+r)^t}\right] - \sum_{t=0}^{n} C_t \left[\frac{1}{(1+r)^t}\right] \tag{11-6}$$

where the terms are the same as those defined for the internal rate of return. In this instance, r is the minimum acceptable rate of return. If the sum of the discounted cash flows is equal to or greater than zero, the proposal is beneficial; if not, the proposal is rejected.

Figure 11-7 illustrates the cash flows for two projects being considered. The assumed value for the minimum acceptable rate of return is 10 per cent, and the cash flows of benefits and costs are adjusted using the discount factors presented in Figure 11-4. The present value of benefits and costs is tabulated in Figure 11-8. The net present value is \$30,486.02 for project A and \$7,960.20 for project B. Both projects are acceptable as long as the allocated budget is adequate. If budget limitations exist, project A should receive priority over project B.

When using the net-present-value method for project evaluation, the discount rate selected becomes highly significant, since the attractiveness of proposals varies with changes in the discount rate. As the discount rate increases, the proposal becomes more attractive with early benefits. On the other hand, a low discount rate will not affect the timing of cash flows as significantly. Figure 11-9 serves to illustrate the importance of the discount rate used in the calculations. Different answers will evolve, depending on the discount rate employed. In fact, project B would be rejected using a minimum acceptable rate of return of 20 per cent. Proposals where the benefits and the costs are nearly eqivalent (e.g., project B) should be checked closely to ensure that the estimates of the magnitude and timing of all cash flows are realistic.

3. ***Payback method.*** The payback method is used to determine the length of time required for a project to become profitable. Undiscounted benefits and costs are estimated, and the point in the life cycle when the benefits exceed the initial investment plus sustaining costs is the *payback point.* This point is also known as the *investment-recovery point* or the *break-even point.* Figures 3-2 and 11-7 show the payback points for three project profiles.

Referring to the cash profile for project A in Figure 11-7, the initial investment of \$15,000 and the additional cost factors of \$6,000 and \$3,000 in years 2 and 3, respectively, represent a \$24,000 total cost. The benefits of \$5,000 in year 2, \$12,000 in year 3, and a proportion of the \$16,500 in year 4 are added and compared to the cost. The point in time when the incremental amount of benefits accumulated during year 4 is equal to the total cost is the

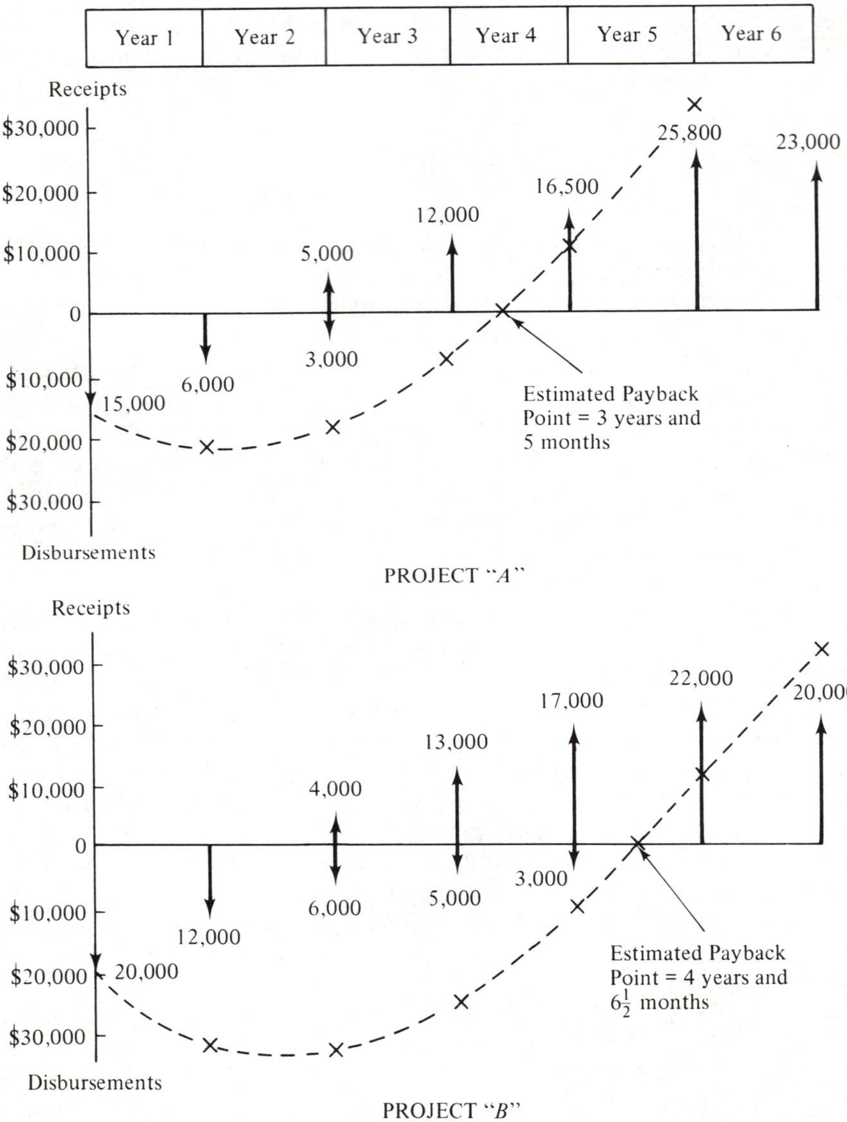

Figure 11-7. Comparison of cash flows and payback points.

payback point, or approximately 3 years and 5 months after initiating the project.

When evaluating investment proposals using the payback method, one has to determine the maximum allowable payback time period considered feasible for project acceptance. This value is dependent on several factors. First, how quickly from a financial standpoint does management wish to

PROJECT "*A*"

Year	Cash Flow		Discount Factor	Net Present Value	
	Benefits	Costs		Benefits	Costs
0		$15,000	0.0000		$15,000.00
1		6,000	0.9091		5,454.60
2	$5,000	3,000	0.8264	$ 4,132.00	2,479.20
3	12,000		0.7513	9,015.60	
4	16,500		0.6830	11,269.50	
5	25,800		0.6209	16,019.22	
6	23,000		0.5645	12,983.50	
Total	$82,300	$24,000		$53,419.82	$22,933.80

PROJECT "*B*"

Year	Cash Flow		Discount Factor	Net Present Value	
	Benefits	Costs		Benefits	Costs
0		$20,000	0.0000		$20,000.00
1		12,000	0.9091		10,909.20
2	$ 4,000	6,000	0.8264	$ 3,305.60	4,958.40
3	13,000	5,000	0.7513	9,766.90	3,756.50
4	17,000	3,000	0.6830	11,611.00	2,049.00
5	22,000		0.6209	13,659.80	
6	20,000		0.5645	11,290.00	
Total	$76,000	$46,000		$49,633.30	$41,673.10

Note: These values represent the cash flows in Figure 11-7. A 10% discount rate is assumed.

Figure 11-8. Present value of cash flows (*two projects*).

recover all costs? The answer to this question is based on the methods of financing and the cash-flow position of the corporation. Second, if a corporation is to develop and produce a product, what is the required time frame for subsequent sales and the accumulation of enough revenue to make the venture profitable? Competition and technical obsolescence are major factors in response to this question. For instance, current technology advances in the field of electronics causes certain products to become obsolete in a relatively short period of time, whereas the competition or obsolescence

PROJECT "*B*"

Year	Cash Flow		Present Value					
	Benefits	Costs	Benefits			Costs		
			6%	10%	20%	6%	10%	20%
0		$20,000				$20,000	$20,000	$20,000
1		12,000				11,320	10,909	9,999
2	$ 4,000	6,000	$ 3,560	$ 3,306	$ 2,778	5,340	4,958	4,166
3	13,000	5,000	10,915	9,767	7,523	4,198	3,757	3,762
4	17,000	3,000	13,466	11,611	8,199	2,376	2,049	1,447
5	22,000		16,441	13,660	8,842			
6	20,000		14,100	11,290	6,698			
Total	$76,000	$46,000	$58,482	$49,634	$34,040	$43,234	$41,673	$39,374

Net Present Value (6% discount rate) = $15,248

Net Present Value (10% discount rate) = $ 7,961

Net Present Value (20% discount rate) = – $ 5,334

Figure 11-9. Present value of Project "B" with different discount rates.

possibilities may not be as great in other product areas. Thus, in some instances a maximum acceptable payback period of 3 years is considered feasible, and in other situations 10 to 15 years may be allowable. Assuming that a 5-year payback period is considered feasible, both projects A and B in Figure 11-7 are accepted; however, project A should be ranked higher, since the payback period is shorter and the project is less risky on a relative basis.

The payback method is used quite often since the emphasis is on early cash flows and the payback point is easy to calculate. In addition, where certain proposals are ranked over others when different evaluation methods are used for life-cycle comparisons, the payback point becomes significant when considering outside competition and obsolescence, and the ranking of proposals may change if the payback period is too long. On the other hand, use of the payback method frequently ignores the cash flows beyond the payback point. Thus, when using the payback method as the sole criterion for proposal evaluation, the magnitude of the ultimate benefits derived from an investment may not become readily apparent.

4. Additional criteria. Although the criteria discussed thus far constitute popular financial approaches used in proposal evaluation, there are a few additional and related measures that deserve mention at this point. A brief definition of three of these are noted below.

a. *Benefit-cost ratio.* As the name implies, a benefit-cost ratio is simply the ratio between the sum of the benefits and the costs of the project. Benefits and costs are expressed in monetary terms. The sum of the benefits is divided by the costs, and proposal acceptance is based on a ratio equal to or greater than unity. Reference to different financial textbooks will indicate slightly different approaches in expressing the benefit-cost ratio quantitatively.[5]

b. *Return on investment.* This measure is similar to the rate-of-return method described earlier, except that the expected return in this instance usually relates to one's individual investment in stock, securities, bonds, and so on. A way of expressing return on investment is

$$P_o = \sum_{t=1}^{n} \left[\frac{D_t}{(1 + r)^n}\right] + \ldots \left[\frac{P_n}{(1 + r)^t}\right] \tag{11-7}$$

where P_o is the purchase price of the stock at time zero, D_t the expected dividend at the end of period *t*, P_n the expected market value of the stock after the dividend is paid in the last period, *r* the return on investment, and *n* the last period in which a return is expected.

Investments will vary greatly as different investors follow individual patterns when dealing with the stock market. The holding period of stock will range anywhere from one day to a lifetime, depending on the return expected. If an individual considers 10 per cent as a good return on investment, he will invest in and hold on to the stock; if not, he will not invest in the venture, or he will sell the stock that he already owns when the yield falls below 10 per cent.

c. *Cost effectiveness.* Cost effectiveness relates to the measure of a system in terms of some degree of mission fulfillment (i.e., system effectiveness) and total life-cycle cost. Mission fulfillment may address one or a combination of performance and/or effectiveness parameters, such as range, accuracy, speed, capacity, availability, reliability, and others, representing the inherent characteristics of system development and production. Life-cycle cost includes all anticipated future costs associated with research and development, production/construction, system operational use and logistic support, and material phase-out and disposition. These costs, spread over the life cycle, are discounted to the present value. This relationship of system design characteristics with cost can be expressed in terms of a variety of figures of merit, as discussed in Section 5.3A1.

Cost-effectiveness figures of merit are utilized extensively in the evaluation of alternative engineering design proposals as discussed in Chapters 2,

---

[5]Discussions of benefit-cost ratios, presented in a slightly different manner in each instance, are included in: W.J. Farbrycky and G.J. Thuesen, *Economic Decision Analysis,* Prentice-Hall, Inc., Englewood Cliffs, N.J., 1974; J. C. Van Horne, *Financial Management and Policy,* 3rd ed., Prentice-Hall, Inc., Englewood Cliffs, N.J., 1974; and G. David Quirin, *The Capital Expenditure Decision,* Richard D. Irwin, Inc., Homewood, Ill., 1968.

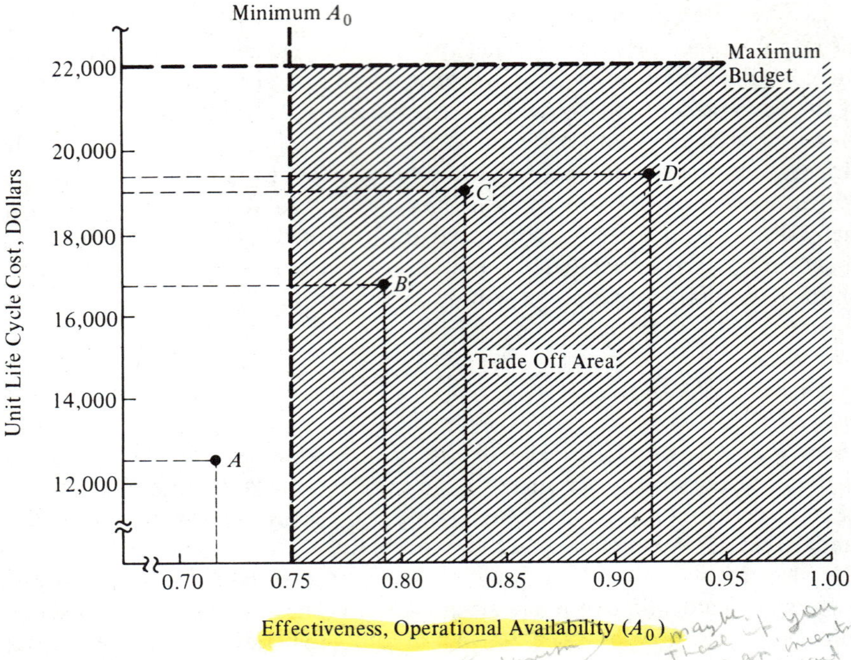

Evaluation Criteria	Conf. "*A*"	Conf. "*B*"	Conf. "*C*"	Conf. "*D*"
Operational Availability	0.72	0.79	0.83	0.92
Unit Life Cycle Cost ($)	12,500	16,600	19,100	19,400
Cost Effect. Ratio (10^{-5})	5.76	4.76	4.35	4.72

Figure 11-10. Evaluation of cost effectiveness measures.

3, and 5. An example of a comparison of four different system design configurations is presented in Figure 11-10. In this instance, operational availability (A_o) and life-cycle cost are used for comparative purposes. Configurations B, C, and D will each meet the criteria of the established availability and cost budgets. However, configuration B is preferred, since these requirements are met at the least overall unit-life-cycle-cost expenditure.

The use of cost effectiveness figures of merit, although not purely a financial criterion per se, relates closely to the methods of evaluation discussed earlier with respect to the application and treatment of life-cycle cost.

11.2 METHODS OF FINANCING

Management acceptance of proposals is based not only on the feasibility of the proposal itself in terms of technical criteria and one or more of the financial figures of merit discussed earlier, but on the status of the corporation relative to profitability and risk. Projects require the acquisition and utilization of manpower, raw material, and fixed assets, such as real estate, facilities, and capital equipment. The availability of these resources is dependent on the current cash position of the corporation and the ability to acquire additional sources of funding for project support. Outside sponsors obviously simplify the decision process; however, investment decisions to expend effort now in hopes of realizing significant benefits in the future may require an initial capital outlay of considerable magnitude. Thus, as part of the investment decision process, management should (a) review the corporation's profitability and current cash position, (b) forecast future cash needs, and (c) determine possible methods for attaining additional funds through long-term and/or short-term financing. The first step evolves around the funds flow of a corporation, as illustrated in Figure 11-11.[6]

The flow of funds within a corporation is basically a continuous process, particularly if the corporation has been in business for a period of time. Products are developed and sold either for cash or on credit. Credit sales constitute accounts receivable, which when collected convert to cash. If the total cash inflow exceeds all costs (including depreciation on assets) for a period of time, there is a profit for that period; if not, there is a loss.[7] Figure 11-11 illustrates a typical cash-flow diagram with the focal point being the *reservoir of cash*. The amount of available cash in the reservoir that can be invested today, which fluctuates over time, will influence management's accept-reject decisions on investment proposals.

Given that the decision is to accept a proposal, a cash budget or a projection of future cash receipts and disbursements is accomplished following the method(s) discussed in Section 11.1 (refer to Figures 11-7 and 11-8). The results, in terms of the overall requirements for funds, may be presented for any project or a group of projects as shown in Figure 11-12. The figure illustrates a variable requirement for funds and the possible methods of attaining funds not readily available in the cash reservoir.

The acquisition of additional funds is realized through a combination of long-term and short-term financing, which is discussed briefly below. One of the functions of financial management is to develop an optimum mix of each,

---

[6] A more in-depth analysis of the funds flow of a corporation may be found in: J. C. Van Horne, *Financial Management and Policy*, 3rd ed., Prentice-Hall, Inc., Englewood Cliffs, N.J., 1974, Chapter 27.

[7] The aspect of depreciation is discussed in Chapter 12.

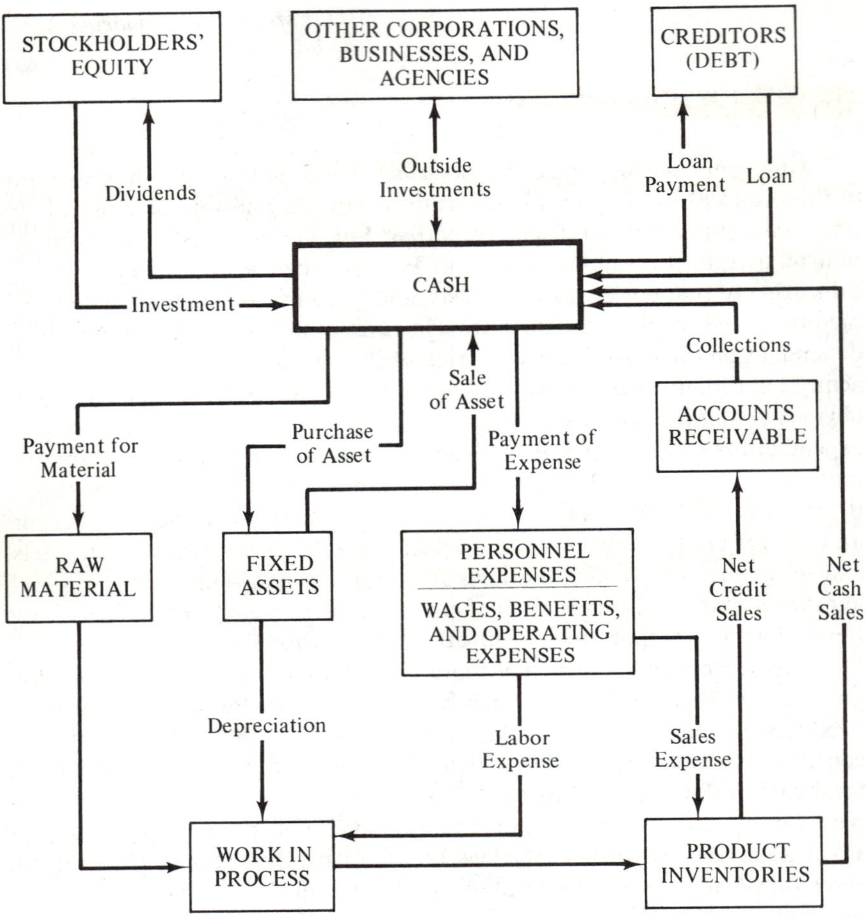

Figure 11-11. Funds flow of a corporation.

considering profitability and risk to the corporation. This mix may be presented as a level of financing illustrated by the dashed line in Figure 11-12.

A. Long-Term Financing

Long-term financing usually refers to the borrowing of money for a long period of time (e.g., 5 years or longer) in order to invest in fixed assets relatively permanent in nature with long life. Interest charges on the loan become available through increased earnings over time from the use of these assets. In raising funds to support projects, management may choose between equity capital and borrowed money. *Equity capital* refers to ownership money acquired through the sale of common and preferred stock discussed in Section 10.1B, and *borrowed money* (in this instance) refers to money attained from the sale of bonds.

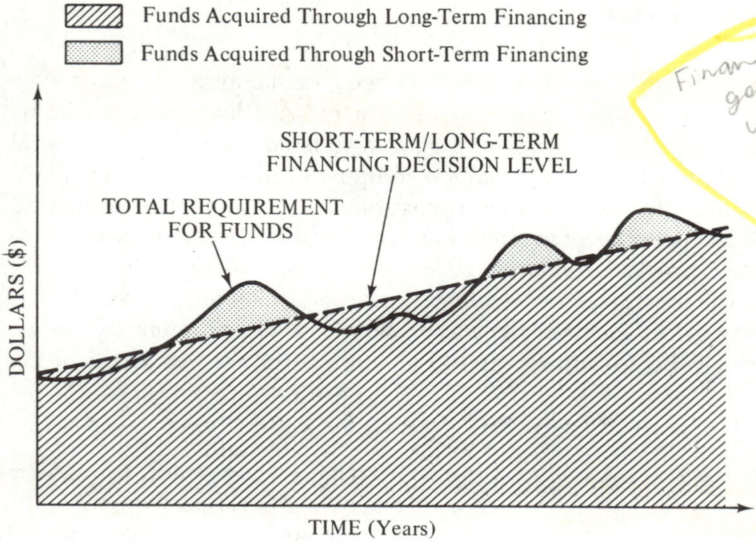

Figure 11-12. Requirement for funds.

1. Equity capital. The corporation charter specifies the number of authorized shares of stock that can be issued. Frequently, a corporation will not issue some of its authorized allotment, to allow flexibility in granting stock options, splitting the stock, pursuing mergers, and so on. Additional financing can be attained through the sale of unissued stock. In pursuing this form of financing, the corporation will usually provide its existing stockholders with the first opportunity to obtain additional shares on a privilege-subscription basis, allowing the stockholder to purchase at a reduced price a designated number of shares determined by the rights established from the number of shares that he already owns. After the initial option to its stockholders is offered, the corporation will offer the remaining shares to the public at the going market price.

2. Corporate bonds. Corporations, in search for outside funds to support operation and growth, may initiate a bond issue with the promise of paying the investor(s) a designated interest on his money at certain scheduled intervals of time. Unlike stockholders, investors in bonds are willing to allow the use of their funds for business purposes but do not assume ownership participation in the business. Two basic categories of bonds are noted.

a. *Unsecured bonds.* Unsecured bonds, often referred to as "debentures," have no claim on any specific asset of the corporation but are backed by the earning power and general credit of the corporation as viewed by the investor. Usually, only corporations with a strong credit rating will be able to sell unsecured bonds, as investors will require security in some form from weaker corporations.

b. *Secured bonds.* Secured bonds are tied to some form of claim on the assets of the corporation in the event that the corporation fails to fulfill its promise to the investor. Examples are mortgage bonds, where the collateral is real property (i.e., real estate, capital equipment, inventories), and collateral trust bonds, where the collateral constitutes a pledge of other securities.

There are many specific varieties of bonds within these categories, each with different provisions relative to maturity date, interest rate and the frequency of interest payment, callable versus noncallable features, and so on. The financial manager of the corporation, as well as the investor, must carefully weigh the advantages and disadvantages of each in terms of the objectives.

3. General revenue bonds. In many areas revenue bonds are used to finance public works projects such as highways, toll bridges, power plants, water works, and sewerage systems. Bonds are sold and the accumulated revenue is used to pay the interest and repay the principal involved. Revenue bonds provide the necessary funds for facilities without adding to bonded indebtedness, and are usually tax exempt.

B. Short-Term Financing

Short-term financing usually includes loans that mature within a year or two. Such loans are frequently used to raise temporary funds to cover seasonal or other special funding needs involving a short time frame (refer to Figure 11-12). Short-term loans are generally self-liquidating, in that the assets acquired with the borrowed money should be easily convertible to cash with a high degree of certainty associated with the conversion ratio or price. Primary sources of short-term financing include (1) trade credit, (2) bank loans, (3) loans from finance companies, (4) sale of commercial notes, and (5) loans from insurance companies. While short-term loans adequately respond to a temporary need, the cost of such loans is usually high. Interest rates are higher, maturity dates are constantly in the picture, and refinancing is difficult, owing to changing conditions in the money market.

11.3 DIVIDEND DECISIONS

As indicated in the beginning of the chapter, the financial function of a corporation is concerned with maximizing the wealth of its stockholders. The stockholders not only expect the corporation to grow in value but anticipate a return on investment in the form of cash dividends (dollars per share of stock), stock dividends, stock splits, or a combination thereof. If the stockholder feels that his investment is worthwhile, he will hold onto the shares that he already owns and may even purchase additional shares. At the same time, the "good word" will spread, and others will invest, also. On the other hand, if the stockholder believes that his investment is poor, he will quickly sell his shares, and the reputation and stability of the corporation may suffer through his actions and those of other stockholders who follow in a like manner. Thus, the corporation must adopt a policy that provides the right mix of retained earnings (for direct reinvestment into the corporation for operations and growth) and dividend distributions.

From the viewpoint of many stockholders, two considerations are important relative to dividend decisions. Initially, an investor may look at the magnitude of the dividend in terms of the per cent of total earnings and whether future projections reflect growth. A number of corporations will establish a target dividend pay-out ratio over the long run. When earnings increase, dividends are increased only when corporate management believes that the new level of earnings can be maintained. As earnings decrease, the dividend distribution remains constant until the business declines or a recession forces a reduction in dividends. The established per cent of earnings distributed as dividends is a key factor to relative investment.

As a second point, the stability of dividend distributions is important, since a stable dividend policy intuitively increases the investor's confidence in the corporation and its management. Many individual investors desire a periodic income that they can depend on without having to rely on their principal. Unstable dividend distributions often induce concern and insecurity (whether warranted or not) and lead to an increasing number of stock transactions and the possible financial instability of the corporation.

Finally, the corporation may wish to offer a stock dividend (i.e., an extra share of stock) or initiate a stock split, which does not increase the stockholder's proportionate ownership of the corporation but does create a favorable psychological effect on the stockholders. Stock dividends are often issued when the corporation decides to reduce dividend distributions and increase retained earnings. Actually, benefits to the stockholder must be evaluated in terms of the combined effects of cash dividends, stock dividends, and stock splits.

11.4 FINANCIAL MANAGEMENT

Financial management involves decisions pertaining to investment proposals, methods of financing, and dividend policies. Decisions in each of the three areas significantly impact the other areas as well as the financial position of the corporation, as indicated from the interactions depicted in Figure 11-1. The financial manager of the corporation is responsible for the trade-offs required in arriving at the best possible mix of investments, cash flow, and benefits to the stockholders.[8]

Financial management requires the establishment of budget levels and determining the maximum amount of funds that can be invested during a specified period of time (except in instances where projects are financed incrementally by outside sponsors). Investment decisions are made based on the established budget levels, projects are initiated, and financial data are collected and analyzed to assess the corporation's business posture.

Financial analysis and assessment requires the use of certain figures of merit for comparative purposes with other similar businesses and for the determination and evaluation of business trends. A few representative and commonly employed financial measures are noted below.

A. Current Ratio

The *current ratio* is a measure of liquidity used to judge the corporation's ability to meet short-term obligations, and is expressed as

$$\text{current ratio} = \frac{\text{current assets}}{\text{current liabilities}} \tag{11-8}$$

Here current assets include cash, inventories, and accounts receivable; and current liabilities include accounts payable and other short-term liabilities. Theoretically, as the ratio increases, the corporation is better able to pay its bills. Usually, corporations that have a small inventory and accounts receivable that are easily collectible can operate safely with a lower current ratio than those firms with large inventories and many fixed assets.

B. Earnings/Common Share Ratio

The *earnings-per-share ratio* is a popular financial performance measure of particular interest to the stockholders, and is expressed as

$$\text{earnings per share} = \frac{\text{net available earnings for the year}}{\text{number of common shares of stock}} \tag{11-9}$$

The net available earnings is the profit for the year less dividend requirements on preferred stock.

---

[8]When making trade-offs that involve investment decisions and the financial position of the corporation, it should be noted that in some instances the best engineering proposal may not be acceptable in terms of financial criteria (i.e., budget constraints, anticipated consumption rate of available funds, the timing of required external funds relative to money market conditions, etc.). Engineering proposals must be evaluated not only in terms of technical feasibility but on the basis of financial objectives.

C. Debt/Equity Ratio

The *debt-to-equity ratio* is often used to determine the corporation's ability to meet long-term obligations, and can be expressed as

$$\text{debt/equity} = \frac{\text{long-term debt} + \text{current liabilities}}{\text{total stockholders equity}} \tag{11-10}$$

D. Profitability Ratio

The *profitability ratio*, an indication of the corporation's efficiency of operation, can be expressed in relation to sales and investment. These expressions are

$$\text{profitability (sales)} = \frac{\text{net profit after taxes}}{\text{total sales}} \tag{11-11}$$

$$\text{profitability (investment)} = \frac{\text{net profit after taxes}}{\text{total tangible assets}} \tag{11-12}$$

E. Turnover Ratio

The *turnover ratio* indicates the relative efficiency with which the corporation utilizes its resources in order to generate output, and is often expressed as

$$\text{turnover ratio} = \frac{\text{net sales}}{\text{total tangible assets}} \tag{11-13}$$

There are many such ratios employed, both individually and collectively, in financial analyses. Financial ratios are of special significance to the financial manager in the evaluation and control of the business condition and performance of the corporation. Certain measures are more valid for a given industry than for others. Proper application suggests that these ratios be employed for comparison purposes with other corporations involved in the same industry or in related product lines. In addition, such ratios are appropriately used in the evaluation of year-to-year trends for a given corporation.

The input factors used to compute these figures of merit are derived from the corporation's financial statements (i.e., balance sheet and income statement). Discussion of financial statements is deferred to Chapter 12, since familiarity with accounting methods is necessary for the preparation of these statements. Financial management depends on accounting data for analysis purposes.

QUESTIONS AND PROBLEMS

1. Describe the general relationship between the financial function of a corporation and engineering. Construct a flow diagram that illustrates the major interfaces.
2. How does the financial function become involved in the decision as to whether to accept or reject an engineering proposal?
3. Describe the relationship in financial management between investment decisions, decisions involving methods of financing, and the dividend policy of a corporation.
4. Assume that your were given the responsibility to evaluate engineering proposals for the corporation and to make recommendations to corporate management for acceptance or rejection. How would you proceed with the evaluation effort and what criteria would you use in arriving at a decision?
5. Determine the present value of the following cash flows, using a 10 per cent discount rate:

Year	*Cash Flow*	*Year*	*Cash Flow*
1	$ 7,500	6	$12,000
2	6,000	7	16,000
3	12,000	8	18,500
4	11,500	9	22,000
5	10,750	10	19,000

6. Refer to problem 5. What is the discounted value of the cash flows? Assume a decision point at the end of year 4.
7. Calculate the internal rates of return for the following projects:

	Project A	*Project B*	*Project C*
Initial cost	$15,000	$11,000	$18,000
Benefits			
Year 1	7,000	3,000	4,000
Year 2	8,000	4,000	6,000
Year 3	8,500	5,000	6,000
Year 4	9,000	6,500	7,000
Year 5	10,000	7,000	7,000
Year 6	10,000	9,000	8,000

8. Assume that the cost of capital is 15 per cent. Calculate benefit-cost ratios for the projects listed in problem 7. Which project is preferable?
9. Calculate the benefits and cost of the projects identified in problem 7. Use an assumed discount rate of 15 per cent. Graph the present-value profiles for each project and determine the payback point. Which project is preferable?
10. What factors should be considered in determining whether a payback point is realistic?

11. What is the difference between the "minimum acceptable rate of return" and the "cost of capital"? Which would you use for project selection purposes? Why?
12. Examine the operations of several large corporations and define the functions of the financial manager. What are the major objectives of these corporations?
13. An investment of $2,000 today will return $1,500 at the end of 1 year, $500 at the end of 2 years, and $150 at the end of 3 years. What is the internal rate of return?
14. The stock of the XYZ Corporation is currently selling at $30 and is expected to pay a $3 dividend at the end of the year. If an investor purchased the stock now and sold it for $35 after receiving the dividend, what rate of return would he earn?
15. Graph the present value of $5 per year for 5, 10, 15, and 20 years at 0, 10, 15, and 30 per cent rates of discount. Explain the difference in the slopes of the curves.
16. Examine the annual reports of two large corporations. What sources of long-term and short-term financing are used?
17. Refer to Figure 11-8. What is the NPV of project A for a discount rate of 10 per cent? 15 per cent? 30 per cent?
18. Refer to Figure 11-10. If another design configuration (i.e., configuration E) with the values of 0.77 for A_O and $15,000 for the unit-life-cycle cost was being considered, how would this configuration rank with the others? Of all the design configurations being considered, which one would you select? Why?
19. Refer to Figure 11-11. Describe briefly the impact of engineering activities on the funds flow of a corporation. How could the cash position of the corporation affect decisions on whether to accept or reject engineering proposals?
20. Refer to Figure 11-12. What factors would you consider in establishing the mix of short-term and long-term financing (e.g., the decision level indicated in the figure)?
21. What is equity capital and how is it acquired?
22. What is the difference between unsecured and secured bonds?
23. How does the corporation's dividend policy affect its ownership?
24. As a prospective investor, what criteria would you employ in determining whether or not to invest in stock of a corporation?
25. As a financial manager of a corporation, what approach would you use in selecting appropriate measures for the evaluation of the corporation's business posture?
26. When evaluating the business position of a corporation, financial measures (i.e., ratios) are often used for comparative purposes. What cautions should be observed in the selection and application of financial measures? (Refer to Section 11.4.) Can the same measures be employed universally for all businesses? Why?
27. How are budgets developed for approved engineering proposals?

12

BUDGETING AND ACCOUNTING FUNCTION

All of us are faced constantly with making decisions related to both business operations and the management of our personal affairs. In making such decisions, a great deal of reliance is placed on initial budgeting and on the availability of valid information covering both current financial status and past transactions. Much of the information required is derived from a formalized accounting function which provides a systematic means for data collection and reporting. Of particular interest is the initial budgeting and funding for acceptable projects, and the subsequent accounting and evaluation of project costs.

The accounting function described herein is oriented to reporting the economic affairs of an organization, and the material presented deals primarily with cost accounting versus bookkeeping. The budgeting and accounting function discussed directly supplements the financial function included in Chapter 11.

12.1 INTRODUCTION

The budgeting and accounting function for a corporation, business, or government agency cuts across all facets of the organization and is designed to:

A. Provide information for use in planning, decision making pertaining to the acceptance or rejection of proposed projects, and budget allocations for those projects that are approved.

B. Provide information for use in the review, evaluation, and control of in-house operations and for the identification of problem areas for which the need for corrective action is indicated.

C. Provide information for use in reporting to the stockholders, creditors, the government, and other outside parties interested in the financial status of the organization for one reason or another.

These three purposes are closely interrelated and can be addressed in the context of the illustration in Figure 12-1. When determining whether or not to accept individual proposals for investment purposes, the ultimate decision will depend on the creditability of the proposal from a technical standpoint and on the financial position of the corporation or business (as discussed in Chapter 11). Given the decision to invest, the estimated funds required for the approved project are budgeted and allocated to individual work packages and the work breakdown structure as defined in the project proposal (refer to Figures 3-15 and 3-16). Total dollar allocations and expenditure rates are collected in terms of incremental and total costs. These data are compared with the initial projections, as illustrated in Figures 3-19 and 3-20, to ensure that the actual expenditures and the budgeted amounts are compatible. If incompatibilities exist, the responsible manager should investigate the cause(s) and initiate corrective action as appropriate.

Evaluating the status of the business enterprise as a whole requires accounting information that covers all projects and other activities which either directly or indirectly support the operations of the enterprise. Projects may be staggered as indicated in Figure 12-2, and there are numerous support activities which are continuous as long as there is work in-house. In any event, *all* activities within the organization are budgeted and must be controlled on a sustaining basis.

Adequate monitoring and control of business operations are highly dependent on the availability of the *right* type of information at the time needed. Certain information may be required for managerial purposes on a weekly or monthly basis, whereas it may be sufficient to report other information once a year. Also, data covering individual projects are generally conveyed only through *internal* reports, whereas information reflecting the overall financial status of the organization is presented both internally and through *external* reporting, usually in a formal annual report or at specific times when a review of the current business status of the organization is necessary for making decisions relative to large investments. Thus, the accounting function must be objective, consistent and yet flexible, and must be responsive to a wide variety of needs.

To facilitate this, the accounting function (together with the financial function discussed in Chapter 11) is usually organized as a separate activity supporting all project and staff engineering activities, technical and management service functions, centralized test laboratories, production operations, and so on, throughout the corporation, business, or agency. Accounting

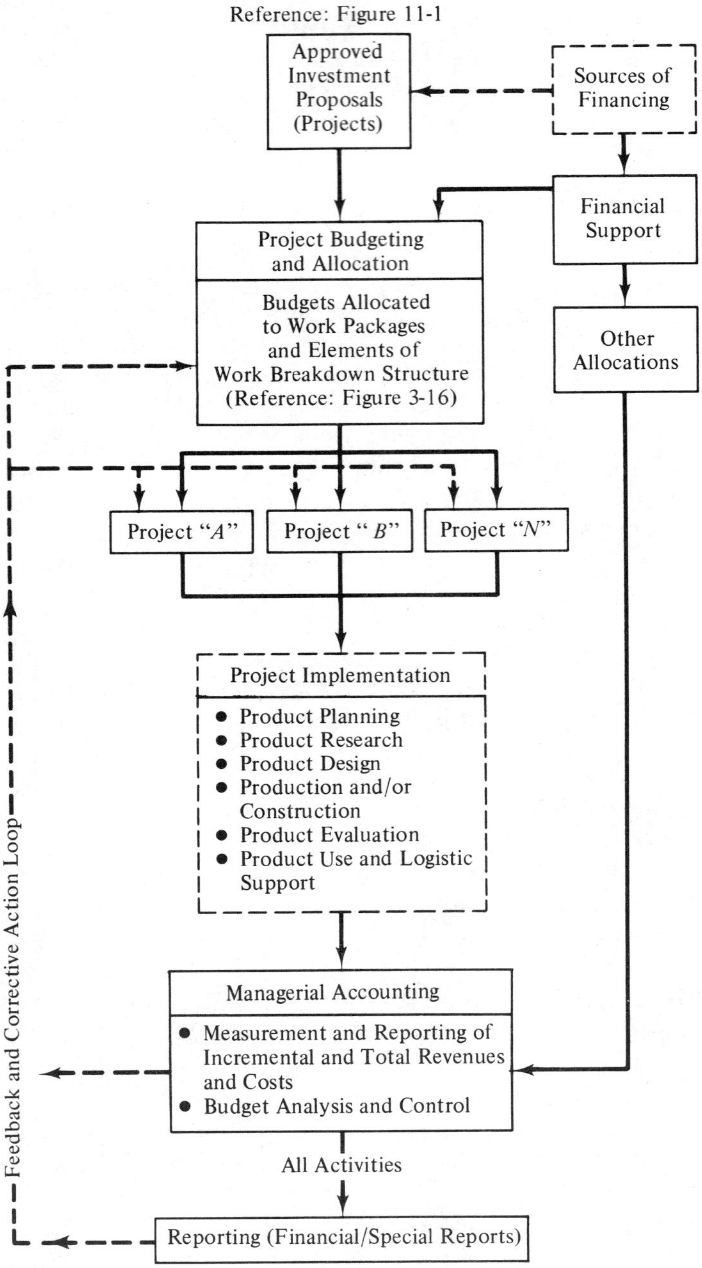

Figure 12-1. Budgeting and accounting cycle.

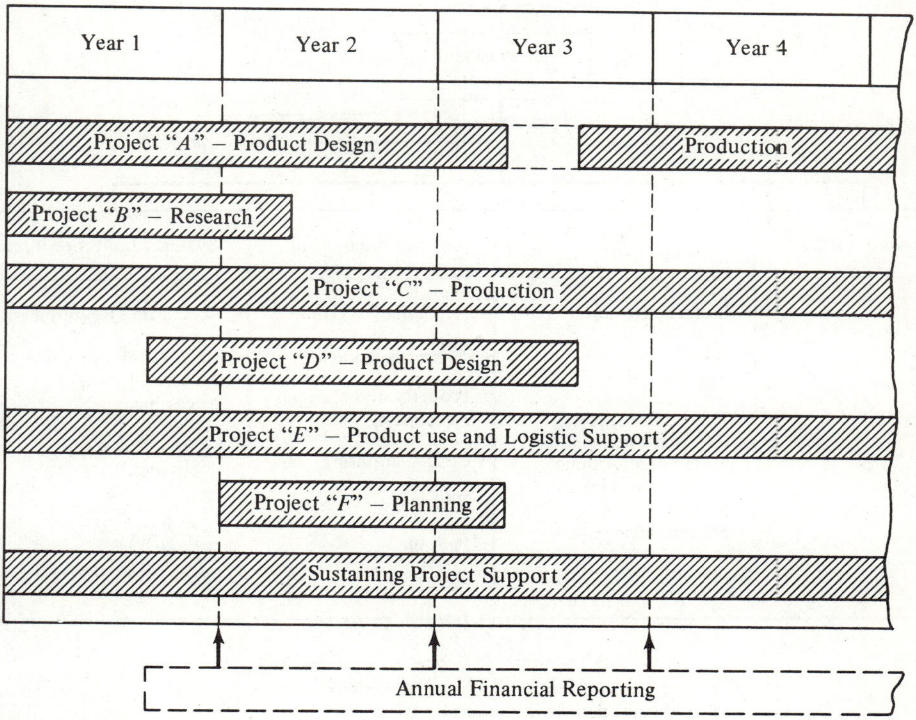

Figure 12-2. Scheduled activities of a firm.

should identify the individual needs of each organization element and develop a data-processing and information capability that will provide the proper data at the time needed to each level of management as appropriate. Common accounting tasks are identified in Figure 12-3.[1]

Accounting is a *service* activity and must support total organizational needs. As needs change, the reporting requirements will change (i.e., types of data and frequency of need), and since changes occur frequently, management must continually assess the extent to which the accounting function is responsive. Experience in the past has indicated that on numerous occasions the accounting data provided managers in many organizations are not

[1]Several good references that cover accounting functions/tasks in depth are: C. T. Horngren, *Cost Accounting—A Managerial Emphasis,* 3rd ed., Prentice-Hall, Inc., Englewood Cliffs, N.J., 1972; and M. J. Gordon and G. Shillinglaw, *Accounting—A Management Approach,* 3rd ed., Richard D. Irwin, Inc., Homewood, Ill., 1966. A reader wishing to study accounting principles per se should refer to either or both references.

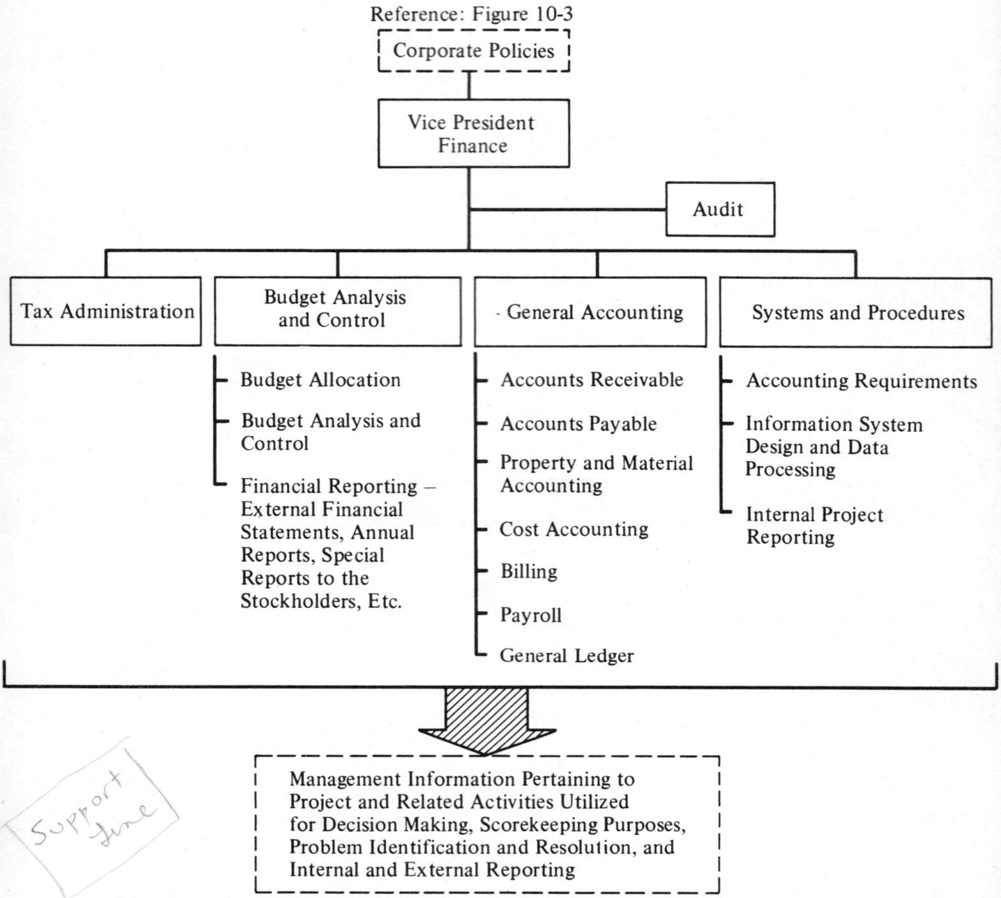

Figure 12-3. Basic finance/accounting functions.

adequate for one reason or another, and the awareness of inherent business problems is not apparent until it is too late and the business is in trouble.[2]

12.2 ACCOUNTING REQUIREMENTS

As stated, a well-designed accounting capability serves a variety of purposes, with the ultimate objective of providing information that will aid management in making decisions on investment opportunities and in assess-

[2]Inadequacies in accounting may develop because (a) the using organization does not adequately identify its needs, (b) the manager of the accounting function does not understand the needs of the total organization, (c) the manager of the accounting function refuses to respond to the requirements of the individual organization elements, and/or (d) the manager of the accounting function incorporates changes in the data-processing and communications loop without considering the effects on using organization elements. The wrong type of data, delays in data reporting and processing, and so on, will not allow for adequate project control.

ing the financial position of the organization. Investment decisions are highly dependent on the prediction of revenues and costs, and the ability to make accurate predictions is necessary to yield beneficial results. The availability of good accounting data that reflect actual historical experience is a key input requirement in assuring the accuracy of predictions, particularly when estimating the *cost* of doing something. Managers are constantly faced with questions concerning the cost of developing a new product, the cost of producing a unit of something or a group of items, the cost of operating and maintaining a plant, the cost of operating and supporting a product in the field throughout its life cycle, and so on. Cost is a major factor in decision making, and as a result emphasis will be placed on cost accounting.

The accounting of costs should be identified with the appropriate project work breakdown structure elements and each facet of the corporation, business, or agency's organization. Costs should be broken down to the depth necessary to ensure compatibility with the level of management control applied to various activities. Activities that are repetitive in nature and essentially standard do not require the extent of accountability and control that is necessary for variable functions involving numerous unknown quantities. Cost accounting should be somewhat tailored to the specific needs of the organization.

In an attempt to further understand the requirements for cost accounting, some additional discussion of a few of the more commonly utilized cost categories is considered appropriate at this time, since terminology often becomes a major problem when discussing cost.

A. Total Cost and Unit Cost

When dealing with cost, the *total cost* concept should serve as a starting point to ensure that *all* factors are appropriately considered. The decision maker can easily overlook a significant facet of cost unless he intuitively addresses the overall cost spectrum before selecting the specific elements of cost that are central to the decision. Total cost includes all life-cycle costs associated with each product, plus any composite costs generated from a mix of products. Product-life-cycle cost may be represented by a cost breakdown structure as illustrated in Figure 12-4. There is no set method for breaking down cost; however, cost breakdown structures should exhibit the following basic characteristics:

1. *All* product cost elements must be considered.
2. Cost categories are generally identified with a significant level of activity or with a major item of material. Cost categories must be well defined, and managers, accountants, engineers, and others must all have the *same* understanding of what is included in a given category and what is not! Cost doubling (i.e., counting the same cost in two or more categories) and omissions must be eliminated.

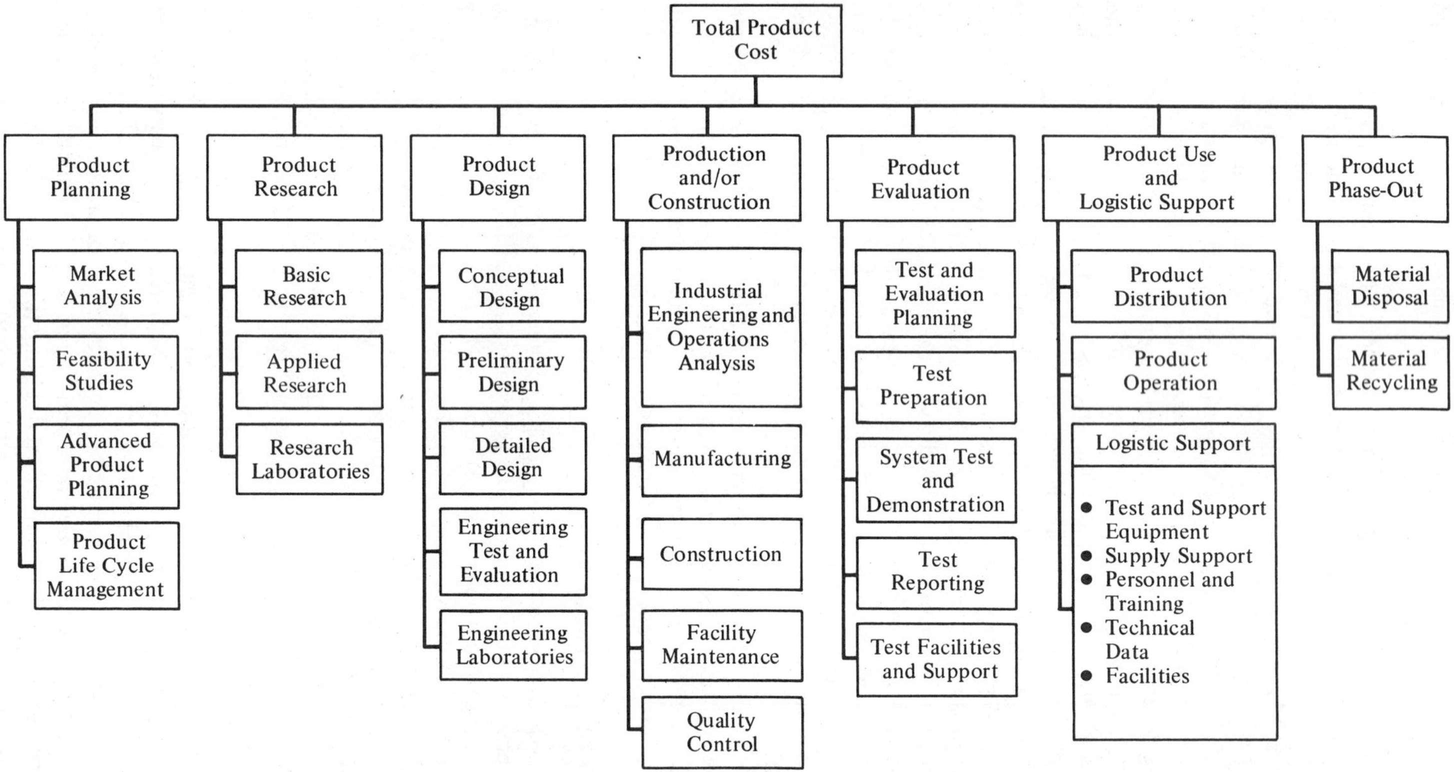

Figure 12-4. Product cost breakdown structure.

3. The cost structure and categories defined should be coded in such a manner as to allow for the analysis of specific areas of interest while virtually ignoring other areas.
4. When related to a specific project, the cost structure should be compatible (through cross-indexing, coding, etc.) with the project work breakdown structure.
5. The cost structure and categories defined should be coded in such a manner as to enable the separation of producer costs, consumer costs, and supplier costs in an expeditious manner.

Obviously, the cost categories identified in Figure 12-4 are too broad to ensure any degree of accountability and control and should be broken down into subcategories that represent the number of indenture levels considered feasible for the intended purpose. These may include levels of specific organization activity representing personnel costs and/or components of a system, such as hardware, software, facilities, or data. In addition, costs associated with producer-controlled operations must be identified as a starting point in determining the accounting requirements for most corporations involved in product development, production, test and evaluation, and other related producer functions.[3] Nevertheless, viewing the overall cost in the context of the illustrated cost breakdown structure (or equivalent) is an excellent starting point.

Unit cost, on the other hand, is often used as a preferred measure for decision making and is usually expressed in terms of cost per item produced, cost per person, cost per increment of effectiveness, cost per capacity output, cost per interval of time, and so on. Preferably, unit cost is derived by dividing the factor of interest into total cost; however, on many occasions, unit cost is expressed using some value less than total cost, such as product development cost or manufacturing cost. Actually, unit costs should be expressed in the terms that are most meaningful for the purpose(s) at hand as long as complete understanding exists relative to what is included in both the numerator and denominator of the unit-cost figure of merit used. Also, caution must be exercised, since unit cost represents an average that may change with the magnitude of the numerator, denominator, or both.

B. Direct Cost and Indirect Cost

Within the total cost spectrum, some costs are identified as being *direct* and others as being *indirect.* The difference relates to traceability, where "direct" refers to costs that can be traced easily to a given cost object. An

[3]Accounting requirements may vary considerably, depending on the producer-consumer interfaces and the areas of control desired. Ideally, accounting data will cover both producer and consumer operations such that total costs can be readily determined.

attempt to define some of the basic differences in these categories is noted below.

1. *Direct labor.* All labor that is obviously related to a given product is classified as direct labor. This includes engineers assigned to the planning, design, development, manufacture, test, and support of a product. Direct production labor includes machine operators, assemblers, and test personnel involved in the fabrication, manufacture, assembly, and test of a product. For projects sponsored by outside agencies through a contracting arrangement, all personnel charging their time against external project contractual funding represent direct labor.

2. *Direct material.* All materials that are included as an integral part of the finished product output constitute direct material. This includes units, assemblies, modules, and component parts of a system.

3. *Indirect or overhead.* Indirect labor or overhead involves those personnel expenses not readily assignable to a specific product output. Examples are top level and middle managers; some engineers pursuing basic research and assigned to staff activities; and organizations, including marketing, finance and accounting, legal, personnel administration and industrial relations, engineering and production laboratories, material handling, and quality control and product inspection. In addition, personnel fringe benefits (i.e., life and health insurance, retirement, social security, pension, etc.) are covered in this category. Indirect material often includes small items, such as fasteners, glue, cleaning compound, and office supplies. Other indirect costs are facility operation and maintenance, property taxes and insurance, utilities, and capital equipment.

Sometimes it is difficult to determine which costs fall into what category since there are many areas that are not very well defined. Thus, the decision as to how to classify costs usually relates back to the allocated budget, which is based on top management's financial decisions and policies.

The relationship of indirect to direct cost is a major factor in the success of a business. Usually an indirect overhead factor is applied to direct labor on a project, and an indirect general administrative charge is applied to the total estimated labor and material costs. Some corporations or businesses use a fixed percentage of the direct cost to establish indirect cost; others use a variable factor to cover indirect costs. The ratio of indirect to direct cost determines in many instances whether or not the corporation will be successful in acquiring new business. If the ratio is too high, the total proposed cost for individual projects will be high, and the competition will prevail. On the other hand, a low ratio may be attained at the expense of reducing the corporation's overall capability and benefits to the employees, which leads to other problems. Thus, the management of the corporation must maintain the proper balance of indirect to direct costs, and accounting data must accurately report these factors regularly and at frequent intervals.

C. Variable Cost and Fixed Cost

Variable and fixed costs are generally defined in terms of how the cost changes in relation to fluctuations in the activity associated with a given cost allocation base. If the cost changes in proportion to activity changes, it is *variable*. A typical example is the variation of total cost when producing different quantities of an item at x dollars per unit. In this case, cost variations occur with variations in volume. On the other hand, if the cost remains constant for a designated period of time despite fluctuations in activity, it is *fixed*. Property taxes, insurance, certain personnl expenses, and depreciation are examples of fixed cost; however, these costs are fixed only within a relevant period of time, since they may ultimately vary in the long term as the overall economy shifts. Different methods commonly used in projecting variable and fixed costs are illustrated in Figure 12-5.[4]

Basically, all costs can be classified as either variable or fixed, although few costs fit clearly into one category or the other. In some instances, interactions between the two categories may exist, since fixed costs may be deliberately increased or decreased to obtain more benefits in terms of sales and profits. For example, if labor rates increase, more sophisticated computerized (automated) equipment may be introduced to reduce the unit variable cost. Conversely, as volume decreases, it may become feasible to reduce the fixed cost and increase the variable cost by selling in-house capital equipment and purchasing the necessary items on the outside instead of manufacturing them internally. There are numerous alternative possibilities of this type that are subject to evaluation in decision making and an understanding of the variable-fixed cost relationship, along with the concurrent impact on revenue from sales, is essential.

The discussion of cost is a necessary requisite in determining accounting requirements. Initially, the aspect of total life-cycle cost must be considered, followed by a determination of the categories of cost (within the total cost spectrum) that are essential for decision making and for the evaluation and control of the business. Next, it is necessary within the financial and budgeting structure of the firm to identify direct and indirect costs and variable and fixed costs and to ensure that these categories are readily accessible through accounting data. Finally, it is essential that the proper elements of data be aligned with organization requirements such that managers at each level of the organization can effectively evaluate and control their respective activities.

---

[4]It is realized that the straight-line representations of cost projected in Figure 12-5 are perhaps an oversimplification of real-world conditions, particularly from the economist's point of view, where nonlinear relationships are considered more valid. However, the accounting simplifications are usually quite adequate in establishing variable and fixed-cost relationships when the acquisition of more accurate data is often more difficult and expensive, and the results may or may not lead to better decisions.

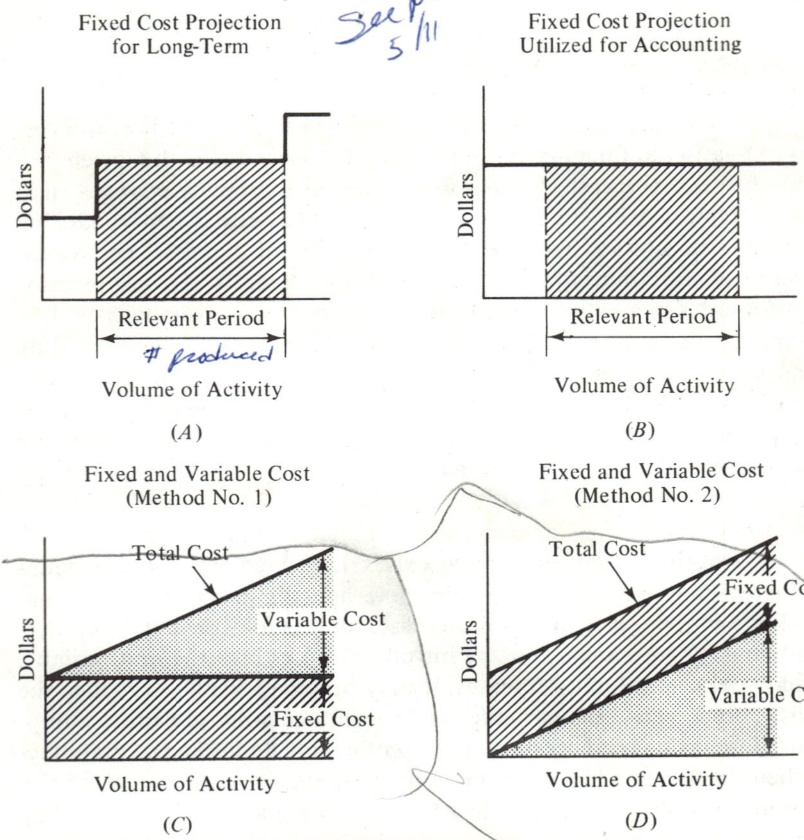

Figure 12-5. Variable and fixed cost relationships.

The determination of accounting requirements is not an easy task, and the ultimate results may not be ideally suited for all purposes. Too much data are costly and time-consuming from the standpoint of data processing; too few data will lead to decisions that may be costly as a result of not having the right information available. Thus, the optimum approach of developing a capability somewhere in the middle is challenging but necessary.

12.3 OVERVIEW OF THE ACCOUNTING PROCESS

The detailed aspects of accounting are quite comprehensive and beyond the scope of this book; therefore, discussion will be limited to a brief overview of some of the key points that are considered pertinent from an engineering interface standpoint. The sections below will touch on design of the accounting process, data collection and handling, and reporting.

A. Design of the Accounting Process

The nature and organization of the enterprise and the particular needs of management will dictate accounting requirements. These requirements will vary considerably depending on whether the needs are associated with a single proprietorship, a business, a government agency, or a large corporation. However, since the requirements are considerably more extensive for a corporation, the discussion will be oriented in this direction.

The definition of accounting requirements should commence with an identification of what is essential in terms of data, the format used in presenting the data, and the frequency of data reporting. The need for data was discussed in Sections 12.1 and 12.2 while the formatting and frequency will vary with individual reporting requirements within a corporation. The objectives include the reporting of information internally for *project control* and *product costing*. For control purposes, large and complex projects, where the type of activity is quite extensive and varied, will require accounting reports on a weekly or biweekly basis, whereas monthly reports will be sufficient for smaller projects. On the other hand, the monthly or bimonthly time interval is probably sufficient for product costing. In addition to internal reporting requirements, corporations are obliged to prepare annual reports covering the business posture and the corporation's financial status for submission to the stockholders, creditors, the government, and others who have a vested interest in the enterprise.

With these output requirements in mind, the accounting process should be viewed in terms of data input and the steps necessary in converting these data to the format desired. Figure 12-6 is presented for discussion. Since internal corporate objectives are oriented to both project control and product costing, the first step is to establish a method for coding and collecting cost information at all apropriate levels of management (i.e., total cost, indirect and direct costs, variable and fixed costs). Initially, a complete listing of accounts by organization element and cost category is generated and used in conjunction with separately assigned project numbers to establish a framework for cost collection.

Figure 12-7 illustrates a sample listing which is often referred to as the *chart of accounts.* Using the assigned codes, a given organization such as systems engineering may expend certain direct labor resources against project 814763. Accounting will assign a number, such as 4010-211-814763, to collect all applicable costs in this area. If electrical design were to incur direct material costs against the same project number, the accounting charge number would be 4020-221-814763. The chart of accounts must be comprehensive and flexible enough to cover all required contingencies. Flexibility is incorporated by allowing for additional subindentures of cost categories through an expansion of the chart of accounts illustrated in the figure.

The next facet for consideration is that of input source documentation necessary for the recording of all events, direct and indirect, that contribute to costs. One might start with the basic question: What source data elements are required to meet the control and reporting needs of the corporation, and which activities should be involved in the completion of such data elements?

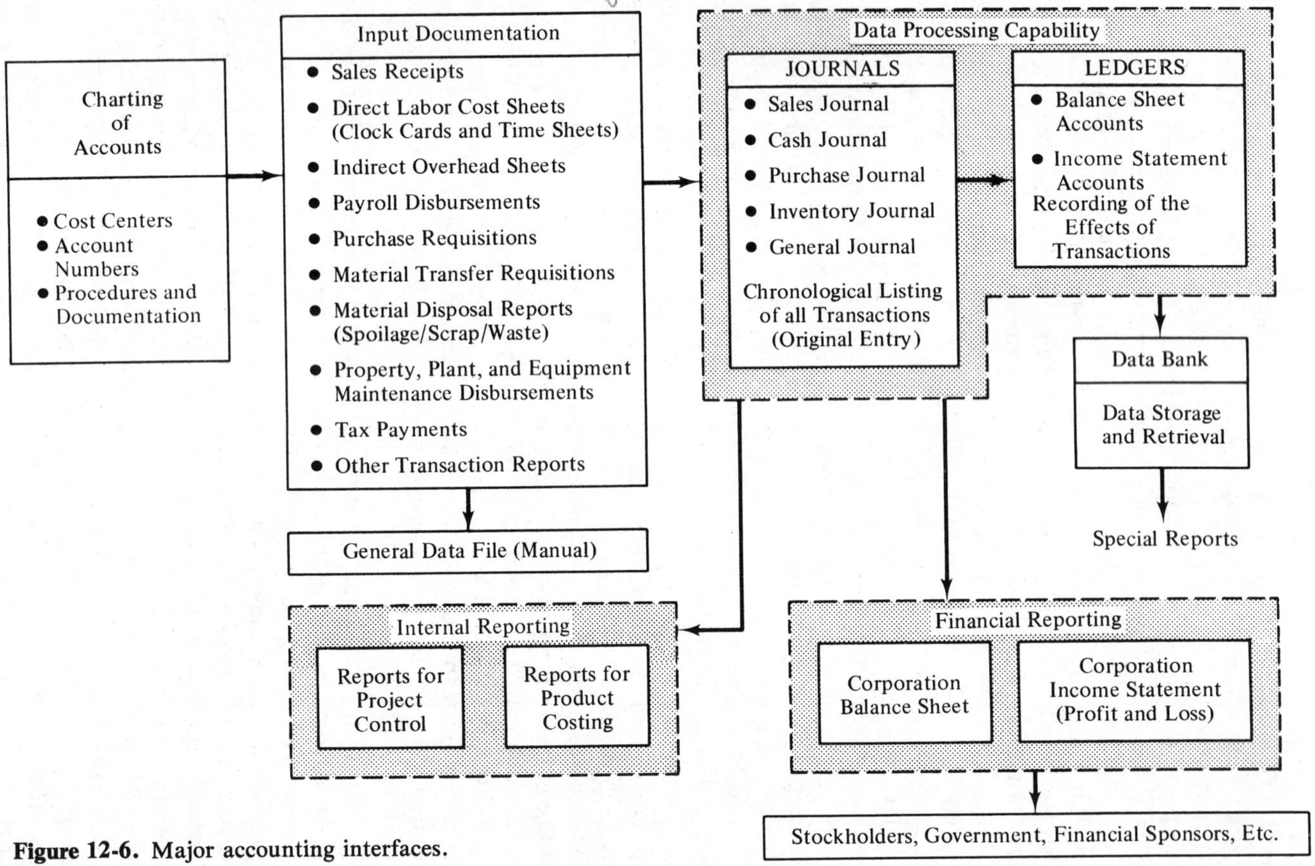

Figure 12-6. Major accounting interfaces.

ACCOUNTING NUMBERS FOR ORGANIZATION ELEMENTS

Department Number	Department Nomenclature	Department Number	Department Nomenclature
1000	Advanced Product	5000	Technical Services
2000	Marketing	5010	Drafting
3000	Research	5020	Publications and Reproduction
3010	Basic Research	6000	Management Services
3030	Applied Research	6010	Data Processing
4000	Engineering Design	6020	Centralized Test Laboratories
4010	Systems Engineering	6021	Components Test Lab.
4020	Electrical Design	6022	Model Shop
4030	Mechanical Design	7000	Industrial Engineering
4040	Reliability Engineering	8000	Plant Engineering
4050	Maintainability Engineering	9000	Manufacturing Engineering
4060	Human Factors	1100	Production Operations
4070	Logistics Engineering	1200	Quality Control
4080	Product Safety	1300	Product Support
4090	Prototype Development	1400	Project Management

ACCOUNTING NUMBERS FOR SALES AND COST

Account Number	Budget Item	Account Number	Budget Item
100	Sales Revenues	220	Material Cost
200	Cost	221	Direct Cost
210	Labor Cost	222	Indirect Cost
211	Direct Cost	223	Disposal Cost
212	Indirect Cost	?	Other Items

Figure 12-7. Chart of accounts.

Data that cover sales and costs are required, with costs being incurred through individual transactions involving labor and material expenditures. Of particular relevance to engineers is the recording of their labor hours in the weekly or monthly time report for the cost center in which the activities are accomplished. Also, engineers are involved in the initiation of purchase orders for items of material and technical data and for the completion of expense reports covering travel, reproduction, computer usage, and the acquisition of office supplies. These transactions are generally recorded through manual means and should be initiated with care to ensure that accuracy is maintained. A haphazard approach to data recording should obviously be avoided if the output results are to be meaningful. It is relatively easy to use an improper accounting charge number or to make errors in the recording of one's time. The proper recording of data is essentially a function of personal motivation, and accounting personnel and management must be sensitive to possible errors.

Individual input data transactions initiated in the organization where personnel are physically located are generally recorded in a *journal*. A journal constitutes a chronological listing of all transactions and is the formal mechanism for the *original entry* of data in many accounting processes. Journal information is usually converted to a *ledger*, where the effects of all transactions are recorded in a form compatible with the requirements for financial reporting (i.e., the balance sheet and the income statement).[5] Financial reports are discussed further in Section 12.4.

B. Data Processing and Project Reporting

In most corporations, the mechanics of processing accounting data from the manually intiated time report or the purchase order to the formal project report and financial statement are relatively complex and should be adapted for computerized data-processing techniques where possible. For larger projects, in particular, it is necessary that input data be submitted promptly on the date required and that the turnaround time for completion of the project status report be minimized. If the manager is to be held responsible for the complete control of all assigned project activities, accurate accounting information must be made available to him on a timely basis. Late transaction entries and untimely reporting may be costly, resulting in delays in initiating prompt corrective action, which might otherwise have been avoided. Thus, it becomes equally important that the engineer complete the input data *accurately* and that the accounting function be *responsive* to project needs.

---

[5]The reader is advised to refer to an accounting textbook for a discussion of the details concerning journal and ledger entries, traditional bookkeeping (T-accounts and debit-credit procedures), and transactions analysis. Refer to the bibliography in Appendix D.

Cost Center 4010		Department Name: Systems Engineering						Date: 4/10/76	
		Status This Period			Total Status to Date			Projected Status At Completion*	
WBS No.	Project No.	Budget this Period ($)	Cost this Period ($)	Period Overrun (Underrun) ($)	Project Budget ($)	Actual Cost To Date ($)	Total Overrun (Underrun) ($)	Overrun (Underrun) ($)	Percent (%)
3A1100	814763	1,200	1,150	(50)	23,000	10,100	(12,900)	50	– 0.2
2A3100	824823	6,700	6,200	(500)	37,500	21,400	(16,100)	1,005	– 2.7
3A5400	833110	15,500	16,250	700	64,750	53,530	(11,220)	(12,100)	+ 18.7
6B3200	712001	500	610	110	9,400	1,330	(8,070)	115	– 1.2

* Projected status is based on a continuation of expenditures at the current average rate for the remaining time to project completion.

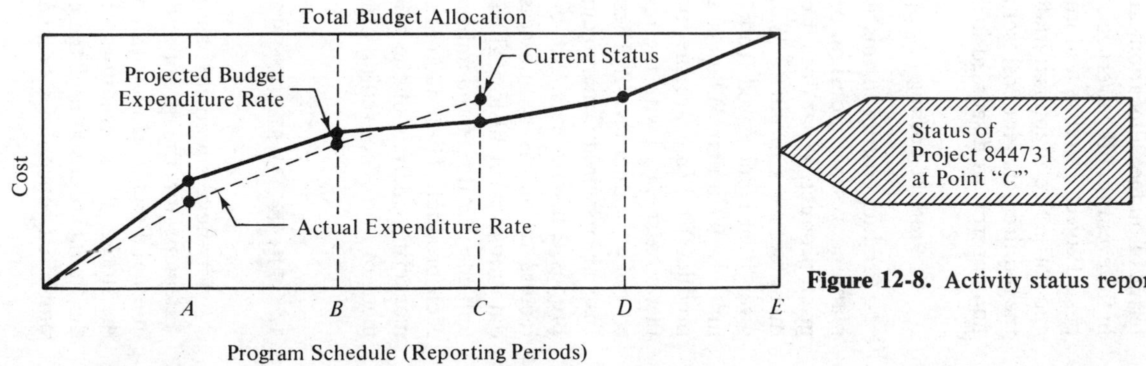

Figure 12-8. Activity status reporting.

Figure 12-8 represents an abbreviated example of a cost center report, indicating the status of each applicable project for the reporting period (e.g., weekly, biweekly, or monthly), the overall status for the project(s) to date in terms of budget allocation, and the anticipated status at project completion in the event that the average expenditure rate continues as experienced in the past. Project overruns and underruns are noted, the causes are investigated, and corrective action is initiated as appropriate. Reports of this type support the requirements for cost/schedule/performance/effectiveness measurement and control, as discussed in Section 3.4C4.

C. Product Costing

Product costing in this instance refers to the costs associated with the planning, research, design, production, test and evaluation, and support of a given product. What does it cost to develop a product? What does it cost to manufacture a product? What is the cost if the production quantity of a given item is increased from 100 units to 500 units, or decreased from 100 units to 99 units? What is the overhead cost associated with developing product X? What are the material requirements and scrapage or spoilage factors associated with the manufacture of product Y? What is the average cost of a product recall? These and other questions associated with product costs are required if management is to make intelligent decisions concerning proposed investments. Not only are total costs required, but marginal cost factors are necessary in assessing alternative possibilities. Thus, accounting data must cover historical experience related to (1) total product cost and unit cost, (2) direct and indirect costs, and (3) the fixed and variable costs for each product. These factors should be readily available to enable a proper marketing analysis of the sales-volume-cost relationship and the establishment of a product selling price.

12.4 FINANCIAL STATEMENTS

Financial statements usually include the balance sheet and the income statement (or the profit-and-loss statement) and are prepared for both internal dissemination within the corporation and for external distribution to the stockholders, creditors, the government, and others with an interest in the financial position of the corporation. Figures 12-9 and 12-10 present a simplified version of a balance sheet and income statement for review.

A. Balance Sheet

A *balance sheet* is a statement giving the value of the assets owned by a corporation, the liabilities of the corporation, and the ownership equity at a particular time. Balance sheets are prepared at least annually but may be

ABC CORPORATION

Balance Sheet (As Of Some Date)

Assets		Equities	
Current Assets		Current Liabilities	
Cash	$ 300,000	Accounts Payable	$ 315,000
Marketable Securities	650,000	Provision for Taxes	135,000
Accounts Receivable	230,000	Total Current Liabilities	$ 450,000
Inventories	820,000	Other Liabilities	
Other Current Assets	200,000	Long-Term Debt	850,000
Total Current Assets	$ 2,200,000	Total Liabilities	$1,300,000
Other Assets		Ownership Equity	
Deferred Receivables	150,000	Capital Stock	775,000
Property, Plant, and Equipment (Net)	950,000	Retained Earnings	1,225,000
Total Assets	$ 3,300,000	Total Equities	$3,300,000

Figure 12-9. Abbreviated balance sheet.

XYZ CORPORATION

Income Statement (For Some Period Of Time)

Revenue from Sales	$4,500,000
Other Income	250,000
Total Income	$4,750,000
Costs and Expenses	
Costs of the Product Sold	$3,550,000
Costs of Selling the Product	200,000
Costs of Administration	330,000
Depreciation of Property, Plant, and Equipment	75,000
Other Expenses	30,000
Total Costs and Expenses	$4,185,000
Net Income (Revenue Less Cost)	$ 565,000

Figure 12-10. Abbreviated income statement (*profit and loss*).

prepared more often and on specific occasions when the need exists in connection with making decisions concerning large project investments, the sale or exchange of securities, dividend distributions, and so on. The balance sheet is divided into two sections with *assets* being stated on the left side and *liabilities* plus *stockholders' equity* shown on the right side.

1. Assets. *Assets* are usually classified as current assets, long-term investments, and fixed assets, such as property, plant, and capital equipment.

a. *Current assets.* Current assets include cash and other assets that, under normal business conditions, can be converted into cash within a short time frame (i.e., 1 year or less). Examples of current assets are (1) funds in checking accounts; (2) short-term investments, such as savings accounts, marketable securities, short-term government securities and treasury notes that mature within 1 year or less; (3) accounts receivable or credit extended by the corporation to its customers in conducting business; and (4) inventories.

Inventories may include raw-material inventories, purchased-item inventories, and finished product inventories (refer to Figure 6-2). The valuation of these inventories can vary considerably and may have a significant effect on both the balance sheet and income statement. Four commonly used inventory valuation methods are noted.

(1) The *first in-first out* (*FIFO*) method attempts to determine the cost of a "lot" of items based on the assumption that the first units entering the inventory are the first units to be issued in response to a given demand and that the last units to enter the inventory are still in the inventory at the end of the valuation period. Valuation of the inventory is based on the acquisition costs of those specific items actually remaining on hand.

(2) The *moving average* procedure is to add the cost of each new inventory acquisition to the total cost of the inventory previously at hand, to determine an average cost per unit by dividing this number by the total quantity of units, and to price all anticipated items that will be issued prior to the next acquisition at the resulting figure. Thus, the inventory is always valued at a single moving average cost.

(3) The *lower of cost of market* rule assumes that inventory valuation is based on replacement cost (or the current "market" price) rather than at the initial acquisition cost. This value in itself may vary considerably as market prices rise and fall.

(4) The *last in-first out* (*LIFO*) method is to assign inventory costs on the basis of purchases initiated within the *current* accounting period. In other words, current item acquisition costs are used in the valuation of the entire inventory.

It should be noted that these assets are *working assets* in the context that they are in a constant cycle of being converted into cash. Inventories when sold become accounts receivable, which in turn become cash when collected. Cash is used to pay debts and for operating expenses.

b. *Long-term investments.* Investments in bonds and stock designated for future long-term needs, deferred receivables, cash value of insurance policies, and so on, are included in this category.

c. *Fixed assets.* Fixed assets have a relatively permanent existence and are not readily converted into cash. These include *tangible* items, such as property, facilities, and capital equipment, plus *intangible* nonphysical items, such as goodwill, franchises, patents, and the like, which influence business conditions. Tangible items represent those assets not intended for sale that are used continuously in support of product design, test, manufacture, transportation, storage, etc.

Facilities and capital equipment are subject to depreciation, the subtraction of which from the initial acquisition cost results in the figure on the balance sheet representing the current value of the asset. Depreciation refers to an accounting method devised for the purpose of systematically distributing the cost of a tangible asset over the estimated useful life of the asset. Depreciation is not an out-of-pocket expense, but a bookkeeping charge that represents a reduced value of the asset with time. Depreciation also appears on the income statement as part of the costs associated with operating the business. Figure 12-11 is included as a simplified illustration of the concept of depreciation.

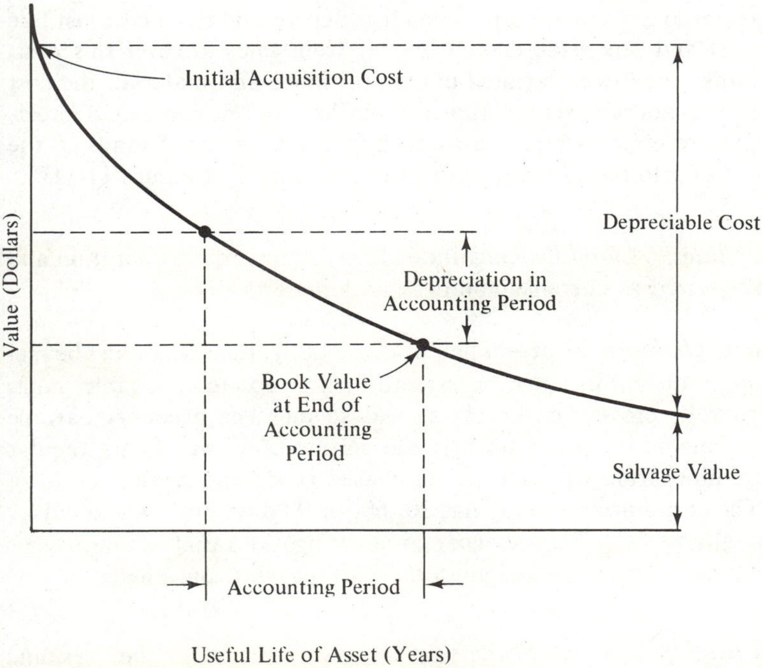

Figure 12-11. The concept of depreciation.

Three basic methods are commonly used in determining depreciation cost.

(1) The *straight-line* method of depreciation assumes that the fixed asset loses the same amount of value each year during its projected useful life. The estimated final salvage value of the asset is subtracted from the initial acquisition cost, and the balance is divided equally over the useful-life time frame to determine the annual depreciation charge.

(2) The *declining-balance* method requires that a fixed percentage of the asset's value at the beginning of the accounting period be charged off each year. The same percentage is used throughout; however, since the value of the asset is declining each year, the amount of the depreciation is higher in the earlier years and lower later.

(3) The *sum-of-the-digits* method is somewhat more complicated and constitutes an accelerated depreciation during the early years of an asset's life. The depreciation charge each year is determined by applying a changing percentage to the acquisition cost less the salvage value of the asset. The percentage charged is a fraction, where the numerator represents the remaining years of the asset's life at the beginning of the accounting period, and the denominator is the combined sum of all the useful years.

Depreciation is recorded in financial statements as a cost along with labor and material costs. The corporation lists depreciation as a cost just like any other cost and sets aside enough money from sales to cover this cost. However, unlike the direct payment of cash for labor and material, the cost of depreciation is not represented as a cash outflow. In this context, depreciation charges are often referred to as an important source of funds to the corporation (refer to the funds-flow diagram illustrated in Figure 11-11).

2. Liabilities. *Liabilities* constitute claims against the corporation and may be categorized as current liabilities and long-term debt.

a. *Current liabilities.* Current liabilities are obligations that must be met by the corporation within 1 year or less and include accounts payable, notes payable, payroll, provisions for taxes, and so on. The accounts payable category represents the amounts that the corporation owes to its regular business creditors from whom it has purchased goods and services on open account. The corporation usually has 30, 60, or 90 days in which to pay.

b. *Long-term debt.* This category covers obligations that extend beyond the 1-year time frame, such as long-term notes payable and bonds.

3. Ownership equity. Ownership equity represents the residual amount left after subtracting liabilities from assets. This amount is generally divided into capital stock (i.e., outstanding issues of stock) and retained earnings. Retained earnings are the profits earned by the corporation during the previous accounting period but not paid out to the stockholders. This surplus is often used to support new projects.

B. Income Statement

The *income statement,* or *profit-and-loss statement* as it is often called, is a report of the financial position of a corporation over a period of time (e.g., from January 1 through December 31, 19xx) that indicates primary sources of income during that period, costs, and the net profit or loss. Major income-statement entries are noted below (refer to Figure 12-10).

1. Total income. Income is acquired through direct product sales, the product sales generated by owned subsidiaries, the leasing of equipment, interest payments from loans and other investments, and/or the sale of portions of the business and other tangible assets.

2. Costs and expenses.

a. *Costs of the product sold.* This category generally relates to all direct costs associated with product planning, research, design, production, test and evaluation, product-life-cycle support, and material phase-out and disposal as applicable.

b. *Costs of selling the product.* Indirect costs in this area usually include salaries of sales personnel, sales commissions, travel, advertising and promotion, the packing and shipping of goods, office supplies, telephone and telegraph, and so on.

c. *Costs of administration.* All miscellaneous indirect costs are covered, such as the salaries of top-level and middle managers; some engineers pursuing basic research or assigned to staff functions; such organizations as accounting and finance, legal, personnel administration and industrial relations, laboratories; personnel fringe benefits; patent licensing costs, special taxes and rent payments, and facility maintenance and utilities.

d. *Depreciation of property, plant, and equipment.* As discussed earlier, depreciation costs are recorded in the income statement as well as on the balance sheet.

3. Net income or loss. Net income (or loss) is simply the difference between total income and all costs. In Figure 12-10, the net income value of $565,000 represents the net ordinary income before taxes. At an income tax rate of 30 per cent plus an additional surtax rate of 20 per cent of all earnings above $300,000, the estimated total income tax is $222,500, and the net income after taxes becomes $342,500. This figure is then used as a basis for decisions concerning the possible distribution of dividends to the stockholders. There is no set rate covering dividend distributions; however, the amount set aside is usually decided by a vote of the board of directors. The residual is retained within the corporation for the purposes of expansion, new project investments, and so on.

Financial statements in general will vary from corporation to corporation. Several actual examples, extracted from annual reports, are presented

General Motors Acceptance Corporation

and Subsidiaries

Condensed Consolidated Balance Sheet

December 31, 1974 and 1973

Assets	1974	1973
Cash	$ 271,292,640	$ 207,789,543
Marketable Securities	237,504,900	247,762,460
Finance Receivables (including instalments maturing after one year: 1974—$4,708,802,838; 1973—$4,678,807,737; less unearned income: 1974—$865,035,983; 1973—$784,428,384 and reserves for losses: 1974—$131,578,539; 1973—$126,803,446)	14,531,312,314	13,748,055,565
Insurance Receivables	25,860,086	27,143,040
Unamortized Debt Expense	19,941,136	19,516,614
Other Assets	67,236,814	39,404,324
Total Assets	$15,153,147,890	$14,289,671,546

Liabilities, Reserves and Stockholder's Equity				
Notes, Loans and Debentures Payable Within One Year (less unamortized discount: 1974—$22,450,728; 1973—$24,338,825)			$ 7,473,280,458	$ 6,198,390,324
Accounts Payable, Accrued Liabilities and Reserves				
General Motors Corporation and affiliated companies (including noninterest bearing loan of $500,000,000 at December 31, 1973)			1,359,792,896	1,696,627,830
Dealers			116,537,508	115,006,639
United States and foreign income and other taxes			50,242,062	29,250,424
Interest			119,994,977	106,137,047
Unearned insurance premiums			137,871,123	147,095,655
Insurance losses and claims			39,137,386	39,678,422
Other			59,982,252	51,765,140
Total Accounts Payable, Accrued Liabilities and Reserves			1,883,558,204	2,185,561,157
Notes, Loans and Debentures Payable After One Year (maturing prior to 2000—less unamortized discount: 1974—$10,693,486; 1973—$10,890,113)			3,843,976,817	3,952,330,172
Subordinated Indebtedness (maturing prior to 1993)			765,000,000	810,000,000
Stockholder's Equity				
Preferred stock, $100 par value (authorized and outstanding, 1,100,000 shares):				
6% cumulative			75,000,000	75,000,000
7¼% cumulative			35,000,000	35,000,000
Common stock, $100 par value (authorized and outstanding, 5,150,000 shares)			515,000,000	515,000,000
Net income retained for use in the business:	Year 1974	Year 1973		
Balance at beginning of the year	$518,389,893	$473,002,595		
Net income	105,980,018	88,424,798		
Total	624,369,911	561,427,393		
Cash dividends	62,037,500	43,037,500		
Balance at end of the year			562,332,411	518,389,893
Total Stockholder's Equity			1,187,332,411	1,143,389,893
Total Liabilities, Reserves and Stockholder's Equity			$15,153,147,890	$14,289,671,546

Figure 12-12. Sample balance sheet. (Annual Report, GMC, 1974.)

in Figures 12-12 and 12-13.[6] Although the major data entries are compatible with those indicated in Figures 12-9 and 12-10, variations will occur in the form of disclosing different information that is peculiar to each corporation but is considered necessary for reporting to the public. In any

[6]It is recommended that the reader review the annual reports for several large corporations and attempt to analyze them from the standpoint of the major considerations discussed in this chapter.

Standard Oil Company of California
Consolidated Statement of Income

	1974	1973
Revenues:		
Sales and other operating revenues (including excise taxes)	$17,924,357,000	$8,479,978,000
Equity in net income of certain affiliated companies	707,196,000	410,576,000
Other income	143,654,000	36,269,000
	18,775,207,000	8,926,823,000
Costs and other deductions:		
Purchased crude oil and products	12,766,039,000	3,675,515,000
Operating expenses	1,620,582,000	1,469,277,000
Selling, general and administrative expenses	537,416,000	518,702,000
Depreciation, depletion and amortization	510,352,000	406,278,000
Taxes other than taxes on income	1,521,320,000	1,459,125,000
Interest and debt expense	64,580,000	68,949,000
	17,020,289,000	7,597,846,000
Income before taxes on income	1,754,918,000	1,328,977,000
Provision for taxes on income	784,900,000	485,400,000
Net income	$ 970,018,000	$ 843,577,000
Net income per share of common stock	$5.71	$4.97

Consolidated Statement of Retained Earnings

	1974	1973
Balance at beginning of year	$ 3,941,711,000	$3,361,103,000
Net income for year	970,018,000	843,577,000
Cash dividends on common stock— $1.92½ per share in 1974 and $1.55 in 1973	(326,613,000)	(252,969,000)
Balance at end of year	$ 4,585,116,000	$3,941,711,000

(See Accompanying Notes to Financial Statements)

Figure 12-13. Sample income statement. (Annual Report, Standard Oil Company, 1974.)

event, whatever information is conveyed should be reported in an objective and consistent manner. Objectivity, consistency, and comparability are key factors in presenting financial information. Objectivity and consistency support the requirement for *reliable* information, and comparability is extremely important to investors, since decisions as to invest or not to invest depend heavily on the ability to compare the status of one corporation with the next.

12.5 AUDITS

Practically all business enterprises require, by law, periodic audits by an independent certified public accountant (CPA) to ensure fair and accurate financial reporting practices and to verify the traceability of data from the transactions recorded in journals to the entries noted in financial statements. The public accountant investigates the measures used in reporting revenues and costs as well as the methods employed in the valuation of assets. Finally, the accountant verifies that the practices used are consistent and that full disclosure is followed in the reporting of financial information to the public. Confirmation of formal audits is noted in financial reports through the inclusion of a statement to this effect signed by an authorized public accountant or public accounting firm.

QUESTIONS AND PROBLEMS

1. What purpose(s) does the accounting function serve?
2. Why is the availability of good accounting information essential?
3. Why should the manager of accounting (and the accounting staff) have a good understanding of the organization structure of the firm?
4. Assume that you are an engineering manager responsible for the initiation, organization, and management of a new project. How would you determine and specify the accounting needs that are required for project control? What elements of data would you likely select?
5. Assume that you are assigned as the manager of the accounting function in a newly established division of XYZ Corporation. How would you proceed in determining the overall accounting requirements for the division? How would you decide on what data are to be included and what should not be included?
6. What is he purpose of project budgeting and allocation? Why is it important?
7. The accounting function in the organization of a corporation is usually considered as a staff function. Why?
8. How do accounting data support project management and control?
9. How do project budgeting and control functions relate to the work breakdown structure?
10. Describe what is meant by "total cost." Why is it preferable to address the total-cost concept in all costing exercises?
11. What is meant by "unit cost"? Why is the unit-cost concept important?
12. Describe the basic differences between direct and indirect cost.
13. Define: direct labor, indirect labor, direct materials, indirect materials, and overhead. Include examples under each category.

14. Describe the basic differences between variable and fixed cost.

15. The reporting of a given cost item can be considered variable under certain conditions and fixed under other conditions. True or false? Why?

16. Define "product cost." What is meant by "marginal cost"?

17. Select a corporation or business of your choice, and attempt to identify the specific items of cost that are included under the direct and indirect categories.

18. Assume that you are responsible for preparing a cost proposal for a prospective project covering the development and production of system A. Identify by line item the estimated costs for the proposed scope of work. Include direct labor and direct material, the supplies and services of outside subcontractors, indirect factors, profit, and so on. Lay out the cost proposal as if you were about to submit it to a potential customer. Use hypothetical cost values if you wish.

19. Select a corporation or business of your choice, and construct a flow diagram (similar to the illustration in Figure 12-6) of the accounting process and major interfaces.

20. When determining an accounting capability, what considerations would apply in determining the application(s) for automatic data-processing techniques.

21. In reviewing the problems associated with many accounting procedures in existence today, the cause is likely to be associated with the computer. True or false? Why?

22. Describe the difference(s) between a journal and a ledger.

23. What is a chart of accounts? Select a large project of your choice and develop a project numbering scheme that covers all elements of work.

24. Refer to Figure 12-8. The status of project 844731 at point C indicates an overrun condition. As project manager, what would you do, if anything?

25. Describe the difference(s) between a balance sheet and an income statement.

26. Why is financial reporting to the stockholders, creditors, and the government necessary?

27. What is the difference between current assets and fixed assets? Current liabilities and long-term debt? How would you classify a computer used for the purposes of computer-aided design?

28. Describe four commonly used methods for inventory valuation. How does the selection of any one of these methods affect financial reporting?

29. Define "depreciation." Describe three commonly used methods for determining depreciation cost. How does the selection of any one of these methods affect financial reporting?

30. Objectivity, consistency, flexibility, and comparability are key attributes in accounting. True or false? Why?

31. What is the purpose of periodic audits? What benefits may be derived?

13

PURCHASING FUNCTION

The purchasing function is established within a corporation or business to handle all transactions that involve the outside procurement of materials and/or services through the implementation of purchase orders. In many corporations, this means that while large-dollar-value transactions in excess of x dollars are handled through formalized contracts, there are many instances when materials and services are procured through individual purchase orders.

13.1 PURCHASING PROCESS

Purchasing is one of the key functions in the successful operation of any business enterprise. The sources selected for the acquisition of materials and services and the methods used in procurement can have a significant impact on costs to the enterprise, particularly if the items procured are of inferior quality and the quantity delivered is not compatible with the needs. Purchasing objectives are to ensure that:

A. The materials and/or services needed are provided in the right quantity, at the right time, and at the right place.
B. The performance features, reliability characteristics, and quality of the materials and/or services needed are in accordance with the specified requirements.
C. The materials and/or services needed are procured from the right source and at the right price.

In most instances, the purchasing function is not responsible for determining what materials and services are to be procured, since the needs evolve from research, design engineering, production, test, and support activities throughout the organization. Like accounting, the purchasing function provides a *service* across many organization elements throughout the corporation. On the other hand, the purchasing function does include the necessary expertise to combine individual requirements where appropriate and to ensure that the best overall deal is made, on behalf of the corporation, in the acquisition of these requirements in total.

Figure 13-1 illustrates the basic functions and interfaces applicable in most instances where outside procurements are handled through purchase orders. A discussion of some key points in the purchasing process follows.

A. Purchase Requisitions

Purchase requisitions are initiated in the organization where the need exists and may cover the acquisition of materials, personnel services, and/or data. Purchase requisitions are usually accompanied by a procurement specification, a material specification, and/or drawings that cover the technical characteristics of the item (refer to Figure 3-7) and include such information as the quantity of units required, the delivery date, and the location where the delivery is to be made.

Purchase requisitions initiated by engineering functions often cover limited quantities of items, such as capital equipment to support laboratory research and testing needs, component parts for the fabrication of engineering models and prototype equipment, spare parts for equipment already in the inventory, and data to support design tasks. Quite often the descriptive material accompanying these requisitions is quite vague, as the engineer initiating the requisition either cannot adequately define what he needs or is unwilling to commit himself for one reason or another. This, of course, places quite a burden on the purchasing function, which should respond but in doing so, may end up by procuring the wrong item. Hence, it is incumbent that engineers initiating requisitions work closely with purchasing agents to ensure that the item(s) desired is adequately defined prior to proceeding to the next step.

On the other hand, purchase requisitions initiated in support of manufacturing operations often involve large quantities of one or more given items. Where large quantities of an item are required, the economic order principle is applied. The principle relates to optimization between the placing of many orders for an item, resulting in high material acquisition costs, and placing orders less frequently (with more items per order), maintaining a higher level of inventory, causing increasing inventory maintenance and carrying costs. In other words, ordering creates procurement cost, and inventory creates carrying cost. The economic order principle equates the *cost to order* to the *cost to hold,* and the point at which the combined costs are at a

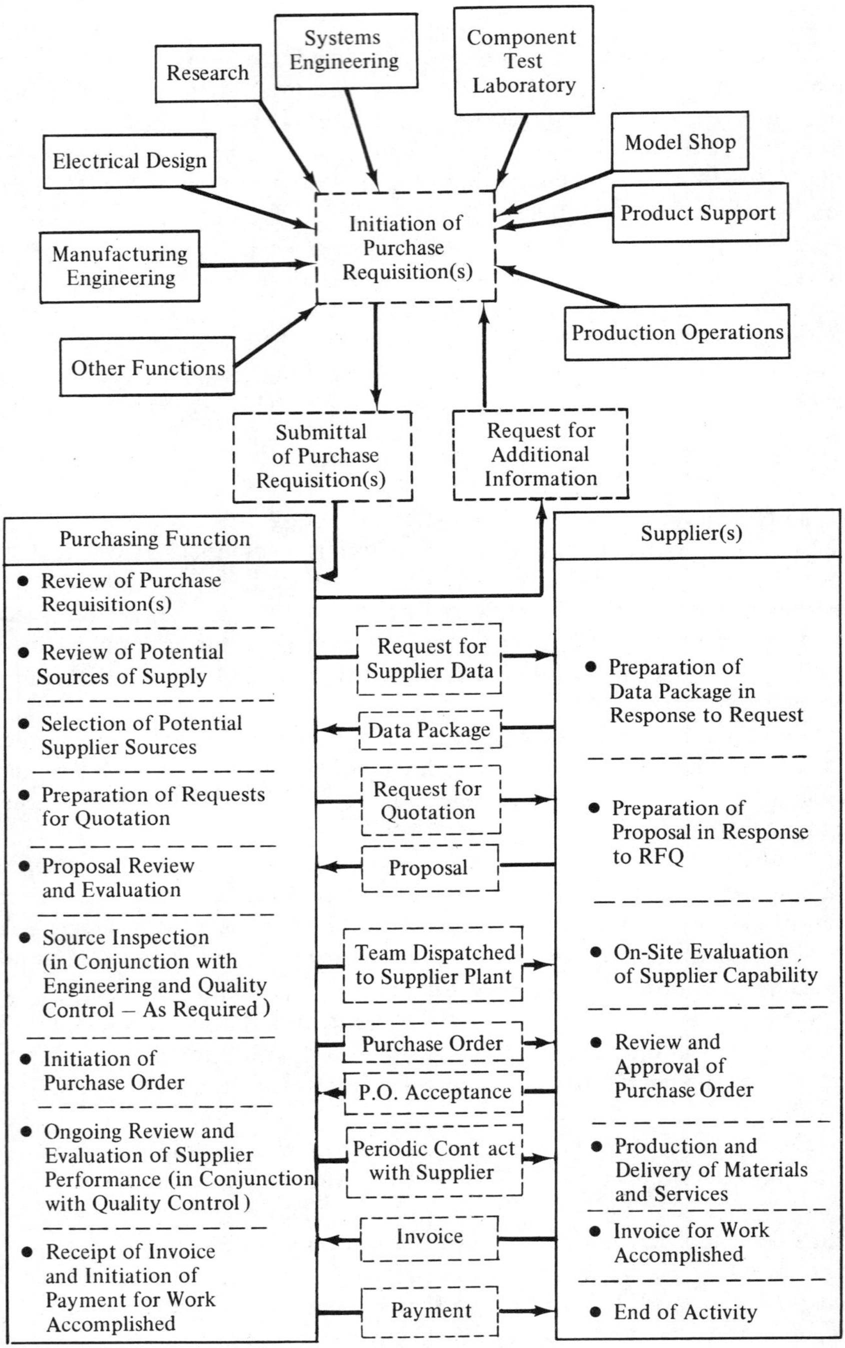

Figure 13-1. Major purchasing functions and interfaces.

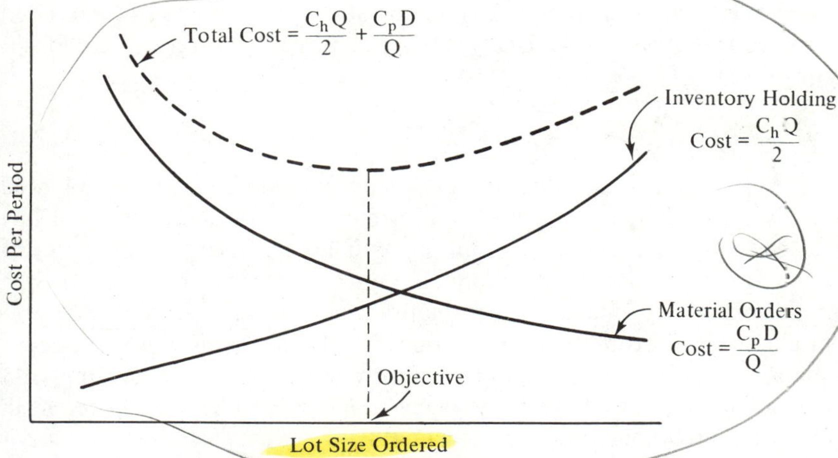

Figure 13-2. Economic order quantity relationships.

minimum indicates the desired amount to order. Figure 13-2 illustrates this concept. The illustration forms the basis for the size of orders placed, which is known as the *economic order quantity* (EOQ).[1] The total cost curve in Figure 13-2 is expressed as

$$\text{total cost} = \frac{C_h Q}{2} + \frac{C_p D}{Q} \tag{13-1}$$

Differentiating and solving for Q, which represents the quantity of units ordered, gives the following equation for EOQ:

$$Q^* = EOQ = \sqrt{\frac{2C_p D}{C_h}} \tag{13-2}$$

where C_p is the average procurement cost of ordering, in dollars per order (includes setup costs, processing orders, receiving, etc.), C_h the cost of carrying a unit in inventory (e.g., per cent of the unit price multiplied by the total unit price in dollars), D the annual demand in units (assumed as being constant throughout the time period), and Q^* the most economical quantity.

The purchasing function receives requisitions from many organizational groups throughout the corporation, evaluates the requirements in terms of commonalty, and considers the EOQ principle in initiating procurement actions. On occasion, requisitions will be held up until additional requisi-

---

[1]The EOQ concept presented is a simplified representation of the major considerations in establishing an inventory within a firm. However, there are many variations and additional material that should be reviewed prior to making a decision on a specific application.

tions covering the same items are received; however, care must be exercised to ensure that delays are not costly relative to scheduled project accomplishments.

B. Sources of Supply

The effective operation of the purchasing function depends on how well the individual purchasing agents are informed as to the best places to buy items. Maintaining a status quo philosophy is not sufficient, since the business conditions of individual suppliers may vary considerably from year to year. Suppliers may reduce product quality to cut costs or may shift management emphasis in promoting different product lines. In any event, the reputation of a supplier may change significantly in a short period of time, and products, although listed by the same part number, may be quite different in terms of material characteristics and quality of workmanship.

The efficient purchasing agent will maintain a continued surveillance of the activities and status of the various supplier sources for the items needed. The agent should solicit updated specifications and catalogs from current suppliers and should initiate an historical record to verify that (1) the items ordered are being delivered on schedule, (2) the items received are in accordance with the procurement or material specification, and (3) the supplier lives up to the negotiated contractual and warranty provisions associated with the product. In addition, the purchasing agent should attempt to acquire data from new suppliers entering the market. The net results of the agent's knowledge concerning the suppliers of various products can be summarized by a weighted evaluation and ranking method, listing the suppliers for each major product line in order of preference, with some suppliers being completely disqualified from consideration. The ranking approach must be dynamic, as preferences will shift from time to time.

C. Request for Quotation

With the identification of an acceptable list of suppliers for a given item, the purchasing agent initiates a *request for quotation* (RFQ), soliciting cost proposals from each of the suppliers selected.[2] The RFQ usually constitutes a data package that includes a complete specification describing the item (e.g., procurement or material specification), the quantity of items required, the delivery date, and the place where the delivery is to be made. In addition, the RFQ will specify a time limit within which the supplier must respond. Generally, the purchasing agent will attempt to solicit proposals from at least three suppliers in order to provide some flexibility in source selection.

---

[2]The terms *invitation for bid* (IFB) and *request for proposal* (RFP) are also used in soliciting proposals from potential suppliers.

Completion Contract negotiation

D. Source Selection

Upon receipt of supplier proposals, Purchasing in conjunction with engineering (or the organization that originally initiated the purchase requisition) will review each proposal to ensure compliance with the RFQ and will evaluate those which are acceptable against predetermined criteria. The approach used in the evaluation process is described in Section 3.4F4, and the criteria imposed will vary with the nature of the item being procured. For large items the criteria may be comparable to that which is specified in Figure 3-22, while smaller items may be rated in terms of performance and effectiveness characteristics, product quality, acquisition price and utilization cost, and supplier capability. The goal is to select a product that will meet all RFQ requirements and a supplier that is dependable. This does not necessarily mean that the supplier with the *lowest* proposed initial acquisition price should be selected, since product quality may be marginal relative to desired standards, and the management integrity of the supplier might be questionable. It is often more beneficial in the long run to invest a little extra at the start in order to realize greater overall effectiveness over the life cycle.

Supplier evaluation and source selection are usually accomplished as a team function through review of the proposals submitted (both those that are solicited and unsolicited), combined with a trip to the facilities of each potential supplier for the purpose of assessing supplier capability.. Since many proposals convey a certain degree of eloquence and salesmanship that may or may not reflect the *true* capability of the supplier, a visit to the supplier's facility is recommended in instances when prior knowledge in this area is lacking. On-site visits generally include representation from purchasing, engineering, and quality control. On occasion, the quality control organization may actually represent the design engineering functions on such visits.

E. Initiation of Purchase Order

Upon completion of source selection, Purchasing will initiate a *purchase order* (PO) covering the quality, quantity, delivery, and price for the item(s) desired. In addition, a PO, which upon acceptance by the supplier constitutes a contractual agreement, usually includes the specification of both state and federal laws that influence purchasing agreements. The PO is a legal document that must be signed by an authorized representative of the contractor and by his counterpart in the supplier's organization.

F. Review and Evaluation of Supplier Performance

The approved PO will legally obligate the selected supplier to provide X quantity of items, produced in accordance with specification ABC, at Y dollars per unit for delivery on a given date and at a designated location. Since compliance with these requirements is critical to the contractor's needs, it is incumbent on the contractor to ensure that supplier performance is in accordance with the PO. For relatively simple items, the contractor may send a quality control representative to visit the supplier's facility on a periodic basis to conduct spot on-site inspections and to assess progress. In the procurement of highly complex equipment, the contractor may assign a quality-control representative at the supplier's facility on a full-time basis. The intent is to assure that all subcontracted work is progressing in a satisfactory manner, as planned.

In the event that the assigned quality control representative detects problems in supplier performance, he will report his findings both to supplier management and to his own management for the purpose of initiating corrective action. Problems of a technical nature may require engineering support. Failure on the part of the supplier to correct a confirmed deficiency may necessitate legal action and will certainly influence the supplier's reputation relative to possible future business.

13.2 PURCHASING ORGANIZATION

Purchasing, like the financial and accounting functions, usually reports in a staff capacity and provides a service for many organization elements within a corporation or business. In providing this service, Purchasing must respond in some manner to numerous purchase requisitions initiated as a result of varying needs. Purchasing agents work closely with engineering and other functions to ensure that the identified needs are first validated and then fulfilled. The fulfillment aspect is accomplished through the processing of purchase requisitions, the acquisition of supplier data, the evaluation of potential suppliers and source selection, the issuance of purchase orders, and the processing of invoices and payments upon completion of work. In performing these functions, purchasing agents are acting on behalf of the corporation (as a whole) from an investment standpoint and on behalf of the organization group initiating the requisition in response to a technical need.

Purchasing agents must work with suppliers on a continuous basis in the procurement of specified products. As products are often quite varied in terms of complexity and function, the purchasing organization usually includes a number of purchasing agents who are knowledgeable in specific product lines and are assigned to handle purchasing actions in these areas. The purchasing organization can vary considerably in size and degree of specialization, depending on the depth and versatility of the product lines undertaken by the corporation.

On the other side of the spectrum, individual suppliers in their marketing roles will actively solicit business through the purchasing function of a corporation. Solicitations often result in attempts to attain a favorable position with respect to potential future procurements. The offering of personal gifts and the entertainment of purchasing agents are not an uncommon practice, and in some instances favoritism toward certain suppliers develops as a result of aggressive marketing tactics of this nature. Consequently, biases on the part of purchasing agents may inhibit the source selection process to the detriment of the corporation by choosing an inferior product. On occasion, purchasing agents have been known to actually change purchase requisition requirements to meet the qualifications of favored suppliers. Thus, extreme caution must be exercised to ensure that the personnel hired as purchasing agents are above reproach when faced with attractive outside influential situations.

QUESTIONS AND PROBLEMS

1. What are the basic objectives of the purchasing function?
2. Select a corporation or business of your choice and describe the purchasing process in that corporation or business.
3. As an engineer you have identified a need for certain equipment. What information would you provide in initiating a purchase requisition for the desired equipment?
4. As an engineer you are working with purchasing in source selection. What criteria would you use in the evaluation of potential suppliers, and how would you select the best source?
5. Assume that you as an engineer were requested to participate as a team member in conducting an on-site review of a supplier. What information would you seek while visiting the supplier's facility?
6. Determine the economic order quantity of an item for inventory replenishment when:
   (a) The cost per unit is $100.
   (b) The cost of preparing a shipment and sending a truck to the warehouse is $25.
   (c) The estimated cost of holding the inventory, including capital tied up, is 25 per cent of the inventory value.
   (d) The annual demand is 200 units.

   Assume that the cost per order and the inventory carrying charge are fixed.
7. Once a given supplier of an item (part 27134) has been approved as a preferred source, subsequent approvals of that supplier for reprocurements of the same item are not necessary. True or false? Why?
8. In source selection, why are visits to the facilities of potential suppliers considered important?
9. Once purchase orders are issued by the contractor and accepted by the supplier, they are considered to be legal contracts. True or false?
10. Identify some of the basic functions of the purchasing organization. Describe the major interfaces between the purchasing function and engineering, and between purchasing and suppliers. What cautions should be observed in maintaining relationships with suppliers?

14

PERSONNEL FUNCTION

The first official contact that engineers make with a corporation, business, government agency, academic institution, or any other organization where they are seeking employment is through the personnel function. When employed, the engineer continues to work closely with Personnel in dealing with formalized training programs, performance reviews and salary adjustments, job transfers, individual counseling, fringe benefits, employee and recreational services, labor relations, health and safety, and many other activities of a similar nature. As the engineer terminates his job to seek employment elsewhere, Personnel conducts exit interviews in an attempt to determine the reason for termination and will subsequently (through the time span of his career) serve as a reference source for other potential employers in their evaluation of him as a candidate for a job. The interfaces between the engineer and Personnel are continuous and varied, as discussed in Chapter 9, and the basic activities of Personnel are noted in Figure 14-1 and discussed throughout this chapter.[1]

Although personnel work per se is conducted throughout the organization (e.g., the daily activities between the manager and his subordinates), there is a need for a centralized group to establish policy and to administer the corporation's personnel program across organization lines. This need

[1] In many corporations, Personnel's responsibilities include personnel administration, wage and salary administration, and labor relations, all under the vice president, director, or manager of industrial relations. Organization structures differ; however, the basic activities are fairly consistent from one instance to the next.

References: Figure 10-3;
Chapter 3

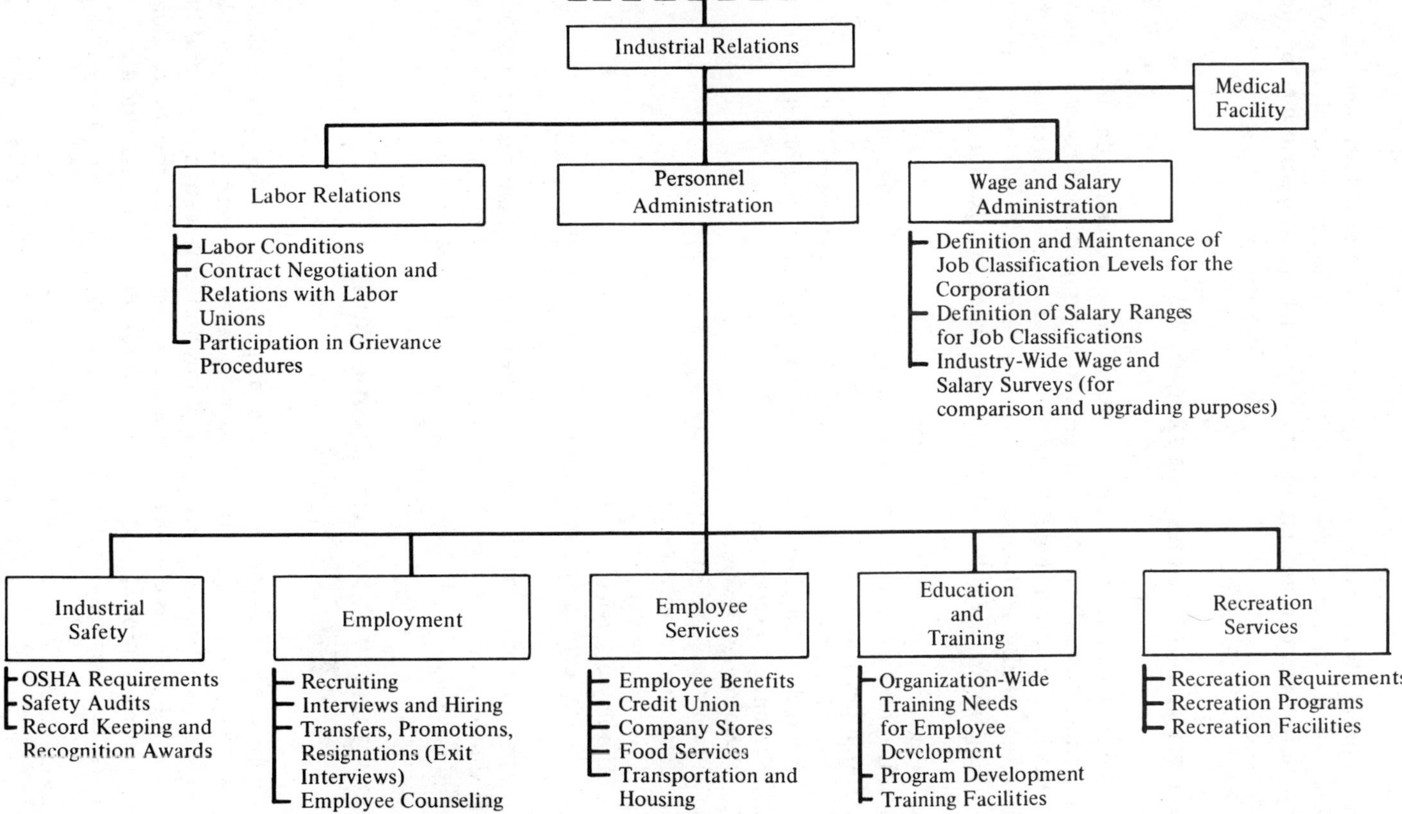

Figure 14-1. Industrial relations and personnel administration.

exists primarily to ensure that all employees are treated fairly and that consistency is maintained in the application of policies that affect all employees. Figure 14-1 presents an example of the functions usually performed by such a group.[2]

14.1 PERSONNEL POLICIES AND PROCEDURES

Personnel policies and procedures generally cover the day-to-day operations that affect all organization groups throughout the corporation and/or its divisions and are included in the *Policy and Procedures Manual* discussed in Section 10.2B4. Although not to be considered as all-inclusive, some of the more significant areas that should be addressed are listed as follows:

A. Personnel group functions and responsibilities.
B. Establishment of job classifications and the preparation of classification descriptions.
C. Establishment of salary ranges for job classifications.
D. Recruiting, employment, and the termination of personnel.
E. Personnel development.
F. Performance reviews.
G. Promotions and merit increases.
H. Personnel transfers.
I. Fringe benefits.
J. Employee services.
K. Health services and medical provisions.

These and other procedures should be readily accessible to all employees and should provide the necessary step-by-step guidance in response to the "what," "when," and "how" questions for the appropriate area.

14.2 ESTABLISHMENT OF JOB CLASSIFICATIONS AND WAGE/SALARY LEVELS

One of the initial steps in developing personnel policies is the establishment of job classifications and a salary or wage-rate scale for each classification. The classification process constitutes an identification of *general* titles for the various levels of expertise that will be required by the corporation to perform both project and staff activities. Personnel classifications are defined to cover jobs that require comparable skill levels throughout the

---

[2]A good reference dealing with personnel in depth is: G. Strauss and L. Sayles, *Personnel—The Human Problems of Management,* 3rd ed., Prentice-Hall, Inc., Englewood Cliffs, N.J., 1972.

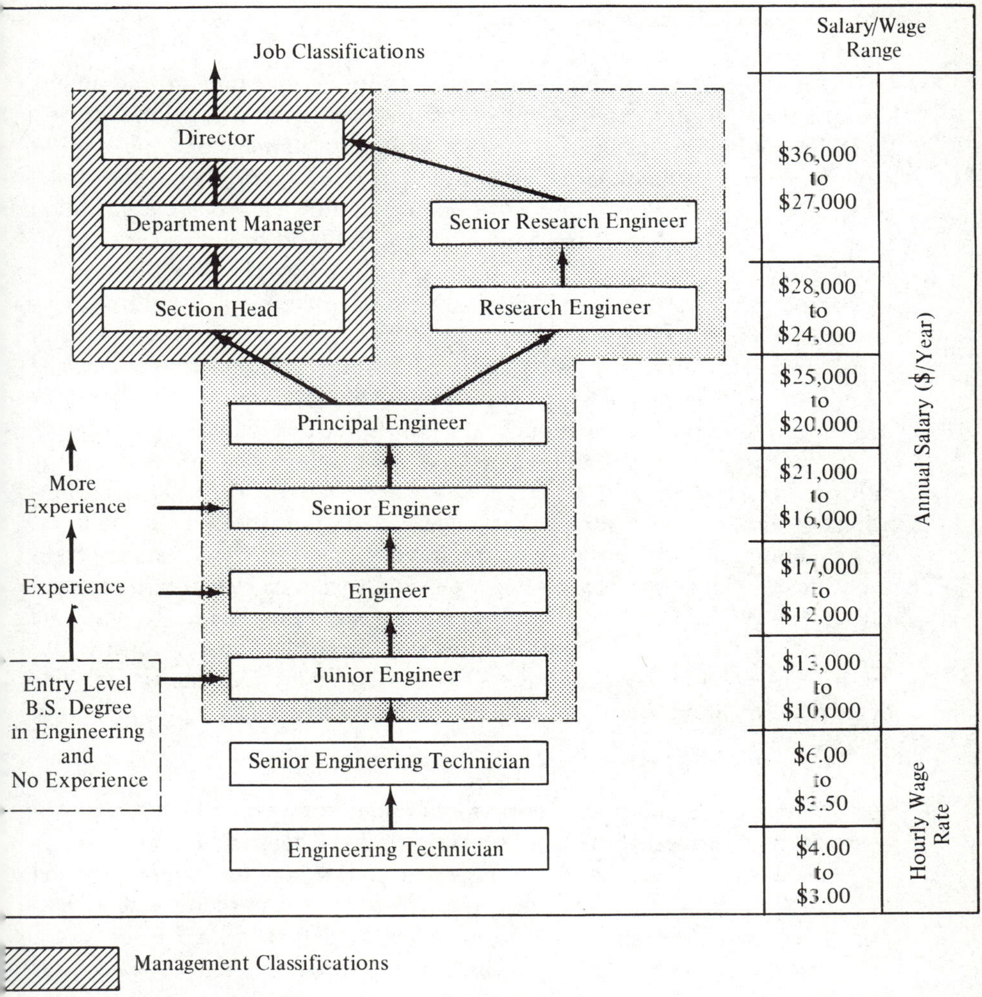

Figure 14-2. Partial job classification scale.

corporation and serve as a frame of reference for initial employment and employee growth through promotion. Although variations occur with the nature of the business and general market conditions, one partial set of job classifications is presented in Figure 14-2.[3]

[3]Figure 14-2 presents a partial hierarchy of position titles, oriented to engineering activities, for a corporation. The ordering of positions as shown does not necessarily reflect a job reporting scheme. For instance, senior engineers may report directly to section heads and junior engineers may report organizationally to principal engineers. It is not necessary that engineers report to next higher level, as shown.

The progression in engineering starts with the technician and moves upward through the principal engineer.[4] If an individual is promoted to a classification above that level, he either assumes a management position or accepts a highly scientific research-oriented job. In any event, Personnel (through wage and salary administration) will establish a broad classification such as "Senior Engineer" and will define the position through a set of minimum entry criteria. For instance, to be qualified for this classification, an individual must have at least a baccalaureate degree in some field of engineering plus 5 to 10 years of experience related to the area of need. Within this general classification, individual engineering functions throughout the corporation will develop position descriptions that identify *specific* objectives and entry requirements for a particular job (refer to Figure 9-11).

Concurrently, wage and salary administration must systematically address a feasible compensation policy for each classification. The combination of *wage and salary surveys,* designed to determine the general pay levels in the community and throughout the industry, and *job evaluations* establishes the relationships between pay levels and various jobs within the corporation. Job evaluations are initiated to equate task complexities to pay. The pay scale should be high enough to permit the corporation to attract and retain employees and yet not so high as to create unnecessary costs. The determination of an overall wage/salary policy is influenced by the following factors:

A. Employment conditions both in the community and nationally will affect pay scales. If the labor market is tight, a corporation may have to pay a little more than the average rate to compete for highly qualified candidates, whereas high unemployment will cause the reverse effect.

B. The corporation may wish to establish a high pay scale to attract a greater number of job applicants and allow for better selection. Of course, once that engineers are hired, pay alone is not sufficient to guarantee motivation and employee retention (refer to the hierarchy of needs discussed in Section 9.4B).

C. The corporation's business posture will influence pay scales. When income and profits are high, the pay scale is apt to be a little higher than at times when the corporation is losing money.

D. Corporations whose employees are highly unionized are often forced to pay high wages/salaries because of labor union pressures and collective

---

[4] In many organizations, the classification Junior Engineer is considered as the first entry level and as a salaried employee requiring an engineering degree for qualification. Technicians, who assist engineers during various phases of the life cycle, are considered as hourly employees, where a formal degree is not necessarily required but a certain amount of experience is desired. However, many technicians will work toward a degree through night school (or equivalent) to allow for future growth into the engineering ranks.

bargaining. In addition, nonunionized enterprises may initiate a high pay scale in an attempt to keep out labor unions. More will be said about labor unions in Section 14.7.

Wage and Salary Administration includes consideration of the above factors in establishing pay scales for different classifications as illustrated in Figure 14-2. The pay range for each individual classification overlaps somewhat with adjacent classifications, to provide some flexibility in employment, promotions, and/or merit increases.

Pay scales associated with corporations and businesses are reviewed periodically and adjusted, as necessary, with the market conditions and inflation. It is often appropriate for representatives from different corporations in the same community or industry to get together and agree on classifications and entry levels in order to minimize tendencies toward personnel piracy and poor recruiting practices, which may turn out to be quite costly in the long run (e.g., Corporation ABC revises pay scales upward for a short period to gain certain short-term recruiting advantages).

For government agencies, pay scales are established by the federal or state governments and are standardized by civil service grade, except for periodic percentage increases to cover inflation. Within each grade (or classification) there are designated incremental steps that are tied to "years of employment." Progression on the job is usually automatic within a given grade and merit increases are fixed, independent of job accomplishments.

14.3 EMPLOYMENT

Employment, as discussed herein, refers to the steps necessary in filling one or more vacant positions in the organization as identified by individual position descriptions initiated by the project or staff function with the need. In response, Personnel has two choices:

A. The identified position requirement(s) may be filled from within the corporation through promotion, provided that qualified individuals are available.
B. The identified position requirement(s) may be filled from outside through recruiting.

Promotion from within usually allows for a better selection, since employees that have been in the organization for awhile are known quantities, and their knowledge of internal operating procedures is certainly advantageous. In addition, opportunities for promotion from within stimulate individual motivation and overall job satisfaction and result in less cost to the corporation, as outside recruiting is more expensive. On the other hand, outside recruiting allows for the introduction of new blood into the organization, promoting new ideas and reducing possible internal stagnation. Also, it allows for the direct employment of an individual already trained for a given position, whereas selection of an internal employee may necessitate formalized training to qualify him for job entry.

Actually, there should be an optimum mix of newcomers and old-timers, and this mix is dependent somewhat on the extent to which personnel development is taking place within the organization (i.e., formal training, upgrading through participation in technical societies and industry associations, self-study, etc.). This is particularly true for engineering and related scientific fields, where the state of the art relative to technology is moving ahead at a rapid pace. If engineers within the organization are remaining current in the technical specialties being pursued by the corporation, management should encourage promotion from within and limit the need for outside recruiting to new areas, where internal expertise is not available. Conversely, if complacency exists and engineers are not abreast of the technology, management is obliged to promote the external recruiting option.

Unfortunately, in most corporations this optimum balance does not exist. First, there are the trade-offs between external recruiting versus internal personnel development. Sometimes it is difficult for supervisors to encourage personnel development, since this may be considered a hindrance to the accomplishment of daily project tasks. Thus, the employee concentrates his energies on short-term ventures and may soon find himself to be obsolescent. Second, qualified personnel working within the organization are often designated by their immediate supervisors as being "unavailable," even though it would be beneficial for the corporation to promote from within. Supervisors frequently consider their own activities on a higher priority basis than the welfare and growth of an individual; thus, highly qualified candidates are detained (unless top management intercedes) while less competent persons are put on the availability list. Hence, Personnel will initiate the outside recruiting approach in order to satisfy the requirement. This, in turn, may cause increased dissatisfaction within the corporation when qualified personnel learn that they have been, or are likely to be, passed by relative to possible opportunities for promotion. The net result will be resignations, a higher personnel attrition rate, and higher costs to the corporation, owing to the necessary increase in recruiting activity.[5]

---

[5]Past experiences in this area have led to the accepted practice of "job hopping," particularly by engineers and executives. Unfortunately, it has been proved in many instances that an individual can progress further and at a faster pace by changing jobs. He or she usually remains in his profession but practices in different organizations. Perhaps this trend will be reversed somewhat as management becomes more aware of individual needs, the needs of the organization, and the matching of needs in both areas.

Whatever the case may be, there is always the need for employment in both categories. Internal promotions are discussed further in Section 14.6, and outside recruiting is covered below.

A. Sources of Technical Personnel and Recruiting Plan

In outside recruiting, Personnel must consider the nature of the labor market and respond to the following questions:

1. What are the possible sources for the technical skills required?
2. What is the probability of being able to attract the necessary skills from these sources?
3. What geographical boundary areas should be pursued in terms of active recruiting?
4. What method should be used in contacting possible candidates?

Sources of technical personnel may include new engineers and scientists graduating from college, college faculty members, engineers currently employed by other organizations, unemployed engineers who are seeking employment, and retired technical personnel. The identification of a source must consider *all* qualified individuals, independent of sex, race, and creed. In other words, fair employment practices dictate that equal opportunities be provided for women and minority groups (e.g., blacks, orientals, American Indians, Latin Americans). Further, because these practices have not been commonly followed in the past, federal and state agencies have enforced legislation that provides guidelines to ensure that equal opportunities are made available for all technically qualified individuals relative to recruitment, selection, training, promotion, and so on.[6] In compliance with the intent of these guidelines, corporations and agencies are being required to prepare an *affirmative action plan,* indicating the positive action that will be taken to reduce discrimination, even at the risk of increasing costs and reducing productivity in the organization. The affirmative action goals are multiple; however, one example is to equate the total minority workforce within a corporation (as a percentage) to the surrounding community and/or national population makeup; another is to ensure that minorities are proportionately represented at each level in the organization.

---

[6]There are a number of federal government agencies responsible for enforcing equal employment opportunity laws, including the Office of Federal Contract Compliance in the Labor Department and the Equal Employment Opportunity Commission. State agencies are also responsible for enforcing equal employment opportunity practices. Without the enforcement of these laws, little progress would have been made toward improving fair employment for women and minorities.

With the identification of personnel resources, one must assess the probability of being able to attract qualified individuals. The attractiveness of a corporation or agency to potential candidates is certainly a key factor in determining the extent of recruiting effort required and the geographical boundaries where active recruiting is to be pursued. This factor is highly dependent on the corporation's reputation concerning starting salaries and salary growth potential, work challenge, fringe benefits, stability of employment, facilities, and location. If the corporation enjoys a good reputation, the relative effort expended in recruiting should be minimal, and the geographical areas in which recruiting is to be accomplished will be limited. On the other hand, a poor reputation is difficult to overcome and will require an extensive effort on the part of the corporation to generate the necessary interest. Thus, on the basis of potential resources and the ability to attract, Personnel will develop a strategy and a formal plan to meet organization employment needs.

B. Contacting Technical Personnel

There are numerous sources of technical personnel and many ways of reaching them. Some of the more common methods for communicating needs are noted below.

1. Direct advertising. Personnel may place formal advertisements directly with newspapers, technical magazines, professional journals, and so on, describing the specific position needs of the corporation. Personnel will select the media used based on the highest expected return, which requires some knowledge of market conditions and the possible sources of technical expertise.

2. Working through employment agencies. Employment agencies are in business for the purpose of placing individuals in jobs, and as a result are knowledgeable of available technical personnel resources. The personnel function in a corporation may specify its position needs to an employment agency, which in turn will initiate communications through direct advertising or via some other means. Generally, Personnel will pay the employment agency a fee if an individual is hired as a result of this effort.[7]

---

[7]Since employment agencies make their profit through personnel placements, the organization with the need is likely to receive many résumés and/or calls from individuals seeking employment. Although a match between the advertised position and a qualified individual may result, the majority of résumés submitted will not in all likelihood be responsive. Also, the employment agency, in attempting to attain its objective, will often exert pressure for employment confirmation. Personnel and the corporation/agency manager responsible for recruiting should be aware of this possibility.

3. *Informal personal contacts.* One of the most effective means of communicating job needs is accomplished informally with a personal contact over a cup of coffee, at technical meetings, during vacations, or through telephone conversations. Engineers deal all the time with other engineers in different organizations, and after awhile they are well aware of who's who in their respective fields of expertise and in related areas.

Informal contacts, involving inquiries concerning job interest, are made frequently, although these types of contacts are not officially sanctioned by the Personnel function. It is in the interest of Personnel to keep recruiting activities for the corporation out in the open; otherwise, personnel piracy takes place, and distrust among various organizations in the community builds up. However, what usually occurs is that a personal contact results in an interested individual approaching Personnel on his own to inquire about job possibilities. When this happens, the personnel function can respond officially.

4. *Formal recruiting in the field.* Depending on the extent of corporation needs, Personnel may contact prospective candidates at a time and location coincident with a large convention, symposium, or trade fair. Personnel representatives will usually rent a room at or near the convention site and will advertise such beforehand.

5. *College recruiting.* Personnel, management, and/or engineering representatives from many corporations and agencies will visit college campuses to discuss possible job opportunities after graduation.[8] Contact is made with the placement office at the college(s) selected, a date for an on-campus visit is set, and the placement office will notify students of the prospective visit and make appointments accordingly. When the recruiters arrive on the campus, they will usually have a full slate of informal interviews.

Probably the most common means of reaching technical personnel of all categories are through direct advertising and the use of employment agencies. A few examples of direct advertising both by corporations and employment agencies are illustrated in Figures 14-3 and 14-4. Job advertisements of this type usually appear in the Sunday edition of large city newspapers or in professional journals with wide circulation. However, the ultimate strategy employed by the corporation will usually include a mix of formal advertising, college and field recruiting, and responding to unsolicited individual contacts.

---

[8]Some corporations will select engineering managers or high-level engineers to visit college campuses, since they can usually relate quite well with young prospective engineering graduates. In addition, the college student will observe "success in the outside world" on a first-hand basis.

EMPLOYMENT OPPORTUNITIES

TODAY'S ENGINEER....

. . . we need you to design tomorrow's aerospace products

Sundstrand Aviation, with headquarters in Rockford, Illinois, is growing. Currently, we need 40 more engineers. A division of Sundstrand Corporation, Sundstrand Aviation is a leading supplier of mechanical, hydraulic and electrical components and systems for aircraft throughout the world. Right now we have openings in these areas:

Research

—3-10 years' experience
—Degrees in mechanical, aeronautical, control or electrical engineering
—To do preliminary design, development and testing of advanced turbomachinery, fans and compressors; hydraulic and pneumatic control systems; and electrical system controls

Quality Assurance

—1-3 years' experience
—Degree in metallurgy or metallurgical engineering
—To function as project manager on a wing sweep actuation system

Design

—3-10 years' experience
—Degrees in mechanical engineering or engineering mechanics
—To do mechanical design of aircraft accessory equipment, including hot gas and pneumatic turbomachinery; environmental control systems; accessory gearboxes; actuation systems, etc. Specialists in analytical and experimental stress analysis, fracture mechanics and damage tolerant design also needed.

Marketing

—Minimum 3 years' experience
—Degree in mechanical engineering
—To handle product line marketing for flight control equipment, aircraft actuation systems, pneumatic turbomachinery, and fuel systems

Test Equipment

—3-5 years' experience
—Degree in mechanical or electrical engineering
—To design turbine, hydraulic, and electronic test equipment

Project

—3 or more years of experience
—Degree in mechanical, aeronautical or electrical engineering
—Mechanical—to design, develop and qualify turbine systems or their components; develop organic rankine cycle total energy systems.
—Electrical—To design solid state, $400H_2$ power control circuits; to design and package $400H_2$ electromagnetic components

Sundstrand Corporation provides competitive salaries combined with an excellent benefit program. Qualified candidates from within the Company are given ample opportunities for advancement. Rockford is a pleasant community near the Wisconsin border with a population of 150,000. Our community offers abundant recreational and cultural activities, plus easy access to Chicago and southern Wisconsin's recreational areas.

Send your resume or request a professional employment application.

Mike Kofoed
Employment Manager

SUNDSTRAND
Corporation
4751 Harrison Avenue, Rockford, Illinois 61101
An Equal Opportunity Employer

Taken From Aviation Week & Space Technology

Figure 14-3. Classified advertising.

ENGINEERS F/PD TO $16,000

COMPUTERS

New Prod Design From Boards To Production Graphics Industry-Mech-E/M-Circuits Hardware Software 1 Specialist Each Category-Excellent Co N.Y.Metro Area-Top Benefit Package.
ROSS personnel agency
20 rte 110 melville LI 516-549-5656

ENGINEER-CHEMICAL

Degree and exper. in process development or research. Call 301—821-6633.

ENGINEER MECH-PROJECTS

Our client a Fortune 500 Co. is avidly seeking a degreed individual with 4-7 yrs exp in high speed design machine fabrication. Exp in any of the following desirable, tape slitting, plastic blow & injection molding. Please send resume including salary history in strict confidence to
PO BOX 2066
Central Station, E. Orange, NJ 07019
Our Client Is An
Equal Opportunity Employer

CAREER OPPORTUNITIES

Several well known companies, locally and elsewhere in the US have retained us to locate top caliber, professional personnel for both current and projected long term assignments. Areas of interest include (but are not limited to):

Energy R & D	$14-$30K
Coal Liquefaction Systems	$15-$28K
Environmental Systems	$15-$28K
ELINT Systems	$14-$25K
Law Enforcement Systems	$15-$28K
Satellite Communications	$12-$28K
Digital Communications	$12-$27K
Digital-Computer Systems	$12-$25K
Weapons Systems	$12-$28K
ECM/EW Systems	$12-$29K
Electronics Systems Test	$12-$22K
Systems Intergration	$12-$26K
Mini-Computers	$12-$27K
Configuration Management	$14-$26K
Intergrated Logistics Support	$14-$25K
GFCS, UWCS, NTDS, JPTDS, WDE, WDS, & ASMD Systems	$14-$25K
Ecological Impact Analysis	$17-$25K
Systems Effectiveness	$14-$25K
Avionics System	$14-$26K
Radar Systems	$14-$28K
Highway Safety Analysis	$17-$25K
Intergrated Circuit Design	$12-$22K
Digital Analog Circuit Design	$12-$24K
Digital Logic Design	$12-$24K
Receiver/Transmitter Design	$12-$25K
Micro Wave Circuit Design	$12-$22K
RF Circuit Design	$12-$23K
RFI/EMC Test & Analysis	$12-$22K
Signal Processing	$12-$25K
Computer Systems Analysis	$11-$28K
Software/Systems Devel.	$12-$25K
AN/UYK-7 Or AN/UYK-20, O.S.	$14-$25K
Programmers—All Types	$9-$22K
Management Consulting	$16-$28K
Economists/Econometricians	$15-$32K

Our fees are entirely assumed by our client companies at NO COST WHATSOEVER TO YOU! Candidates must have at least a Bachelor's Degree, be a U.S. Citizen and have at least one years experience since receiving Degree. Act Now! Send detailed resume, in confidence, (including salary history) to: Donald M. Wallach, President or call him at (301) 762-1100.

WALLACH ASSOCIATES

PROFESSIONAL STAFFING CONSULTANTS
1010 Rockville Pike, Rockville, Md. 20852
(A Licensed, Affirmative Action Agency)

ENGINEERS — DESIGNERS

PROJECT MANAGERS
Cut and cover subway projects

SUPERVISING & PROJECT ENGINEERS
Rail transit and railroad projects

BRIDGE ENGINEERS & DESIGNERS
Rail experience helpful

ELECTRIC TRACTION POWER & SIGNAL ENGINEERS

Positions available in Philadelphia, Baltimore, Boston & Buffalo. Your resume will be treated with complete confidence.

C. JONES

URBAN ENGINEERS, INC.
19th Street & Delancey Place
Philadelphia, Pennsylvania 19103
An Equal Opportunity Employer M/F

ELECTRONIC ENGINEER

Ground floor opportunity for talented Servo Mechanisms Engineer with DC Drive Design experience. Digital knowledge would be a plus. Beautiful Virginia location, challenging technical environment. Lucrative profit sharing program. Write in complete confidence to:
M25868 Electronic News
An Equal opportunity employer

ENGINEER/Civil Fee Pd $Open
Some drafting experience.
ACCURATE agency 41E42st., 986-5805

UPSTATE NEW YORK

We are a multi-man electronic rep organization selling to all major OEM's in the area. We are in need of:
Transformers
Stepping Motors
Resistor - film - power
for major programs now in design. We'll be attending IEEE. Please arrange for appointment, but we could discuss now. Send particulars to:
M34993, Electronic News

ENGINEERS

PROJECT
(Food Processing)
FACILITIES
(Pharmaceuticals)
PROJECT MGRS
(Power or Process)
CHIEF ESTIMATOR
(Power-MidWest)
MAINTENANCE
(Process Industry Exp)
COST ENGRS
(Power or Process)
MECH'L/NUCLEAR
(Power Plants)
STRUCTURAL
(Concrete/Steel)
QUALITY ASSURANCE
(Nuclear Codes)
FEE PAID $22-40,000

BENNETT & MASTERS
PERSONNEL ASSOCIATES
101 Park NYC 10017
(212) 683-5154

DEVICE DESIGN ENGINEER

Engineering position available in our Device Development group to design & develop state-of-the-art microwave power transistors.

Experience in RF power, small signal, or low-frequency transistor design & development required.

Reply by resume only,
stating salary requirements to:

LINDA MORGAN

MICROWAVE SEMICONDUCTOR CORP.
100 School House Road
Somerset, N.J. 08873

MSC An Equal Opportunity Employer

Figure 14-4. Classified advertising (*examples*).

C. Personal Résumés, Application Forms, and Interviewing

After initial contacts have been made, each job applicant may follow a one- or two-step process. If there is interest in employment and such interest was generated through an advertisement or as a result of a brief field contact, the corporation or agency soliciting technical personnel may desire to see a *résumé* of the applicant's background and experience, to determine whether there is a possible match between the applicant and the position need before proceeding. The résumé, submitted to Personnel, will be reviewed by the technical functions within the corporation. Résumés are particularly relevant when technical jobs of a specific nature and/or high-level positions are being advertised and prior job experience is required. In the event that the response is favorable, the applicant will be requested to complete a formal *application form* and will be invited to visit the corporation's facility for a formal interview. On the other hand, submission of a résumé may be omitted if enough information was exchanged during an initial contact or if the advertised position(s) does not require prior experience, as is the case for a new college graduate. In these instances, the applicant will complete only the formal application form for submission directly to Personnel.

*1. **Résumés.*** The preparation of résumés in response to an advertised position that requires previous experience is an *art*. First, the résumé must include the following basic information:

a. Name and address of applicant.
b. Current job title or position.
c. Future job objectives.
d. Educational data (formal degrees and continuing education).
e. Employment experience (describing the most recent experience first).
f. Publications and special presentations.
g. Professional affiliations (technical society and industry association accomplishments).
h. Significant honors and awards.
i. Military history.

As a second point, the résumé should be *tailored* to the maximum extent possible (without altering the truth) for the job position being solicited. Experience oriented to job needs should be highlighted, with the inclusion of specifics where possible. Organizational experience (e.g., project and staff activities) should be described in such a manner as to facilitate recognition and a favorable response on the part of those reviewing the résumé and responsible for making decisions as to whether an interest prevails. Tailoring is extremely important and often provides an applicant with a distinct advantage over those who submit a *broad* resume that covers an all-inclusive overview.

Finally, résumés that are too voluminous and contain trivia will in all probability not be read. There are numerous instances when an individual, in attempting to justify his case, will overdo it and list every detail, relevant or not. Such an individual will probably not be considered further. Thus, it is recommended that résumés be limited to three pages or less.

2. Formal job applications. Independent of the preparation and review of personal résumés, most corporations and agencies require that a candidate complete a standard application form. This ensures that certain essential information is made available to the corporation and that the completed data are presented in a consistent format to facilitate review by the personnel function. The application form may be supplemented with a personal résumé when it is appropriate to present additional information under certain categories (e.g., an elaboration of prior employment experience). Application forms will vary from organization to organization; however, much of the information requested is the same. An example of a standard application form is presented in Figure 14-5.

3. Interviewing. After a review of the résumé and the standard application form, Personnel and the technical manager defining the position requirement will make a decision as to whether to invite a candidate to the corporation's facility for a personal interview. If positive, an agreement will be made as to time and date, and in most cases all expenses for the candidate's trip will be paid by the corporation. When the candidate arrives at the corporation's facility, he or she should report to Personnel for a general indoctrination on personnel policies, employee benefits, education and training opportunities, employee and recreation services, and other areas of interest involving the corporation and the community. In many corporations, the conductance of testing is a required practice for candidate selection, and Personnel will have the candidate take some tests during the early stages of the interview process. Testing categories may include the following:

a. *Performance tests,* designed to determine a candidate's proficiency relative to a particular job. These tests are generally given to candidates seeking jobs in manufacturing and/or in administrative areas, where defined skills are required.

b. *Aptitude tests,* designed to measure a candidate's capacity to learn the skills required in a given area.

XEROX EMPLOYMENT APPLICATION

Xerox Personal Data

INSTRUCTIONS

1. Type or print in black ink.
2. Answer each question.
3. Read declaration carefully – then sign and date form.

FEDERAL, STATE AND MUNICIPAL LAWS PROHIBIT DISCRIMINATION BECAUSE OF RACE, COLOR, RELIGION, SEX, NATIONAL ORIGIN, OR AGE.

NAME LAST	FIRST	INITIAL	SOCIAL SECURITY NO.	DATE
PRESENT ADDRESS STREET	CITY	STATE ZIPCODE	HOME PHONE: INCLUDE AREA CODE	ALTERNATE PHONE: INCLUDE AREA CODE
PERMANENT ADDRESS STREET	CITY	STATE ZIPCODE	WHO REFERRED YOU TO XEROX?	

IF NOT A CITIZEN OF U. S., CURRENT VISA STATUS?	HOW WOULD YOU DESCRIBE YOUR GENERAL HEALTH?

WHAT TYPE OF WORK DO YOU PREFER?
1. 2.

GEOGRAPHIC LIMITATIONS	GEOGRAPHIC PREFERENCES	GIVE NAMES OF CLOSE RELATIVES EMPLOYED BY XEROX 1. 2.

HAVE YOU EVER BEEN CONVICTED OF A FELONY? (RECORD OF CONVICTION DOES NOT DISQUALIFY THE APPLICANT FROM EMPLOYMENT CONSIDERATION) ☐ YES ☐ NO — IF "YES" PLEASE EXPLAIN:

MAY WE CONTACT YOUR PRESENT EMPLOYER? ☐ YES ☐ NO	WAGE OR SALARY REQUIRED	DATE AVAILABLE	HAVE YOU EVER APPLIED FOR WORK AT XEROX? ☐ YES ☐ NO IF YES, WHEN AND WHERE	HAVE YOU EVER WORKED AT XEROX? ☐ YES ☐ NO IF YES, WHEN AND WHERE

EDUCATIONAL DATA

CIRCLE HIGHEST GRADE COMPLETED — ELEMENTARY 1 2 3 4 5 6 7 8 — HIGH SCHOOL 9 10 11 12 — COLLEGE 13 14 15 16 — GRADUATE 17 18 19 20 21 22

SCHOOL NAME AND LOCATION	YEARS ATTENDED		GRADUATE?	DEGREE	MAJOR SUBJECT OR TOTAL HOURS
HIGH SCHOOL	FROM	TO			
COLLEGE	FROM	TO			
COLLEGE	FROM	TO			

OVERALL COLLEGE SCHOLASTIC AVERAGE/DEGREE	CLASS STANDING (UPPER ¼, ½, ETC.)

EMPLOYMENT AND MILITARY HISTORY

IN EMPLOYMENT HISTORY SECTION LIST MOST RECENT JOB FIRST. INCLUDE ALL JOBS SINCE GRADUATION FROM HIGH SCHOOL (OR LAST FIVE JOBS, WHICHEVER IS LESS).

FROM MO / YR	TO MO / YR	BRANCH OF U.S. MILITARY SERVICE	RANK AT DISCHARGE	LAST DUTY STATION

DATES FROM	DATES TO	NAME/ADDRESS OF EMPLOYERS	TYPE OF POSITION	SUPERVISOR	PAY RATE		REASON FOR LEAVING
MO. / YR.	MO. / YR.				START	FINISH	
MO. / YR.	MO. / YR.				START	FINISH	
MO. / YR.	MO. / YR.				START	FINISH	
MO. / YR.	MO. / YR.				START	FINISH	
MO. / YR.	MO. / YR.				START	FINISH	

Form 50574 (7/72) Printed in U.S.A.

XEROX IS AN EQUAL OPPORTUNITY EMPLOYER (M/F)

Figure 14-5 A. Employment application (Sheet 1). (*Courtesy* Xerox Corp.)

OUTSIDE ACTIVITIES

PROFESSIONAL MEMBERSHIPS, CERTIFICATES, OR LICENSES, HELD *(EXCLUDE THOSE INDICATING RACE, COLOR, RELIGION, SEX, NATIONAL ORIGIN, OR AGE.)

PATENTS AND/OR INVENTIONS

PRINCIPAL HOBBIES

IN WHAT EXTRA-CURRICULAR, CIVIC, OR CULTURAL ACTIVITIES HAVE YOU BEEN OR ARE YOU CURRENTLY ACTIVE? INCLUDE OFFICES HELD *(EXCLUDE THOSE INDICATING RACE, COLOR, RELIGION, SEX, NATIONAL ORIGIN, OR AGE)

DESCRIBE BRIEFLY THE NATURE OF YOUR LAST TWO **MOST APPROPRIATE** POSITIONS, INCLUDING SUPERVISORY RESPONSIBILITY, IF ANY.
IF YOU HAVE DONE RESEARCH, INCLUDE A BRIEF DESCRIPTION OF ITS NATURE AND SCOPE.

1.

2.

WHAT ARE YOUR PLANS FOR CONTINUING YOUR EDUCATION?

WILL YOU PERFORM SHIFT WORK? ☐ YES ☐ NO	WILL YOU WORK OVERTIME? ☐ YES ☐ NO

PROFESSIONAL/WORK REFERENCES WE MAY CONTACT:

NAME	HOME PHONE NO. (INCLUDE AREA CODE)	NAME	HOME PHONE NO. (INCLUDE AREA CODE)
EMPLOYED BY	EMPLOYER PHONE NO. (INCLUDE AREA CODE)	EMPLOYED BY	EMPLOYER PHONE NO. (INCLUDE AREA CODE)
EMPLOYER ADDRESS		EMPLOYER ADDRESS	

Persons employed at Xerox have access to confidential information regarding various phases of Company business. Therefore, the Company follows the usual practice of requiring new professional employees at the time of employment to sign a proprietary information and conflict of interest agreement. Information concerning competitor's operations, products, designs or other proprietary information will not be solicited from an applicant for employment, or from the Company's employees. Xerox will honor any valid post employment restrictions contained in an applicant's employment contract and fully respects the applicant's duty of loyalty and non-disclosure to his former employer.

I certify that the answers provided above are accurate to the best of my knowledge and belief. I am aware that failure to complete this application, intentional omissions or misstatements may result in refusal of employment or discharge. If employed, I understand that continued employment may be subject to issuance and continuance of bond by the bonding agent of Xerox Corporation.

SIGNATURE OF APPLICANT | DATE

FOR EMPLOYMENT USE ONLY

INTERVIEWER'S SIGNATURE | DATE

Figure 14-5 B. Employment application (Sheet 2). (*Courtesy* Xerox Corp.)

c. *Intelligence tests,* designed to determine the candidate's ability to deal with abstract ideas, relationships, numbers, and so on.

d. *Personality tests,* designed for the purpose of assessing the candidate's ability to adjust emotionally to specific situations, self-confidence, originality, and other features of a related nature.

e. *Interest tests,* intended to measure the candidate's preferences for certain work-related activities.

The benefit of testing pertains more to eliminating those candidates whose chances for success are minimal than as a selection technique in choosing the best of several candidates with demonstrated capabilities.[9] Testing provides only indicators, which are not always correct, and the tests per se should not be considered as a replacement for the personal interview.

Once Personnel has completed its portion of the interview process, the candidate will meet with the technical manager and if more than one group is interested, the candidate may have several interviews during the visit. Technical interviews should be conducted in private, and the interviewer should give full attention to the candidate. The candidate should be made to feel at ease and not rushed. At the same time, the interviewer should control the interview and attempt to seek out pertinent information in response to the following general questions:

a. Is the candidate technically up to date and qualified to do the job as advertised? Does the candidate seem to be motivated in this direction?
b. What are the candidate's reasons for seeking this job? Are they valid?
c. If currently employed, why does the candidate wish to leave his present job?
d. What are the candidate's personal goals?
e. What is the candidate's feeling toward continuing education and training?
f. How well will the candidate be able to get along with others in the organization?
g. Is the candidate's family life compatible with a change at this time? An unhappy wife or husband can ruin morale, individual initiative, and job output in almost every instance.

The interviewer will not always ask such direct questions. However, an overall impression and some insight in these areas can be gained by asking a few general questions and allowing the candidate to talk freely. Sometimes the staging of work situations and what the candidate would do under certain circumstances offers a great deal of information. Also, the manager may ask

---

[9] With the ever-increasing sensitivities toward the rights of individuals, the aspect of personnel testing could bring up some legal questions. Therefore, consulting with a legal expert is advisable prior to instituting a testing program.

several lead engineers in his group to interview the candidate to gain some additional perspective. In any event, the technical interview may take various forms—a private conversation in an office, conversation during a facility tour, comments during a luncheon, and so on. Sometimes the manager will take the candidate out to dinner the evening before, or to his or her home for a social visit, as much information can be conveyed in an informal setting.

After the technical interview, Personnel will once again meet with the candidate for the purpose of determining the candidate's overall impression of the corporation's facility, job interest, and interview(s). The candidate will be given brochures that describe the corporation's organization and product lines and instructions pertaining to the submission of an expense report to cover the costs of the trip. Personnel will also inform the candidate that he or she will be contacted soon relative to a possible job offer.

D. Employment Decisions

To fill a vacant position, the corporation may wish to invite several candidates in for an interview, and the process outlined briefly above is repeated in each instance. Subsequently, both Personnel and the technical manager(s) will evaluate each candidate against preestablished criteria. These will generally include a spotcheck with one or more of the references or previous supervisors listed on the application form. Personnel will usually initiate a form for evaluation purposes. Refer to Figure 14-6 for an example.

After all evaluations are completed, a joint decision between Personnel and the technical manager will be made as to whether to make an offer and, if so, what job classification and salary will be offered. Sometimes these decisions are quite difficult, since the technical manager may be quite impressed with a candidate and recommend that he or she be offered a high starting salary, whereas Personnel, responsible for maintaining consistency throughout the corporation, may recommend a lower salary. If after further discussion and review of the candidate's credentials an impasse still exists, a decision is made at a higher level of authority.[10]

The results of the decision are conveyed to the candidate selected in the form of a job offer. The job offer will specify position classification, salary, fringe benefits, desired reporting date, provisions for moving, and will usually indicate that the offer is contingent upon acceptance within a reasonable period of time and on the successful completion of a physical examination.

---

[10] The problem of haggling over salary does not exist in federal or state agencies, since each individual employed must start at the lowest step in the designated classification for which he or she is qualified.

Applicant's Last Name: ______________________

INTERVIEW SUMMARY

Cooper Tire & Rubber Company

(COMPLETE THIS FORM IMMEDIATELY AFTER YOU FINISH YOUR INTERVIEW)

NAME OF APPLICANT: ______________________

POSITION APPLYING FOR: ______________________

INTERVIEWER: ______________________ DATE: ______________________

1. Does the applicant have the academic qualifications for the position? (Explain)

2. Does the applicant have the job qualifications required? (Explain)

   Technical Skills: ______________________

   Experience: ______________________

3. Does the applicant have the personal qualifications the job requires? (Explain)

   Personality: ______________________

   Appearance & Manner: ______________________

   Initiative & Ambition: ______________________

4. Does the applicant have the communications skills the job requires? (Explain)

   Verbal: ______________________

   Written: ______________________

Form No. 20-124 Rev. 1-72

Figure 14-6 A. Interview summary (Sheet 1). (*Courtesy* Cooper Tire & Rubber Co.)

E. Job Acceptance and Reporting to Work

Upon receipt of an offer, the applicant will weigh the pros and cons of that offer (e.g., salary, benefits, potential working conditions, location), along with other opportunities, and make a decision as to whether to accept.

INTERVIEW SUMMARY (Con't.) -2-

5. Does the applicant have the potential or qualifications to work in another job or area of work? (Explain) ______________________________

____

____

____

____

____

6. Do you recommend we hire this applicant? Why or why not? ______________________________

____

____

____

____

7. If hired, what do you think his initial assignment should be? Why? ______________________________

____

____

____

____

____

Additional Comments or reservations about this applicant:

____

____

____

____

____

____

____

____

Figure 14-6 B. Interview summary (Sheet 2). (*Courtesy* Cooper Tire & Rubber Co.)

Actually, it is quite probable that a great deal of thought has already been given to the possibility of receiving an offer and what to do about it in the event that it becomes a reality. Thus, the final decision boils down to the specific terms and conditions proposed. If there is dissatisfaction, the applicant may wish to negotiate for a "better deal." With small businesses, the negotiation process may prove to be successful; however, for large corporations and government agencies such negotiations usually prove to be fruitless. If the applicant ultimately decides to accept the offer, he or she will notify the corporation accordingly and will indicate confirmation of the planned reporting date.

When starting a new job, the employee will initially report to Personnel for the purpose of filling out forms covering tax deductions, benefits, insurance, the corporation investment program, and so on. In addition, in large corporations the new employee will in all likelihood be required to take an in-plant physical exam to officially verify health records with Personnel. This is accomplished to ensure that full knowledge of the employee's physical condition is known at the start of employment.

14.4 PERSONNEL DEVELOPMENT

Sections 9.4D and 9.5C emphasize the need for individual self-improvement and personnel development. Self-improvement is *vital* not only to the individual but to the organization that he or she serves if ultimate overall growth is to occur.

When a newly employed engineer reports to work, the appropriate manager should meet with the engineer and establish some goals and objectives. These goals should not only cover the day-to-day project/staff activities but should address areas that lead to personal growth. The combination should be included in a personnel development plan against which the engineer is evaluated in periodic performance reviews.

Personnel development usually commences with some formalized training to familiarize the engineer with the policies and procedures applicable to the overall corporation/agency. Subsequently, an individual's growth is attained through a combination of on-the-job experience and formal education and training programs designed to upgrade the engineer in the theory and application of new methods in his technical field of interest. The needs for education and training, specified by various organization groups and individuals throughout the corporation, are indicated to the education and training function of Personnel, which in turn is responsible to (1) evaluate these needs on a total integrated basis and identify education/training program requirements; (2) develop, coordinate, and schedule programs in response to the identified needs; (3) work with the various engineering managers throughout the corporation to arrange for selected engineers to attend these programs; and (4) evaluate the results and effectiveness of each program after completion.

Formal education and training programs may fall into a variety of categories, a few of which are identified below.

A. Undergraduate and Graduate Engineering Courses

Many engineers employed in corporations, businesses, and government agencies are interested in taking formal engineering courses for credit, either with the intent of working toward a degree (baccalaureate, master's, or doctoral) or for the material content included in the course(s). In response, the education and training function of Personnel should initiate contact and establish a mutually compatible working relationship with one or more of the universities or colleges in the area. The trend in many academic institutions is to directly support local industry and government organizations by providing the desired courses either on campus if the institution's campus is located in the geographical area of need, or off campus if the institution's campus is remotely located. Off-campus courses may be provided at the corporation or agency's facility if the need is sufficient, and such courses are usually conducted in the evenings after work. In addition, in some fields of academic pursuit, an engineer can obtain a degree without the requirement for residency on campus.

Engineers interested in actively pursuing course work should, with their supervisor's cognizance, contact Education and Training and request assistance. If a local academic institution is offering a program or course of interest, the engineer may proceed to enroll, and his tuition and book expenses are usually paid in full (or at least partially) by the corporation, provided that (1) the engineer has applied to and been accepted by the academic institution, (2) the course taken is somewhat job-related, and (3) the engineer successfully passes the course. Failure to pass the course usually requires that the expenses incurred be refunded to the corporation. Engineers pursuing formal university/college programs and courses must comply both with personnel policies and the rules of the academic institution.

B. Long-Term Education Programs

As a supplement to university/college programs offered locally, some corporations and government agencies sponsor a long-term training program, in which individuals with proved capability, who are highly motivated, are selected to participate for a designated period of time (e.g., an academic year, semester, or quarter) in an approved program on the campus of a university with all expenses paid by the corporation. In addition, the corporation may also retain the individual on part-time salary while he is in school. The ultimate objective is to reward individual initiative by allowing an engineer to complete work for a degree in a relatively short period of time and to realize benefits for the corporation through a more effective utilization of the added expertise acquired by the selected individual. Usually, an individual has to work hard and demonstrate tremendous growth potential to be eligible for such a program. He or she must also have the support of supervision and must be accepted by the applicable academic institution.[11]

C. Workshops, Short Courses, and Seminars

There are many instances when an engineer feels the need for some short-term formalized continuing education to provide him with the necessary tools to facilitate the accomplishment of an immediate job assignment. He is not interested in degree-granting programs or university/college courses, but would like something *now* in a specific area. For instance: a three-day workshop covering "the application of microelectronics in design" may be appropriate for engineers in electrical design; a two-day seminar on "the use of PERT/CPM techniques in project planning" would be applicable for an engineer recently assigned to a product planning activity; a five-day short course in "systems engineering and management" or "logistics engineering" may be timely prior to initiating a system development project; a four-day workshop covering "linear and dynamic programming" may be beneficial for an engineer assigned to perform an operations analysis task; and so on.

Short-duration training programs of this nature may be organized internally by the education and training function using the capabilities available within the corporation/agency, by soliciting assistance in continuing education from a university or college, or by responding to one or more of the many widely advertised programs sponsored by job-shop business operations throughout the nation. If there is enough interest within the corporation for a specific workshop or seminar, it would in all probability be economically feasible to contact either an inside or an outside source (or a combination) to conduct a program in the corporation's facility and *tailored* to the specific needs of that corporation. On the other hand, if the interest is limited to a few individuals only, the corporation may decide to send those few to a remote location where a program covering the general subject matter desired is being conducted for people from all over. In this case, the program would constitute more of an *overview*, since tailoring to a specific need is not always practicable.

Many programs of this type offer the continuing education unit (CEU) as a standard form of measure for the successful completion of continuing education work. CEUs are being awarded on the basis of the quantity of

---

[11]When an employee is away at school, there is a natural tendency on the part of his supervisor to lose contact and not convey an active interest relative to how the employee is doing or what his interests or problems are. The employee often feels somewhat lost and needs to be reinforced relative to his value to the organization. Also, when the employee returns to work, some supervisors will exert very little effort to provide the employee with new challenges and interesting job assignments. Thus, a once-motivated employee may soon lose interest, productivity will decrease, and he may actively seek another job, unless mnagement makes the effort to effectively utilize his capabilities.

contact hours of high-quality instruction received (e.g., one CEU per 10 contact hours of instruction) and are being recognized more and more by academic institutions, industry, and government agencies as a measure of achievement.

Other means for attaining personal growth are through self-study, the completion of correspondence courses, the participation in technical society and industry association activities, and the presentation of papers at conventions and conferences. These activities are generated through the initiative of the individual.

In summary, personnel development refers to the upgrading of engineers on a continuing basis, through the various means discussed above, to ensure that they are current in their respective fields of technology and that they are able to apply their knowledge to the benefit of the corporation. Hopefully, as an ultimate objective, the engineer will seek certification in a technical society and/or professional registration in the state in which he is employed.[12] Personnel development must be encouraged by the technical manager (i.e., the engineer's supervisor), and the opportunities must be available through the education and training function. These opportunities include not only the provision of education and training programs but the facilities and resources necessary to support these programs.

14.5 PERFORMANCE REVIEWS AND COMPENSATION

Among the questions most significant in the minds of engineers are:

A. How well am I performing on the job?
B. What are my chances for a promotion or a merit increase?
C. What are my chances for reassignment or transfer to another organization that will provide better opportunity for advancement?

The engineer is interested in feedback from his supervisor on a regular basis, which is attained through informal communications and formal job performance reviews. Performance reviews, in conjunction with compensation and reward policies, are initiated to ensure that:

A. Individual personnel are recognized and rewarded in proportion to the degree of demonstrated performance. Highly qualified individuals with distinguished accomplishments should progress at a more rapid pace than those whose performance has been mediocre or below.

---

[12] Certification and professional registration are becoming more and more important to ensure that engineers maintain a defined level of expertise in their respective technical fields. This is particularly true in view of the legal trends in the area of consumer protection, and engineers involved in product design and consulting are being held legally accountable for their actions and practices. More discussion of this area is presented in Chapter 15.

B. Personnel compensation is equitable with respect to others in the organization, as well as similar employees in other corporations/agencies throughout the industry.
C. Personnel who are obviously misplaced in their current jobs are considered for work for which they are better qualified.

Performance reviews are intended not only to officially praise an individual for a job well done, but to inform an employee of marginal or inadequate performance when necessary. Such reviews are usually prepared and conducted by the engineer's supervisor on a semiannual or annual basis, discussed between the employee and the supervisor, and then submitted to the personnel function for approval and inclusion in the employee's permanent file. Salary adjustments and other compensation, based on the performance review, are initiated shortly thereafter.[13]

A. Preparation of Performance Reviews

The preparation of performance reviews is generally a multistep process. Initially, the supervisor prepares a written review on each employee covering such performance and qualification factors as quality and quantity of work accomplished, job knowledge, dependability, creativity, socialness, analysis and judgment, initiative, leadership, verbal ability, and personal development (refer to Figure 9-9). The employee is evaluated in terms of these or comparable criteria using a numerical rating method along with a qualitative description that supports major achievements (or lack thereof) in each area. The numerical rating approach is necessary for the subsequent ranking of employees and may take the following form:

Evaluation Criteria	*Poor*	*Average*	*Excellent*
Quality of work accomplished	1 2 3	4 ⑤ 6	7 8 9
Quantity of work accomplished	1 2 3	4 5 6	⑦ 8 9
Job knowledge	1 2 3	4 5 ⑥	7 8 9
Dependability	1 2 3	4 5 6	7 ⑧ 9
Creativity	1 2 3	④ 5 6	7 8 9
Socialness	1 2 3	4 5 ⑥	7 8 9
Analysis and judgment	1 2 3	4 ⑤ 6	7 8 9
Initiative	1 2 3	4 5 6	⑦ 8 9
Leadership	1 2 3	④ 5 6	7 8 9
Verbal ability	1 2 3	④ 5 6	7 8 9
Personal development	1 2 ③	4 5 6	7 8 9
Total score—59 of a possible 99			

[13]The discussion presented is primarily oriented to a reward policy that is based on actual accomplishment and not to the practice of initiating rewards as a function of time and grade on the job, independent of achievement.

The ranking of employees is a systematic and logical approach for relating job performance with salary. Each supervisor ranks the engineers in his organization. Ranking is accomplished for each engineering classification. The next-higher-level manager then integrates this and other group rankings into a composite ranking by meeting with his supervisors and attempting to get agreement. Often he will get agreement only on a few key individuals and then will have to rank the remaining personnel on his own. The ranking is used later as the basis for salary determination.

As part of the performance review, the supervisor should include positive recommendations to the employee which, if followed, will lead toward improved performance in the future. These recommendations, along with the individual performance review form, are discussed jointly between the supervisor and the employee. The employee may wish to comment on the review and will record his comments on the performance review form. Upon completion of the evaluation process, the performance review form is signed by the responsible supervisor, his manager, and a representative in the personnel function. An example of a performance review form used by one corporation is presented in Figure 14-7.

B. Salary Compensation and Reward Policy

Figure 14-8 conveys the overall approved salary structure as established for the corporation by the wage and salary administration. The employee ranking, derived from the performance review process, is compared with the salary structure to determine if the highest-ranked employees are those with the higher salaries. Such comparisons sometimes indicate that top performers are at a lower salary level than some of the less productive performers in the organization. This may have occured for several reasons:

1. As a result of market conditions and the entry levels required to attract engineering personnel, some new employees are hired at a higher starting salary level than the salaries of productive engineers who are already employed and have been with the corporation for a number of years. In other words, starting salaries have increased at a higher rate than internal salary growth within some corporations, and the new employees are probably not high in the ranking, since they are still in the job-learning stage.
2. Individual employees with a corporation who were at one time highly productive and on top in the ranking have subsequently fallen off in effectiveness, and their respective salaries are higher than justified. In essence, their initial salary growth does not reflect current productivity.

When comparing salary levels with the performance ranking, the objective is to adjust individual salaries such that top performers are

PERFORMANCE APPRAISAL
CONFIDENTIAL

COOPER TIRE & RUBBER COMPANY
FINDLAY, OHIO 45840

LAST NAME ______________________

NAME ______________________ DEPARTMENT ______________________

JOB TITLE ______________________ JOB CLASS ______________________

HIRE DATE ____________ APPRAISAL PERIOD: FROM ____________ TO ____________

INSTRUCTION:

Before rating this person, carefully think about their daily on the job performance and overall contribution to the Company.

PERFORMANCE FACTORS	PERFORMANCE RATING				
	EXCELLENT No improvement needed. Outstanding performance. Works far beyond expectations.	VERY SATISFACTORY Work is well above average. Has opportunity for improvement.	SATISFACTORY Work meets basic requirements and expectations demanded by the job. Performance is adequate.	IMMEDIATE IMPROVEMENT NEEDED Less than expected of required performance. Performance demands immediate corrective action.	UNSATISFACTORY Performance is far less than required or expected. Performance must be improved immediately.
QUALITY OF WORK					
QUANTITY OF WORK					
JOB KNOWLEDGE					
JOB SKILLS					
JOB ATTITUDE					
SUMMARY RATING					

(Each factor is followed by: Reason for Rating: ______________________)

FORM NO. 17-135-3 REV. 12-71 (This form will become a permanent part of the Employee's Personnel file.) (OVER)

Fig. 14-7 A. Performance appraisal (Sheet 1). (*Courtesy* Cooper Tire & Rubber Co.)

recognized and rewarded accordingly. Ultimately, the salaries for all individuals in the organization should be scaled to be compatible with the performance ranking.

WHAT POSITIVE CONTRIBUTION HAS THE EMPLOYEE MADE TO THE COMPANY SINCE THE LAST APPRAISAL?

DOES THIS EMPLOYEE HAVE THE QUALIFICATIONS AND POTENTIAL TO BE CONSIDERED FOR POSITION WITH MORE RESPONSIBILITY? (IF YES, WHEN AND WHAT TYPE POSITION?) (IF NO, WHAT ARE YOUR PLANS FOR HIM OR HER?)

SUMMARY PERFORMANCE RATING (FROM FIRST PAGE) ____________

TOTAL COMPENSATION INCLUDING BONUS ____________
PRESENT WITH INCREASE

QUARTILE POSITION ____________
PRESENT WITH INCREASE

LAST INCREASE ____________ ____________
DATE AMOUNT TYPE

☐ I RECOMMEND THE APPROVED BUDGET INCREASE OF ____________

☐ I RECOMMEND A MERIT INCREASE OF ____% EQUAL TO $ ____________ PER WEEK.

☐ I DO *NOT* RECOMMEND A MERIT INCREASE AT THIS TIME BECAUSE ____________

IF NO INCREASE—SHOW DATE REVIEWED WITH EMPLOYEE ____________

IMMEDIATE SUPERVISOR ____________ DATE ____________

This section to be completed only if you are recommending a deviation from the budget:

NEXT HIGHER SUPERVISOR'S RECOMMENDATION AND COMMENTS:

☐ I APPROVE THE INCREASE (COMMENTS) ____________

☐ I DO *NOT* APPROVE THE INCREASE BECAUSE ____________

SUPERVISOR ____________ DATE ____________

Fig. 14-7 B. Performance appraisal (Sheet 2). (*Courtesy* Cooper Tire & Rubber Co.)

There are various methods for rewarding professional personnel:

1. *Merit increases* or salary increases within a given personnel classification. These are usually awarded in recognition of above-average performance and are normally associated with the scheduled performance review.

2. *Promotions* from one classification to another (e.g., from "engineer" to "senior engineer"). Promotions are not always tied in with the scheduled performance review but may be initiated at any time. Also, salary adjustments associated with promotions are usually higher than normal merit increases. Supporting information justifying the promotion should be provided through the preparation of a special performance review and a letter of recommendation.

3. *Special bonuses,* which include cash disbursements, profit sharing, stock options, and increased retirement benefits.

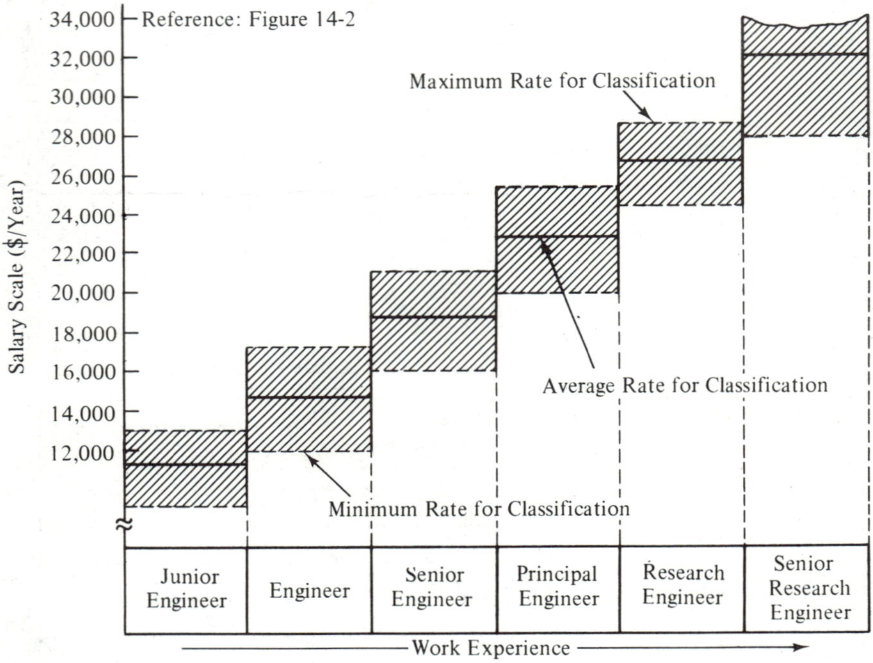

Figure 14-8. Salary structure.

4. Cost of living

Adjustments for inequities in the salary-performance-ranking relationship are made through one or a combination of these methods of compensation. The magnitude of the rewards will depend somewhat, of course, on business conditions, as salary increases and bonuses are likely to be slim if many corporation profits are low. However, if business conditions are good, many corporations will reward their top performers with substantial salary increases plus bonuses. Over the long term, the salary growth for an engineer will tend to follow one of the curves in Figure 14-9.

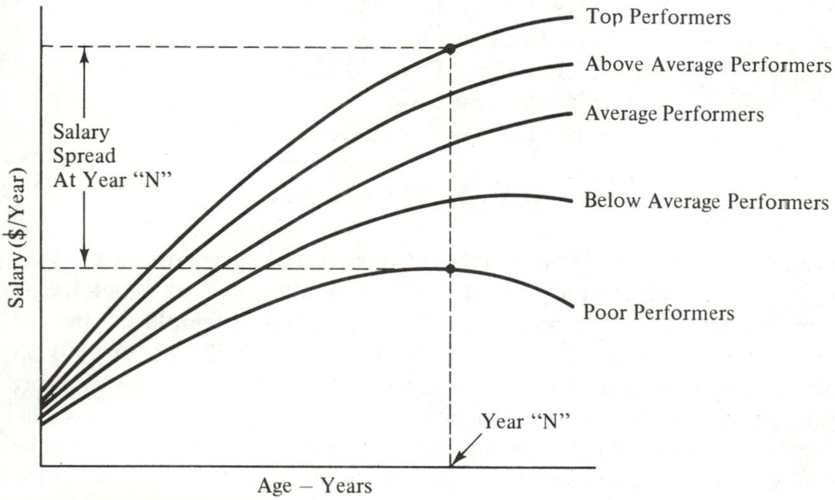

Figure 14-9. Growth curves.

14.6 FRINGE BENEFITS

Fringe benefits include those elements of compensation other than salaries and wages which are significant when initially considering employment and in the ongoing evaluation of one's welfare on the job. They may be of a monetary nature (e.g., retirement, insurance) or nonmonetary (e.g., position title, nice office, furnished automobile, good parking space). However, the monetary aspect constitutes a sizable portion of the total compensation received by the employee and is the main point of discussion. Some prime elements of the total fringe-benefit package are discussed below.[14]

A. Retirement Benefits

The normal retirement benefit constitutes the monthly cash payment, above and beyond social security, which the corporation or agency pays to the individual after retirement at age 65 (or at the age designated as the official retirement age). Retirement is paid until death, and the amount paid may be calculated any number of ways. One method used is to determine the highest average annual salary for 5 consecutive years of employment and multiply the value by some factor to obtain the benefit per year of service. For instance:

---

[14]The fringe-benefit package varies from corporation to corporation, agency to agency, industry to industry, state to state, and country to country and should be evaluated as part of the total compensation received when considering a job.

salary for year H	=	$21,000
salary for year I	=	23,500
salary for year J	=	25,200
salary for year L	=	26,600
salary for year K	=	26,600
average salary	=	$24,598

The total annual benefit is the product of the average salary over the 5-year period, a benefit factor of x percent, and the number of years of service. If the benefit factor is 2 percent and the individual is employed by the corporation for 22 years, his annual benefit will be $10,823.12, or $901.93 monthly. In the event that an employee wishes to retire early, the monthly retirement pay will be reduced somewhat.

The retirement benefit is often based on a joint contribution between the employee and the corporation, with the employee's contribution being deducted from the weekly or monthly paycheck. The engineer, in considering current employment or a future change in employment, should determine his retirement benefits and evaluate the gains or losses, whichever the case may be. At some point in his lifetime, as he gets older and the remaining years of eligible employment are limited, his losses could be substantial if he changes employment and loses either partial or full retirement benefits. Whether an employee loses full retirement benefits in the event of a job change sometimes depends on whether or not the employee has attained *vested rights* with the corporation. Vesting usually refers to the minimal number of consecutive years that an employee must work for a corporation to be eligible for retirement and other benefits.

B. Insurance Benefits[15]

Insurance benefits may be categorized in terms of either group life insurance or medical benefits. *Group life insurance* constitutes the protection that an employee receives throughout his employment with the corporation, with the benefits going to his beneficiary (or beneficiaries) in the event of death prior to his normal age of retirement. Group life insurance may assume a variety of forms, depending on whether the corporation contracts on the basis of "whole life," "paid-up life," or "term" insurance. The amount of insurance available to the employee is usually based on his salary level (e.g., two times the annual salary of Engineer X, or $50,000 of life insurance for an engineer with an annual salary of $25,000). The rates and the employee-corporation sharing ratio may be different, depending on

[15]The variety of and stipulations concerning insurance benefits are extremely complex and cannot be realistically covered in this text. The reader is advised to review additional literature and/or seek further instruction in the subject area.

the type of insurance contracted by the corporation. The employee's contribution is deducted from his paycheck.

Medical benefits, on the other hand, may be classified in terms of (1) policies, such as Blue Cross-Blue Shield, that cover limited hospital services (i.e., up to x days) and certain outpatient expenses; (2) maternity expenses; (3) major medical, for extensive surgery and hospital services; and (4) disability, which provides some income to the employee during the period that an illness or injury prevents him from working. Most corporations and agencies provide various forms of coverage for the employee and his (or her) dependents at a reduced rate, with the employer and employee sharing the cost. The engineer should review carefully the medical benefits offered, to ensure that he is adequately covered in the event of need, particularly in view of the current high costs of medical services in general.

C. Investment Plans

Many corporations and businesses provide the employee with the opportunity to invest in bonds, stock, or a combination thereof. Stock investments may include shares of the corporation's stock or a diversified mix of stock such as that represented by mutual funds. Corporations that sponsor investment plans will often match, on a one-for-one basis, the employee's investment. In other words, for every share of stock (or a portion of a share) that the employee purchases, the corporation will contribute an equal amount to the employee.

If the employee remains with the corporation long enough to attain vested rights, the employee will ultimately receive not only his or her investment but that which has been contributed by the corporation along with the accumulated interest on investment. If a change in employment occurs prior to the designated time for vesting, the employee will receive his or her own contributions upon termination with the corporation, but will not necessarily receive the matching contributions made by the corporation.

D. Additional Benefits

Although retirement benefits, insurance benefits, and investment plans are significant relative to the total fringe-benefit package offered and have a long-range effect on the employee, there are many other benefits that are important in making decisions pertaining to employment.

1. *Vacation and sick leave.* Most corporations will support a given amount of paid vacation and sick leave, which is accrued with time on the job. Of concern to the employee is the annual amount of earned vacation and sick leave available to him, the rate of accrual, whether or not he loses it over a period of time if unused, and whether he is compensated for unused vacation and sick leave if he decides to terminate with the corporation.

2. *Tuition-refund plans.* Tuition refunds are often made available to employees who are enrolled in undergraduate and graduate courses, long-term education programs, and workshops, short courses, and seminars (refer to Section 14.4). The employee may wish to evaluate the extent of tuition-refund coverage and the conditions associated with such coverage.

3. *Overtime salary/wage benefits.* Employees may or may not, depending on their classification, be entitled to overtime pay if they work a number of hours in excess of the normal workweek (e.g., 40 hours) or if they work during regular holidays. The overtime may be compensated on the basis of "time-and-a-half" or "double-time" pay. The employee should determine whether or not he is eligible for overtime compensation.

4. *Paid holidays.* Corporations will allow for x number of paid holidays per year and will designate the applicable dates. In many instances, some of the identified holidays will fall on a Friday or a Monday in order to provide the employee with a long weekend.

5. *Moving expenses.* When an employee is initially hired, many large corporations will pay for the employee's moving expenses plus an allowance for a limited period of time while the employee and his family are getting settled. Small businesses and government agencies usually do not provide such funds. On the other hand, most organizations will pay all expenses associated with an internal transfer from one geographical location to another, particularly if the transfer is initiated at the convenience of the organization.

6. *Interest from inventions.* Some organizations will pay to the employees responsible interest on the benefits received for successful inventions, as a reward for creativity.

7. *Cost-of-living increases.* With the inflationary trends of the past several decades, a number of corporations and agencies have periodically initiated an across-the-board cost-of-living increase to all employees. Promotions and merit increases alone have not adequately covered the rate of inflation in recent years.

8. *Unemployment compensation.* Sudden layoffs through reduced employment conditions can have a significant impact on the individual(s) affected. Some organizations will provide unemployment benefits (e.g., x per cent of the prior earnings) for a specified period of time after termination from the organization.

The fringe-benefit package per se, as generally recognized by most organizations, includes supplementary pay (paid vacations, holidays, etc.), insurance, and retirement provisions. Costwise, the fringe package comprises from 25 to 30 per cent of the total cost of employment, and these costs have been rising at a rapid rate. The fringe-package cost, along with the costs associated with the other benefits mentioned above, are included as overhead expenses in the organization's accounting structure.

14.7 LABOR RELATIONS

The personnel function covers many facets of the total activity of an organization. One of these, not yet discussed, is concerned with the employment conditions as arranged between the corporation's management and its employees. These conditions evolve and are sustained, or changed, through the process of labor relations. Labor relations activities deal with the formalized working relationships and conditions between management and labor. The process for establishing these relationships is generally accomplished through "collective bargaining" between management and a recognized group of individuals representing the employees. Collective bargaining usually implies that an established labor union is present (e.g., American Federation of Labor-Committee for Industrial Organizations) and that the labor union represents the employees of the corporation, particularly those employees on an hourly wage, such as machine operators, manufacturing assemblers, technicians, maintenance specialists, and clerical employees.[16] In a number of corporations and in certain geographical areas, labor unions have been established to represent engineering and comparable professional activities as well as hourly employees.

Labor unions have evolved primarily as a result of dissatisfactions with working conditions and the desires of employees to assert themselves through a formalized mechanism of some type. Such dissatisfactions may stem from legitimate complaints concerning a low salary/wage or poor fringe benefits, or from complaints developing as a result of human emotions, having little factual basis. Labor relations deals with management and the labor unions in arriving at a satisfactory resolution of problems of this nature and is primarily involved in the activities of collective bargaining and the handling of grievance procedures.

A. Union Organization and the Corporation[17]

Unions may be nationally organized with local affiliations or strictly local in scope, without any higher-order connections. The nationally organized union is controlled by a group of elected officials, a constitution, and rules and regulations that govern the local affiliations. The local affiliation has its own charter and bylaws (prepared in accordance with the national union's constitution) and maintains direct contact with management and the employees within the corporation. In fact, the elected officers

[16]Corporations not faced with labor union representation still should address and pay considerable attention to the collective reactions of its employees and the working conditions established.

[17]A detailed discussion of labor relations and the history and organization of labor unions is presented in: H. T. Amrine, J. A. Ritchey, and O. S. Hulley, *Manufacturing Organization and Management*, 3rd ed., Prentice-Hall, Inc., Englewood Cliffs, N.J., 1975.

of the local affiliation are usualy regular employees of the corporation, and their objective is to act on behalf of the employees represented by that union. Union representation at the local level may vary somewhat from state to state and in terms of whether or not corporations are organized for a union shop or an open shop, which is a function of the negotiated union contract.

1. *Union shop.* A union shop is one which specifies that all employees in the bargaining unit must become dues-paying members of the union within a certain period of time after employment (e.g., 30 days). In a situation where all employees in designated job classifications are required to become members of the union eventually, it is obvious that the local union acquires a great deal of strength, improving its bargaining potential with management.

2. *Open shop.* The open shop allows the employee an option of either joining or not joining the union. Corporation management does not discriminate against union and nonunion members, and individuals in both categories can bargain directly with management. In this situation, the local union is usually not as strong as in the case of the union shop.

Whether a union shop or an open shop exists, the local union is organized in a manner somewhat parallel to that of other organizations, with representation at all levels in the corporation where employees in the bargaining unit work.

B. Collective Bargaining

Collective bargaining constitutes the negotiations between the management of a corporation and the union (representing labor), with the ultimate objective of agreeing on a written contract covering the terms and conditions of employment. Collective bargaining is basically a continuous give-and-take process involving proposals and counterproposals. Even though contracts are eventually negotiated and implemented, collective bargaining actually continues through contract maintenance and renewal. In most instances, collective bargaining deals with negotiations pertaining to salary and fringe benefits.

In the negotiation process, the union position is to attain the best benefits possible on behalf of labor, while management's position is to maximize corporation profits and keep costs to a minimum. Meetings between management representatives (i.e., the labor relations function in Personnel) and union officials are conducted in an attempt to arrive at an agreement. If this occurs, a contract is signed and implemented. If not, a mediator from the outside is often brought in to review the situation from both sides in an objective manner and to make suggestions that hopefully will lead to contract settlement. In the event that the outside mediator is unsuccessful, the union officials will probably call for a strike.

When negotiating contracts, the position of the union can be appreciated from the standpoint of attaining an equitable salary/wage and benefit package for members of the bargaining unit. However, union proposals can also push the wage and benefit package up to a level where the corporation loses its competitive position because of the high costs of employment. If this occurs, the corporation may lose business, resulting in employee reduction through layoffs. In some instances, the corporation may be forced to relocate to another geographical area where the personnel labor rates are more competitive. Thus, for the benefit of both the corporation and its employees, during the contract negotiations, a reasonable approach must be maintained by both sides.

C. Grievance Procedures

A *grievance* may be defined as any feeling on the part of the employee, whether real or imagined, of personal injustice during the performance of his or her job. In other words, if an employee does not like the job conditions or feels that he or she is being discriminated against by the supervisor, the employee can initiate a protest on the basis of "violation of the labor contract." Such a protest is known as a grievance and is handled through a formalized process that involves the following basic steps:

1. The unhappy employee, as a member of the bargaining unit, verbally discusses the complaint with his or her supervisor.
2. If the grievance is not settled, the employee presents the grievance in writing to his supervisor.
3. If a settlement does not evolve, the employee presents the grievance to his designated union representative (i.e., steward), who in turn takes the grievance to the next-higher level of management.
4. If the situation still remains unsettled, a union official, or committee, then approaches top management for a resolution. At this stage, the labor relations function of Personnel usually gets involved.
5. If the situation is still unsettled, the grievance is usually presented to an outside arbitrator for a final decision as to the action required.

Starting the grievance procedure at the level of the employee's immediate supervisor is preferable in view of the direct contact between the two on a day-to-day basis, and many grievances can be settled at this stage with a little understanding on both sides. As grievances are presented at higher levels, arbitration becomes a group function, involving several union and management representatives. In addition, the process becomes more formalized, and established time limits are allotted for each step to prevent unnecessary delays and to keep the costs to a minimum.

The engineer's involvement in labor relations may take several forms. If he or she is in a corporation and a geographical area where engineering unions prevail, the engineer may be a member of the bargaining unit. Collective bargaining and contract negotiation will affect his or her salary level and the fringe benefits received. On the other hand, if the engineer is not directly represented by a union, it is quite likely that he or she will be supervising technicians and clerical personnel who are dues-paying members of the union. In this capacity, the engineer may become involved in the grievance procedure as a supervisor or as a representative of management. Finally, the engineer's work is certainly affected when strikes are called. During a strike, the engineer may be able to continue on the job without the assistance of a technician or clerical personnel, or may be barred from entering the facility because of picketing. In any event, the effectiveness and efficiency of job accomplishment will be impaired.

14.8 EMPLOYEE SERVICES

In many personnel organizations, there is a function entitled employee services (or equivalent), which is responsible for a number of activities. Employee Services may:

A. Provide assistance to employees in reviewing employment status and growth, explaining personnel benefits, and in the processing of changes in personnel records (e.g., change in the beneficiary status of insurance, change in number of dependents, change in payroll deductions). Employee Services may also provide free expertise relative to the preparation of income tax statements.
B. Coordinate local campaigns involving the solicitation of funds (e.g., United Fund, Scouts, Chamber of Commerce, bond drives).
C. Operate internal stores responsible for the acquisition and disbursement of office supplies and materials.
D. Operate food services to include in-plant cafeterias and vending machines.
E. Operate the corporation's credit union, where employees can invest money and obtain low-interest loans (often at better rates than are provided by local banks).
F. Operate a corporation-wide *suggestion system.* Suggestion systems are designed to allow employees to directly submit recommendations for the improvement of operations, and serve a twofold purpose:

    1. The suggestion system gives management the benefit of employee suggestions relative to improving the efficiency of the corporation. The average employee may think of a number of ways to improve operations, cut cost, prevent safety hazards, and so on. Some corporations will monetarily reward employees who submit good suggestions that are accepted and followed by management.

2. The suggestion system improves employee morale by giving him or her a chance to express ideas relative to job activities and work conditions. The employee can take pride in knowing that these suggestions are being reviewed and considered for implementation.

G. Provide transportation for distinguished visitors.
H. Assist new employees and those being transferred to other locations in the buying and selling of real estate and housing.

Although some of these activities may appear to be relatively minor, there are a number of instances when such a function can effectively respond to employee needs.

14.9 RECREATION SERVICES

Many organizations will encourage their employees to participate in after-hours recreational activities and will sponsor athletic programs, social functions, clubs, and so on. The intent is to promote a greater esprit de corps and feeling of belonging within the organization, promote greater communications between groups throughout the organization, and to promote goodwill in the community through activities that involve outside participation (e.g., sponsoring a team in the community baseball league, joint sponsorship of a social event). It is not uncommon to find a wide variety of activities available to the employee, and if not, the opportunity to start a new activity is usually encouraged if there is enough interest.

14.10 SUMMARY

In summary, the personnel function is a multifaceted operation which works closely with all organizational elements throughout the corporation, business, or agency. The working interfaces involve both management actions affecting entire groups of people and individual interactions in response to an employee need. Personnel is responsible for the establishment of job classifications and salary/wage levels, the contacting and employment of personnel, maintaining personnel records and counseling employees while on the job, and for the processing of personnel terminations as a result of resignations, layoffs, or discharges. Personnel is also responsible for the ongoing assessment of total organizational effectiveness in terms of turnover rates, absenteeism and dissatisfaction, accidents, requests for transfer, and so on. Personnel attempts to deal effectively relative to the proper mix of internal promotions versus external recruiting. Finally, Personnel provides a major interface relationship with other organizations in the community and throughout the industry to insure that personnel policies and procedures are up to date and are realistic in relation to the market conditions.

QUESTIONS AND PROBLEMS

1. Select a corporation, business, or agency of your choice. Describe the personnel organization (construct or organization chart), and list the major responsibilities and functions of that organization.
2. Why is it necessary to establish job classifications and salary/wage ranges for each classification?
3. How are personnel requirements developed? Who is responsible for determining personnel requirements?
4. How are engineering salaries determined? What factors influence engineering salaries?
5. Identify some of the sources of technical personnel, and describe the recruiting approach that you would employ in contacting these sources.
6. Assume that you are an engineer looking for a job. Prepare a résumé covering your background and experience.
7. Assume that your are an engineer currently employed by Corporation ABC and have been requested by your supervisor to review three résumés, interview each of the candidates who have been invited to the facility for a personal interview, and make a recommendation as to whether an offer should be extended. What criteria would you use in evaluation of the résumés? What factors or characteristics would you look for in conducting a personal interview?
8. What are the benefits of personnel testing? Name some of the disadvantages.
9. When deciding on whether to accept a position for employment, what factors would you consider? List these factors in order of priority.
10. Why is personnel development important? In your own career, what are your objectives relative to personal growth, and how do you intend to accomplish these objectives?
11. If you were assigned the responsibility of reviewing currently employed engineers relative to job performance, what criteria would you use in the review and evaluation process? Develop a performance review form of your own.
12. Assume that you are an engineer working for Corporation XYZ. How would you evaluate your own progress (i.e., growth)?
13. What are fringe benefits? What do you consider to be important in evaluating the fringe-benefit package? List the factors in order or priority.
14. Assume that you are an engineer employed in Corporation AJAX. Where would you go to review and evaluate your benefit package?
15. As a class project, research and prepare a brief report on the following legislation: Davis-Bacon Act, Social Security Act, Walsh-Healey Public Contracts Act, Fair Labor Standards Act, and Taft-Hartley Act. What provisions are covered by each act?
16. Describe the basic functions of the Office of Federal Conract Compliance and the Equal Employment Opportunity Commission. How do these functions influence employment in corporations, businesses, and agencies?

17. As a class project, investigate and describe the job classifications and salary levels for engineers employed by the state and federal governments.

18. What is a labor union, and how is it organized? Identify several of the more active unions. Can you identify an engineering union?

19. What is collective bargaining? Describe the grievance procedure. Define the engineering interfaces with labor relations.

20. What factors tend to strengthen a labor union's position in a corporation? What are some of the benefits of a labor union? Identify some of the disadvantages.

21. What is the difference between a union shop and an open shop?

22. What are the benefits of a suggestion system within an organization?

23. What are the responsibilities of Personnel in terms of relations with the community?

15

LEGAL FUNCTION

There are three major areas that should be addressed in terms of engineering involvement with legal activities. These are (1) product liability; (2) patents and copyrights; and (3) contract types, negotiations, and incentive-penalty provisions. Although there are other instances where engineering—legal interfaces exist, these are considered as being the most pertinent.

15.1 PRODUCT LIABILITY [1]

Federal laws aimed at consumer protection, the legal interpretations and acceptance of *strict liability,* and the proliferation of litigation combined with large court awards in favor of the consumer are having a significant effect on industry.[2] Trends toward consumer protection impact all stages of product design, production, test and evaluation, and logistic support, with emphasis being placed primarily on product reliability and safety. Product manufacturers are being held liable if:

---

[1]A good discussion on product liability is found in: A.S. Weinstein, A.D. Twerski, H. R. Piehler, and W. A. Donaher, *Product Liability: An Interaction of Law and Technology,* Carnegie-Mellon University, Pittsburgh, Pa.; reprinted from *Duquesne Law Review,* Vol. 12, No. 3, Spring 1974.

[2]The 1970 President's Commission on Product Safety indicated that some 20 million Americans are injured each year as a result of accidents connected with consumer products. Refer to the *Final Report of the National Commission on Product Safety,* Government Printing

A. Their products malfunction and personal injuries occur as a result.
B. Their products are misused, personal injuries occur, and the manufacturer has failed to warn the consumer against misuses and/or failed to instruct the consumer as to proper use.
C. Their products fail to live up to the effectiveness (e.g., reliability), adequacy, and performance factors as initially advertised.

The extent of liability may take the form of a court settlement resulting in the reward of a large sum of money to the plaintiff, or in the form of *product recall,* where products are either modified in the field or returned to the manufacturer's plant and modified for the purpose of precluding a safety hazard or correcting a detected deficiency. In either situation, the potential costs to the manufacturer may be quite extensive.

In response, some manufacturers have initiated positive steps toward accepting responsibility for consumer safety and product reliability, but others have accomplished little in this direction. Initially, a primary objective in any organization is to create an awareness on the part of management and engineering of the new challenges associated with product liability and the consequences if a haphazard approach to product liability is followed. Second, procedures must be implemented in the design, production, test and evaluation, and logistic support facets of the life cycle to ensure that future product liability risks are minimized. A few specific considerations are noted.

A. Design

The engineer must be concerned with the potential *uses* as well as *abuses* to which the product may possibly be subjected. Reliability, maintainability, human factors, and safety are key design characteristics, along with product performance. Concerning reliability and safety in particular, failure-modes-and-effects analysis (FMEA), used as a reliability design aid, is an excellent approach in determining the "what if" questions in this area. For instance, through a functional breakdown of a system or equipment item, the designer can simulate malfunctions at the component level and assess the consequences in terms of complete or partial system failure, effects relative to mission success, effects relative to operator safety,

---

Office, Washington, D.C., LCC No. 76-600753, June 1970. The President's Report on Occupational Safety and Health of May 1972 indicated that incidents connected with industrial products account for an additional 7 million injuries each year. Department of Labor data indicate that for 1973 one of every 10 workers experienced a job-related injury or illness. National Safety Council data for 1974 indicate that the total cost of accidents for that year was approximately $15 billion, covering lost wages, medical expenses, and administrative costs (to include legal fees). Because of these statistics, product liability and safety have become major concerns.

and so on. As part of maintainability and human factors design, detailed operator and maintenance task analyses are accomplished, with the objective of analyzing the anticipated functions to be performed by the operator and the maintenance technician after the product has been delivered to the consumer and is in use. These activities, performed in a timely manner at the early stages of design, can prove to be extremely beneficial in detecting and subsequently preventing possible hazardous conditions later.

A second significant design activity is that of testing, as discussed in Section 5.3C. Testing must deal with the man-machine interfaces and the impact of possible environments on the system. Are the test requirements designed such that the product will be subjected to all anticipated environmental extremes? Will the environments to which the product is to be exposed result in hazardous conditions? These and comparable questions must be addressed through test planning.

Finally, the formal design review process described in Section 5.4 serves as a check function to ensure that the necessary design characteristics have been incorporated and to assure that the technical data for the product are adequate. Data to include operating procedures and maintenance instructions must be detailed to the extent necessary to enable the consumer to properly use the product, and must include warning notices and safety precautions where possible hazardous conditions could develop. In addition, the preparation of such data must be tailored to the consumer's level of understanding.

B. Production

As noted in earlier sections, the best design in the world can be ruined if the production process is less than adequate. Quality control must establish, in conjunction with industrial engineering, the necessary minimum standards for production processes, fabrication and testing techniques, inspection methods, and so on, to ensure that product reliability and safety are maintained throughout the production/construction process. In addition, the maintenance of proper records throughout production is essential in the event of follow-up problems after the product is distributed in the field. Such records should tie production operations with product serialization in the event that the manufacturer faces a product-recall situation and needs to identify products by part numbers and serial numbers when planning for product modification programs.

C. Advertising

All product advertising, sales, and promotional material must be reviewed both from a technical and a legal standpoint to ensure that statements or claims relative to product performance and effectiveness are indeed accurate. Exaggerated claims, which cannot be technically substantiated, may encourage the consumer to abuse the product in such a manner as to create a hazardous situation.

The discussion above relates primarily to preventive measures; however, in the advent that legal action is initiated against a product manufacturer in a liability suit, the engineer may find himself involved as an expert witness called into court to testify on a technical matter or assigned to an investigation team to determine the cause(s) of product failure, perform failure analyses, and prepare supporting documentation.

If a liability suit is instituted, the party suing the product manufacturer (i.e., the plaintiff) employs an attorney, who, after some investigation, may call on a technical expert to give an opinion as to whether the applicable product is faulty in any way. The product manufacturer (i.e., the defendant), on the other hand, may utilize the services of engineering employees to investigate details concerning design, production, test and evaluation, and so on, in preparation for the defense. Attempts at an out-of-court settlement are usually made; however, if this fails, a date is set for a formal trial.

The engineer's appearance in court is usually as an expert witness in the areas where technical expertise is needed. The witness may or may not be permitted to hear prior testimony, and he or she is interrogated by attorneys from both sides. Since the attorneys usually do not possess the equivalent technical background of the expert witness, there will be attempts to assess the competence of the witness in providing such testimony. In general, the witness should reply briefly only to the questions asked and should not elaborate unless requested to do so. The testimony provided by the expert must be geared to the average nontechnical layman, whether judge or jury, since the objective is to raise the level of comprehension concerning the technical material being presented.

Under strict liability, the plaintiff must prove that the product was defective and consequently dangerous. Also, it must be proved that the defect existed when the product left the manufacturer's plant or warehouse and that the defect caused harm. Interpretations as to what constitutes a dangerous situation may depend highly on the testimony of the expert witness. In addition, there is always the question as to whether strict liability or negligence applies. This is an extremely interesting and important area, and the reader may wish to approach a legal counsel for discussion relative to the results of product liability suits, legislative trends, and the increasingly important role of the technical expert witness in court decisions.

Another major role of the engineer in product liability, which is directly supportive of being qualified as an expert witness, is that of being assigned as a member of a special investigation team initiated to conduct failure analyses and laboratory testing. The objective is to reconstruct and/or simulate product malfunctions that have occurred or are likely to occur in the future. Advanced technology has made it possible to determine failure modes, cause-and-effect relationships, and to reconstruct how failures have occurred. For instance, the use of an electron microscope has made it possible to evaluate material fractures, crack growth, and so on, and hardness testing can be used to measure material strength characteristics.

The results of failure analyses and laboratory testing are usually presented in formal reports, often used as supporting documentation in court cases. Sometimes formal reports providing the right type of evidence carry more weight in court than verbal testimony alone given by an expert witness. On the other hand, such reports made available to opposing attorneys prior to trial may provide enough ammunition for discounting the evidence developed unless these reports are thorough and accurate in content. Again, the engineer's role is highly significant.

Product liability is receiving considerable attention in industry and government alike, and the results are impacting all facets of the life cycle as covered in Part Two. Hopefully, through improved communication between the technical and legal communities, better product designs will evolve, and the quantity of consumer accidents will significantly decrease.

15.2 PATENTS AND COPYRIGHTS

Creativity results in ideas, which in turn are converted into inventions. An invention constitutes the creation of a tangible item that did not exist previously. A patent is a legal document that describes and establishes claim to an invention and serves as a contract between the inventor (or the organization initiating the patent request) and the government. The patent provides the right to exclude others from producing, using, or selling the item for a period of 17 years from the date of patent issue. After that period of time, the general public has free access to the invention unless the effective life of the patent has been extended through patent improvements.

Generally, a patentable invention is an item that produces a new result, or something where the results are the same but the method used in producing the item is new. Patents may be obtained either for new products or processes, or for improvements on old products or processes. Some basic conditions of patentability are noted below.

A. The invention must have utility and must not have been previously known or described in any publication. The engineer, in applying for a patent, must be up to date in his field in order to know what has and has not been accomplished to date.
B. The invention must be more than an obvious advance to someone possessing ordinary skills in the technology. Of course, it is often difficult (if not impossible) to determine what constitutes "ordinary skills."

C. The invention must evolve as a result of *continuous* effort. Discontinuing work on an invention and finishing it later infers abandonment, which is not allowed. This does not preclude working on an invention part-time; however, it must be verified that the work is continuous.

D. The invention must not have been described in any other patent description. What may be considered as a new invention might already have been described as part of an existing patented item.

These conditions require a complete and accurate recording and verification of events, providing suitable evidence to prove inventorship. Initially, a written document fully describing the invention must be prepared by the inventor, dated, and witnessed. The witness will generally sign a statement verifying the date when the invention was first explained by the inventor(s). Second, dated documents leading to the invention are necessary to substantiate work progress and continuity. Such documentation may be supported by a working model in some instances. In many corporations, each engineer involved in research and product development is required to complete an engineering notebook. All good ideas, supported by sketches when appropriate, should be described and included in the notebook as soon as conceived. Patentable material is reviewed and signed by the engineer and a witness (generally his supervisor) on a regular basis.

Finally, with the necessary backup material available, the inventor will initiate a patent application which will include those claims considered significant for patent award and protection. Claims must be stated to the extent required to prove item utility and that the item does indeed represent a significant advancement; however, claims should not be stated in such detail as to allow someone else to design around the patent. The intent is to protect one's claims and not enable another individual to produce a comparable item and reduce the number of claims identified in the patent application.

A patent application is usually initiated by the inventor, with the assistance of a patent attorney, and submitted to the U.S. Patent Office for processing. One of the first actions of the patent office is to review the application, identify through investigation any known prior work that may be related, and to indicate to the inventor which (if any) claims it will allow. The inventor and the patent attorney must review the material as supplied by the patent office, make corrections and/or additions when required to further support the claims, and resubmit the application within a specified period of time (e.g., 6 months). This process continues until a patent is either issued by the patent office or disallowed. In the interim, patentable items are often produced with the identification "Patent Pending" inscribed on them. This is done to discourage prospective competitors from producing a comparable item.

While patents are basically intended to protect inventors of new product or process designs, other forms of creativity are covered by copyright registration. Copyright registration provides protection for a 28-year period, and usually applies to the results of an individual's personal intellectual effort, such as published textbooks, special reports, and articles. Registration of copyrights follows the same general process as patent requests but is the responsibility of the Library of Congress instead of the U.S. Patent Office. An individual author, his legal representative, or the corporation that publishes the work will initiate the copyright application. The registrar of copyrights is officially empowered to initiate certificates of copyright registration to those who comply with the requirements.

When considering the legal aspects of patents and copyrights, an engineer should not only investigate the requirements as specified by the U.S. Patent Office and/or the Library of Congress, but should look into the policies imposed by the corporation or agency where he or she is employed. Many corporations will lay claim to all or a part of the benefits received as a result of a patent or copyright. Such policies are usually stated through patent agreements between the engineer and the corporation as a condition of initial employment. In dealing with government agencies, procurement policies often require that any patents/copyrights resulting from work accomplished using federal or state funds assume government ownership as developed. On the other hand, the results of creative work accomplished with private funding revert back to the corporation or the individual responsible. In any event, an engineer may wish to investigate the terms and conditions relative to patents and copyrights prior to accepting job employment.

15.3 CONTRACT TYPES, INCENTIVE/PENALTY PROVISIONS, AND NEGOTIATIONS

The engineer needs to have some understanding of contracts since he or she is often affected by the type of contract imposed. Fixed-price contracts are tightly controlled, with the contractor (e.g., producer) assuming most of the risk. Engineering design should be fairly well established, as changes subsequent to contract negotiation may be quite costly. On the other hand, cost-reimbursement type of contracts (i.e., cost plus fixed fee, cost plus incentive fee) are more flexible in terms of making changes after initial contract negotiation, and the bulk of the risk is assumed by the customer or contracting agency. In any event, the engineer should have some feeling relative to the extent of design definition required and what can and cannot be done by virtue of various contractual requirements.

As a second point, the engineer participates from both a technical and a cost-estimating standpoint in the initial preparation of proposals that lead to contract negotiations (refer to Section 3.4F). When negotiations actually take place, the engineer often participates once again relative to the interpretation of specifications and the technical aspects of task accomplishment. Throughout negotiations, the intended scope of work may change, and these changes must be evaluated for their impact on other proposed line-item tasks. In essence, engineers directly support the contract negotiations in many different respects.

The sections below discuss briefly some of the salient points of contracts and their effects on engineering. This discussion presents an overview of the most common contract types, contract negotiations, progress payments and incentive/penalty provisions, and warranties. The discussion should be viewed in terms of risks to the engineer's organization and how engineering decisions affect the magnitude of the risk assumed.

A. Description of Major Contract Types

Major categories of contracts include fixed-price, cost-reimbursement, time and material, and letter agreements.

1. *Firm-fixed-price contract.* A firm-fixed-price (FFP) contract is a legal agreement to pay a specified amount of money when the items called for by the contract have been delivered and accepted. No price adjustments are allowed for the contracted work after award regardless of the actual costs experienced on the part of the contractor. At a specified price, the contractor assumes all financial risks for performance, and his profits depend on his ability to initially predict cost, negotiate, and to subsequently control costs. In the application of this type of a contract, engineering design should be fairly well defined through appropriate specifications.

2. *Fixed-price-with-escalation contract.* This type of contract is virtually the same as a FFP contract, with the addition of an escalation clause to cover uncontrollable price increases or decreases. Escalation can be applied to both labor and material. Since there are many uncertainties relative to predicting the magnitude of escalation, an escalation ceiling is usually established with the contractor and customer sharing the risks up to that point. Escalated costs above the specified ceiling value are assumed by the contractor.

3. *Fixed-price-incentive contract.* Fixed-price-incentive contracts are appropriate where some cost uncertainties exist and there is an excellent possibility that cost reduction can be attained through good contractor management and by providing the contractor with some profit incentive. A target cost, a minimum cost, and a ceiling price are negotiated, together with a

final profit-adjustment formula. Profit adjustment, from the initial targeted profit, can be made based on total cost performance. Costs above the established ceiling price are assumed by the contractor.

4. *Cost-plus-fixed-fee contract.* Cost-plus-fixed-fee (CPFF) contracts are of a cost-reimbursement type, where the contractor is reimbursed for all allowable costs expended on the project. A negotiated fixed fee (e.g., 10 per cent of the estimated cost) is paid to the contractor upon completion of work. Although this fee is fixed in terms of a percentage of the total cost, fee increases or decreases may occur as the work scope of the contract changes. This is particularly true after initial contract negotiation, when the customer will accept contractor-generated change proposals or in instances where the customer will ask the contractor to perform work beyond the scope of the current contract. CPFF contracts are ordinarily applied to (a) research and advanced development work; (b) projects in which the scope and nature of the work cannot be specifically defined or accurately costed; and (c) programs in which production/construction specifications are incomplete.

5. *Cost-plus-incentive-fee contract.* Like the CPFF contract, the cost-plus-incentive-fee (CPIF) contract is intended to cover uncertainties in project performance. Allowable costs are paid to the contractor, together with additional incentive fee payments on designated accomplishments. At the time of negotiation, individual factors such as task schedules and specific product characteristics may be identified as items where incentives are to be specified in order to motivate the contractor to excel in these areas. Contract negotiation will result in a defined target cost, target fee, a minimum and maximum fee, and a fee-adjustment formula. At the end of the contract, the contractor's project performance will serve as the basis for fee adjustment.

6. *Cost-sharing contract.* Cost-sharing contracts are primarily designed for research and development and are most commonly applied to work accomplished with educational institutions and nonprofit agencies. Such work is jointly sponsored, and reimbursement to the contractor is in accordance with a predetermined sharing agreement. No fee is awarded; however, in lieu thereof, the contractor anticipates that the work accomplished will derive other benefits (e.g., a patentable item, acquisition of technical know-how).

7. *Time and material contract.* A time-and-material contract allows for the payment of actual services and materials expended in the performance of designated tasks. Services constitute incurred direct labor hours reimbursed on the basis of a negotiated fixed hourly or daily rate, and materials are reimbursed at cost. The rates used include direct and indirect

labor, overhead, and profit. This type of contract is employed when the extent and duration of work cannot be determined and when costs cannot be estimated to any degree of accuracy. Appropriate applications include specific subcontracted research and engineering tasks, maintenance repair and overhaul services, and so on.

8. *Letter agreement.* Letter agreements are often used as a preliminary contractual document initiated with the intent of authorizing the contractor to start work immediately. These agreements serve as an interim means for providing a rapid response to an identified need that otherwise might be delayed pending the negotiation of a definitive contract. Letter agreements do not include total pricing arrangements; however, an upper-limit dollar amount is usually specified to preclude excessive spending. Under letter agreements, all costs incurred by the contractor for work accomplished are fully reimbursed by the customer. Ultimately, letter agreements should be converted to a formalized contract.

Associated with each major type of contract are questions concerning the schedule of payments. When will the contractor be reimbursed for the successful completion of contracted work? What is the magnitude of expected reimbursement(s)? Are payments to be made at the completion of each line--item task? For incentive contracting, what is the incentive/penalty payment plan? These and comparable questions are significant, particularly for large contracts, since the customer is generally tied to a specific budgeting cycle, and the contractor must offset operating costs without going too far into debt. Thus, a payment schedule of some type should be developed.

Figure 15-1 indicates one approach, which ties progress payments to the successful completion of four significant project milestone accomplishments (i.e., formal design reviews).[3] From the standpoint of the customer, this approach motivates the contractor to place emphasis in these areas with the intent of ensuring success. On the other hand, the contractor will argue for more frequent payments earlier and state that such payments should not be related to milestones that may be difficult to meet.

If incentive contracting is used, an incentive/penalty plan should be established as a supplement to the schedule for progress payments. Such a plan will specify the application of incentive and penalty payments to significant project milestones and/or to demonstrated product performance and effectiveness characteristics. In other words, incentive/penalty factors may be ap-

[3]Another common approach is for the customer to initiate payment upon the successful completion and acceptance of each line item specified in the contract. Thus, the identification of significant line items becomes important when determining the schedule and the size of each payment.

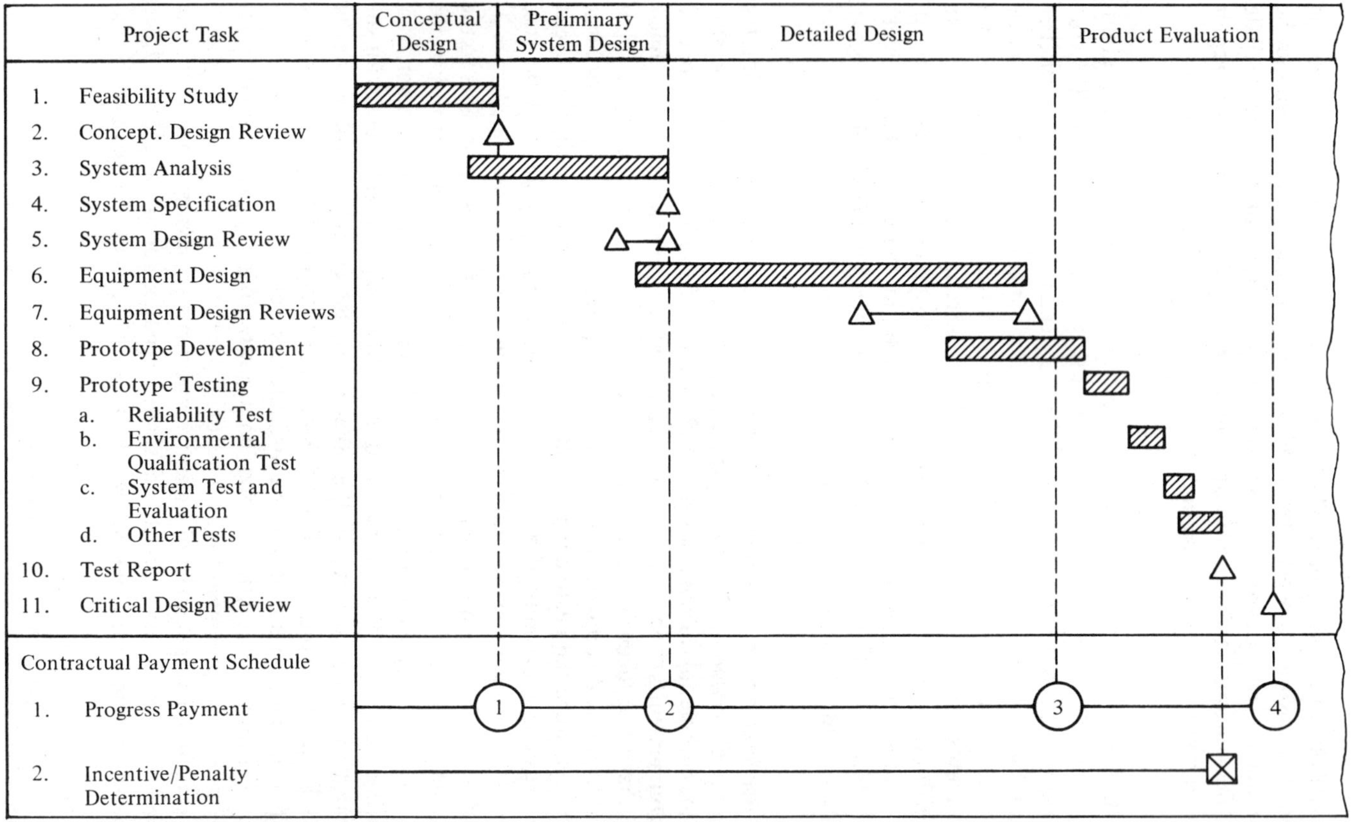

Figure 15-1. Schedule of contractual payments.

plied to both the performance of the contractor in project completion and to the performance and effectiveness of his product in field. Figure 15-1 shows the application of an incentive/penalty plan to the successful completion of prototype testing.

When applying incentive factors, which emphasize the positive aspects from the contractor's standpoint, one must also deal with the negative aspects (i.e., penalty factors), which must be just as strongly stated in order to influence the contractor's decision-making process. The next step is to determine the magnitude of the incentive/penalty values imposed. Certainly the incentive(s) should be large enough to promote a conscientious effort to meet the desired objective(s). On the other hand, incentives must be within reason in view of budget constraints and the total value of the contract. In addition, the ultimate budgeted funding for the customer must not only cover the cost of the completed contracted line-item tasks but an additional sum of money designated for the contractor should the maximum incentive payment be justified.

In developing an incentive/penalty payment plan, it is necessary to identify the parameters to which incentives and penalties are to be applied. In many instances, more than one parameter is indicated, resulting in a multiple structure. The appropriate sum of money for each incentive is difficult to determine and will depend on the type of product and the importance of the item to which the incentive is to be applied. It is unlikely that all selected parameters will be equally important; therefore, it will be necessary to assign an *importance value* or *weighting* for each parameter and to estimate the magnitude of the incentive/penalty values accordingly. An example of a multiple approach, involving two product characteristics, is illustrated in Figure 15-2. A target value is established based on specification requirements. This target value may also be considered as a contract or warranty value. If, after test and evaluation, the measured value is an improvement over the target value, an incentive fee is awarded to the contractor at a designated time in the project life cycle (refer to Figure 15-1). On the other hand, if the measured value represents a degraded condition, the contractor is liable and a penalty payment is due.

The incentive/penalty amounts are reflected in the individual payment plans. Weighting is inherent in the designated amounts indicated by the ordinate values and the established sharing ratios (SRs). Nearly all incentive plans assume the form of a sharing arrangement, generally expressed as a percentage ratio. For example, the 40/60 SR for the MTBM plan indicates that if the measured MTBM were greater than the target figure of 200 hours, the value of dollars indicated on the ordinate would be split, with 40 per cent going to the customer and 60 per cent being paid to the contractor. Added incentive would be provided if the MTBM exceeded an upper confidence limit of approximately 238 hours, with a split of 20 per cent going to the customer and 80 per cent going to the contractor. Conversely, if the measured

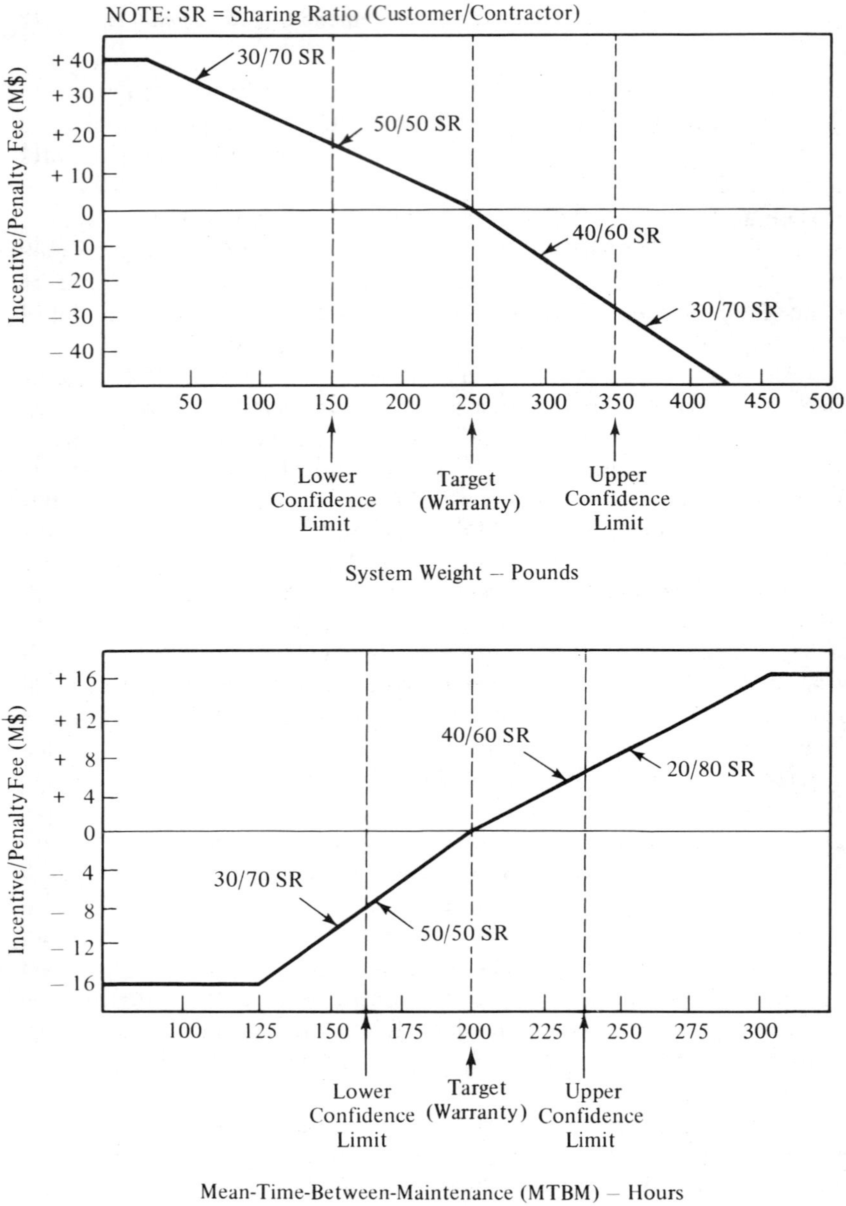

Figure 15-2. Multiple incentive/penalty plans.

MTBM fell below the target objective, a penalty payment of 50 per cent of the indicated ordinate value would be paid by the contractor. Incentive/penalty fees are directly applied to one or more variables, which are controlled by the contractor through the proper application of management and engineering skills.

The aspect of incentive/penalty payment plan development and implementation is a comprehensive subject area. There are all kinds of variations possible, and the selected approach and application are highly dependent on the type of product, the program functions involved, product development risks, budget limitations, and the formal negotiated contractual structure.

B. Contract Negotiations

The objective of contract negotiation is to achieve the most advantageous contract agreement from the standpoint of technical requirements, pricing, the type of contract imposed, and the payment schedule. Obviously, the customer and the prospective contractor each views this objective relative to his own individual position in terms of the risks associated with the numerous options that are available to him. At one extreme is the firm-fixed-price contract, where the risks are primarily assumed by the contractor; the cost-plus-fixed-fee contract covers the other extreme, where the customer assumes most of the risk. Between these two extremes there are a number of relatively flexible options.

Within the contract structure itself, there may be a variety of terms and conditions relating to:

1. *Management and technical requirements*—program planning and specification requirements, line-item tasks to be accomplished, task schedules, product delivery requirements, and so on.
2. *Pricing information*—total price and price per line-item task.
3. *Contractual payment schedule*—progress payments, schedule, and payment amounts.
4. *Incentive/penalty factors*—application and magnitude of incentives and/or penalties.
5. *Warranty/guarantee provisions*—time period, product-use stipulations, liability, and consumer/producer risk.

The ultimate achievement of definitive contract terms and conditions is accomplished through formalized negotiations between the customer and the prospective contractor. Negotiations per se can assume a simplified approach involving several representatives from each side meeting on a given date to discuss requirements in general, or the negotiation process may be relatively complex, with a team of specialists meeting for an extended time period to discuss each proposed line item in detail. In a more formalized negotiation, the customer, in response to a contractor's proposal, will interrogate the contractor relative to the validity of his proposed technical approach and the price. Questions along a technical line will attempt to ascertain whether or not the contractor has justified that his technical approach is the best (based on design trade-offs) and that he has the technical expertise and experience to follow through in developing and producing the proposed item. On a cost basis, the object is to determine that the contractor's price is fair and reasonable and that it was developed through a logical cost analysis. From the contractor's standpoint, the negotiation takes the form of defending the proposal as initially submitted to the customer. The contractor may be required to provide any amount of supporting material to help convince the customer that he is thorough, honest, and offering the best deal possible.

Negotiation in general is an art and usually requires some strategy on both sides. Initially a plan is established which identifies the location where negotiations are to be held and the agenda. The contractor and the customer will each identify the personnel who will participate in the negotiation. Prior to negotiation, individual meetings will be held to determine what should and what should not be brought up. Also, decisions are made concerning the feasibility of possible counteroffers in the event that disagreements cannot be resolved. During the formal negotiations at the bargaining table, both sides will assume the minimum-risk position, considering the contractual terms and conditions mentioned above. Interruptions will occur, short strategy meetings will be held to discuss events that have occurred, attempts will be made to gain some sympathy from the opposition, and hopefully an agreement will be made after some compromises on both sides. This process may evolve through a number of iterations consuming more time than initially anticipated.

The engineer's role in contract negotiation is to provide the necessary technical support, as required. He or she should respond only to the questions asked and should take direction from the lead contract negotiator representing his organization. Conveying the wrong information and/or providing too much elaboration in any given area may prove to be harmful to the negotiator's bargaining position.

QUESTIONS AND PROBLEMS

1. What are the implications of product liability to a manufacturer?
2. What is meant by "strict liability"?
3. As a design engineer, what steps would you initiate to ensure that future liability risks associated with your product are minimized?
4. How does product liability relate to design? Production? Test and evaluation? Logistic support? Advertising and marketing?
5. What roles are engineers likely to fill in litigation?
6. What benefits are derived through the use of the reliability FMEA in the early design process?

7. Identify some products that have been recalled by the manufacturer for the purpose of correcting a deficiency and/or averting a possible safety hazard.
8. What is a patent? Identify some of the basic requirements for patentability.
9. What steps should a design engineer follow when dealing with potential patentable material?
10. What function does the U.S. Patent Office perform in processing patent applications?
11. What is a copyright registration? What are the requirements for copyrighting material?
12. Assume that you have been assigned as a proposal manager in a corporation and have been asked to participate in contract negotiations with the customer. If given the choice, what type of contract would you seek through negotiation? Why?
13. Assume that you are employed by a government agency and have been assigned to negotiate a rather large contract with a potential contractor. What type of contract would you try to impose? Why?
14. Assume that you are responsible for a major subcontract. Develop a hypothetical program plan for a product of your choice, describe the contractual requirements that you would impose, and identify a payment schedule. Discuss the reasons behind your decisions.
15. As a customer representative negotiating a contract with a potential contractor, you have identified certain product characteristics (i.e., product range, effectiveness, etc.) that are considered extremely critical to your needs. Through contract negotiation, how would you help to ensure that the contractor will apply the necessary emphasis in fulfilling your needs?
16. When establishing multiple incentives under incentive contracting, what steps would you follow?
17. Define what is meant by an incentive/penalty sharing ratio. How is it applied? How does the SR relate to producer/customer risk?
18. Describe the engineer's role in contract negotiation. How can the engineer help in negotiations? How can he hinder his organization's bargaining position in negotiations?

16

PRODUCT SAFETY FUNCTION

Safety can be broadly defined as "freedom from those conditions that can cause injury or death to personnel, damage to or loss of equipment or property."[1] Safety includes both *personal safety,* protecting the individual against death and physical injury, and *equipment safety,* the protection of equipment against damage. Equipment damage may in turn result in personal injury; so the interactions between personal safety and equipment safety are significant.

16.1 BACKGROUND AND INTRODUCTION

Until fairly recently, safety has not received a great deal of emphasis; however, with the ever-increasing numbers of personal accidents and product losses, government legislation and controls have been implemented to create and maintain a safer environment overall. The Williams-Steiger Occupational Safety and Health Act (OSHA) of 1970 is a prime example of legislation that not only imposes industrial safety requirements per se but shifts the responsibility for safety from the individual employee of a firm, or the consumer of a product, to the product manufacturer. This fact is supported by the discussion covering product liability in Section 15.1.

OSHA, which became an official part of national labor law on April 28, 1971, was designed to ". . . assure so far as possible every working man and

---

[1] W. P. Rodgers, *Introduction to System Safety Engineering,* John Wiley & Sons, Inc., New York, 1971.

woman in the Nation safe and healthful working conditions and to preserve our human resources . . . ."[2] Compliance with the Act imposes requirements on the part of both the employer of personnel and the employee. The employer is required to provide a work place free of safety and health hazards and to comply with certain standards directed toward this end. National standards have been prepared and applied across the board to many industrial lines, covering such areas as personal protection; the handling of hazardous materials; fire protection; electrical practices; housekeeping and sanitation; maintenance practices; medical examinations and first aid; protective clothing; and environmental conditions relating to air and water pollution, noise, illumination, radiation, temperature, and humidity. In addition, individual standards have been prepared to cover requirements applicable to particular industries, including construction, pulp and paper, textiles, maritime products, and mining. The employer is responsible for initiating and maintaining the necessary facilities and resources to allow for compliance with all applicable standards. This is often accomplished through the establishment of a formalized safety program as an official ongoing organization function.

The employee, on the other hand, must become thoroughly familiar with OSHA standards and must comply with the requirements of the organization's safety program. This includes not only following the recommended procedures in the day-to-day aspects of job accomplishment, but pertains to designing a product in such a manner as to preclude the possibility of future accidents. It can be stated that accident prevention begins at the early stages of product development, and this implies that the design engineer is responsible for many accidents. Hence, the engineer must incorporate the necessary features in the design that will in the future prevent accidents and protect personnel. Such features must take into account *Murphy's law* which addresses the following basic premises:

A. No matter how remote the possibility, any task that *can* be done incorrectly *will* eventually be done that way.
B. No matter how difficult it is to damage equipment, it will occur.
C. Equipment that *can* fail *will* fail at the most inappropriate time.
D. Operating and maintenance instructions in task accomplishment will be ignored.

In other words, the engineer must not only avoid possible hazardous conditions in the product itself, but must also view the operator and the maintenance man as potential hazards.

---

[2]Information concerning OSHA is included in the following publication: OSHA 2056, "All About OSHA—The Who, What, Where, When, Why and How of the Occupational Safety and Health Act of 1970," Occupational Safety and Health Administration, U.S. Department of Labor, Government Printing Office, Washington, D.C.

OSHA standards, combined with the emphasis on consumer protection, have imposed rather stringent requirements on product manufacturers. Many of these requirements can be related directly to product design and the responsible engineer. Failure to meet such requirements often results in legal actions initiated against the manufacturer, which can turn out to be quite costly. Thus, many organizations (large and small) have established formal safety programs specifying organization responsibilities, tasks to be accomplished, training courses, related product design analyses, review and reporting procedures, and so on, to create the necessary awareness of safety throughout all organization levels.

16.2 ESTABLISHING A SAFETY PROGRAM

In many organizations, safety programs have been somewhat bifurcated because of different degrees of emphasis on *industrial safety,* dealing with in-plant operations and *system/product safety* oriented to product design for consumer use. As a result, separate program efforts have often evolved where the former has been placed under the direction of a safety manager (or coordinator) responsible for internal employee safety, and the aspects of safety dealing with consumer product design have been handled to varying degrees through different engineering disciplines (i.e., reliability engineering, maintainability engineering, human factors, logistic support, etc.). With the advent of the new emphasis on *total* safety (as discussed above), it is appropriate to ensure that all safety activity is integrated across organizational lines and coordinated through a single functional entity. This integration and coordination is accomplished by the establishment of a formal safety program and the preparation of an integrated safety plan.

A. Integrated Safety Plan

The integrated safety plan includes both management and engineering activities and constitutes the mechanism through which the responsible safety manager monitors and controls organization-wide industrial safety practices affecting all in-house programs as well as the applicable safety-oriented functions associated with individual projects.[3] The following outline provides some guidelines to the content of a safety plan.

1. *Objectives.* State the objectives of the safety program in terms of both industrial safety in response to OSHA standards and system/product design safety as applied to in-house projects.

---

[3]Safety activities for individual projects should be included, in detail, in the applicable product plans discussed in Section 3.4 (Advanced Product Planning). Such activities should directly support the organization-wide integrated safety plan discussed herein.

2. *Definitions.* Owing to the inherent problems of communications, it is appropriate to define such terms as hazard (and hazard categories), safety, safety and hazard analysis, safety testing, damage, and risk.[4]
3. *References.* List the specifications and standards applicable to the activities of the entire organization.
4. *Management Requirements.*
   a. Policies and procedures.
   b. Safety organization and responsibilities.
   c. Safety tasks and milestone schedules.
   d. Program reviews and facility inspections.
   e. Problem identification, evaluation, and corrective action.
   f. Safety program status reporting.
5. *Technical Requirements*
   a. Safety criteria.
   b. Safety and hazard analysis.
   c. Related activities (e.g., reliability, maintainability, human factors).
   d. Safety testing.
6. *Training Programs.* Identify the organization's plans for management and engineering personnel training.
7. *Appendices.* Include support requirements as appropriate.

The integrated safety plan follows the same overall approach as the planning process accomplished for product research, design, production/construction, test and evaluation, and product use and logistic support (refer to chapter 3). The basic elements of policies and procedures, organization, tasks and schedules, design support, program reviews, and status reporting are included. Because of the relative significance of safety and the consistency required throughout the organization, the integrated safety plan receives top management support and assumes a higher level of importance than normally provided an individual product plan. All engineering and other personnel in the organization should be thoroughly familiar with the content and requirements specified in this plan.

B. Safety Criteria

Criteria refer to the qualitative and quantitative input characteristics that specify the requirements for both industrial and system/product safety. For industrial safety, such criteria are derived from many occupational and health standards. For instance, one such standard dealing with general industry specifies requirements for the following:

---

[4]A good reference for terms and definitions, safety program requirements, and so on, is: W. Hammer, *Handbook of System and Product Safety,* Prentice-Hall, Inc., Englewood Cliffs, N.J., 1972.

1. *Walking surfaces*—holes in floors, open-sided platforms, ladders, stairwells, aisles and passageways, and handrails.
2. *Means of egress*—pathways from the work place or desk to the facility exits, lighting, and signs.
3. *Powered platforms, man-lifts, and vehicle-mounted work platforms*—external platforms for building maintenance, elevators, and pulley arrangements.
4. *Occupational health and environmental control*—air contaminants, noise, exposure to dust and dangerous chemicals, radiation, and water pollutants.
5. *Handling of hazardous materials*—compressed gases, acetylene, hydrogen, oxygen, flammable and combustible liquids and materials, and explosives.
6. *Personnel protective equipment*—equipment and clothing for eyes, face, respiratory, hands, head, and foot protection against environmental hazards of a chemical, radiological, physical, or mechanical nature.
7. *General housekeeping and sanitation controls*—waste disposal, sewage provisions, drinking water, cleaning materials, and eating facilities.
8. *Medical and first aid*—emergency first aid, clinic or infirmary facilities, and requirements for employee physical examinations.
9. *Fire protection*—portable fire extinguishers standpipe and hoses, automatic sprinklers, and chemical provisions.
10. *Machinery and machine guarding*—safety requirements for the operation and maintenance of all in plant machinery and hand tools.
11. *Electrical provisions and equipment*—installation of wiring and equipment, and protective provisions for equipment usage.

Criteria more directly applicable to system/product design (i.e., the inherent characteristics of an equipment item) cover requirements concerning the incorporation of fail-safe provisions, the size and weight of an item, elimination of extruding pieces and sharp edges, and so on. Some of the major safety-oriented features that should be considered in product design are noted in the checklist in Appendix C.

Many of the safety standards currently in use have been developed by various technical societies and industry associations and convey the minimum criteria that should be observed in product design, production/construction, test and evaluation, and product use and support in the field.[5] Since such standards are developed with safety in mind, the engineer in his

---

[5]Some specific examples of technical society/industry association standards are the National Electric Code (No. 70), Lightning Protection Code (No. 78), and Flammable and Combustible Liquids Code (No. 30) generated by the National Fire Protection Association (NFPA); Boiler and Pressure Vessel Code of the American Society of Mechanical Engineers (ASME); and Impact Protective Design of Occupant Environment—Transport Aircraft (ARP 767) developed by the Society of Automotive Engineers (SAE).

goal toward accident prevention should become thoroughly familiar with these prior to project implementation.

C. Safety Program Tasks

Tasks oriented to industrial safety and system/product safety may be accomplished through various types of program activities. In other words, although the corporation or agency may basically address the subject of safety through an integrated safety plan, the many specific aspects of safety are usually handled through a variety of organizational elements of both a staff and project nature. For instance, industrial safety, which deals primarily with OSHA requirements across all organizational lines, may fall under the cognizance of the industrial relations and personnel administration function, as illustrated in Figure 14-1. On the other hand, system/product safety is usually accomplished through a combination of functional design and design support activities for various projects. Such activities concerned with different aspects of safety are functional design (i.e., electrical design, mechanical design, etc.). reliability engineering, maintainability engineering, human factors engineering, logistics engineering, and test and evaluation. These activities may be accomplished through a project organization, a staff function, or a combination thereof (refer to Chapter 10) and are oriented to the objectives outlined in Figure 16-1. Upon close examination of the activities identified in the figure, it is assumed that major safety program tasks can be combined and categorized as follows:

1. Identify and establish safety design criteria (i.e., the initial definition of design requirements as discussed in Section 16.2B).
2. Accomplish design reviews and analyses to determine and support design decisions, and to subsequently assess a chosen design configuration in terms of compliance with the established safety requirements.
3. Establish a testing program to verify that safety requirements have been satisfied.
4. Implement a procedure for corrective action to resolve problem areas dealing with noncompliant situations.
5. Conduct periodic program reviews to evaluate the status of safety activities against the integrated safety plan.

Whether we are discussing industrial safety or system/product safety, there are still the aspects of criteria determination, design analysis, testing, program reviews, and the requirements for a feedback and corrective-action procedure that must be addressed. These aspects are an inherent part of the process discussed in Chapter 5 and illustrated in Figure 5-1, independent of the type and nature of the system.

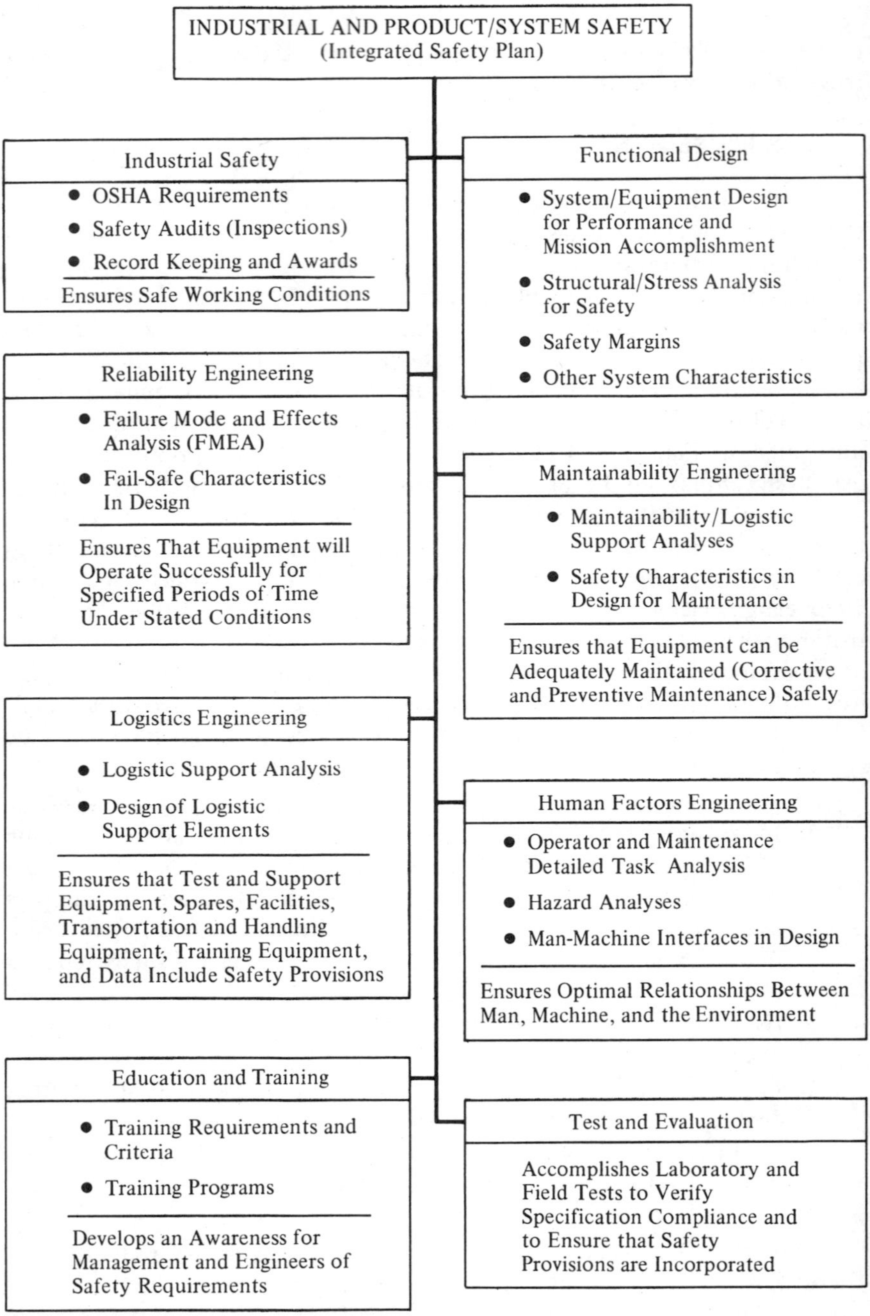

Figure 16-1. Major safety activities.

When reviewing Figure 16-1 and the preceding sections, there are a number of activities where the engineer is directly involved, along with others who have a significant impact on his or her day-to-day accomplishments. Three of these activities, of particular significance to safety, are discussed next.

*1. **Hazard analysis.*** During conceptual design or early in the preliminary system design phase, a hazard analysis is initially conducted with the intent of identifying potential hazardous conditions which could at some future time cause personnel death or injury as well as equipment damage. As the design progresses, the hazard analysis is updated to reflect the system/product configuration represented through formalized design documentation. The hazard analysis, conducted as part of the overall systems analysis effort, reflected in block 2 of Figure 5-1, includes the following basic information:

a. *Description of hazard.* Mission requirements, system operational requirements, and the system maintenance concept are reviewed to identify possible hazardous conditions. Past experience on similar systems or products utilized in comparable environments serves as a good starting point. Hazardous conditions may include acceleration and motion, electrical shock, chemical reactions, explosion and fire, heat and temperature, radiation, pressure, moisture, vibration and noise, and toxicity.

b. *Cause of hazard.* Possible causes should be described for each identified hazard. In other words, what events are likely to take place in creating the hazard?

c. *Identification of hazard effects(s).* Describe the effects of each identified hazard on both personnel and equipment. Personnel effects may include injuries such as cuts, bruises, broken bones, punctures, heat exhaustion, asphyxiation, trauma, and respiratory or circulatory damage.

d. *Hazard classification.*[6] Hazards may be categorized according to their impact on personnel and equipment:

(1) Negligible hazard (category I)—such conditions as environment, personnel error, characteristics in design, errors in procedures, or equipment failures that will not result in significant personnel injury or equipment damage.

(2) Marginal hazard (category II)—such conditions as environment, personnel error, characteristics in design, errors in procedures, or equipment failures that can be controlled without personnel injury or major system damage.

---

[6]Hazard classifications are identified in: W. P. Rodgers, *Introduction to System Safety Engineering,* John Wiley & Sons, Inc., New York, 1971; and MIL-STD-882, "System Safety Program for Systems and Associated Subsystems and Equipment: Requirements For," Government Printing Office, Washington, D.C.

(3) Critical hazard (category III)—such conditions as environment, personnel error, characteristics in design, errors in procedures, or equipment failures that will cause personnel injury or major system damage, or that will require immediate corrective action for personnel or system survival.

(4) Catastrophic hazard (category IV)—such conditions as environment, personnel error, characteristics in design, errors in procedures, or equipment failures that will cause death or severe personnel injury, and/or complete system loss.

Depending on the effect, hazards will be categorized in one of the four classifications.

e. *Anticipated probability of hazard occurrence.* Through statistical means, estimate the probability of occurrence of anticipated hazard frequency in terms of calendar time, system operating cycles, equipment operating hours, or equivalent.

f. *Corrective action or preventive measures.* Describe the action(s) that can be taken to eliminate or minimize (through control) the hazard. Hopefully, all hazardous conditions will be eliminated; however, in some instances it may only be possible to reduce the hazard level from category IV to one of the lesser critical categories.

The hazard analysis basically serves as an aid in initially establishing design criteria and as an evaluation tool for the subsequent assessment of design for safety. Although the format may vary somewhat, the hazard analysis may be applied in support of both industrial safety and system/product safety requirements.

2. Failure-mode-and-effects analysis. A failure mode-and-effects analysis (FMEA) is performed as a reliability engineering function during the early stages of preliminary system design to identify possible problems that could develop as a result of system failures.[7] The FMEA is oriented to equipment only and does not generally cover the effects of human actions on equipment. The objective is to determine the ways in which equipment can fail and the effects of such failures on other elements of the system. The FMEA includes the following information:

a. *Item identification.* Identify each significant system component that is likely to fail. This can be accomplished in conjunction with the functional

---

[7]Sometimes the FMEA is known as the "failure modes, effects, and criticality analysis (FMECA)."

analysis described in Section 5.2A, where functional flow diagrams are employed to identify subsystems, units, assemblies, and components.

b. *Description of failure modes.* Define the most probable modes of failure for each identified item. Failure modes may be classified as ruptures, fractures or cracks, voids, short and open circuits, physical separations, and so on, and are related to the operational modes that the system experiences through the performance of its designated mission. In other words, how will the equipment fail, and under what operating or environmental conditions is the equipment subjected to when the failure occurs?

c. *Cause of failure.* The anticipated cause of failure should be described for each instance. Typical causes might include abnormal equipment stresses during operation, aging and wear-out, poor workmanship, defective materials, damage due to transportation and handling, or operator and maintenance induced faults.

d. *Possible effects of failure.* Describe the most probable effects as a result of each identified failure. Effects may range from complete system destruction to partial system operation. To what extent does the failure impact on successful accomplishment of the mission of the system?

e. *Probability of occurrence.* Through statistical means, estimate the probability of failure occurrence. Probabilities of occurrence initially may evolve from experience factors or through reliability allocation (refer to Section 5.2A) and will be based on reliability prediction data as system development progresses.

f. *Criticality of failure.* Failures may be classified in terms of criticality as follows: [8]

(1) *Minor failure*—any failure that does not degrade the overall performance and effectiveness of the system beyond acceptable limits.

(2) *Major failure*—any failure that will degrade the system performance and effectiveness beyond acceptable limits but can be controlled.

(3) *Critical failure*—any failure that will degrade the system beyond acceptable limits and could create a safety hazard if immediate corrective action is not initiated.

(4) *Catastrophic failure*—any failure that could result in significant system damage, such as to preclude mission accomplishment, and could cause deaths and personnel injuries.

Failures are classified in any one of the four categories, depending on the defined failure effects.

---

[8]Reliability criticality classifications are similar in nature to the hazard categories identified earlier. Refer to: W. Hammer, *Handbook of System and Product Safety,* Prentice-Hall, Inc., Englewood Cliffs, N.J., 1972.

g. *Possible corrective action or preventive measures.* Describe the action(s) that can be initiated to reduce the probability of failure occurrence or to minimize the effects of failure.

The FMEA can be effectively utilized to provide a logical review of system failure modes, to ensure that the results of any given failure will cause minimal damage to equipment and will not create hazardous conditions for system operators and maintenance personnel. The FMEA and the hazard analysis, described earlier, provide an excellent means to assure the incorporation of system/product safety characteristics in design.

3. Accident/incident investigation and analysis. The hazard analysis and FMEA are methods that are primarily employed throughout system/product design to preclude or minimize possible safety problems that may occur later. However, it is unrealistic to assume that *all* problems will be eliminated, because accidents *will* happen. The occurrence of an accident should lead to an investigation and an analysis of the cause-and-effect relationships. The analysis results will reveal deficiencies that should cause the initiation of corrective action. A description of the accident, the cause and effects, and the corrective action taken should be recorded in an accident/incident report for historical purposes.

D. Program Reviews, Inspections, and Status Reporting

Enforcement of safety requirements and implementation of the integrated safety plan are assured through formal program reviews and inspections. Management reviews should be conducted on a regularly scheduled basis to ensure that the overall objectives of the corporation/agency's safety program are met. Formal design reviews (discussed in Section 5.4) are scheduled at appropriate times during the life cycle of a project to cover the design aspects of safety through review of hazard analyses, FMEAs, and related tasks. Both scheduled and unscheduled inspections of facilities and premises are conducted to ensure compliance with OSHA industrial safety requirements.[9] The combination of these reviews, supported by periodic status reports covering safety program activities, will create the emphasis and awareness necessary to meet current industrial and system/product safety requirements.

---

[9]OSHA requirements specify that inspections of corporation/business/agency facilities and premises may be conducted at any time (without delay) and that both employer and employees may be questioned either privately or publicly concerning the enforcement of safety standards. If violations are noted, inspectors may issue citations, and the applicable violation(s) must be corrected within a specified period of time or the responsible employer will be penalized. If the employer disagrees with the citation, a protest may be initiated, resulting in further review and discussion. There are formalized OSHA procedures covering industrial safety practices.

QUESTIONS AND PROBLEMS

1. Why is safety important? Describe the basic differences between industrial safety and system/product safety. How do they relate to each other?
2. In what way is the design engineer responsible for safety?
3. What are some of the implications of OSHA on a corporation, business, or agency?
4. How do OSHA requirements relate to product liability as discussed in Chapter 15?
5. Why are safety standards necessary? List at least five universally applied safety standards.
6. Identify the various design activities associated with system/product safety and briefly describe the objectives of each.
7. What is the FMEA and its purpose? When is it applied?
8. Select an equipment item of your choice and accomplish a FMEA.
9. What is a hazard analysis and its purpose? When is it applied?
10. Select an area of your choice and conduct a hazard analysis.
11. Why is it important to conduct training programs in safety?
12. Select a corporation, business, or agency of your choice and describe the safety program being implemented. What organizational group is responsible for safety?
13. How are OSHA requirements enforced?
14. What is meant by "safety criteria"?
15. What is the purpose of conducting laboratory and field tests for safety?
16. Select a system or product of your choice. Develop a safety checklist and evaluate the system in terms of safety characteristics in design.

part five

Engineering and the Environment

The most nebulous (and yet important) aspect of engineering pertains to its interactions with ecological, economic, political, social, and technology considerations. This part discusses briefly the impact of these considerations on the engineering and related activities covered in parts One through Four and is represented by the outer ring in the conceptual illustration of the text content.

Conceptual Outline

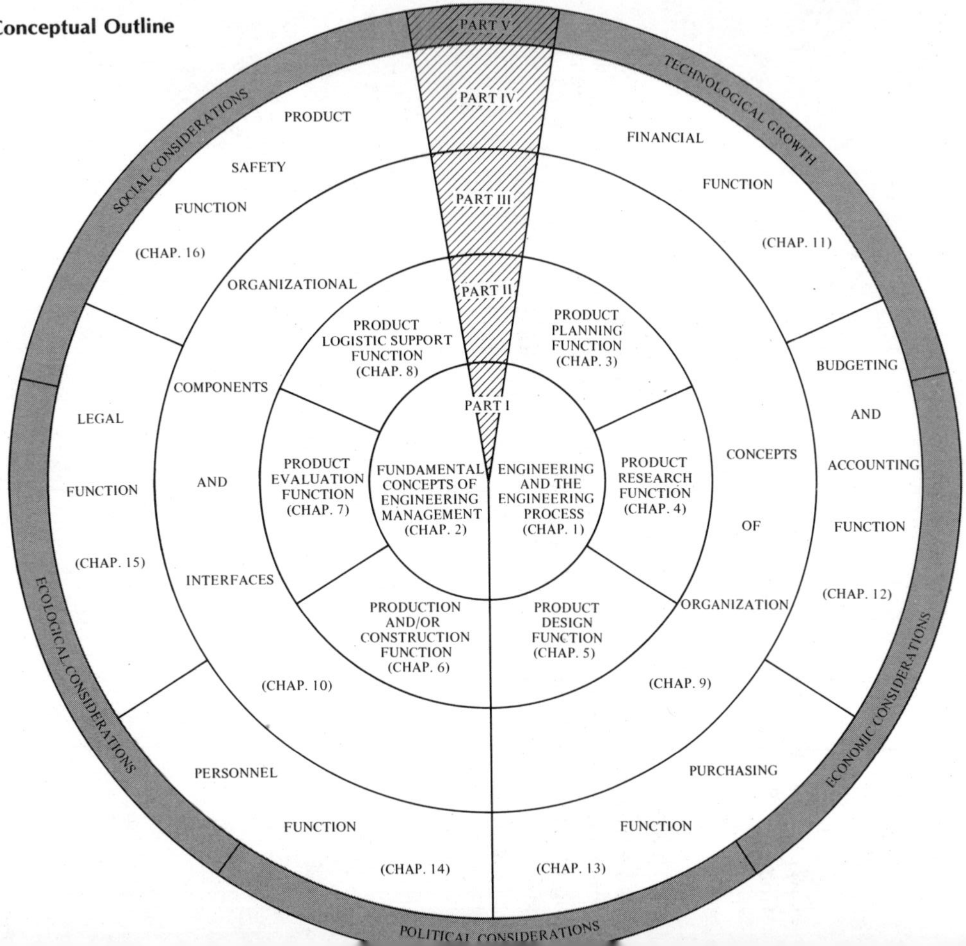

PART I	ENGINEERING IN MODERN TIMES
PART II	THE PRODUCT LIFE CYCLE
PART III	ORGANIZATION AND THE PRODUCT LIFE CYCLE
PART IV	RELATED FUNCTIONS AND INTERFACES
PART V	ENGINEERING AND THE ENVIRONMENT

17

ENGINEERING AND ENVIRONMENTAL CONSIDERATIONS

The previous chapters primarily dealt with the engineering process and related project activities. In the interest of attaining some additional perspective, engineering activities must now be addressed in the context of the total environment in which we live. Environment in this instance refers to the ecological, economic, political, social, and technology considerations that impact on engineering and in turn are impacted by the results of engineering. These considerations are closely related, and the interactions are many, as conveyed in Figure 17-1. Each of the environmental categories noted in the figure constitutes a rather complex subject in itself, with conditions changing continually. Changes in any one category will usually impact each of the other areas, and engineering decisions will be affected accordingly.

Discussion in the paragraphs to follow will touch briefly on the relationships between engineering and ecological, economic, political, and social considerations, with the objective of introducing some additional concerns that the engineer will have to address in system/product development, production, test and evaluation, operational use, and logistic support. Future engineering activities as related to these facets of the total environment will indeed include further scientific and technology challenges, which are discussed in Chapter 18.[1]

[1]The discussion presented herein is obviously extremely cursory in nature. However, it is hoped that the student will acquire an awareness relative to some of the external (and not too well defined) factors that will influence engineering decision making. The more relevant aspects of ecology, economics, political influences, and societal functions are covered briefly.

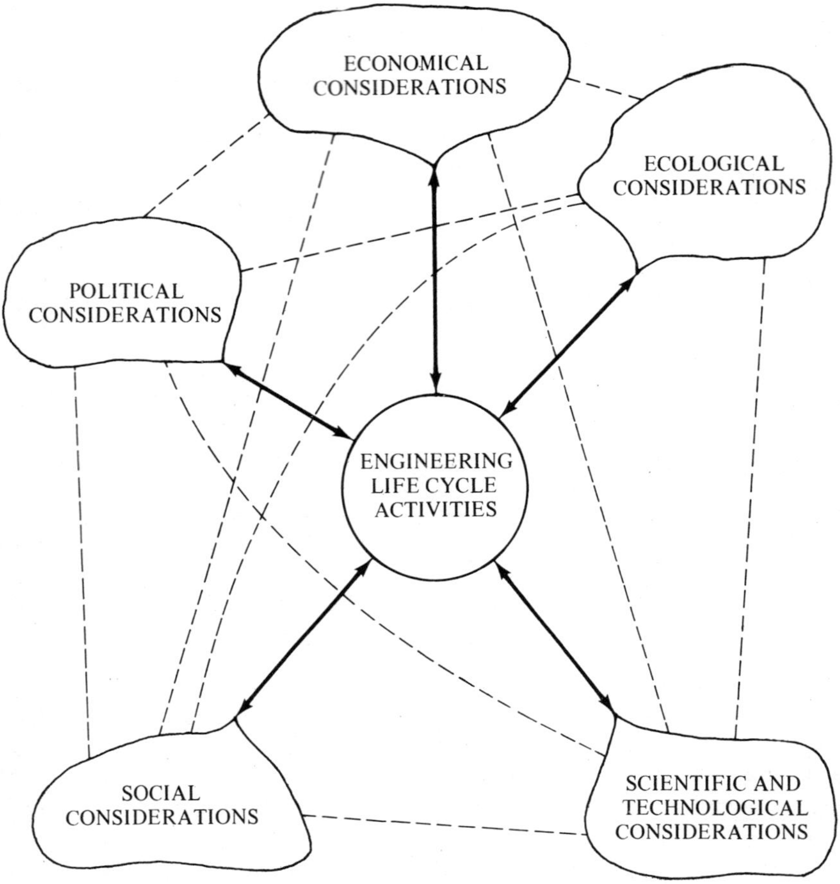

Figure 17-1. Engineering and the environment.

17.1 ECOLOGICAL CONSIDERATIONS

Ecology generally pertains to the study of the relationships between various organisms and their environment. This includes consideration of plant, animal, and human populations in terms of rate of population growth, food habits, living habits, reproductive habits, and ultimate death. In other words, we are basically talking about the conventional biological process as viewed in the broad sense.

In recent decades, the world population growth, combined with the technological changes associated with our living standards, has created a greater consumption of our resources, resulting in potential shortages which in turn has stimulated shifts toward establishing other means for accomplishing objectives. Concurrently, the amount of waste has increased significantly. The net effects of this have caused alterations to the basic biological process, and to some extent these alterations have been harmful. Of particular concern are those problems dealing with air pollution, water pollution, noise pollution, radiation, and solid waste management.

A. Air Pollution and Control

Air pollution may be defined as any gaseous, liquid, or solid material suspended in the air which creates an undesirable effect. Air pollution can result in health hazards to humans, which may be classified as episodes and chronic diseases.[2] During *episodes*, large numbers of individuals with particular susceptibilities will become ill or succumb and die as the atmospheric levels of pollution build up. *Chronic diseases* refer to the long-term individual effects from pollution and primarily relate to respiratory-system problems (e.g., emphysema, pneumonia). In addition to the direct effects on humans, air pollution can also cause considerable damage to plant life, which influences land use and results in losses through corrosion.

Human exposure to air pollution may be categorized as (1) personal exposure resulting from social habits such as smoking, (2) exposure under occupational or job conditions, and (3) exposure of small or large segments of the population to man-made or natural sources of environmental pollution. Personal exposure associated with social habits can be controlled only by the individual himself; however, occupational and general exposure to the environment can be altered by controlling the source where the contamination occurs or through the movement of the individuals involved.

Air pollutants known to cause harmful effects generally fall into one of the following five categories:[3]

1. *Particulate matter.* Particulates are small solid (or liquid) substances in the air resulting primarily from fuel combustion, incineration of waste materials, or industrial process losses. Individuals who breathe such substances into their lungs may suffer from respiratory disorders or may find that the incidence of respiratory ailments is increasing. In addition, the presence of particulate matter decreases visibility and often causes eyestrain.

2. *Sulfur oxides.* Sulfur oxides, which result form the burning of high-sulfur fuels, can irritate the upper respiratory tract and cause injury to delicate lung tissue.

3. *Carbon monoxide.* Carbon monoxide, which primarily comes from motor vehicle exhaust, is of course a poisonous gas. In low concentrations it

---

[2] L. G. Rich, *Environmental Systems Engineering,* McGraw-Hill Book Company, New York, 1973.

[3] The five categories of pollutants are included in the document: U.S. Department of Health, Education, and Welfare, U.S. Public Health Service, "The Effects of Air Pollution," Government Printing Office, Washington, D.C., 1966.

may cause headaches and nausea, impair mental processes, produce dizziness, and reduce the oxygen supply to the body. In high concentrations, it may cause death.

4. *Nitrogen oxides.* Most nitrogen oxides come from stationary combustion sources or from motor vehicle exhausts and may damage vegetation and contribute to photochemical smog. In addition, nitrogen oxide will corrode metals.

5. *Hydrocarbons.* Although there are numerous sources of hydrocarbon pollution, the major source constitutes the emissions as a result of the incomplete burning of gasoline in automobiles and in other vehicles with gasoline engines. Hydrocarbon pollution contributes (with nitrogen oxides) to photochemical smog, lung and eye irritation, and may inhibit vegetation growth.

At one time, it was estimated that the total mass of air pollutants emitted from man-made sources exceeded 200 million tons per year, and it was anticipated that this figure would rise with increased industrial activity, more automobiles on the road, and so on. As a result, the federal government initiated the Clean Air Act of 1963, the Air Quality Act of 1967, and the Federal Clean Air Act Amendments of 1970. The Environmental Protection Agency (EPA) was assigned the responsibility of establishing emmission standards and to ensure that such standards are adhered to. Air quality standards are in two parts, primary and secondary. Primary standards apply exclusively to human health; while secondary standards apply to property damage and the general welfare of the public.[4]

With the advent of the Federal Clean Air Act Amendments of 1970, engineers are required to ensure that newly developed products are in compliance with the EPA primary and secondary standards. This imposes some design restrictions that may be costly to the producer, particularly if the producer ignores the air quality standards initially and is required later to modify the product to attain acceptance. Thus, the engineer should become thoroughly familiar with these standards and ensure that proper consideration is given toward air quality requirements early in the product life cycle.

B. Water Pollution and Control

Water pollution constitutes any contaminating influence on a body of water brought about by the introduction of materials that will adversely affect the organisms living in that body of water. A significant measure of water quality is the dissolved oxygen content in the water, which is

[4]Specific coverage of air quality standards and the quantitative emission values for particulate matter, sulfur oxides, and so on, is included in: P. A. Vesilind, *Environmental Pollution and Control,* Duke Environmental Center, Duke University, Durham, N.C., 1974.

fundamental to aquatic life (i.e., fish, plant life, etc.). Population and industrial growth in the past have degraded the water quality of many lakes, rivers, and streams through the discharge of raw sewage, organic wastes, mercuric compounds, and municipal wastes. As a result, the federal government initiated the Water Quality Act of 1965, requiring water quality standards for (1) drinking water; (2) bodies of water used for water sports and fishing, but not for drinking: (3) bodies of water used for fish and wildlife propagation, but not for water sports and fishing; and (4) bodies of water used for agricultural and industrial purposes. Different standards have been established based on the degree of beneficial use of the body of water in question.[5]

The impact of the Water Quality Act on product/process manufacturers is significant, as it forces a measurement and control mechanism to ensure that wastes being discharged into bodies of water are not contaminating the water or that such wastes are treated prior to discharge. As a further constraint, the Clean Water Act of 1972 establishes a nationwide policy of *zero discharge* by 1985 (i.e., all wastes must be removed before discharges are made into bodies of water).[6]

C. Noise Pollution and Control

Noise is another environmental pollutant in a similar category with air and water pollution. Noise, usually measured in decibels, is definitely a health hazard and could, if not properly controlled, cause an individual to lose his hearing. Noise sources include (1) industrial noise caused by machines, handling equipment, and so on; (2) community noise caused by aircraft, automobiles and trucks, and construction; and (3) domestic noise resulting from machines and appliances around the home.

Noise control is being accomplished through various means. The Noise Control Act of 1972 directs the Environmental Protection Agency (Office of Noise Abatement and Control) to investigate the broad aspects of noise problems, identify products that are major contributory sources of noise, and to establish noise emission standards for various product categories. The Occupational Safety and Health Act has identified maximum permissible noise levels for industrial operations. The Federal Highway Administration has established noise-level design criteria for various land uses associated with highway routing and transportation modes. The Federal Aviation Administration has established noise controls for airport locations by diverting aircraft flight paths away from populated areas, altering aircraft operating procedures to reduce unnecessary power thrusts in takeoffs and

---

[5]P. A. Vesilind, *Environmental Pollution and Control,* Duke Environmental Center, Duke University, Durham, N.C., 1974.

[6]It may turn out that the zero-discharge requirement is too costly to implement and that a compromise of some sort is necessary (i.e., 80 or 90 per cent cleanup rather than 100 per cent).

landings, and by imposing limits on aircraft engine noise. Although these controls are a start in the right direction, noise still remains a major problem in our society.

D. Radiation

Radiation can be defined as energy transmitted through space and may be classified as "natural radiation" and "man-made radiation." Natural radiation includes cosmic rays bombarding the earth's atmosphere on a continuing basis and radioactive isotopes found in the mineral deposits and water. Man-made radiation stems from medical applications (e.g., x-rays), reactors in nuclear power plants, reactors used in the propulsion of aircraft or ships, and radioactive wastes. Radiation can cause damage to human cell tissues and may in certain instances cause skin cancer, leukemia, and/or mutations in the sex cells.[7]

Radiological health, or the protection of man and his environment from unacceptable exposure to radiation, is controlled by external and internal standards. External standards are based on maximum exposure to the external portions of the human body, while internal standards are based on maximum exposure to internal tissues and organs of the human body as a result of breathing air and drinking water contaminated with radioactivity. The Atomic Energy Commission (AEC) has initiated external standards for individuals employed at AEC plants, and the U.S. Public Health Service has generated drinking water standards for radiation protection.

E. Solid Waste Management

Solid waste, another form of pollution, includes garbage and refuse (i.e., paper, wood, cloth, garden waste materials, metals, plastics, ceramics, and glass) which have been discarded for one reason or another. Roadside dumps, piles of industrial debris, junk car yards, and so on, are good examples of solid waste. Although it is difficult to directly correlate solid waste with public health problems, it is believed that improper solid waste disposal is a true health hazard in view of the fact that flies, rats, and other disease-carrying rodents are attracted to areas where there are solid wastes. In addition, there may be a significant impact on air pollution if windy conditions prevail or on water pollution if the solid waste is located near a lake, river, or stream.

The solid waste problem covers three aspects. First, products must be initially designed with disposability in mind, as discussed in Chapter 5. New applications of natural resources, concepts in packaging, and so on, are addressing this issue. Second, discarded garbage and refuse must be

---

[7]L. G. Rich, *Environmental Systems Engineering*, McGraw-Hill Book Company, New York, 1973.

collected and transported to a designated facility or location. How is the solid waste to be collected (e.g., garbage trucks, grinders, pneumatic pipes)? The collection task currently constitutes about 80 per cent of the total solid waste management cost. Here again, solid waste collection must be considered early in the product life cycle to minimize the problem(s) encountered later. Finally, once that solid waste has been collected, there is the problem of disposal. How is the disposal function to be accomplished? Can the material be recycled for other uses, to alleviate disposal problems? Can the material be compacted or shredded, or should incineration be considered to attain volume reduction prior to disposal? If incineration is used, how will it affect the air quality index? Can sanitary landfills be used for material disposal?

These and comparable questions remain unanswered at the present. As a result, the federal government established the Solid Waste Disposal Act of 1965 and the Resource Recovery Act of 1970 (as an amendment) to initiate research and development programs to determine new and improved methods for proper economic solid waste disposal. These programs are directed toward the conservation of natural resources by reducing the amount of waste and unsalvageable materials and toward the possibilities of potential uses for solid waste and the methods of recycling. The responsibility for solid waste programs of this nature is assigned to the Environmental Protection Agency at the federal level with various state agencies involved in supporting activities at the local level.

17.2 ECONOMIC CONSIDERATIONS

Economic considerations related to engineering activities within corporations, businesses, and government agencies have been covered to varying degrees throughout this text. However, such considerations are obviously a function of the total economic status of the country and state in which we live. In periods of rising prosperity, consumer expenditures rise, product development and production/construction activities increase, national employment rises, and so on. In other words, the gross national product (GNP), consumption, government spending, and investment trends react in a certain manner and all reflect our total economic posture.[8] During prosperous times, corporations and agencies are readily willing to invest in research and engineering functions related to new product developments and/or modifications of existing products to improve performance characteristics. In addition, along with the advent of prosperity comes waste, which results in unnecessary costs.

---

[8]The reader may wish to review some of the principles of macroeconomics to gain further insight relative to GNP, investment, total consumption, and other measures of our national economy and their effects on our day-to-day activities. A good reference is: T. F. Dernburg and D. M. McDougall, *Macroeconomics,* 3rd ed., McGraw-Hill Book Company, New York, 1968.

At the other end of the spectrum, during periods of decline (i.e., a slight depression or a recession), the economic ailments at the national level affect us all. Consumer expenditures decline, corporation profits fall off, unemployment increases, and related societal problems evolve. Under such conditions, corporations and agencies are more selective in terms of investing in proposed research and engineering projects. Sometimes, particularly toward the end of the fiscal budgeting year, investments in engineering projects will not be considered at all, regardless of the credibility of the proposal. During periods of economic decline, there is less spending, and a tendency to become more cost effective in our day-to-day operations becomes a reality.[9]

Over the past several decades, we have been experiencing prosperity in general; however, there are undoubtedly some economic declines in the future. The challenge to engineering is to respond to these various conditions by developing products that will not only meet the ever-increasing needs of our society but will do so in an effective and economical manner.

17.3 POLITICAL CONSIDERATIONS

Political considerations pertain to policies applied in the day-to-day administration of the national government, state government, corporation, or the organization with which we are associated. Policies may be established through formal legislation (e.g., the legislation implemented to control environmental pollution) or through the initiative of individuals who occupy key management functions. Those policies established through formal legislation usually evolve as a result of societal pressures and certainly have an impact on engineering and related areas of activity. For instance, the standards imposed by the OSHA and the environmental standards discussed in Section 17.1 significantly affect the design of future products/processes.

Policies established through individual initiative generally apply directly to specific functions of a corporation, business, or government agency. Individual managers may react to various perceived trends in a number of ways, the results of which often become operating policies that directly affect the activities within the organization. Many of these policies turn out to be extremely beneficial to the operations of the organization. On the other hand, there are some instances when such policies are detrimental. The following examples are presented as an illustration of what is implied in the latter case.[10]

---

[9]Unfortunately, it is quite often true that a slight depression or recession has to occur to emphasize the need for cost effectiveness. Why can't we emphasize a cost-effective approach during prosperous times?

[10]Policies as discussed here are based more on political motivations rather than on the establishment of standard operating procedures within a corporation, business, or agency, as discussed in Section 10.2B4.

A. The equal employment opportunity (EEO) practices oriented toward the employment of women and minorities have created situations where in some instances a minority-group person will be hired (or retained) for a job position over an individual who may be more qualified for the position. The objective here, to improve the balance of minorities in the organization, may result in a reduction in job accomplishment and efficiency. In other words, an *overreaction* to EEO practices may create a reverse effect and a degrading situation.

B. A manager may perceive that his organization's workload will be declining in the future and may attempt to counter this trend by:

1. Generating unnecessary tasks, resulting in redundancy and waste.
2. Extending the time duration for existing tasks by initiating unnecessary design changes.
3. Producing marginal design data such that only *his* organization possesses the knowledge and expertise to accomplish follow-up tasks.
4. Delaying decisions to the point where additional follow-up work will be required.
5. Blocking requests for personnel transfers.
6. Unofficially soliciting work from other organizations through various means.
7. Inhibiting communications to other organizations with the intent of gaining an "upper hand" on a relative basis with the other organization elements.

The intent is to increase or maintain organization status regardless of the consequences to the corporation, business, or agency's objectives and best interests.

There are, of course, numerous other examples where the reaction (or overreaction) to administrative trends and the internal politics within the lower levels of the organization significantly influence day-to-day decision making. Hopefully, such decisions will affect the organization in a positive manner.

17.4 SOCIETAL CONSIDERATIONS

Society can be defined as "the system of community life in which individuals, ordinarily in a territorial establishment, form a continuous and regulatory association for their mutual benefit and protection."[11] A society

---

[11] *Standard Dictionary of the English Language,* international ed., Funk & Wagnalls, Inc., New York, 1960.

constitutes a group of individuals who have common interests. These interests may be directed toward community development, urban renewal, industrial growth, or may assume a status quo posture. Social trends generally follow economic trends. Activities of individuals and organizations within the society should be compatible with the objectives and interests of the society; otherwise, the society as an entity will reject these activities. In other words, if a corporation involved in product design and development is located in a given community, common interests must be developed between the corporation and the community. The community should support the objectives of the corporation, and the corporation in turn must provide benefits to the community. In a similar capacity, individuals employed in the corporation must be accepted by the community. This degree of acceptance, which is a function of corporation-community relationships and individual behavior, will determine on-the-job effectiveness. In essence, engineering activities are highly influenced by the social environment in which they are accomplished. If the community shows pride and demonstrates support for the corporation, individual employee performance is likely to be well above average. If not, employee dissatisfaction will occur, family life will be affected, and job performance will deteriorate. Thus, the society's and engineer's social welfare are both critical to the accomplishment of engineering objectives.

17.5 SUMMARY

This chapter reviews briefly some of the environmental considerations that influence, and are influenced by, engineering. Scientists and engineers develop technology that impacts the environment in which we live. Sometimes the impact effects are beneficial, and in other instances they are detrimental. Detrimental effects are controlled through the implementation of environmental specifications and standards, which in turn places more stringent constraints on future technology. Thus, the relationships between engineering and the environment are interactive.

Because of the detrimental effects that have resulted in the past, future projects of significant magnitude (e.g., proposed construction and land development) must be supported through the preparation of an *environmental impact statement* prior to commencing with the project.[12] Environmental impact statements provide a detailed assessment of the probable effect of the project on the environment and include the following basic information:

A. A description of the major project objective and the alternative possibilities for achieving this objective.

---

[12]Environmental impact studies are required by the Environmental Policy Act of 1969.

B. An evaluation of the characteristics and conditions of the *existing* environment.
C. An analysis of the various alternative methods for achieving the objective, and a detailed description of the most cost-effective approach.
D. An assessment of each step of the recommended approach to determine environmental impact.
E. A recommendation for further action.

QUESTIONS AND PROBLEMS

1. How is engineering generally impacted by politics? Macroeconomic conditions? Social conditions? Ecological requirements? Identify some specific examples in each instance.

2. How does engineering impact on social conditions? Ecological conditions? Identify some specific examples.

3. Why is ecology important? Describe a situation where the ecological balance of nature has been unintentionally altered by man.

4. Research the literature and identify the quantitative requirements for air quality and water quality in your state. (That is, what is the maximum allowable pollution index?)

5. Which of the air pollutants appear to be most dangerous relative to man's longevity? Why?

6. What is the objective of solid waste management? Why is it important? What is being done in your community relative to solid waste management?

7. What technological innovations would you propose to help solve the solid waste management problem?

8. Research the literature and plot a noise-level spectrum (i.e., zero decibels and on up). At approximately what noise level does human injury occur? From a technology viewpoint, what are the basic causes of noise in buildings? What are the noise levels that you are personally subjected to in accomplishing your day-to-day activities?

9. What types of radiation hazards exist? What harmful effects can be caused by radiation?

10. Describe how the engineering process can be detrimentally affected by political decisions within your organization. Give some examples, if possible.

11. What is the purpose of an environmental impact statement? What factors are included?

12. Identify an industry of your choice and assume that your community has been selected as a possible site for a new plant. Develop a checklist of the factors that you would use in assessing the environmental impact of the new plant.

13. Identify some products that have been designed, or redesigned, to meet environmental standards (be specific in terms of design characteristics).

14. What can you as an engineer do that would help improve the general standard of living in your community?

18

CHALLENGES THAT FACE ENGINEERING MANAGEMENT

Review of current national and international trends indicates that (1) population growth is increasing at a rapid pace and will continue to increase almost exponentially if current growth rates remain unchanged,[1] (2) population mobility is increasing, (3) our civilization is maturing in terms of technology wants and needs, (4) and the quality and quantity of our natural resources are dwindling in certain areas as a result.[1] Combined with these trends is the public concern for environmental quality and the implementation of quality standards and procedures.

When we analyze these trends on an *integrated* basis, it readily becomes apparent that many challenges face engineering and engineering management. Of particular concern are the needs for new transportation systems, health and medical products, communication systems, automation, buildings and structures, and so on, in light of limited or constrained energy, environment, and material resources. Figure 18-1 illustrates the technology-resource relationship.

---

[1]Where it has taken approximately 2 million years for the world population to reach over 3 billion persons, it will take only 35 years to add the next 3 billion persons if present growth rates remain unchanged. Human life spans have been extended as a result of health and medical provisions while birth rates continue to climb, particularly in the developing countries. Refer to G. Beakley and H. Leach, *Engineering—An Introduction to a Creative Profession,* 2nd ed., Macmillan Publishing Co., Inc., New York, 1972.

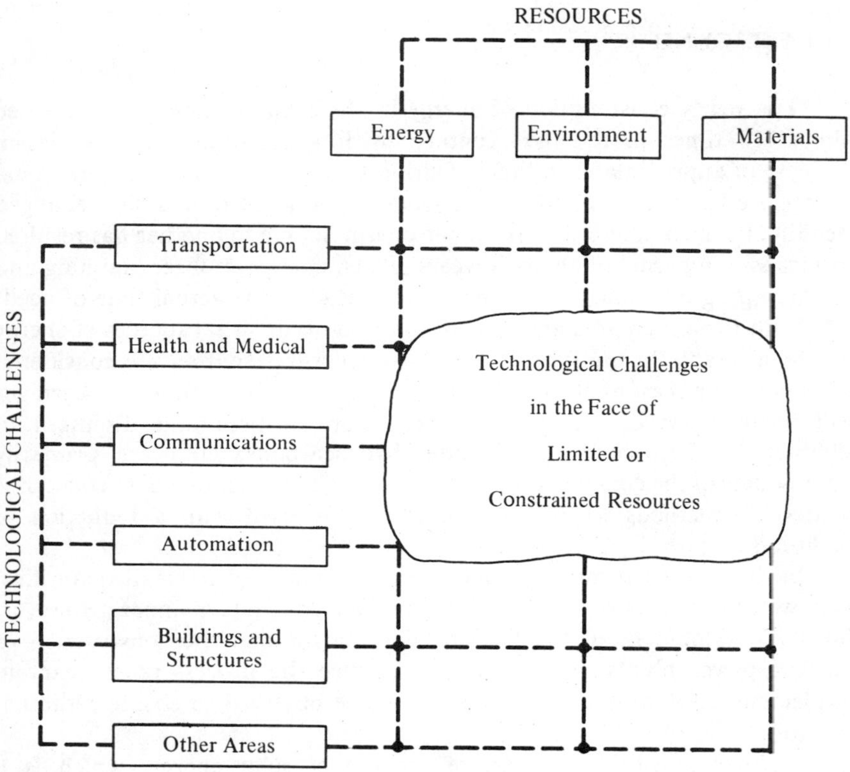

Figure 18-1. Future technology considerations.

Engineering activities must be directed toward finding more effective ways to utilize our current resources by providing alternative means for accomplishing certain objectives. Thus far, new technologies and management strategies have been developed in a timely manner to counter some of the resource shortages; however, the demands in the future will likely increase and become more stringent. Although a sample of projects directed toward future technology is noted in Figures 1-6 and 18-1, emphasis here is placed on the energy, materials, and environmental concerns.[2]

[2]The interrelationships of these concerns are discussed in: C. Starr, "Energy, Materials, Environment—The Triangle," *Research/Development Magazine,* Technical Publishing Company, New York, April 1975.

18.1 ENERGY NEEDS [3]

The yearly consumption of energy in the United States has increased almost 10 times in the past century, with a corresponding population increase of approximately 6 times. During this period, where the per capita energy use has almost doubled, the types of energy resources have changed steadily. Each complete shift from one energy source to another has resulted in a transition period of about 50 years. Thus, energy resource planning and research programs must commence years ahead of the actual time of need.

With the anticipated energy consumption needs, what sources of energy will be available in the future? Fossil-fuel energy resources are considered adequate (at a reasonable price) until around the year 2000, when shortages will occur. However, the five most common air pollutants discussed in Chapter 17 (i.e., carbon monoxide, hydrocarbons, etc.) are primarily by-products of the combustion of fossil fuels. Thus, an immediate concern is to develop methods where fossil fuels can be used with a reduction in pollution output.

In the long term, nuclear fuel resources (now used in small quantities) may well meet the needs of the future in an adequate manner. However, further development of the fast breeder reactor and the construction of nuclear power plants must begin soon, since the process takes years to implement. In addition, precautions must be observed to ensure radiation control.

Another possibility consists of the use of solar energy, which is a continuous energy source currently being employed to a very limited extent, primarily because of the large investment cost associated with the energy collection station. An obvious challenge is to develop alternative methods of solar energy collection.

There are, of course, other possible sources of energy, including gas, wood fuel, hydropower, and wind power. These, along with fossil fuels, nuclear reactors, and solar systems, must be analyzed on an integrated basis to determine the best approach in the pursuit of future technology.

---

[3] The data covering energy are taken in part from: C. Starr and C. Smith, "Energy and the World of A.D. 2000," *Engineering for the Benefit of Mankind,* Third Autumn Meeting of the National Academy of Engineering, published by the National Academy of Sciences, Washington, D.C., 1970. Another source covering energy technology is: H. C. Hottel and J. B. Howard, *New Energy Technology—Some Facts and Assessments,* The MIT Press, Cambridge, Mass., 1973.

18.2 MATERIALS[4]

Greater quantities and varieties of mineral resources are in demand, and this trend is expected to increase on a continuing basis. Concurrently, the depletion of established domestic sources of minerals is accelerating. Although the total output from our mines has risen, the mineral values in the extracted ores have diminished. Such trends would suggest the following:

1. Further exploration is required to locate new sources of mineral deposits. There are many subterranean areas, as well as areas under the ocean floor, that have not yet been explored.
2. More revolutionary methods for mining and material extraction are required. Research is being directed toward improving the techniques for tunneling into and through the earth and toward developing new and rapid excavation systems that will realize a far greater efficiency in the extraction of minerals.
3. In the interest of stretching our resources, effort is required to develop methods for recycling materials for new applications.

18.3 ENVIRONMENT

Environmental characteristics are covered in Chapter 17 and include ecology, polotical, social, and economic considerations. Of particular relevance are the ecological impacts on our life, community, and the world today. Air and water pollution are increasing, and controls must be established to assure ultimate survival. Technology challenges concerning the development of new products or systems that will not cause further degradation to our environment are apparent.

Solid waste is another critical problem. On the average, each person in the United States generated approximately 5.5 pounds of solid waste per day in 1970, and the estimated value for 1980 is 8.0 pounds per person per day.[5] As indicated in Chapter 17, many problems exist in the collection and disposal of solid wastes. How can solid wastes be transported? How is solid

---

[4]Further discussion of materials is presented in the article: W. R. Hibbard, Jr., "Materials Today and Tomorrow," *Engineering for the Benefit of Mankind,* Third Autumn Meeting of the National Academy of Engineering, published by the National Academy of Sciences, Washington, D.C., 1970.

[5]G. Beakley and H. Leach, *Engineering—An Introduction to a Creative Profession,* 2nd ed., Macmillan Publishing Co., Inc., 1972.

waste disposal to be accomplished? Can the amount of solid waste be reduced? Can some of it, particularly scrap metal, be recycled for other uses? Solid waste management is a problem in many urban and rural areas throughout the nation, and the technological challenges in arriving at solutions are great.

All new products and systems impact these resources in one way or another, both in their development and in their operational use and maintenance support. The engineering and engineering management challenge is to introduce new products/systems with resource conservation as a major goal.

QUESTIONS AND PROBLEMS

1. Identify and discuss some of the technological challenges in today's society. You may wish to address the areas identified in Figure 18-1 (i.e., transportation, health and medical communications, use of automation techniques, buildings and structures). How is engineering involved? What are the responsibilities of engineering management?
2. How does technology assessment (discussed in Chapter 3) relate to technology development? What considerations are necessary in technology assessment?
3. Estimate the amount of energy consumed by your activities in 1 year.
4. Analyze the transportation needs of your community and describe what you would recommend for a transportation system. How does your recommended approach impact energy, environment, and material resources?
5. Identify some problems in your community that require engineering solutions. Select one problem and prepare a report describing the problem and a proposed approach to solving the problem.

Appendices

A

SELECTED TERMS AND DEFINITIONS

Throughout this text there are numerous references to different terms and definitions that are significant in the engineering design process but may not be too familiar to the engineer. Such terms as cost effectiveness, system effectiveness, availability, reliability, maintainability, logistic support, and life-cycle cost are as important as range, accuracy, size and weight, volume, capacity, speed of performance, and so on. All these factors, applicable to each system or product, must be considered in the initial establishment of design criteria, the evaluation of alternative design configurations and the selection of a preferred approach, and in the subsequent assessment of system performance and effectiveness. The author has identified a few of these, presented in alphabetical order, which directly support the material presented in Parts One and Two (Chapters 1 through 8).[1]

A. Availability (A)

Availability, considered as a measure of system/product effectiveness, may be expressed four ways:

---

[1] A variety of definitions has been selected for the purpose of introducing common terms used in design. In some instances, the definitions presented are directly extracted from an identified source, and in other instances a definition has been modified with the intent of providing added clarity.

1. *Operational Availability* (A_o) is the probability that a system or piece of equipment, when used under stated conditions in an *actual* operational environment, will operate satisfactorily when called upon. It is expressed as

$$A_o = \frac{\text{MTBM}}{\text{MTBM} + \text{MDT}} \tag{A-1}$$

where MTBM = mean time between maintenance (i.e., uptime) and MDT = mean maintenance downtime. MTBM reflects the frequency of maintenance, which in turn is a significant factor in determining logistic support requirements. MDT includes active maintenance time, logistics supply time, and administrative delay time.

While a system or product may perform well when operating, it must be available when needed. For example, you may own a high-priced automobile, but what good is it when it fails to start on a cold morning? Availability is a necessary effectiveness consideration in design along with performance and cost.

2. *Achieved Availability* (A_a) is the probability that a system or piece of equipment, when used under stated conditions, in an *ideal* support environment (i.e., available tools, spares, manpower, etc.) will operate satisfactorily at any point in time. It excludes logistics supply time and administrative delay time. Achieved availability (A_a) is more directly relatable to the early design process (as a means of measuring equipment reliability and maintainability characteristics) and is expressed as

$$A_a = \frac{\text{MTBM}}{\text{MTBM} + \overline{\text{M}}} \tag{A-2}$$

where $\overline{M}$ = mean active maintenance time. This includes corrective (unscheduled) and preventive (scheduled) maintenance times.

3. *Inherent Availability* (A_i) is the probability that a system or piece of equipment, when used under stated conditions in an *ideal* support environment (i.e., available tools, spares, manpower, etc.) will operate satisfactorily at any time. It excludes preventive maintenance actions, logistics supply time, and administrative downtime and is expressed as

$$A_i = \frac{\text{MTBF}}{\text{MTBF} + \overline{M}ct} \tag{A-3}$$

where MTBF = mean time between failures and $\overline{M}ct$ = mean corrective maintenance time.

4. *Availability* is a measure of the degree to which an item is in the operable and committable state at the start of the mission when the mission is called for an unknown (random) point in time. [2]

B. Cost Effectiveness (CE)

Cost effectiveness relates to a measure of a system or equipment in terms of mission fulfillment (system effectiveness) and total life-cycle cost. Mission fulfillment may be expressed in different terms (one or more figures of merit), depending on the type of system/product and its mission requirements. Refer to Section 5.3A1.

C. Dependability

Dependability is a measure of the item-operating condition at one or more points during the mission, including the effects of reliability, maintainability, and human factors, given the item condition(s) at the start of the mission. It may be stated as the probability that an item will (1) enter or occupy any one of its required operational modes during a specified mission; and (2) satisfactorily perform the functions associated with those operational modes.

D. Life-Cycle Cost (LCC)

All costs associated with the product/system life cycle to include the following (refer to Figure 12-4):

1. *Product planning*— market analysis, feasibility studies, advanced product planning, and life-cycle management.
2. *Product research*—basic research, applied research, and research laboratory support.
3. *Product design*— conceptual design, preliminary design and analysis, detailed design (e.g., design engineering, design documentation, design support, design aids, fabrication of engineering and prototype models), engineering test and evaluation, and engineering laboratory support.
4. *Production and/or construction*— industrial engineering and operations analysis (e.g., manufacturing engineering, plant engineering, methods engineering, production control), manufacturing, construction, facility maintenance, utilities, and quality control.
5. *Product evaluation*— test and evaluation planning, test preparation, system test and demonstration, test reporting, test facilities, and support.

---

[2] MIL-STD-721, "Definitions of Effectiveness Terms for Reliability, Maintainability, Human Factors, and Safety."

6. *Product use and logistic support*— product distribution, product-life-cycle operations, and logistic support (e.g., test and support equipment, supply support, personnel and training, technical data, and facilities).
7. *Product phase-out*—material retirement, disposal, and recycling.

Life-cycle costs include labor and material costs, recurring and non-recurring costs, direct and indirect costs, variable costs, and fixed costs. Life-cycle costs may be categorized in many different ways, depending on the type of system or product and the sensitivities desired in cost-effectiveness measurement.[3]

E. Logistics

Logistics is the art and science of management, engineering, and technical activities concerned with requirements, design, and supplying and maintaining resources to support objectives, plans, and operations.[4] The elements of logistic support include personnel and training, test and support equipment, supply support (provisioning, distribution, spare/repair parts, inventories, etc.), facilities, transportation and handling, and technical data.

F. Logistic Support Analysis (LSA)

An iterative analytical process by which the logistic support necessary for a new or modified system/product is identified. It is a tool that aids in (1) the initial determination and establishment of logistic support criteria or constraints that impact on the design; (2) consideration of those criteria in the design of the system or product; (3) the provisioning and acquisition of logistic support elements; and (4) in the final analysis of the design to validate its feasibility in terms of total logistic support effectiveness. An output of the LSA is the identification and documentation of all specific logistic support resources required to support the system/product throughout its programmed life cycle (e.g., test and support equipment, personnel, spare/repair parts). The LSA often includes maintenance analysis, life-cycle cost analysis, and logistics modeling.

G. Maintainability ($\underline{M}$)

An inherent characteristic of equipment design and installation which is concerned with the ease, economy, safety, and accuracy in the performance of maintenance actions. It is concerned with maintenance times, sup-

---

[3] A different and more in-depth breakdown of life-cycle cost is included in: B. Blanchard, *Logistics Engineering and Management*, Prentice-Hall, Inc., Englewood Cliffs, N.J., 1974, Appendix A.

[4] This definition was established by the Society of Logistics Engineers (SOLE).

portability characteristics in design, and maintenance costs. Maintainability is the *ability* of a piece of equipment to be maintained, versus maintenance, which constitutes a series of actions to be taken to restore or retain an item in an effective operational state. Maintainability is a design parameter and maintenance is a result of design.

Maintainability can also be defined as "a characteristic of design and installation which is expressed as the probability that an item will be retained in or restored to a specified condition within a given period of time when maintenance is performed in accordance with prescribed procedures and resources."[5]

Maintainability can be specified, predicted, and assessed both on a qualitative basis and on a quantitative basis. The measures of maintainability are in terms of a combination of maintenance times, supportability factors, and projected maintenance cost. Those measures most commonly employed are as follows:

1. MTBM—mean time between maintenance, which includes both preventive (scheduled) and corrective (unscheduled) maintenance requirements. It includes consideration of reliability MTBF and MTBR. MTBM is also considered as a reliability parameter.
2. MTBR—Mean time between replacement of an item due to a maintenance action (usually generates a spare-part requirement).
3. $\overline{M}$—mean active maintenance time (a function of $\overline{M}ct$ and $\overline{M}pt$).
4. $\overline{M}ct$—mean corrective maintenance time. Equivalent to mean time to repair (MTTR).
5. $\overline{M}pt$—mean preventive maintenance time.
6. $\widetilde{M}ct$—median active corrective maintenance time.
7. $\widetilde{M}pt$—median active preventive maintenance time.
8. M_{max}—maximum active corrective maintenance time (usually specified at the 90th and 95th per cent confidence level).
9. MDT—maintenance downtime (total time during which a system/product is not in condition to perform its intended function). MDT includes active maintenance time ($\overline{M}$), logistics supply time, and administrative delay time. Logistics supply time is that portion of nonactive maintenance time during which maintenance is delayed solely because a spare/repair part, test and support equipment item, or similar element is not immediately available. Administrative delay time or waiting time is that portion of nonactive maintenance time during which maintenance is delayed for reasons administrative in nature (personnel assignment priority, inadequate manning, organizational constraint, etc.).

---

[5]This definition is included in MIL-STD-721. The definition deals with the time element only. As there are other significant characteristics to consider (i.e., cost and supportability), it is felt that this definition is too limiting.

10. MMH/OH—maintenance man-hours per equipment operating hour.

11. Cost/OH—maintenance cost per equipment operating hour.

12. Cost/MA—maintenance cost per maintenance action.

13. Turnaround time (TAT)—that element of maintenance time needed to service, repair, and/or checkout an item for recommitment. This constitutes the time that it takes an item to go through the complete cycle from operational installation through a maintenance shop and into the spares inventory ready for use.

14. Self-test thoroughness—scope and accuracy of testing.

15. Fault isolation accuracy—accuracy of equipment diagnostic routines, in per cent.

An understanding of specified maintenance times (i.e., $\overline{M}$, $\overline{M}ct$, $\overline{M}pt$, $\widetilde{M}ct$,) requires some appreciation for basic statistics and the distribution of maintenance times over a designated system operational cycle. In most instances, maintenance times follow either the normal or log-normal distribution.

H. Maintenance

All actions necessary for retaining an item in, or restoring it to, an operational condition. Maintenance includes servicing, repair, remove and replace, modification, inspection, calibration, overhaul, and condition verification.

I. Maintenance Concept

A series of statements and/or illustrations that define criteria covering maintenance levels or echelons, support policies, effectiveness factors (e.g., maintenance time constraints), and basic logistic support requirements. The maintenance concept is a prerequisite to system/equipment design and development. Refer to Section 3.3A2.

J. Maintenance, Corrective

Unscheduled maintenance actions performed, as a result of failure, to restore an item to a specified condition. A typical maintenance cycle applicable to each unscheduled maintenance action is illustrated in Figure A-1. The elapsed time for each maintenance action is Mct_i (refer to item O below).

K. Maintenance Downtime (MDT)

Maintenance downtime constitutes the total elapsed time required to repair, restore, and/or retain a system/product in full operating status.

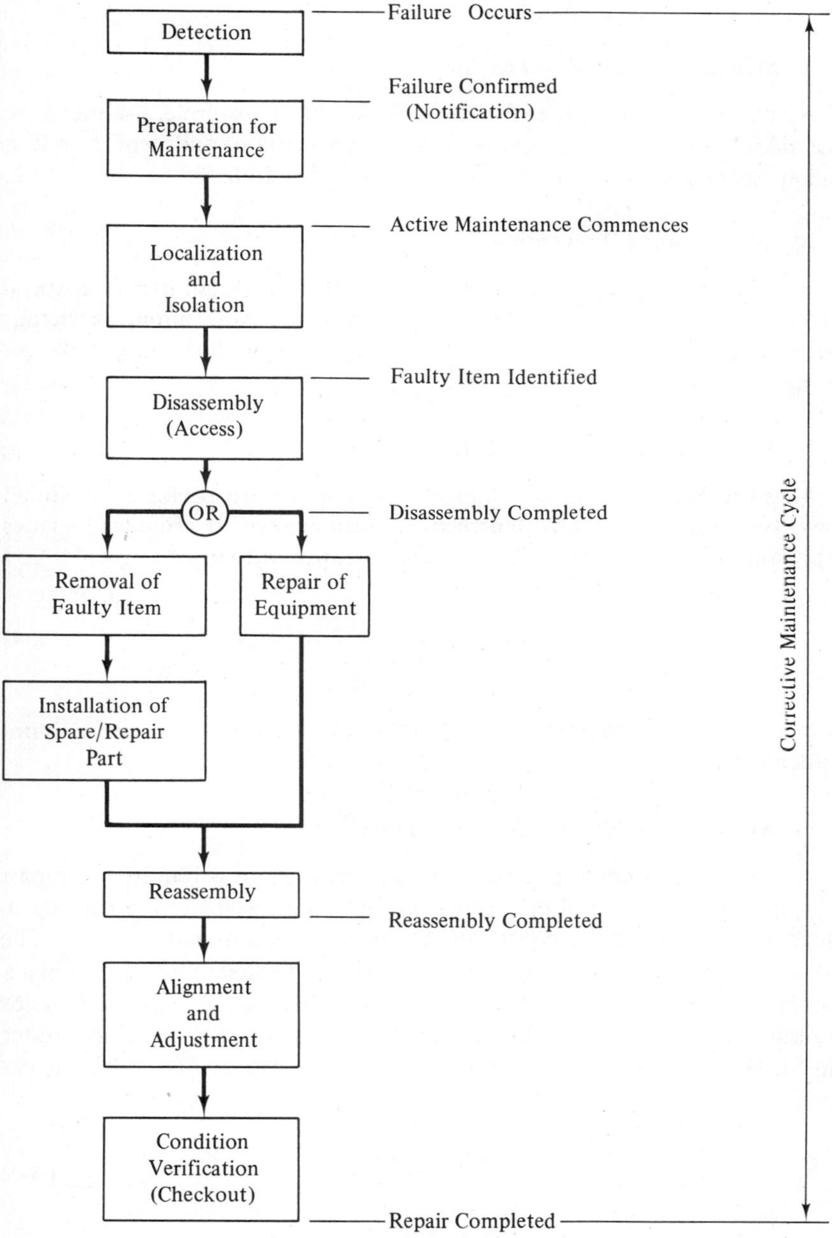

Figure A-1. Corrective maintenance cycle.

MDT includes $\overline{M}$, logistics supply time, and administrative wait time. The mean or average value is calculated from the elapsed times for each function and the associated frequencies (similar to the approach used in determining $\overline{M}$).

L. Maintenance Level or Echelon

A division of maintenance functions/tasks. As an example, maintenance levels may be classified as organizational, intermediate, and depot; or first echelon, second echelon, third echelon, and so on. Refer to Section 3.3A2.

M. Maintenance, Preventive

Scheduled actions performed in an attempt to retain an item in a specified condition by providing systematic inspection, calibration, servicing, selected item replacements, and comparable actions with the intent of preventing incipient failures.

N. Mean Active Maintenance Time ($\overline{M}$)

$\overline{M}$ is the mean or average elapsed time required to perform scheduled (preventive) and unscheduled (corrective) maintenance. It excludes logistics supply and administrative wait time, and is expressed as

$$\overline{M} = \frac{(\lambda)(\overline{M}ct) + (fpt)(\overline{M}pt)}{\lambda + fpt} \tag{A-4}$$

where λ = corrective maintenance or failure rate and fpt = preventive maintenance rate.

O. Mean Corrective Maintenance Time ($\overline{M}ct$)

Each time a system or product fails, a series of steps is required to repair and restore the item to full operational status. These steps are illustrated in Figure A-1 and together constitute a corrective maintenance cycle. The elapsed time to complete a maintenance cycle is represented by Mct_i. When a number of maintenance cycles occur (e.g., the six maintenance cycles required as a result of the failures indicated in Figure A-4 described under item S), the time for all events are combined to compute $\overline{M}ct$, which is expressed as

$$\overline{M}ct = \frac{\Sigma\, Mct_i}{N} \text{ or } \frac{\Sigma (\lambda_i)(Mct_i)}{\Sigma \lambda_i} \tag{A-5}$$

where λ_i = failure rate of the individual (ith) element of the item being measured, usually expressed in failures per equipment operating hour. Note that $\overline{M}ct$ considers only active maintenance time or that time which is spent working directly on the equipment. Logistics supply time and administrative wait time are not included. Although all elements of time are critical, the $\overline{M}ct$ factor is primarily oriented to a measure of the suppportability characteristics in the equipment design.

P. Mean Preventive Maintenance Time [$\overline{M}pt$]

$\overline{M}pt$ is the mean (or average) elapsed time required to perform scheduled or preventive maintenance on an item. This may include calibration, servicing, inspection, and overhaul, and can be accomplished while the system is in full operating condition, or could result in downtime.

$$\overline{M}pt = \frac{\Sigma\,(fpt_i)\;(Mpt_i)}{\Sigma\,fpt_i} \tag{A-6}$$

where fpt_i = frequency of the individual (*ith*) preventive maintenance action in actions per equipment operating hour and Mpt_i = average elapsed time required for the *i*th preventive maintenance action.

In this instance, the concern is for preventive maintenance actions that result in system downtime. Again, $\overline{M}pt$ includes only *active* equipment maintenance time and not logistic supply and administrative wait times.

Q. Mean Time Between Maintenance Actions [MTBM]

MTBM is the mean or average time between all maintenance actions (corrective and preventive) and is calculated as

$$\text{MTBM} = \frac{1}{1/\text{MTBM}_c + 1/\text{MTBM}_P} \tag{A-7}$$

where MTBM_c = the mean interval of corrective (unscheduled) maintenance and MTBM_P = the mean interval of preventive (scheduled) maintenance. MTBM_c, representing the frequency of maintenance, must consider the following factors:

1. *Inherent reliability characteristics.* This category covers primary failures based on the physical makeup of the item and the stresses to which the item is subjected. Part catastrophic failures and out-of-tolerance conditions are included. Primary failures are derived through reliability prediction data and are usually based on a constant failure rate.

2. *Manufacturing defects.* Quite often when an equipment item first comes off the production line, there are a rash of failures until the system is operated for a short period of time and a constant failure rate is realized. This is particularly true for electronic equipment when certain corrective actions are necessary to attain system stabilization. The extent of difficulty at this early point in the operational use phase is dependent on the amount of equipment operation and the type of testing accomplished in the production/construction phase. If enough hours of equipment operation are

attained through testing, the quantity of failures after delivery will probably be less. In any event, when new equipment first enters the inventory, there may be more initial failures than indicated by the predicted reliability failure rate, and this factor must be considered in determining the requirements for initial system support.

3. *Wear-out characteristics.* After equipment has been in operational use for a period of time, various individual components begin to wear out. Some components (i.e., mechanical linkages, gears) wear out sooner than others. When this occurs, the resultant frequency of failure and corrective maintenance will increase.

4. *Dependent failures* (*chain effect*). This category includes those secondary failures which occur as a result of a primary catastrophic failure. In other words, the failure of one item may cause other items to fail. The frequency of dependent failures is based on the extent of fail-safe characteristics inherent in the design. In electronic equipment, circuit protection provisions may be incorporated to reduce the probability of dependent failures. The reliability failure-mode-and-effect analysis is the best data source for reflecting dependent failure characteristics.

5. *Operator-induced failures.* Overstress of equipment due to operator human error is a possibility and is very significant in determining frequency factors. Unplanned stresses can be placed on the equipment as a result of different operating methods (by different operators or by the same operator), and these stresses can accumulate to the extent that the equipment will fail more frequently. Hopefully, conditions such as this can be minimized or eliminated through the proper emphasis on human factors in the design process.

6. *Maintenance-induced failures.* Damage to equipment caused by human error during maintenance actions may result from not following the proper maintenance procedures, improper application and use of tools and test items, losing components or leaving parts out when reassembling items after repair, forcing replacement items to fit when physical interferences are present, causing physical damage to components lying adjacent to or near the item being repaired, and many others. Induced faults of this type may occur due to improper maintenance environment (inadequate illumination, uncomfortable temperatures, high noise level), personnel fatigue, inadequately trained personnel performing the maintenance tasks, equipment sensitivity or fragility, poor internal accessibility to components, and not having the right elements of logistic support available when needed. Maintenance-induced faults are possible both during the accomplishment of corrective maintenance and preventive maintenance. For instance, during a scheduled calibration when performing a fine adjustment, a screwdriver slip may cause damage in another area, resulting in corrective maintenance.

7. *Equipment damage due to handling.* Another important aspect is the probability of equipment damage due to bumping, dropping, shoving, tossing, and so on. These factors are particularly relevant during transportation modes and often contribute to subsequent equipment failures. The extent to which consideration is given toward transportability in the equipment design will influence the effects of handling on the equipment failure rate. One should evaluate anticipated transportation and handling modes, review the design in terms of effects, and assign an appropriate frequency factor.

$MTBM_c$ should be generally equivalent to the reliability MTBF factor.

R. Producibility

Producibility is the extent to which an item can be reproduced in multiple quantities through normal production operations. This is a system/product design characteristic.

S. Reliability

Reliability is the probability that a system or product will operate in a satisfactory manner for a specified period of time when used under stated conditions. Where related to a specific mission, reliability may be defined as the probability of a successful mission of given duration under specified use conditions. Reliability, sometimes equated to the probability of survival, is expressed as

$$R = e^{-\lambda t} \tag{A-8}$$

where e = Napierian or natural logarithm base (2.7183), λ = failure rate in failures per hour ($\lambda = 1/\text{MTBF}$), and t = total operating time in hours. The reliability exponential function is illustrated in Figure A-2. Assuming that an item has a constant failure rate (λ), the reliability can be determined from the exponential distribution curve. Test and field data covering a variety of products have indicated that if the design of a product is mature, the failure rate (λ) will be relatively constant throughout a defined period of the product's operational life. In addition, when a product is produced and first introduced into the inventory, there are usually more failures during a "debugging" period. Likewise, when the product reaches a certain age, there is a "wearout period" when failures increase. A typical failure-rate curve is depicted in Figure A-3. The object is to predict the constant failure-rate

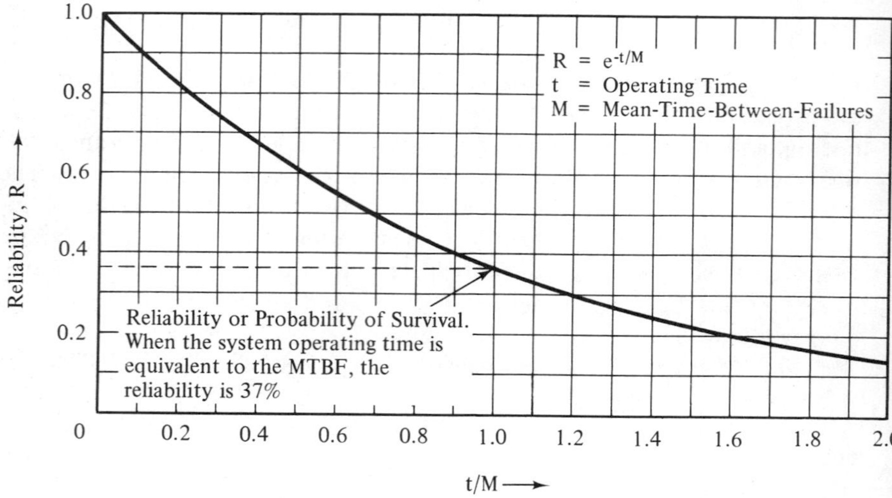

Figure A-2. Basic reliability curve.

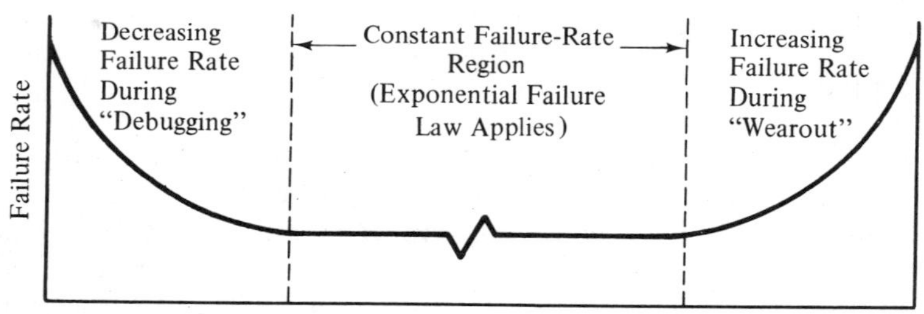

Figure A-3. Typical failure-rate curve.

region for different items and use the predicted data in the estimation of corrective and/or preventive maintenance requirements. However, if the product is in operational use early in the debugging period or when wearout occurs, the impact on logistics may be significant as the number of maintenance actions will be higher than initially anticipated.

As indicated earlier, reliability is also expressed in terms of MTBF, which is the reciprocal of the failure rate (λ). Assume that a system or product operational cycle is 169 hours and that six failures occur during that time; the failure rate (λ) or corrective maintenance frequency is

$$\lambda = \frac{\text{number of failures}}{\text{total mission time}} = \frac{6}{169} \tag{A-9}$$

0.036 failures or maintenance actions/hour

A failure is defined as an instance when the system or product is not operational within a specified set of parameters, and failures occur at the points indicated in Figure A-4. The system mean life or the mean time between failure (MTBF) is

$$\text{MTBF} = \frac{1}{\lambda} = \frac{1}{0.036} = 27.78 \text{ hours} \tag{A-10}$$

A reliability nomograph is presented in Figure A-5 to simplify calculations.

The failure rate and MTBF are the primary reliability quantitative factors used in the logistic support analysis for determining specific support requirements (e.g., the quantity of corrective maintenance actions and the test and support equipment, spare/repair parts, etc., required in each instance). Failure rate and MTBF factors used for logistic purposes are derived from reliability allocations and/or predictions.

T. System Effectiveness

System effectiveness may be defined as the probability that a system can successfully meet an overall operational demand within a given time when operated under specified conditions, or the ability of a system to do the job for which it was intended (function of availability, dependability, system performance, etc.). System effectiveness relates to total mission fulfillment.

U. Trade-Off Study (Analysis)

An evaluation tool used in decision making. As related to this text, a trade-off is concerned with decisions involving the best mix of desired system/product performance parameters, effectiveness characteristics, and cost.

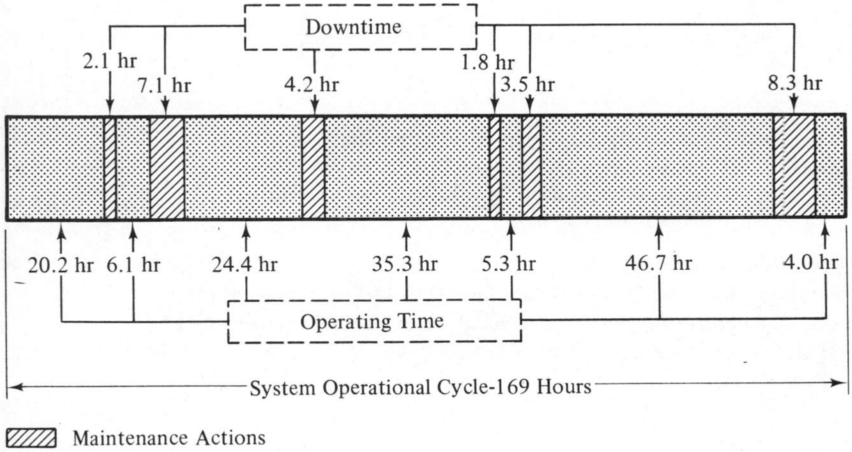

Figure A-4. System/product operating cycle.

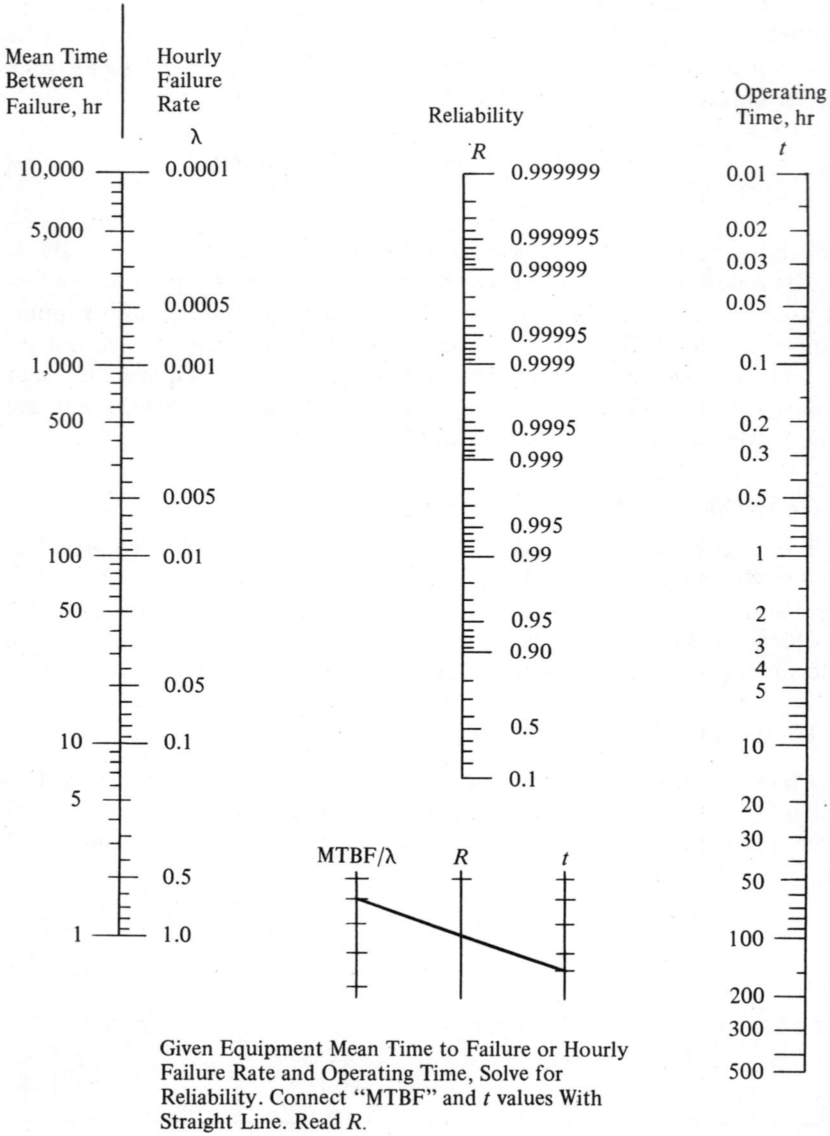

Figure A-5. Reliability nomograph for the exponential failure distribution. (*Source:* NAVAIR 00-65-502/NAVORD OD 41146, *Reliability Engineering Handbook,* Naval Air Systems Command and Naval Ordnance Systems Command, Revised March 15, 1968, Chart 3-1, Page A3-3.)

B

ENGINEERING EVALUATION METHODS

Chapters 2 and 5 introduce trade-off studies and the use of analytical methods in engineering decision making. Trade-off studies are made to evaluate alternative configurations, with the ultimate intent of selecting a preferred approach. As a means of simply illustrating the process, five examples of typical problem applications have been selected. The topics discussed are:

A. Cost-effectiveness analysis, involving evaluation of two alternative design configurations using reliability and life-cycle cost as decision criteria.
B. Level of repair analysis, comparing the options of designing an item for discard at failure, repair at the intermediate level of maintenance, or repair at the depot/supplier facility.
C. Optimum system/equipment packing design, using the branch-and-bound technique for seeking a feasible solution among a large number of alternatives.
D. Evaluation of three alternative test and support equipment configurations in terms of performance, operability, effectiveness, design characteristics, logistic support, schedule, cost, and so on. Weighting factors are employed in the scoring process.
E. Reliability and maintainability trade-off evaluation (variation of MTBF and $\overline{M}ct$ in terms of a given inherent availability).

There are numerous examples of a related nature involving all facets of design and support. However, it is hoped that the reader will acquire the insight desired through a careful in-depth review of what is presented herein. Through such a review, many additional questions should arise that will encourage further pursuit of the subject matter.

A. COST-EFFECTIVENESS ANALYSIS (Example 1)

This example illustrates a cost-effectiveness evaluation emphasizing life-cycle costing and the systems approach.

1. Definition of the Problem

A ground vehicle currently in the preliminary system design phase requires the incorporation of radio communication equipment. A decision is needed as to the type of equipment deemed most feasible from the standpoint of performance, reliability, and life-cycle cost. Budget limitations suggest that the equipment unit cost (based on life-cycle cost) should not exceed $20,000.

2. Evaluation Approach

a. The accomplishment of a cost-effectiveness evaluation requires further expansion of the problem definition. A description of operational requirements and the maintenance concept (discussed in Section 3.3A) are essential. In addition, one needs to know the program schedule.

The communication equipment is to be installed in a light vehicle. The equipment shall enable communication with other vehicles at a range of 200 miles, overhead aircraft at an altitude of 10,000 feet or less, and a centralized area communication facility. The system must have a reliability MTBF of 450 hours, $\overline{Mct}$ of 30 minutes, and an MMH/OH requirement of 0.2. The operation and maintenance concepts and program time frame are illustrated in Figure B-1.

b. Review of all possible supplier sources indicates that there are two design configurations that appear (based on preliminary design data) to meet the specified requirements. Each is evaluated on an equivalent basis in terms of reliability MTBF and total life-cycle cost.

c. The next step is to identify data needs and to structure an analytical model for use in the evaluation process. For each configuration the analyst needs:[1]

[1]Material relating to the reliability and maintainability prediction techniques, logistic support analysis, and life-cycle cost analysis is included in: B. Blanchard, *Logistics Engineering and Management,* Prentice-Hall, Inc., Englewood Cliffs, N.J., 1974.

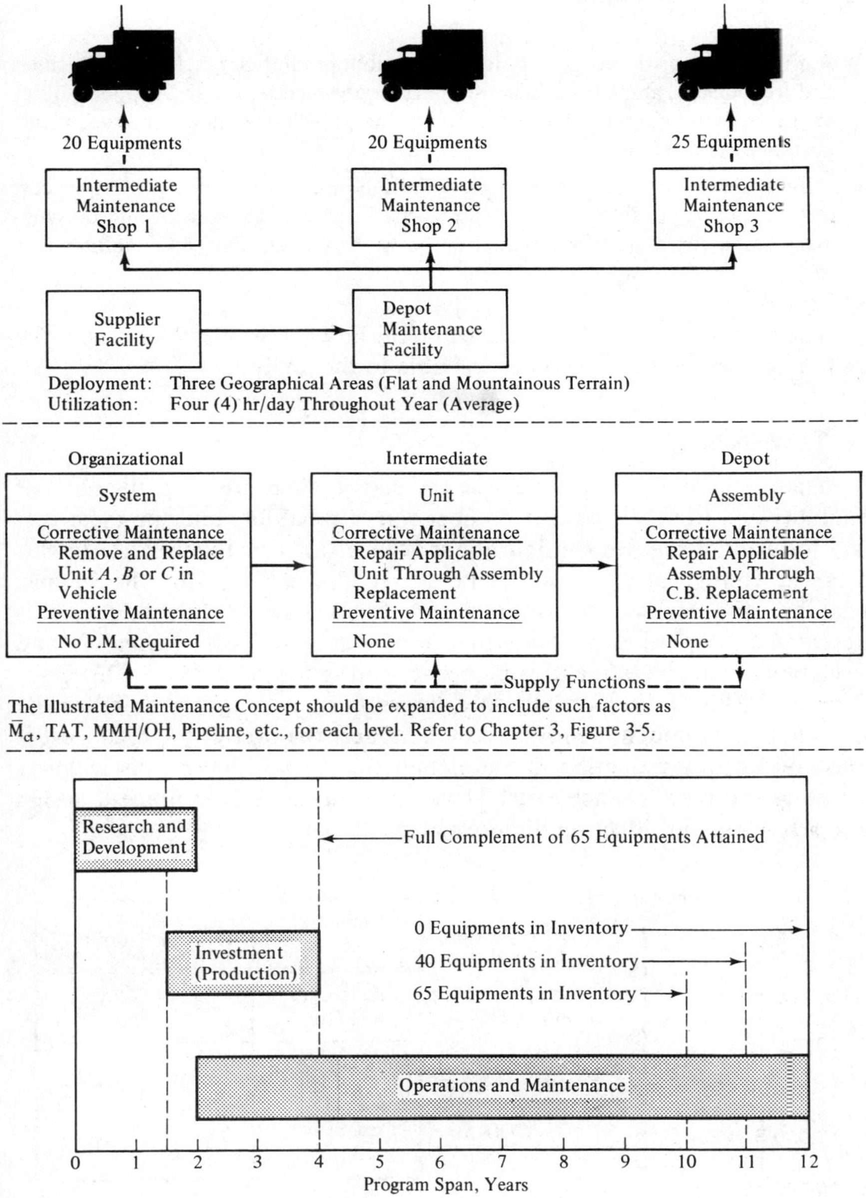

Figure B-1. Basic system concepts.

(1) A reliability allocation or prediction providing estimated component failure rates and an equipment MTBF. The equipment MTBF must be 450 hours or greater.

(2) A maintainability allocation or prediction, providing $\overline{Mct}$ and MMH/OH factors for the equipment and its components.

(3) A gross-level logistic support analysis, describing maintenance tasks, task times and frequencies, and basic logistic support requirements (test and support equipment, spare/repair parts, personnel and training, technical data, facilities, transportation, and handling).

(4) A life-cycle cost analysis, involving a definition of cost categories plus input cost factors. In this instance, costs are categorized in terms of research and development costs, investment or production costs, and operations and maintenance cost.

The analytical model is structured on the basis of problem definition, the evaluation criteria, and the data available to the analyst.

3. Evaluation Results

The problem is to select the best among two alternatives on the basis of reliability and life-cycle cost. A comparison of reliability and life-cycle cost data for each of the two configurations is illustrated in Figure B-2. In this instance, configuration A is the preferred alternative, with the highest reliability and lowest life-cycle cost. A breakdown of life-cycle cost is presented in Figure B-3. Note that the acquisition cost (research and development plus investment) is higher for configuration A ($478,033 versus $384,131). This is due partially to a better design using higher reliability components. Although the initial cost is higher, the overall life-cycle cost is lower, owing to a reduction in maintenance actions which results in lower operating and maintenance costs. These characteristics in equipment design have a tremendous effect on life-cycle cost.

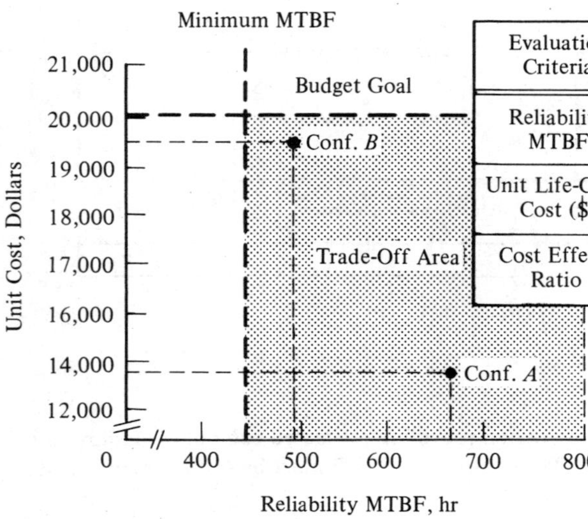

Evaluation Criteria	Conf. *A*	Conf. *B*
Reliability MTBF	675	495
Unit Life-Cycle Cost ($)	13,850	19,505
Cost Effect. Ratio	0.040	0.025

Figure B-2. Reliability versus unit cost.

Cost Category	Configuration *A*		Configuration *B*	
	P.V. Cost	% of Total	P.V. Cost	% of Total
1. Research and development	$ 70,219	7.8	$ 53,246	4.2
(a) Program management	9,374	1.1	9,252	0.8
(b) Advanced *R* and *D*	4,152	0.5	4,150	0.4
(c) Engineering design	41,400	4.5	24,581	1.9
(d) Equipment development and test	12,176	1.4	12,153	0.9
(e) Engineering data	3,117	0.3	3,110	0.2
2. Investment	407,814	45.3	330,885	26.1
(a) Manufacturing	333,994	37.1	262,504	20.8
(b) Construction	45,553	5.1	43,227	3.4
(c) Initial logistic support	28,267	3.1	25,154	1.9
3. Operations and maintenance	422,217	46.9	883,629	69.7
(a) Operations	37,811	4.2	39,301	3.1
(b) Maintenance	384,406	42.7	844,328	66.6
• Maintenance personnel and support	210,659	23.4	407,219	32.2
• Spare/repair parts	103,520	11.5	228,926	18.1
• Test and support equipment maintenance	47,713	5.3	131,747	10.4
• Transportation and handling	14,404	1.6	51,838	4.1
• Maintenance training	1,808	0.2	2,125	Neg.
• Maintenance facilities	900	0.1	1,021	Neg.
• Technical data	5,402	0.6	21,452	1.7
(c) System/equipment modifications	....	....	....	....
(d) System phase-out and disposal	....	....	....	....
GRAND TOTAL*	$900,250	100%	$1,267,760	100%

*The cost values presented are hypothetical but realistically derived. A 10% discount factor was used in determining present value costs.

Figure B-3. Life-cycle cost breakdown.

Figure B-4 projects the life-cycle cost in the form of a profile. The anticipated cost for each year is indicated. These figures (when not discounted) can be directly applied for budgeting purposes.

As stated earlier, configuration A is preferred on the basis of total life-cycle cost. Prior to a final decision, however, the analyst should perform a break-even analysis to determine the point in time that configuration A becomes more effective than configuration B. Figure B-5 illustrates the payback point, which is 6 years and 5 months, or a little more than a year after the equipment is introduced into the operational inventory. This point is early enough in the life cycle to support the decision. On the other hand, if the payback point were much further out in time, the decision might be questioned.

Category	Program Year												Total
	1	2	3	4	5	6	7	8	9	10	11	12	
Research and Development	32,119	38,100	—	—	—	—	—	—	—	—	—	—	70,219
Investment	—	94,110	156,852	156,852	—	—	—	—	—	—	—	—	407,814
Operations and Maintenance	—	—	12,180	32,480	60,492	57,472	53,480	50,484	50,470	50,494	37,480	17,185	422,217
Total	32,119	132,210	169,032	189,332	60,492	57,472	53,480	50,484	50,470	50,494	37,480	17,185	$900,250

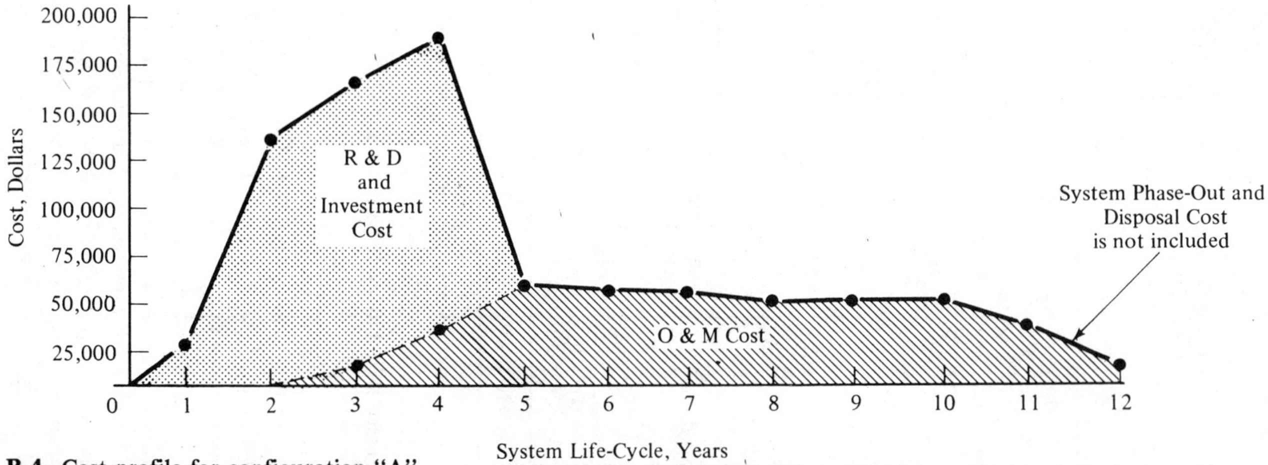

Figure B-4. Cost profile for configuration "A".

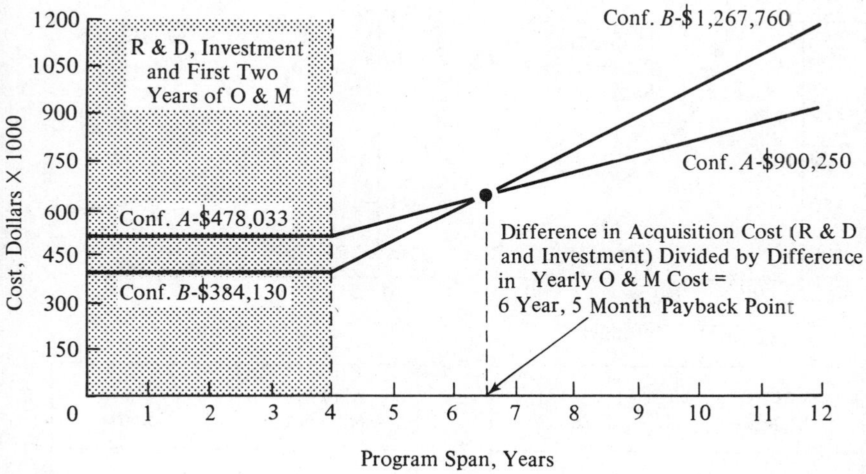

Figure B-5. Investment payback.

Referring to Figure B-3, the analyst can readily pick out the high contributors (those which contribute more than 10 per cent of the total cost). These are the areas where a more refined analysis is required and greater emphasis is needed in providing valid input data. For instance, maintenance personnel and support cost and spare/repair parts cost contribute 23.4 and 11.5 per cent, respectively, of the total cost for configuration A. This leads the analyst to reevaluate the design in terms of impact on personnel support and spares; the prediction methods used in determining maintenance frequencies and inventory requirements; the analytical model to ensure that the proper parameter relationships are established; and cost factors such as personnel labor cost, spares material costs, and inventory holding cost. If the analyst wishes to determine the sensitivity of these areas to input variations, he may perform a sensitivity analysis. In this instance, it is appropriate to vary MTBF as a function of maintenance personnel and support cost and spare/repair cost. Figure B-6 presents the results.

The analyst or decision maker should review the break-even analysis in Figure B-5 and determine how far out in time he is willing to go and remain with configuration A. Assuming that the selected maximum payback point is 7 years, the difference in alternatives is equivalent to approximately $65,000 (the present-value difference between the two configurations at the 7-year point). This indicates the range of input variations allowed. For instance, if the design configuration A changes or if the reliability prediction is in error, resulting in an MTBF as low as 450 hours (the specified system requirement), the maintenance personnel and support cost will increase to approximately $324,000, an increase of about $113,340 above the baseline value. Thus, although system reliability is within the specified requirements, the cost increase due to the input MTBF variation causes a decision shift in favor of

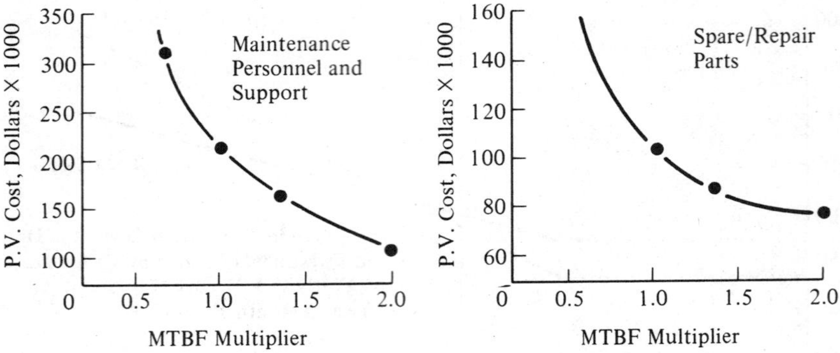

MTBF Multiplier	P.V. Cost, Dollars
0.67	323,140
**1.00	210,659
1.33	162,325
2.00	112,565

**Baseline Configuration *A*

MTBF Multiplier	P.V. Cost, Dollars
0.67	199,576
**1.00	103,520
1.33	92,235
2.00	80,130

**Baseline Configuration *A*

Figure B-6. Sensitivity analysis.

configuration B. The analyst must assess the sensitivity of significant input parameters and determine their impact on the ultimate decision.

B. LEVEL OF REPAIR ANALYSIS [Example 2]

In defining the detailed maintenance concept and establishing criteria for product design, it is necessary to determine whether items should be repaired at the intermediate level, repaired at the depot/producer facility, or discarded in the event of failure. This example evaluates these alternatives through a level of repair analysis.

1. Definition of the Problem

The communication equipment defined in example 1 constitutes three units (unit A, unit B, and unit C). Unit B constitutes 10 operational functions, each of which is to be packaged as a removable assembly. Should the various assemblies be designed for discard at failure or for repair? If the latter, where should repair be accomplished?

2. Evaluation Approach

a. The first step is to define operational requirements in terms of equipment deployment, utilization, effectiveness, environmental factors, and so on. Whether there are 2 or 100 pieces of equipment deployed or whether the equipment is to be utilized 12 hours versus 4 hours per day are significant factors in the decision process. An operational deployment similar to the one presented in Figure B-1 is assumed herein; however, the maintenance concept, as presented, will be reevaluated in terms of repair decisions.

b. Since the repair-discard decision is best supported using maintenance frequency and economic criteria, each applicable assembly of unit B is evaluated in terms of reliability and life-cycle cost. Thus, the analyst must generate reliability and maintainability estimates, a gross-level logistic support analysis, and life-cycle cost data. A typical data format covering each individual assembly is presented in Figure B-7.

Based on the results in the figure, it is economically feasible to accomplish repair of assembly A-1 at the depot maintenance facility. This is primarily due to the large quantity of equipment in the operational inventory, combined with the relatively low reliability. An improvement in reliability would reduce the number of maintenance actions and force the decision toward discard. A lower assembly reliability would tend to force the decision in favor of intermediate maintenance support. When the results are as close as indicated for the two repair options, the analyst should review the data for validity and perform a sensitivity analysis to determine impact on the decision as a result of input-parameter variations.

c. The next step is to evaluate each of the 10 assemblies of unit B in a like manner and record the results in a summary form, depicted in Figure B-8. Review of the values in the figure indicates a mixed policy. However, the analyst should look at the total policy cost (which in this instance reflects depot repair) as a preferred approach. Assuming that the depot repair concept is selected, the analyst should then relate the results to the designer with the intent of instituting design changes for those assemblies designated as "intermediate repair" and "discard." Hopefully, design improvements combined with subsequent evaluations will indicate a more uniform policy at a lower total policy cost. Often this becomes an iterative process of design and evaluation.

3. Evaluation Results

The results of the level of repair analysis, in this instance, lead to a *depot repair policy decision.* Changes in system deployment, utilization, and/or effectiveness will influence this policy. In addition, the variations in data input can have a significant impact.

During the early phases of system/product design, valid input data are often quite limited. Yet, the designer must make decisions concerning item repairability, and some form of repair-discard criteria is needed to support these decisions. In response, the analyst may accomplish the steps above, selecting equipment items that are similar in nature and function and that

Evaluation Criteria	*Repair Cost* Inter.	*Repair Cost* Depot	Discard Cost	Remarks
(a) Estimated unit acquisition cost of *Assembly A-1* (to include R and D and Investment)	\$ 160	\$ 160	\$ 125	Assembly design is simplified in discard case.
(b) Maintenance personnel and support cost	\$ 3,000	\$ 3,600	N/A	Based on 10-year life, 65 deployed equipments, utilization of 4 hours/day and λ of 0.00016, the quantity of MA is 150. Labor rates: intermediate—\$10/hour. depot rate—\$12/hour. $\overline{M}ct = 2$ hours. two technicians.
(c) Spare assemblies	\$ 480	\$ 960	\$18,750	100% spares in discard case. Intermediate = 3 and depot = 6 to cover TAT and pipeline.
(d) Piece parts for assembly repair	\$ 2,250	\$ 2,250	N/A	Assume \$15 per repair action.
(e) Test and support equipment	\$ 4,500	\$ 1,500	N/A	Special test equipment required at an acquisition and support cost of \$1500. Cost is allocated to assembly A-1. Intermediate—3 test sets. Depot—1 test set.
(f) Packing and transportation	N/A	\$ 780	N/A	300 one-way trips @ \$65/100 pounds. One assembly weighs 4 pounds.

Figure B-7. Repair versus discard-at-failure evaluation.

Evaluation Criteria	Repair Cost		Discard Cost	Remarks
	Inter.	Depot		
(g) Maintenance training	$ 192	$ 48	N/A	Delta training cost to cover maintenance of assembly. Assume $160/student week. Intermediate—24 students at 2 hours each. Depot—8 students at 2 hours each.
(h) Maintenance facilities	$ 174	$ 174	N/A	Assume $0.29 per direct labor hour of corrective maintenance.
(i) Technical data	$ 800	$ 800	N/A	Assume 4 pages for diagrams and text covering assembly repair at $200/page.
(j) Inventory maintenance	$ 546	$ 642	$ 3,750	Assume 20% of inventory value (spare assemblies plus piece parts).
Total estimated cost	$12,102	$10,914	$22,625	

Figure B-7. *Continued.*

Assembly Number	Repair At Intermediate	Repair At Depot	Discard	Decision
1	$ 12,102	$ 10,914	$ 22,625	Repair-depot
2	11,312	14,378	12,514	Repair-inter.
3	13,142	15,448	16,798	Repair-inter.
4	15,122	13,992	12,898	Discard
5	12,119	9,810	17,980	Repair-depot
6	14,221	8,992	18,552	Repair-depot
7	16,112	13,142	21,962	Repair-depot
8	19,702	17,129	16,898	Discard
9	12,121	13,136	23,648	Repair-inter.
10	14,200	11,460	19,230	Repair-depot
Policy Cost	$140,153	$128,401	$183,105	

Figure B-8. Summary of repair level costs.

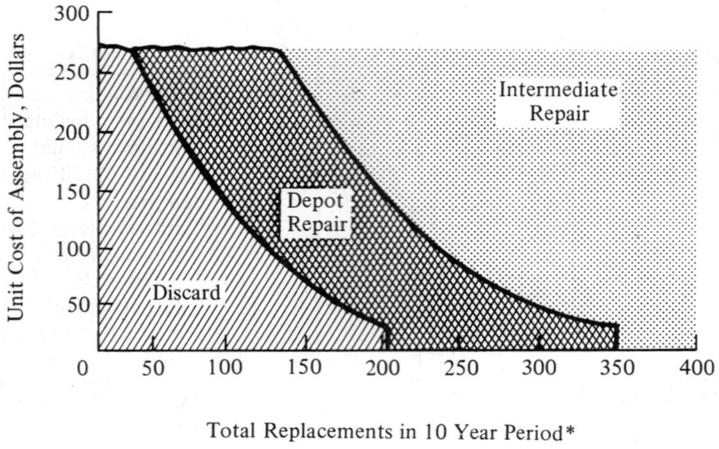

*Can be Related Directly to Failure Rate

Figure B-9. Repair-discard screening criteria.

are deployed and utilized in a like manner. By accomplishing a level of repair and sensitivity analysis, the analyst can establish gross screening criteria, as illustrated in Figure B-9. Numerous points are identified through successive analysis iterations (this can be accomplished quite easily through the use of a computer model). Regression lines are then constructed to establish the approximate crossover points.

With screening criteria developed, reliability and cost goals can be determined for newly designed assemblies (e.g., all equipment assemblies shall be designed for discard at failure using the criteria in Figure B-9). In addition, off-the-shelf items can be evaluated to determine the most economic repair policy. In any event, the process illustrated in Figure B-10 is applicable.

C. OPTIMUM EQUIPMENT PACKAGING DESIGN (Example 3)

In the design of equipment, the engineer is faced with the task of packaging. The basic packaging criteria are identified through the functional analysis in Chapter 5. However, a more detailed approach is required to determine the optimum packaging configuration at a lower indenture level. This example, which relates closely with the level of repair analysis in example 2, illustrates use of the branch-and-bound technique in design.

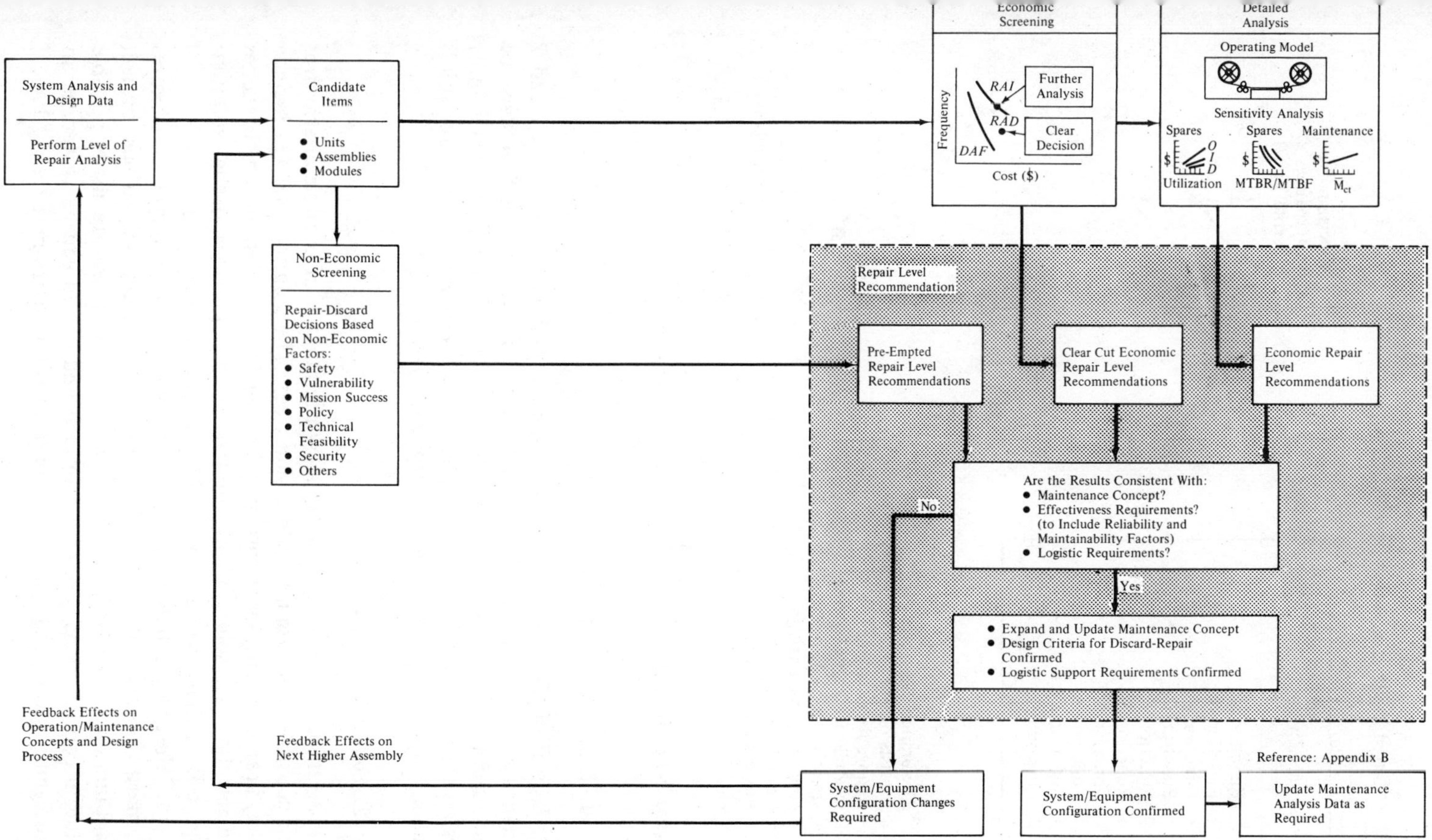

Figure B-10. Level of repair analysis procedure.

Function	Package Assignment
1. Function *A*	1
2. Function *B*	1
3. Function *C*	1
4. Function *D*	2
5. Function *E*	2
6. Function *F*	3
7. Function *G*	4
8. Function *H*	4
9. Function *I*	4
10. Function *J*	5

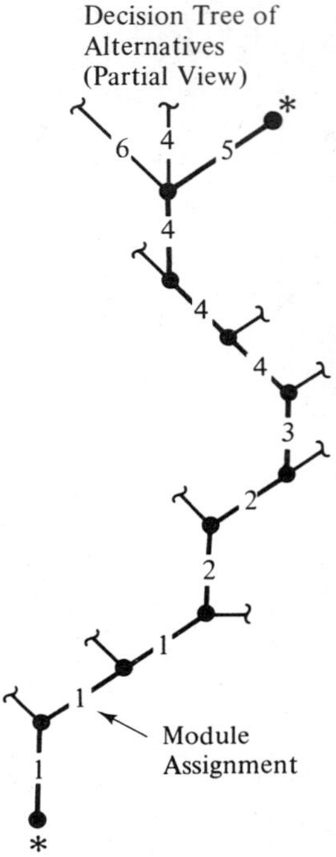

*The branch illustrated consists of five (5) removable assemblies or modular packages. There are other possible branch combinations ranging from all functions in a single package to 10 packages

Figure B-11. Branch and bound technique for optimum packaging.

1. Definition of the Problem

Unit B of the communication equipment defined in examples 1 and 2 constitutes 10 operational functions. How should these functions be packaged (10 removable assemblies, 9, 8, . . .,1) to reflect a high reliability design at the lowest life-cycle cost? The task is to evaluate *n* alternative unit B packaging concepts, where *n* is a large number.

2. Evaluation Approach

a. Initially, the analyst must assess the number of possible alternatives and construct a solution tree (decision tree). The solution tree will include many branches. The nodes correspond to partial solutions, and the branches indicate how the solution was obtained from a previous partial solution. Figure B-11 illustrates a particular branch of the solution tree.

b. The analyst then must determine reliability and life-cycle cost figures of merit for each branch and compare the results of various branch combinations to arrive at a feasible solution. As there are many possible solution combinations, use of the branch and bound can be viewed as an intelligently structured search of groups of solutions from a tree of logical possibilities. In general, the tree of all possible solutions is repeatedly partitioned into smaller and smaller subsets, and a lower bound (for minimization) is determined for reliability and an upper bound for cost within each subset. After each partitioning, the subsets with a bound that exceeds the cost (or is less than the prescribed reliability bound) of a known feasible solution are excluded from all future partitions. The partitioning continues until a feasible solution is found such that its cost is no greater than the bound for any subset or its reliability is no less. The problem can be solved through the use of integer programming.

3. Evaluation Results

From the figure, the function assignment for a given modular package is noted. The total solution in this instance (indicated by the terminal nodes designated by the asterisk) constitutes five packages. This example, like the level of repair analysis, illustrates a technique that can be applied to the design process with the intent of influencing design for optimum packaging.

D. SELECTION OF TEST AND SUPPORT EQUIPMENT CONFIGURATION (Example 4)

This example illustrates the selection of test and support equipment for the depot level of maintenance. The evaluation criteria include a number of parameters, such as performance, effectiveness, design characteristics, schedule, and cost considerations. Through this means, both qualitative and quantitative considerations are covered in the evaluation process.

1. Definition of the Problem

A system requires test and support equipment for maintenance functions at the depot or producer facility. The operational requirements and the maintenance concept have been defined, and a preliminary logistic support analysis has identified items assigned to the depot, arrival rates, and loading factors. A specific configuration of test and support equipment must be selected to fulfill this requirement.

Based on a review of the market, three alternative configurations are being considered. Each configuration is in existence (supporting other equipments in the inventory) and with some redesign could be made compatible with the new requirements. The design and development of new equipment are precluded on the basis of cost and need date.

2. Evaluation Approach

a. The analyst commences with developing a list of major evaluation parameters, as depicted in Figure B-12.

b. Quantitative weighting factors from 0 to 100 are assigned to each parameter in accordance with the degree of importance. The Delphi or some equivalent technique may be used to establish the weighting factors. The sum of all weighting factors is 100.

c. Each of the three alternative test and support equipment configurations is independently evaluated using special checklist criteria. Such criteria shall include desired qualitative and quantitative goals for the 11 evaluation parameters in Figure B-12. Using the criteria, base rating values from 0 to 10 are applied according to the degree of compatibility with the desired goals. If all goals are met, a rating of 10 is assigned.

d. The base rating values are multiplied by the weighting factors to obtain the score. The total score is then determined by adding the individual scores for each configuration. Since some redesign is required in each instance, a special derating factor is applied to compensate for risk and uncertainty. The resultant values are summarized in Figure B-12.

3. Evaluation Results

Referring to Figure B-12, configuration B represents the preferred approach based on the highest total score (730 points). This configuration is recommended in terms of its inherent features relating to performance, operability, effectiveness, design characteristics, logistic support, and so on.

E. RELIABILITY MAINTAINABILITY EVALUATION (Example 5)

This example presents a trade-off evaluation of reliability and maintainability, given a specific inherent availability (A_i). The results impact logistics in terms of frequency of maintenance, maintenance downtime, and cost.

1. Definition of the Problem

There is a requirement to replace existing equipment in the inventory with a new item for the purpose of improving operational effectiveness. The existing equipment meets an availability of 0.961, a MTBF of 125 hours, and a $\overline{Mct}$ of 5 hours. The new system must meet an availability of 0.990, a MTBF greater than 300 hours, and a $\overline{Mct}$ not to exceed 5.0 hours. An anticipated quantity of 200 equipments is to be procured. Three alternative design configurations are being considered to satisfy the requirement.

2. Evaluation Approach

a. Availability, reliability MTBF, and maintainability $\overline{Mct}$ factors are illustrated in Figure B-13, together with allowable area for trade-off.

Item	Evaluation Parameter	Weighting Factor	Conf. A		Conf. B		Conf. C	
			Base Rate	Score	Base Rate	Score	Base Rate	Score
1	PERFORMANCE—Compatibility, maintenance capability, maintenance load, method/mode of testing, versatility of testing	14	6	84	9	126	3	42
2	OPERABILITY—Simplicity of operation	4	10	40	7	28	4	16
3	EFFECTIVENESS—A_O, MTBM, $\overline{M}ct$, $\overline{M}pt$, MMH/OH	12	5	60	8	96	7	84
4	DESIGN CHARACTERISTICS—Reliability maintainability, human factors, safety, transportability, producibility, interchangeability	9	8	72	6	54	3	27
5	DESIGN DATA—Design drawings, specifications, logistics provisioning data, technical manuals, reports	2	6	12	8	16	5	10
6	TEST AIDS—General test equipment, calibration standards, maintenance tapes	3	5	15	8	24	3	9
7	FACILITIES and UTILITIES—Space, weight, volume, environment, power, heat, water, air conditioning	5	7	35	8	40	4	20
8	SPARE/REPAIR PARTS—Part type and quantity, standard parts, procurement time	6	9	54	7	42	5	30
9	FLEXIBILITY/GROWTH POTENTIAL—Test accuracies, performance range, space, reconfiguration, design change acceptability	3	4	12	8	24	6	18
10	SCHEDULE—R and D, production	17	7	119	8	136	9	153
11	COST—Life-Cycle (R and D, investment, O & M)	25	10	250	9	225	5	125
Subtotal				753		811		534
Derating factor (development risk)				113 15%		81 10%		107 20%
Grand Total		100		640		730		427

Figure B-12. Evaluation summary.

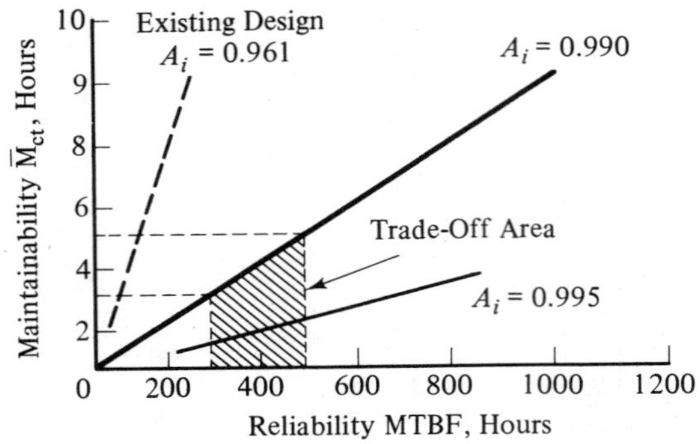

Conf.	A_i	MTBF	$\bar{M}_{ct}$
Existing	0.961	125	5.0
Alt. *A*	0.991	450	4.0
Alt. *B*	0.990	375	3.5
Alt. *C*	0.991	320	2.8

Other systems are eligible for consideration as long as the effectiveness parameters fall within the trade-off areas

Figure B-13. Reliability-maintainability trade-off.

b. Life-cycle cost data are developed for each of the three configurations in a manner similar to procedures described in the previous examples. In this instance, it is feasible to perform a gross-level analysis by selecting categories expected to have a major impact on the decision. For instance, improving reliability and/or maintainability characteristics in design will result in an increase in R&D and investment (manufacturing) cost. In addition, experience has indicated that such improvements will result in lower O&M cost, particularly in the areas of maintenance personnel and support cost and the cost of spare/repair parts. Thus, initially, the analyst looks only at these categories. If the final decision is close, then it may be appropriate to investigate other categories. A summary of partial cost data is presented in Figure B-14.

3. Evaluation Results

Configuration A satisfies the system availability, reliability, and maintenance requirements with the least life-cycle cost.

Figure 2-2 illustrates the engineering evaluation process which constitutes the overall approach utilized in decision making. The abbreviated examples presented above, and many more, basically follow the steps identified in the figure. The evaluation itself is facilitated through the use of models. A model constitutes a simplified representation of a real-world situation and is employed to aid the decision maker in addressing and analyzing certain problem areas. Models can vary from a basic mathematical expression to a complex computer program. The type of model required is a function of the quantity of variables, the number of alternatives, and the complexity of the operation. The decision maker must analyze the problem at hand, select techniques that can be used to resolve individual segments of the problem, and select or develop a model that properly employs these

Category	*Conf. A*	*Conf. B'*	*Conf. C*	*Remarks*
R and D cost				High Reliability Parts, Packaging, Accessibility, etc.
•Reliability design	$ 17,120	$ 15,227	$ 12,110	
•Maintainability design	2,109	4,898	7,115	
Investment cost manufacturing (200 Systems)	$3,422,400	$3,258,400	$3,022,200	$17,112/Equipment A; $16,292/Equipment B; $15,111/Equipment C
O & M cost				
•Maintenance personnel and support	$1,280,000	$1,536,000	$1,800,000	12,800 Maint. Action/Equipment A; 15,360 Maint. Action/Equipment B;
•Spare/Repair parts	342,240	325,840	302,220	18,000 Maint. Action/Equipment C; 10% of manufacturing cost for spares
Total Cost	$5,063,869	$5,140,365	$5,143,645	

Figure B-14. Cost summary (*partial costs*).

techniques. Typical problem applications may fall into any one of the following categories.

A. *Conceptual design and preliminary system design.* Applications in this area deal with the comparison of alternative operational concepts, mission profiles, performance parameters, effectiveness factors, product utilization rates, maintenance concepts, and related system/product functions. The designer in accomplishing feasibility studies (Section 3.3) and preliminary system design (Section 5.2) may utilize models in the evaluation of the many possible alternatives that exist at this stage in the life cycle.

B. *Detailed system/equipment design.* Models may be effectively employed to evaluate equipment, performance factors, packaging schemes, diagnostic and fault monitoring provisions, range and accuracy alternatives, size and weight, standardization, reliability, maintainability, human factors, and supportability characteristics in design.

C. *Determination of specific logistic support requirements.* As system/product design matures, it is necessary to evaluate the various alternative methods for supporting the system in the field. This includes the numerous and wide variety of approaches which become evident through the accomplishment of life-cycle cost analyses. The use of logistic support models facilitates the evaluation of various alternative support postures (i.e., maintenance plans).

D. *Determination of production/construction requirements.* When designing a production capability, there are numerous considerations related to man-machine functions, facility location and needs, number of production lines, inventory requirements, test and inspection stations, and so on. Models are utilized to evaluate to many alternative design possibilities in this area.

E. *Assessment of system/product effectiveness.* After initial deployment of the prime equipment and its associated elements of logistic support in the field, it is necessary to measure overall cost/system effectiveness. Are the elements of logistic support that have been acquired adequate (in terms of type, quantity, and location)? How should logistic support provisioning factors used in reprocurements be changed to better meet the needs? What is the impact of a prime equipment change on the various elements of logistic support? These and other factors need to be evaluated on a continuing basis to ensure that the required degree of system operational effectiveness is maintained, and logistic models may be employed accordingly.

When addressing problems of this type, we may ask, "Why models?" Models are useful for the following reasons:

A. There are numerous interrelated elements of a system/product that must be treated on an integrated basis. The interactions among reliability, maintainability, performance, effectiveness, logistic support, and so on, are many, and a variation in any one will affect the others in differing degrees. A model can be developed to represent these conditions in a realistic integrated manner and will allow for a productive analysis effort.
B. There may be many different alternatives available for consideration that require evaluation. A model may be developed to assess each alternative in a timely manner. The decision maker is provided the opportunity to investigate many more situations than would be possible through other means.
C. The use of models often permits the solution to problems that are impossible to solve by other means.
D. The use of appropriate models aids in rapidly identifying high risk areas (i.e., problems).
E. Models often facilitate the processing and handling of historical and statistical data in an effective manner.
F. Models may be employed to readily assess alternative corrective actions in terms of impact on the total system.

Modeling in general has tremendous educational benefits as well as being an effective tool in decision making. The analyst can readily identify relationships between system/product parameters and is better prepared to respond to the "what if?" questions. For instance, what is the impact on spare/repair parts if the equipment reliability is degraded? What is the impact on reliability if equipment component changes occur? What is the impact on life-cycle cost if equipment performance is increased? What is the cost impact if automation is implemented in place of performing functions through manual means? What is the impact on support equipment if the prime equipment package design changes? What is the impact on overall system maintenance if the equipment is utilized to a greater extent than initially planned? There are many questions of this nature to which an analyst with experience can readily respond, and in many instances these intuitive rapid responses are necessary to avoid possible future problems.

Models may be developed by applying a variety of analytical techniques.[2] For instance, we may employ simulation to analyze the overall behavior of a system over time under a specified set of constraints.

---

[2]For a better understanding of available analytical techniques and their specific applications, the reader should review additional material covering the use of operations research methods.

Functional relationships exist between the solution parameters and the control variables in the model, and in many instances the solutions are stated in terms of a range of values (not point estimates) or one or a set of intervals that contain the correct answer. On the other hand, straight analytical methods may be used to provide a single answer or a unique set of answers for any given set of input variables.

Models are often categorized by the analytical method(s) used to solve the problem. There are reliability models based on the probability of occurrence of various events. Given the configuration, certain basic probabilities, and a distribution, such models may be used in the prediction of failure rates and the quantity of maintenance actions. In a similar manner, maintainability models may be employed to determine anticipated maintenance times.

Network models are useful in evaluating the characteristics of systems/products that involve the movement of material and may be employed to evaluate the movement of prime equipment, spare/repair parts, and repairables shipped to a higher-level maintenance facility (a.g., intermediate shop to depot) for corrective maintenance support. A number of network models use probabilistic considerations in determining which path to take when choosing one from a number of alternatives. PERT/CPM models are also considered in this class. Optimization techniques are used when an optimal system design or support policy is desired and there are a number of constraints. Linear programming can be employed in optimizing with linear constraints while nonlinear programming is used when the applicable functions and constraints are nonlinear. Dynamic programming is a technique utilized if the solution involves a discrete set of possible alternatives with sequentially related constraints. This method is particularly suitable when the problem at hand involves a large number of variables.

Accounting models are used when the functions involved are basically additive. A good example is the determination of life-cycle cost, which is structured for the purpose of adding up numerous individual component costs.

There are many quantitative methods available for application in the development of models. Often, a model will incorporate a number of different analytical methods. For instance, optimization will be used along with linear and nonlinear programming, queuing techniques, probabilistic functions, accounting methods, and so on, in a given model. The selection of techniques is tailored to the specific problem areas being addressed and becomes dominant in the development of a model.

When developing models, certain cautions must be exercised. First, a single, complex, all-inclusive model cannot offer some advantages which are available through the use of multiple models designed to function as an integrated set. Computer storage may be quickly exceeded; it may be difficult to acquire the required time on the computer; the model may not incorporate the desired flexibility and/or growth potential; it becomes difficult and costly to incorporate changes in the model; and the model may not provide timely results. On the other hand, developing a set of models (each model in the set having the capability of evaluating a specific class of problems) does provide the necessary flexibility required to solve the large variety of problems listed above. Individual models can be tailored and are more responsive to evaluating the impact of a design change on spare/repair parts, the impact of a waiting line or repairable-item demands on support equipment loading and utilization, and similar lower-level problems. If changes to the model are required, the necessary changes can be incorporated without affecting other models in the set. Thus, when developing models, the modular approach is preferable. Both individual evaluations and the total system evaluation effort must be considered such that different models in the set may be employed singly or on a combined basis.

As a second point, in developing the model, the usual dangers inherent in abstraction are present. A mathematically feasible model or series of expressions may require gross oversimplifications. There is no guarantee that the time and effort invested in model development will provide the results desired. The analyst must be careful to ensure that the model does indeed reflect a realistic situation.

Finally, there is the danger that the analyst, after experimenting with the model for a long period of time, will become too attached to it. Some analysts will become so attached that they will insist that the model *is* the real world and/or is directly applicable to *all* problems at hand. Again, care must be exercised to ensure that the model selected or developed is compatible with the problem(s) being solved. Models are only a tool used in decision making and cannot be considered as a substitute for experience and judgment. As such, the use of models in a variety of situations has produced successful results. Several excellent references which discuss the use and application of models are:

1. W. J. Fabrycky, P. M. Ghare, and P. E. Torgersen, *Industrial Operations Research,* Prentice-Hall, Inc., Englewood Cliffs, N.J., 1972.

2. A. Schuchman, *Scientific Decision Making in Business—Readings in Operations Research for Non-mathematicians,* Holt, Rinehart and Winston, Inc., New York 1963.

C

SYSTEM/PRODUCT DESIGN REVIEW CHECKLIST

On a periodic basis throughout the system/product design and development process, it is beneficial to conduct one or more informal reviews to assess (1) whether essential project tasks have been accomplished, and (2) the extent to which the required characteristics in design have been incorporated. Questions, presented in a checklist format, have been developed to reflect certain features. Not all questions are applicable in all reviews; however, the answer to those questions that are applicable should be *yes*, to reflect desirable results. These questions directly support an abbreviated checklist presented in Figure 5-18.

A. GENERAL

1.0 System Operational Requirements

1.1 Has the system/product mission been defined? Mission scenarios?
1.2 Have all basic system/product performance parameters been defined?
1.3 Has the planned operational deployment been defined (quantity of equipments per geographical location)?
1.4 Has the method(s) of distribution for various elements of the system been defined?
1.5 Has the system/product life cycle been defined?

1.6 Have system utilization requirements been defined? This includes hours of system/equipment operation or quantity of operational cycles per a given time period. Define an operational cycle if used.

1.7 Has the operational environment been defined in terms of temperature extremes, humidity, shock and vibration, storage, transportation, and handling?

2.0 Effectiveness Factors

2.1 Have the appropriate measures of system/product effectiveness been identified?

2.2 Have quantitative availability, reliability, maintainability, and logistic support factors been specified where appropriate? This includes consideration of A_o, A_a, A_i, MTBM, MTBF, MDT, $\overline{M}$, $\overline{M}ct$, $\overline{M}pt$, MMH/OH, cost/OH, cost/MA, life-cycle cost, TAT, test effectiveness, supply support effectiveness, and pipeline time.

2.3 Are the specified effectiveness factors appropriate for system/product operation and mission requirements?

2.4 Are the specified effectiveness factors sensitive to changes in operational requirements?

2.5 In the event that two or more effectiveness measures are appropriate, are the measures properly weighted (the relative weighting in terms of significance or level of importance of each applicable criterion factor employed)?

3.0 System/Product Maintenance Concept

3.1 Have the levels or echelons of maintenance been specified and defined?

3.2 Have basic maintenance functions been identified for each level?

3.3 Have quantitative parameters been established for turnaround time (TAT) at each maintenance level and logistics pipeline time between levels?

3.4 Has the logistics pipeline time between levels been minimized to the extent feasible considering cost? The lack of adequate supply responsiveness has a major detrimental affect on total logistic support.

3.5 Has level of repair policies been established? All assemblies and subassemblies should be classified as "repairable" or "discard at failure."

3.6 Have the criteria for level of repair decisions been adequately defined?

3.7 Has the level of maintenance (organizational, intermediate, depot, or supplier) been defined for each repairable item?

3.8 Have criteria been established for test and support equipment at each level of maintenance?

3.9 Have criteria been established for personnel quantities and/or skills at each level of maintenance?

3.10 Have criteria been established for inventories at each level of maintenance?

3.11 Have criteria been established for facilities at each level of maintenance?

3.12 Have major transportaton and handling functions been identified?

4.0 Functional Analysis and Allocation

4.1 Have all major system/product operational functions been defined?

4.2 Have all major system/product maintenance functions been defined?

4.3 Have performance parameters and effectiveness factors (e.g., availability, reliability, maintainability, logistics support, weight, size, cost) been allocated to the appropriate system elements or indenture levels (i.e., subsystems, units, assemblies, subassemblies, etc.)?

4.4 Has functional allocation been accomplished to the depth necessary to provide adequate criteria for system/product design?

4.5 Have boundaries, ranges, and/or constraints been established in lieu of discrete quantitative values, where appropriate?

5.0 System/Product Trade-off Studies

5.1 Are all feasible alternatives considered in the trade-off study?

5.2 Are all assumptions adequately identified?

5.3 Are the assumptions reasonable? Do any of the specific assumptions treat quantitative or qualitative uncertainties as facts (which should be avoided)?

5.4 Are trade-off studies adequately documented?

5.5 Are the facts stated correctly and with proper qualification where appropriate?

5.6 Are all applicable reference sources identified?

5.7 Are the conclusions and recommendations logically derived from the material contained in the study?

5.8 Are the conclusions and recommendations really feasible in light of ecological, political, cultural policy or other considerations?

5.9 Have biases been eliminated or minimized in arriving at the stated conclusions?

6.0 System and Supporting Specifications

6.1 Have system, procurement, development, process, and material specifications been prepared as required?

6.2 Has a specification tree been developed for each project?

6.3 Are the specifications that are applicable for each system/product compatible with each other? Have conflicting specification requirements been eliminated?

6.4 Does the system specification cover the information identified in Figure 3-8?

7.0 Product Acquisition Plan

7.1 Has a product acquisition plan been prepared? Does it cover the information content described in Section 3.4C?

7.2 Are all appropriate product research requirements identified (basic research, applied research, research methods, etc)?

7.3 Are all appropriate product design requirements identified (conceptual design, preliminary system design, detailed equipment design, design support, and engineering model/prototype development)?

7.4 Are all appropriate production/construction requirements identified?

7.5 Have all specific research, design, and production/construction tasks been defined, scheduled, combined, and identified with an organization function responsible for their accomplishment? Has a cost estimate been developed for each identified task? Have milestone charts, bar charts, and/or PERT/CPM networks been developed where appropriate?

7.6 Have formal system and equipment design reviews been scheduled?

7.7 Have formal program or project reviews been scheduled?

8.0 Design Documentation

8.1 Has a design documentation tree been developed for each project?

8.2 Have design standards manuals been utilized to the maximum extent possible?

8.3 Has the system/product design been adequately defined through design documentation?

8.4 Can the system/product be reproduced in multiple like quantities from the available design documentation?

8.5 Is the design documentation adequate for procurement purposes and for contracting?

8.6 Are the various items of documentation compatible in content? Has unnecessary redundancy been eliminated?

8.7 Does the design documentation truly reflect the system/product configuration as it currently exists? Are all changes incorporated?

9.0 Logistic Support Requirements

9.1 Has a product use and logistic support plan been prepared? Does the plan cover the activities described in Section 3.4E?

9.2 Has a logistic support analysis been accomplished? Does the LSA justify all logistic support requirements?

9.3 Personnel and Training Requirements

a. Have operational and maintenance personnel requirements (quantity and skill levels) been defined?

b. Are operational and maintenance personnel requirements minimized to the greatest extent possible?

c. Are operational and maintenance personnel requirements compatible with the logistic support analysis? Personnel quantities and skill levels should "track."

d. Are the planned personnel skill levels at each location compatible with the complexity of the operational and maintenance tasks specified?

e. Has maximum consideration been given to the use of existing personnel skills for new equipment?

f. Have personnel attrition rates been established?

g. Have personnel effectiveness factors been determined (actual time that work is accomplished per the total time allowed for work accomplishment)?

h. Have operational and maintenance training requirements been specified? This includes consideration of both initial training and replenishment training throughout the life cycle.

i. Have specific training programs been planned? The type of training, frequency of training, duration of training, and student entry requirements should be identified.

j. Are the planned training programs compatible with the personnel-skill-level requirements specified for the performance of operational and maintenance tasks?

k. Have training equipment requirements been defined? Acquisitioned?

l. Have maintenance provisions for training equipment been planned?

m. Have training data requirements been defined?

n. Are the planned operating and maintenance procedures (designated for support of the system throughout its life cycle) utilized to the maximum extent possible in the training program(s)?

9.4 Test and Support Equipment

a. Have the test-and-support-equipment requirements been defined for each level of operation and maintenance? Are the requirements compatible with the system maintenance concept?

b. Have standard test-and-support-equipment items been selected? Newly designed equipment should not be necessary unless standard equipment is either unacceptable or unavailable.

c. Are the selected test-and-support-equipment items compatible with the prime equipment? Does the test equipment do the job?

d. Are the test-and-support-equipment requirements compatible with the logistic support analysis? Are they justified?

e. Have test-and-support-equipment requirements (both in terms of variety and quantity) been minimized to the greatest extent possible?

f. Are the reliability and maintainability features in the test and support equipment compatible with those equivalent features in the prime equipment? It is not practical to select an item of support equipment that is not as reliable as the item it supports.

g. Have logistic support requirements for the selected test and support equipment been defined? This includes maintenance tasks, test equipment, spare/repair parts, personnel and training, data, and facilities.

h. Is the test-and-support-equipment selection process based on cost-effectiveness considerations (i.e., life-cycle cost)?

9.5 Supply Support (Spare/Repair Parts)

a. Have the distribution channels and inventory points for spare/repair parts been established?

b. Are the types and quantity of spare/repair parts compatible with the system/product packaging design?

c. Are the types and quantity of spare/repair parts compatible with the system maintenance concept, the logistic support analysis, and level of repair analysis data?

d. Are the types and quantity of spare/repair parts designated for a given location appropriate for the estimated demand at that location? Too many or too few spares can be costly.

e. Are spare/repair part provisioning factors (e.g., replacement frequencies) directly traceable to reliability and maintainability predictions?

f. Are the specified logistics pipeline times compatible with effective supply support? Long pipeline times place a tremendous burden on logistic support.

g. Have spare/repair parts been identified and provisioned for properational support activities (e.g., interim producer or supplier support, test programs)?

h. Have spare/repair part requirements been minimized to the maximum extent possible?

i. Have test and acceptance procedures been developed for spare/repair parts? Spare/repair parts should be processed, produced, and accepted on a similar basis with their equivalent components in the prime equipment.

j. Have the consequences (risks) of stock-out been defined in terms of effect on mission requirements and cost?

k. Has an inventory safety stock level been defined?

l. Has a provisioning or procurement cycle been defined (procurement or order frequency)? Have EOQ factors been determined?

m. Has a supply-availability requirement been established (the probability of having a spare available when required)?

n. Are spare/repair parts completely interchangeable with those comparable elements of prime mission equipment which they are designed to support? Have the spare/repair parts been modified to incorporate the applicable changes made in the prime equipment?

9.6 Facilities and Storage

a. Have facility requirements (space, volume, capital equipment, utilities, etc.) necessary for production/construction been defined?
b. Have facility requirements (space, volume, capital equipment, utilities, etc) necessary for system-life-cycle operation been defined?
c. Have facility requirements (space, volume, capital equipment, utilities, etc.) necessary for system-life-cycle maintenance at each level been defined?
d. Have operational and maintenance facility requirements been minimized to the greatest extent possible?
e. Have environmental system requirements (e.g., temperature, humidity, and dust control) associated with operational and maintenance facilities been identified? Have hazardous conditions been eliminated?
f. Have storage shelf-space requirements for spare/repair parts been defined?
g. Have storage environments been defined?
h. Are the designated facility and storage requirements compatible with the logistic support analysis?

9.7 Technical Data

a. Have system/product operating and maintenance procedure requirements been defined for the life cycle?
b. Have the necessary procedures been prepared?
c. Are operating and maintenance procedures compatible with logistic support analysis data? This pertains particularly to the logic troubleshooting flow diagrams, task sequences, and support requirements.
d. Are operating and maintenance procedures as brief as possible without sacrificing necessary information?
e. Are operating and maintenance procedures adequate from the standpoint of presenting simple step-by-step instructions, including appropriate use of illustrations, and tables for presenting data?
f. Are operating and maintenance procedures compatible with the level of activity performed at the location where the procedures are used? Depot maintenance instructions should not be included in manuals that are used at the intermediate level of maintenance. The maintenance procedures should be compatible with the level of repair analysis and the maintenance concept.
g. Are operating and maintenance procedures written to the skill level of the individual accomplishing the functions covered by the procedures?

Procedures should be written in a simple, clear, and concise manner for low-skilled personnel.

h. Do the operating and maintenance procedures specify the correct test and support equipment, spare/repair parts, transportation and handling equipment, and facilities?

i. Do the procedures include special warning notices in areas where safety is a concern?

j. Are the designated operating and maintenance procedures used in system/equipment test programs?

9.8 Transportation and Handling

a. Are transportation and handling requirements defined for both operational and maintenance functions? This includes transportation of prime equipment, test and support equipment, spares, personnel, and data.

b. Are transportation and handling environments (temperature, shock and vibration, exposure to dust and salt spray, storage, etc.) defined?

c. Are the modes (air, ground vehicle, rail, sea, or a combination) of transportation known? A profile or scenario, similar to that accomplished for mission definition, should be developed showing the various transportation and handling requirements.

d. Are the requirements for reusable containers known? Design information should be developed on reusable containers.

e. Are the requirements for packing known? This includes labor, material, preservation, storage limitations, and the processing of an item for shipment.

10.0 Ecological Requirements

10.1 Have the required air quality standards been maintained in spite of the introduction, operation, and logistic support of the system/product? Air pollution must not exceed certain specified environmental standards and must be controlled at all times.

10.2 Have the required water quality standards been maintained in spite of the introduction, operation, and logistic support of the system/product? Water quality control is required by legislation.

10.3 Are noise levels in the environment being controlled to preclude possible human injury (e.g., less than the 85-decibel level)?

10.4 Are solid wastes resulting from the introduction, operation, and logistic support of the system/product being properly managed? Solid waste must be collected, treated, processed for disposal, and/or recycled.

11.0 Societal Requirements

11.1 Does the system/product satisfy societal needs?

11.2 Have the societal effects from introducing a new system/product into the inventory been evaluated? How does system/product introduction, operation, and logistic support affect community life?

11.3 Have all adverse societal effects caused by the introduction of a new system/product into the inventory been minimized, if not eliminated?

12.0 Economic Factors

12.1 Has the system/product been justified in terms of total life-cycle cost?
12.2 Are all cost elements considered? Refer to Figure 12-4.
12.3 Are cost categories adequately defined?
12.4 Are cost estimates relevant?
12.5 Are incremental and marginal costs considered?
12.6 Are variable and fixed costs separately identifiable?
12.7 Are escalation factors specified and employed?
12.8 Are learning curves specified and employed?
12.9 Is the discount rate specified and employed in the economic evaluation of alternatives?
12.10 Are the cost aspects employed in the evaluation of alternatives treated in a consistent manner?
12.11 Are cost-estimating relationships used in economic analyses reasonably accurate?
12.12 Are areas of risk and uncertainty identified?
12.13 Has the sensitivity of cost estimates been properly addressed through a contingency or sensitivity analysis?

B. DESIGN FEATURES (Representative Sample)

1.0 Accessibility

1.1 Is the equipment directly accessible for the performance of both operator and maintenance tasks?
1.2 Is access easily attained?
1.3 Are access requirements compatible with the frequency of maintenance (or need)? Accessibility for items requiring frequent maintenance should be greater than that for items requiring infrequent maintenance.
1.4 Are access doors provided where appropriate? Are hinged doors utilized?
1.5 Are access openings adequate in size and optimally located for the access required?
1.6 Are access doors and openings labeled in terms of items that are accessible from within?
1.7 Can access doors that are hinged be supported in the open position?
1.8 Are access door fasteners minimized?
1.9 Are access door fasteners of the quick-release variety?

1.10 Can access be attained without the use of tools?
1.11 If tools are required to gain access, are the number of tools held to a minimum? Are the tools of the standard variety?
1.12 Are access provisions between modules and components adequate?

2.0 Adjustments and Alignments

2.1 Are adjustment requirements and frequencies known?
2.2 Have adjustment requirements been minimized?
2.3 Are adjustment points accessible?
2.4 Are adjustment-point locations compatible with the maintenance level at which the adjustment is made?
2.5 Are adjustment interaction effects eliminated?
2.6 Are factory adjustments specified?
2.7 Are adjustment points adequately labeled?

3.0 Cables and Connectors

3.1 Are cables fabricated in removable sections?
3.2 Are cables routed to avoid sharp bends?
3.3 Are cables routed to avoid pinching?
3.4 Is cable labeling adequate?
3.5 Is cable clamping adequate?
3.6 Are the connectors used of the quick-disconnect variety?
3.7 Are connectors that are mounted on surfaces far enough apart so that they can be firmly grasped for connecting and disconnecting?
3.8 Are connectors and receptacles labeled?
3.9 Are connectors and receptacles keyed?
3.10 Are connectors standardized?
3.11 Do the connectors incorporate provisions for moisture prevention?

4.0 Calibration

4.1 Are calibration requirements known?
4.2 Are calibration frequencies known?
4.3 Are calibration tolerances known?
4.4 Are calibration requirements minimized?
4.5 Are standards available for calibration?
4.6 Are calibration procedures prepared?
4.7 Is traceability to the National Bureau of Standards possible?
4.8 Have the facilities for calibration been identified?
4.9 Are the calibration requirements compatible with the logistic support analysis and the maintenance concept?

5.0 Packaging and Mounting

5.1 Is the packaging design attractive from the standpoint of consumer appeal (e.g., color, shape, size)?

5.2 Is functional packaging incorporated to the maximum extent possible? Interaction effects between packages should be minimized. It should be possible to limit maintenance to the removal of one module (the one containing the failed part) when a failure occurs and not require the removal of two, three, or four modules in order to resolve the problem.

5.3 Is the packaging design compatible with level of repair analysis decisions? Repairable items are designed to include maintenance provisions such as test points, accessibility, and plug-in components. Items classified as "discard at failure" should be encapsulated and relatively low in cost. Maintenance provisions within the disposable module are not required.

5.4 Are disposable modules incorporated to the maximum extent practical? It is highly desirable to reduce overall support through a no-maintenance-design concept as long as the items being discarded are relatively high in reliability and low in cost.

5.5 Are plug-in modules and components utilized to the maximum extent possible (unless the use of plug-in components significantly degrades the equipment reliability)?

5.6 Are accesses between modules adequate to allow for hand grasping?

5.7 Are modules and components mounted such that the removal of any single item for maintenance will not require the removal of other items? Component stacking should be avoided where possible.

5.8 In areas where module stacking is necessary because of limited space, are the modules mounted in such a way that access priority has been assigned in accordance with the predicted removal and replacement frequency? Items that require frequent maintenance should be more accessible.

5.9 Are modules and components, not of the plug-in variety, mounted with four fasteners or less? Modules should be securely mounted, but the number of fasteners should be held to a minimum.

5.10 Are shock-mounting provisions incorporated where shock and vibration requirements are excessive?

5.11 Are provisions incorporated to preclude installation of the wrong module?

5.12 Are plug-in modules and components removable without the use of tools? If tools are required, they should be of the standard variety.

5.13 Are guides (slides or pins) provided to facilitate module installation?

5.14 Are modules and components labeled?

5.15 Are module and component labels located on top or immediately adjacent to the item and in plain sight?

5.16 Are the labels permanently affixed and unlikely to come off during a maintenance action or as a result of environment? Is the information on the label adequate? Disposable modules should be so labeled.

5.17 In equipment racks, are the heavier items mounted at the bottom of the rack? Unit weight should decrease with the increase in installation height.

5.18 Are operator panels optimally positioned? For personnel in the standing position, panels should be located between 40 and 70 inches above the floor. Critical or precise controls should be between 48 and 64 inches above the floor. For personnel in the sitting position, panels should be located 30 inches above the floor.

5.19 Are drawers in equipment racks mounted on roll-out slides?

6.0 Disposability

6.1 Has the equipment been designed for disposability (e.g., selection of materials, packaging)?

6.2 Have procedures been prepared to cover system/equipment/component disposal?

6.3 Can the components or materials used in equipment design be recycled for use in other products?

6.4 If component/material recycling is not feasible, can decomposition be accomplished?

6.5 Can recycling and/or decomposition be accomplished using existing logistic support resources?

6.6 Are recycling and/or decomposition methods and results consistent with environmental, ecological, safety, political, and social requirements?

6.7 Is the method(s) used for recycling and/or decomposition economically feasible?

7.0 Environment

7.1 Has the equipment design considered the following: temperature, shock, vibration, humidity, pressure, wind, salt spray, sand and dust, rain, fungus, and radiation? Have the ranges and extreme conditions been specified and properly addressed in design? Have the proper environmental profiles been addressed?

7.2 Have provisions been made to specify and control noise, illumination, humidity, and temperature in areas where personnel are required to perform operating and maintenance functions?

7.3 Is the design output compatible with air and water quality requirements?

8.0 Fasteners

8.1 Are quick-release fasteners used on doors and access panels?

8.2 Are the total number of fasteners minimized?

8.3 Are the number of different type of fasteners held to a minimum? This relates to standardization.

8.4 Have fasteners been selected based on the requirement for standard tools rather than special tools?

9.0 Handling

9.1 For heavy items, are hoist lugs (lifting eyes) or base-lifting provisions for forklift-truck application incorporated? Hoist lugs should be provided on all items weighing more than 150 pounds.
9.2 Are hoist and base-lifting points identified relative to lifting capacity?
9.3 Are weight labels provided?
9.4 Are packages, units, components, or other items weighing over 10 pounds provided with handles? Are the proper-sized handles used and are they located in the right position? Are the handles optimally located from the weight distribution standpoint? Handles should be located over the center of gravity.
9.5 Are packages, units, or other items weighing more than 40 pounds provided with two handles (for a two-man carrying capability)?
9.6 Are containers, cases, or covers provided to protect equipment vulnerable areas from damage during handling?

10.0 Human Factors

10.1 Has a system analysis been accomplished to verify optimum man-machine interfaces? Are automated and manual functions adequately specified?
10.2 Has a detailed equipment operator task analysis been accomplished to verify task sequences, task complexities and personnel skills, and so on?
10.3 Has a detailed equipment maintenance task analysis been accomplished to verify maintenance task sequences, task complexities and personnel skills, and so on?
10.4 For manual functions, does equipment design reflect ease of operation by low-skilled personnel? Is the design such that human error rates are minimized?
10.5 Are the detailed operator and maintenance task analyses compatible with system/equipment operating and maintenance procedures (e.g., task sequences, depth of explanatory material based on task complexity)? Are the detailed task analyses compatible with reliability data, maintainability data, and the logistic support analysis?

11.0 Interchangeability

11.1 Are equipment, modules, and/or components that perform similar operations electrically, functionally, and physically interchangeable?
11.2 Can replacements of like items be made without adjustments and/or alignments?

12.0 Maintainability

12.1 Have maintainability requirements for the system/equipment been adequately defined? Are they compatible with system performance and effectiveness factors?

12.2 Have maintainability requirements been allocated to the appropriate level (e.g., MTBM, MDT, MMH/OH, $\overline{Mct}$, $\overline{Mpt}$ to the unit, assembly, and/or subassembly)?

12.3 Have maintainability predictions been accomplished to assess the design in terms of the specified requirements? Do the predictions indicate compliance with the requirements?

12.4 Are anticipated system/product corrective and preventive maintenance requirements identified? Have the proper trade-off studies been conducted to attain the proper balance between corrective and preventive maintenance? Too much preventive maintenance can be costly and can significantly impact correct maintenance requirements.

12.5 Are maintainability data compatible with reliability data, the logistic support analysis, level of repair analysis, and the maintenance concept?

12.6 Have maintainability demonstrations been conducted? Do the results indicate compliance with the requirements?

12.7 Is the system/equipment maintainable in terms of the requirements?

13.0 Panel Displays and Controls

13.1 Are controls standardized?

13.2 Are controls sequentially positioned?

13.3 Is control spacing adequate?

13.4 Is control labeling adequate?

13.5 Have the proper control/display relationships been incorporated?

13.6 Are the proper type of panel switches used?

13.7 Is the control panel lighting adequate?

13.8 Are the controls placed according to frequency of use?

14.0 Producibility

14.1 Does the design lend itself to economic production? Can simplified fabrication and assembly techniques be employed?

14.2 Has the design stabilized (minimum change)?

14.3 Has the design been verified through prototype testing and environmental qualification testing?

14.4 Is the design such that many models of the same item can be produced with identical results?

14.5 Are the production drawings and material lists adequate?

14.6 Are common materials used (rather than special materials)?

14.7 Can standard tooling and existing facilities be used for fabrication, assembly, and test operations?

14.8 Is the design such that rework requirements are minimized? Are spoilage factors held to a minimum?

14.9 Are standard fabrication, assembly, test, and inspection procedures applicable?

14.10 Is the design such that automated manufacturing processes (e.g., numerical control techniques) can be applied for high-volume repetitive functions?

14.11 Is the design definition such that two or more suppliers can produce the equipment from a given set of drawings with identical results?

15.0 Reliability

15.1 Have reliability requirements for the system/equipment been adequately defined? Are they compatible with performance and effectiveness factors?

15.2 Have reliability requirements been allocated to the appropriate level (e.g., MTBM, MTBF, and λ, to the unit, assembly, and/or subassembly)?

15.3 Has a failure-mode-and-effects analysis (FMEA) been accomplished to determine the most probable modes of failure and their effect on the system/equipment?

15.4 Have reliability predictions been accomplished to assess the design in terms of the specified requirements? Do the predictions indicate compliance with the requirements?

15.5 Has the system/equipment wear-out period been defined?

15.6 Are item failure rates known?

15.7 Have parts with excessive failure rates been identified?

15.8 Has equipment design complexity been minimized?

15.9 Is protection against secondary failures (resulting from primary failures) incorporated where possible?

15.10 Has the use of adjustable components been minimized?

15.11 Has the use of friction or pressure contacts in mechanical equipment been avoided?

15.12 Have all critical-useful-life items been eliminated from the equipment design?

15.13 Have cooling provisions been incorporated in design "hot-spot" areas?

15.14 Is cooling directed toward the most critical items?

15.15 Are reliability data compatible with the maintainability data, logistic support analysis, level of repair analysis, and the maintenance concept?

15.16 Have reliability tests been conducted? Do the results indicate compliance with the requirements?

16.0 Safety

16.1 Has an integrated safety plan been prepared and implemented?

16.2 Has a hazard analysis been accomplished to identify potential hazardous conditions? Is the hazard analysis compatible with the reliability FMEA?

16.3 Have system/equipment hazards from heat, cold, thermal change, barometric change, humidity change, shock, vibration, light, mold, bacteria, corrosion, rodents, fungi, odors, chemicals, oils, greases, handling and transportation, and so on, been eliminated?

16.4 Have fail-safe provisions been incorporated in the design?

16.5 Have protruding devices been eliminated or are they suitably protected?

16.6 Have provisions been incorporated for protection against high voltages? Are all external metal parts adequately grounded?

16.7 Are sharp metal edges, access openings, and corners protected with rubber, fillets, fiber, or plastic coating?

16.8 Are electrical circuit interlocks employed?

16.9 Are standoffs or handles provided to protect equipment from damage during the performance of bench maintenance?

16.10 Are tools that are used near high-voltage areas adequately insulated at the handle or at other parts of the tool which the maintenance man is likely to touch?

16.11 Are the environments such that personnel safety is ensured? Are noise levels within a safe range? Is illumination adequate? Is the air clean? Are the temperatures at a proper level? Are OSHA requirements being maintained?

16.12 Has the proper protective clothing been identified for areas where the environment could be detrimental to human safety? Radiation, intense cold or heat, gas, loud noise, and so on, are examples.

16.13 Are safety equipment requirements identified in areas where ordnance devices and the like are activated?

17.0 Selection of Parts/Materials

17.1 Have appropriate standards been consulted for the selection of components and materials?

17.2 Have all component parts and materials selected for the design been adequately evaluated prior to their procurement and application? Evaluation should consider performance parameters, reliability, maintainability, human factors, and cost.

17.3 Have supplier sources for component-part and material procurement been established?

17.4 Are the established supplier sources reliable in terms of quality level, ability to deliver on time, and willingness to accept component-warranty provisions?

17.5 Have alternative supplier sources been identified in the event that the prime source fails to deliver?

17.6 Have the reliability, maintainability, and human factors engineers been consulted in the selection and application of parts? Reliability is concerned with part-failure rates, stresses, tolerances, allowable temperature extremes, and signal ratings. Maintainability and human factors are concerned with the part effects on maintenance times, mounting provisions, man interfaces, and so on.

18.0 Servicing and Lubrication

18.1 Have servicing requirements been held to a minimum?
18.2 When servicing is indicated, are the specific requirements identified? This includes frequency of servicing and the materials needed.
18.3 Are procurement sources for servicing materials known?
18.4 Are servicing points accessible?
18.5 Have personnel and equipment requirements for servicing been identified? This includes handling equipment, vehicles, carts, and so on.
18.6 Does the design include servicing indicators?

19.0 Standardization

19.1 Are standard equipment items and parts incorporated in the design to the maximum extent possible (except for items not compatible with effectiveness factors)? Maximum standardization is desirable.
19.2 Are the same items and/or parts used in similar applications?
19.3 Are the number of different part types used throughout the design minimized? In the interest of developing an efficient supply support capability, the number of item spares should be held to a minimum.
19.4 Are identifying equipment labels and nomenclature assignments standardized to the maximum extent possible?
19.5 Are equipment-control-panel positions and layouts (from panel to panel) the same or similar when a number of panels are incorporated and provide comparable functions?

20.0 Storage

20.1 Can the equipment be stored for extended periods of time without excessive degradation (beyond specification limits)?
20.2 Have scheduled maintenance requirements for stored equipment been defined?
20.3 Have scheduled maintenance requirements for stored equipment been eliminated or minimized?
20.4 Have the required maintenance resources necessary to service stored equipment been identified?
20.5 Is equipment design compatible with the defined storage environments?

21.0 Survivability

21.1 Have system/equipment survivability requirements been defined (e.g., survival in space, high radiation, and comparable environments)?

21.2 Are the survivability requirements compatible with performance, effectiveness, reliability, maintainability, and human factors requirements?

22.0 Test Provisions

22.1 Have on-board monitoring and self-test provisions been incorporated where appropriate?
22.2 Is the extent or depth of self-testing compatible with the logistic support analysis, level of repair analysis, and maintenance concept?
22.3 Are self-test provisions automatic?
22.4 Have direct fault indicators been provided (either a fault light, an audio signal, or a means of determining that a malfunction positively exists)? Are continuous performance monitoring provisions incorporated where appropriate?
22.5 Are test points provided to enable checkout and fault isolation beyond the level of self-test? Test points for fault isolation within an assembly should not be incorporated if the assembly is to be discarded at failure. Test-point provisions must be compatible with the level of repair analysis.
22.6 Are test points accessible? Accessibility should be compatible with the extent of maintenance performed. Test points on the operator's front panel are not required for a depot maintenance action.
22.7 Are test points functionally and conveniently grouped to allow for sequential testing (following a signal flow), testing of similar functions, or frequency of use when access is limited?
22.8 Are test points provided for direct testing of all replaceable items?
22.9 Are test points adequately labeled? Each test point should be identified with a unique number, and the proper signal or expected measured output should be specified on a label located adjacent to the test point.
22.10 Are test points adequately illuminated to allow the technician to see the test point number and labeled signal value?
22.11 Can every equipment malfunction (degradation beyond specification tolerance limits) that could possibly occur in the equipment be detected through a no-go indication at the system level? This is a measure of test thoroughness.

The preceding questions are representative (not to be considered all-inclusive) of what should be considered in conducting a system/product design review. These questions stem from specific qualitative and quantitative design criteria that serve as an initial input into the design process. A few examples of criteria currently used are illustrated in Figures C-1 through C-4. Criteria covering these and other areas of design are followed in the design process, to the extent appropriate, and reviews of the product are subsequently accomplished to assess the results.

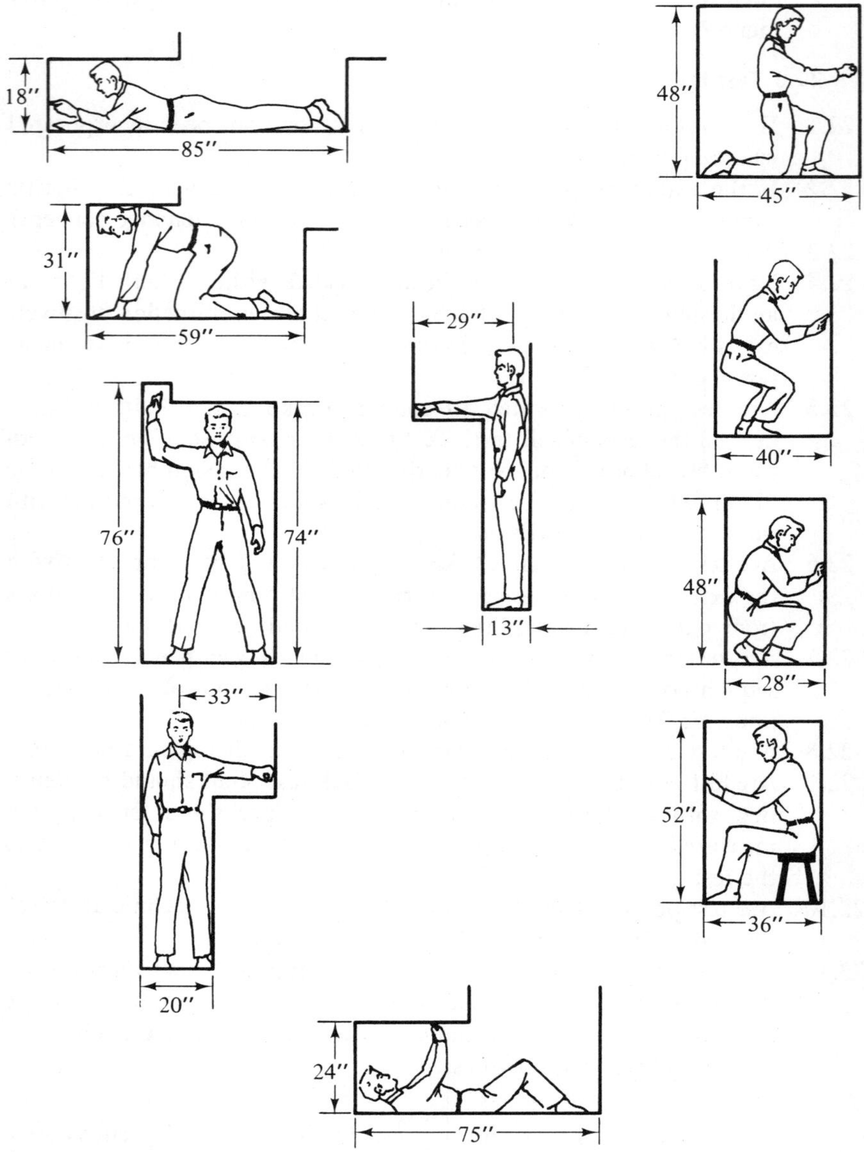

Figure C-1. General criteria. (*Source:* NAVSHIPS 94324, *Maintainability Design Criteria Handbook for Designers of Shipboard Electronic Equipment,* Naval Ship Systems Command, U.S. Navy.)

Minimum Openings For Using Common Hand Tools

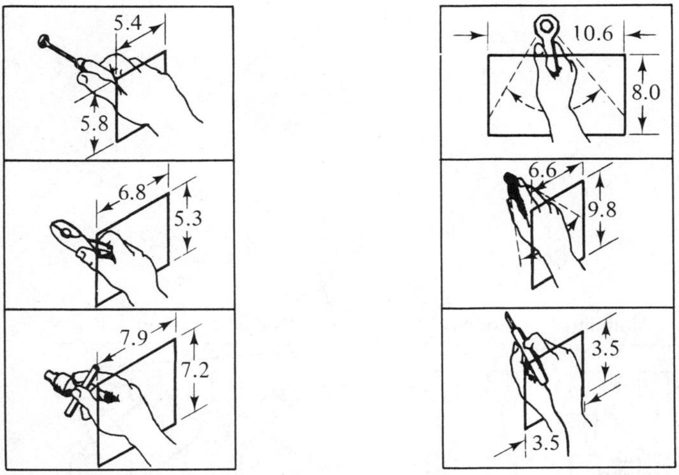

M Criteria
(Accessibility)

Space Required For Using Common Hand Tools

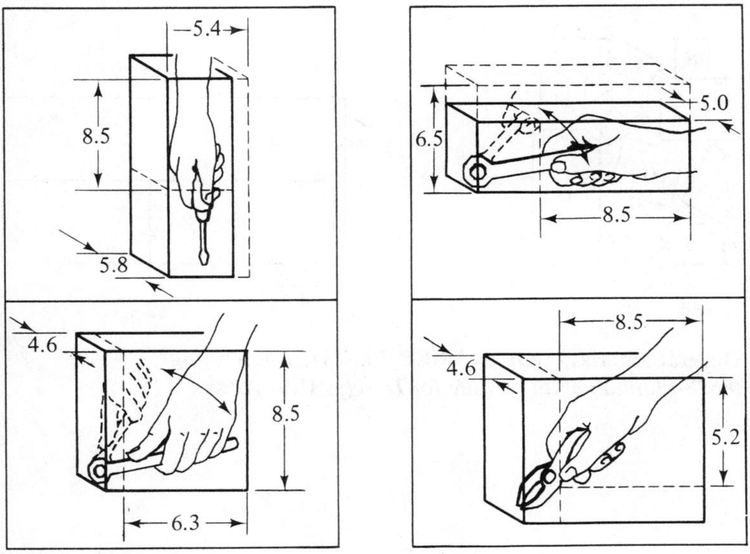

Figure C-2. Design criteria (*accessibility*). (*Source:* NAVSHIPS 94324, *Maintainability Design Criteria Handbook for Designers of Shipboard Electronic Equipment,* Naval Ship Systems Command, U.S. Navy.)

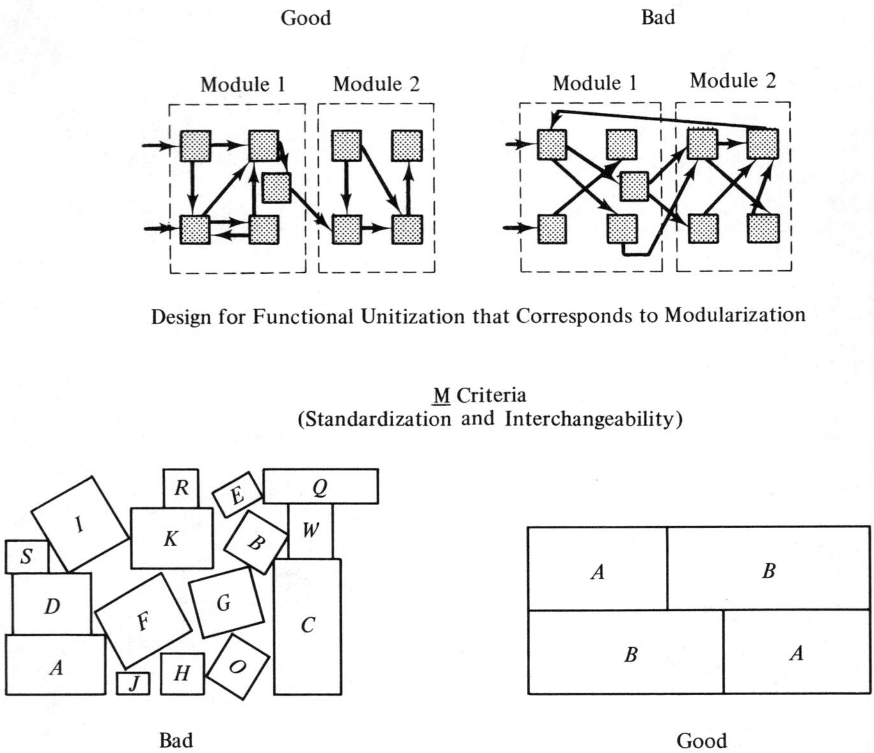

Figure C-3. General criteria. (*Source:* AMCP 706-134, ***Engineering Design Handbook-Maintainability Guide for Design,*** U.S. Army.)

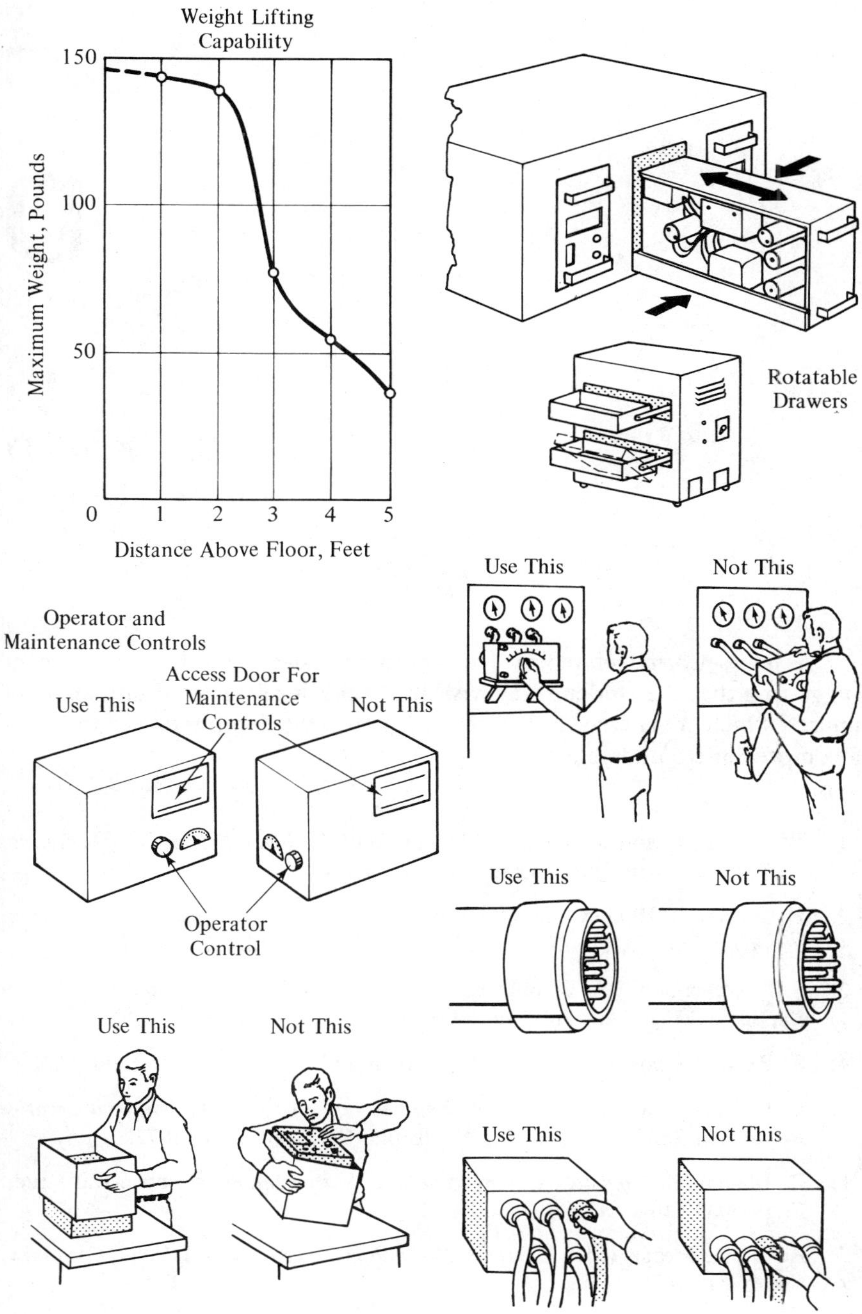

Figure C-4. General design criteria. (*Source:* DH 1-3, *Design Handbook Personnel Subsystems,* AFSC, U.S. Air Force.)

D

SELECTED BIBLIOGRAPHY

A few selected references have been included with the objective of suggesting that the student may wish to further investigate certain areas of interest which are covered in this text but not to the depth required to attain a comprehensive understanding.

1. W. Alderson and P. Green, *Planning and Problem Solving in Marketing,* Richard D. Irwin, Inc., Homewood, Ill., 1966.
2. H. Amrine, J. Ritchey, and O. Hulley, *Manufacturing Organization and Management,* 3rd ed., Prentice-Hall, Inc., Englewood Cliffs, N.J., 1975.
3. S.H. Archer and C.A. D'Ambrosio, *The Theory of Business Finance: A Book of Readings,* Macmillan Publishing Co., Inc., New York, 1968.
4. B. Bass, *Organizational Psychology,* Allyn and Bacon, Inc., Boston, 1968.
5. G. C. Beakley and H. W. Leach, *Engineering—An Introduction to a Creative Profession,* 2nd ed., Macmillan Publishing Co., New York, 1972.
6. B. Blanchard, *Logistics Engineering and Management,* Prentice-Hall, Inc., Englewood Cliffs, N.J., 1974.
7. K. B. DeGreene (ed.), *Systems Psychology,* McGraw-Hill Book Company, New York, 1970
8. T. Dernburg and D. McDougall, *Macroeconomics,* 3rd ed., McGraw-Hill Book Company, New York, 1968.
9. L. Doyle, C. Keyser, J. Leach, G. Schrader, and M. Singer, *Manufacturing*

Processes and Materials for Engineers, 2nd ed., Prentice-Hall, Inc., Englewood Cliffs, N.J., 1969.

10. J. H. Earle, *Engineering Design Graphics,* Addison-Wesley Publishing Company, Inc., Reading, Mass., 1969.

11. J. Gavett, *Production and Operations Management,* Harcourt Brace Jovanovich, Inc., New York, 1968.

12. C. George, *Management in Industry,* Prentice-Hall, Inc., Englewood Cliffs, N.J., 1964.

13. M. J. Gordon and G. Shillinglaw, *Accounting—A Management Approach,* 3rd ed., Richard D. Irwin, Inc., Homewood, Ill., 1964.

14. P. Green and D. Tull, *Research for Marketing Decisions,* Prentice-Hall, Inc., Englewood Cliffs, N.J., 1966.

15. T. Haimann and R. Hilgert, *Supervision: Concepts and Practices of Management,* South-Western Publishing Company, Cincinnati, Ohio, 1972.

16. W. W. Haynes, J. L. Massie, and M. J. Wallace, *Management—Analysis, Concepts, and Cases,* 3rd ed., Prentice-Hall, Inc., Englewood Cliffs, N.J., 1975.

17. C. T. Horngren, *Cost Accounting—A Managerial Emphasis,* 3rd ed., Prentice-Hall, Inc., Englewood Cliffs, N.J., 1972.

18. R. A. Johnson, W. T. Newell, and R. C. Vergin, *Operations Management—A Systems Concept,* Houghton Mifflin Company, Boston, 1972.

19. D. Karger and R. Murdick, *Managing Engineering and Research,* Industrial Press, Inc., New York, 1969.

20. F. Kast and J. Rosenzweig, *Organization and Management—A Systems Approach,* 2nd ed., McGraw-Hill Book Company, New York, 1974

21. H. Koontz and C.O'Donnell, *Principles of Management—An Analysis of Management Functions,* 3rd ed., McGraw-Hill Book Company, New York, 1964.

22. P. Kotler, *Marketing Management—Analysis, Planning, and Control,* 2nd ed., Prentice-Hall, Inc., Englewood Cliffs, N.J., 1972.

23. E. V. Krick, *An Introduction to Engineering and Engineering Design,* 2nd ed., John Wiley & Sons, Inc., New York, 1969.

24. P. Mali, *Managing by Objectives,* John Wiley & Sons, Inc., New York, 1972.

25. J. Martino, *Technological Forecasting for Decision Making,* American Elsevier Publishing Company, Inc., New York, 1972

26. D. Meredith, K. Wong, R. Woodland, and R. Wortman, *Design and Planning of Engineering Systems,* Prentice-Hall, Inc., Englewood Cliffs, N.J. 1973.

27. J. Mize, C. White, and G. Brooks, *Operations Planning and Control,* Prentice-Hall, Inc., Englewood Cliffs, N.J., 1971.

28. G. Odiorne, *Management by Objectives,* Pitman Publishing Corporation, New York, 1965.

29. B. Ostrofsky, *Design, Planning, and Development,* Prentice-Hall, Inc., Englewood Cliffs, N.J., 1977.

30. L. Rich, *Environmental Systems Engineering,* McGraw-Hill Book Company, New York, 1973.

31. E. Schein, *Organizational Psychology,* 2nd ed., Prentice-Hall, Inc., Englewood Cliffs, N.J., 1970.

32. W. Scott, *Organization Theory—A Behavioral Analysis for Management,* Richard D. Irwin, Inc., Homewood, Ill., 1967.

33. A. Shuchman, *Scientific Decision Making in Business,* Holt, Rinehart and Winston, Inc., New York, 1963.

34. W. J. Stanton, *Fundamentals of Marketing,* 4th ed., McGraw-Hill Book Company, New York, 1975.

35. G. Strauss and L.R. Sayles, *Personnel—The Human Problems of Management,* 3rd ed., Prentice-Hall, Inc., Englewood Cliffs, N.J. 1972.

36. H. G. Thuesen, W.J. Fabrycky, and G.J. Thuesen, *Engineering Economy,* 5th ed., Prentice-Hall, Inc., Englewood Cliffs, N.J. 1972.

37. P.E. Torgersen, I.T. Weinstock, *Management: An Integrated Approach,* Prentice-Hall, Inc., Englewood Cliffs, N.J., 1977.

38. J.C. Van Horne, *Financial Management and Policy,* 3rd ed., Prentice-Hall, Inc., Englewood Cliffs, N.J., 1974.

38. P. Vesilind, *Environmental Pollution and Control,* Duke Environmental Center, Duke University, Durham, N.C., 1974.

40. T. T. Woodson, *Introduction to Engineering Design,* McGraw-Hill Book Company, New York, 1966.

INDEX

D

E

O

P

Q

R

U

V

W